Driven by Nature

Plant Litter Quality and Decomposition

———————————————

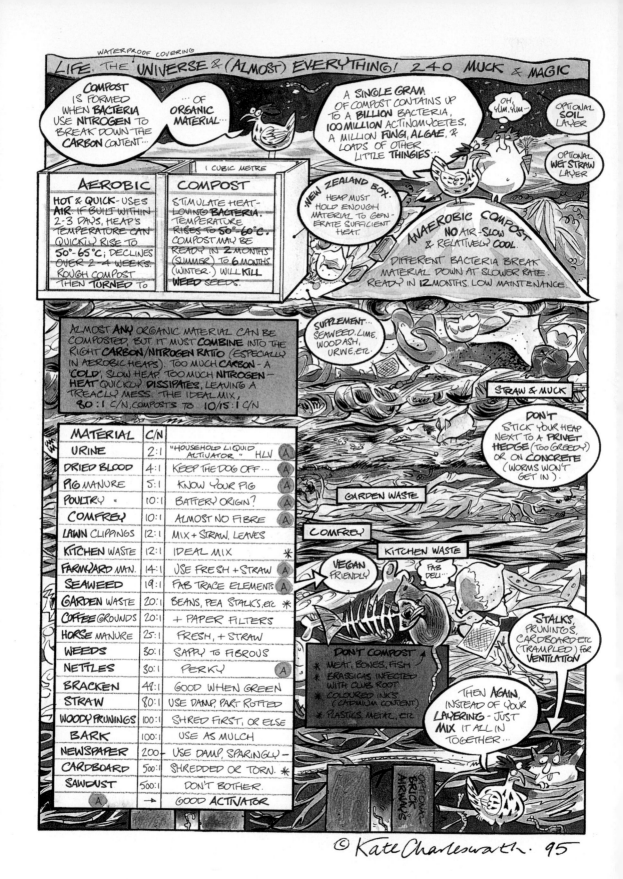

© Kate Charlesworth. 95

Driven by Nature

Plant Litter Quality and Decomposition

Edited by

G. Cadisch and K.E. Giller

Department of Biological Sciences
Wye College, University of London, UK

CAB INTERNATIONAL

CAB INTERNATIONAL
Wallingford
Oxon OX10 8DE
UK

Tel: +44 (0)1491 832111
Fax: +44 (0)1491 833508
E-mail: cabi@cabi.org
Telex: 847964 (COMAGG G)

A catalogue record for this book is available from the British Library.

ISBN 0 85199 145 9

Typeset in Plantin by Columns Design Ltd, Reading
Printed and bound in the UK at the University Press, Cambridge

Contents

Contributors

GORAN ÅGREN
Department of Ecology and Environmental
Research
Swedish University of Agricultural Sciences
Box 7072
S-750 07 Uppsala
Sweden

JO ANDERSON
Department of Biological Sciences
University of Exeter
Exeter EX4 4PS, UK

WIM ARP
Department of Terrestrial Ecology and
Nature Conservation
Wageningen Agricultural University
Bornsesteeg 69
6708 PD Wageningen
The Netherlands

JEFF BALDOCK
Petawawa National Forestry Institute
Canadian Forest Service
Chalk River
Ontario
Canada K0J 1J0

ADRIAN BAVAGE
Institute of Grassland and Environmental
Research
Plas Gogerddan
Aberystwyth
Dyfed, UK

MATHIAS BECKER
West Africa Rice Development Association
(WARDA)
BP 2551
Bouaké 01
Côte d'Ivoire

GRAEME BLAIR
Department of Agronomy and Soil Science
University of New England
Armidale
NSW 2351
Australia

HANS-PETER BLUME
Institute of Plant Nutrition and Soil Science
University of Kiel
Germany

ERNESTO BOSATTA
Department of Ecology and Environmental
Research
Swedish University of Agricultural Sciences
Box 7072
S-750 07 Uppsala
Sweden

LIJBERT BRUSSAARD
Department of Terrestrial Ecology and
Nature Conservation
Wageningen Agricultural University
Bornsesteeg 69
6708 PD Wageningen
The Netherlands

JOHN BURKET
Department of Crop and Soil Science
Oregon State University
Corvallis
OR 97331-7306, USA

GEORG CADISCH
Department of Biological Sciences
Wye College
University of London
Wye
Ashford
Kent TN25 5AH, UK

ANDREW CHESSON
Rowett Research Institute
Bucksburn
Aberdeen AB2 9SB, UK

JOHN CHUDEK
Department of Chemistry
University of Dundee
Dundee DD1 4HN, UK

PHILIP CLARKE
Department of Soil Science
Waite Agricultural Research Institute
The University of Adelaide
Glen Osmond
SA 5064, Australia

IAN DAVIES
Institute of Grassland and Environmental
Research
Plas Gogerddan
Aberystwyth
Dyfed, UK

JAN DIELS
Soil Microbiology
IITA Nigeria
c/o Lambourn and Co.
Carolyn House
26 Dingwall Road
Croydon CR9 3EE, UK

RICHARD DICK
Department of Crop and Soil Science
Oregon State University, Corvallis
OR 97331-7306, USA

BEN DZOWELA
SADC-ECRAF Agroforestry Project
c/o Department of Research and Specialist
Services
PO Box CY594
Causeway
Harare
Zimbabwe

KEN GILLER
Department of Biological Sciences
Wye College
University of London
Wye
Ashford
Kent TN25 5AH, UK

AHMAD GOLCHIN
Department of Soil Science
Waite Agricultural Research Institute
The University of Adelaide
Glen Osmond
SA 5064, Australia

TON GORISSEN
DLO-Research Institute for Agrobiology and
Soil Fertility (AB-DL))
PO Box 14
6700 AA Wageningen
The Netherlands

NOAM GRESSEL
Division of Ecosystem Sciences
Department of Environmental Science
Policy and Management
108 Hilgard Hall
University of California
Berkeley
CA 94720, USA

KENNETH HAMMEL
Institute for Microbial and Biochemical
Technology
Forest Products Laboratory
Forest Service
US Department of Agriculture
Madison
WI 53705, USA

EKO HANDAYANTO
Faculty of Agriculture
Brawijaya University
Jalan Veteran
Malang 65145
East Java
Republic of Indonesia

JEFFREY HARBORNE
Department of Botany
University of Reading
Whiteknights
Reading RG6 6AS, UK

PATRICK HATCHER
Fuel Science Program
Pennsylvania State University
University Park
PA 16802, USA

BILL HEAL
Institute of Ecology and Resource
Management
University of Edinburgh
EH9 3JU, UK

DAVID HOPKINS
Department of Biological Sciences
University of Dundee
Dundee DD1 4HN, UK

ULRICH IRMLER
Ecosystem Research Center
University of Kiel
Germany

LARS JENSEN
Department of Agricultural Sciences
Royal Veterinary and Agricultural University
Thorvaldsensvej 40
DK 1871 FC
Frederiksberg
Denmark

RICHARD JONES
Department of Natural Resource Sciences
Washington State University
Pullman
WA 99164-6226, USA

BIAUW TJWAN KANG
International Institute of Tropical
Agriculture (IITA)
PMB 5320
Ibadan
Nigeria
c/o L.W. Lambourn & Co.
26 Dingwall Road
Croydon CR9 3EE, UK

PETER KUIKMAN
DLO-Research Institute for Agrobiology and
Soil Fertility (AB-DLO)
PO Box 14
6700 AA Wageningen
The Netherlands

JAGDISH K. LADHA
International Rice Research Institute (IRRI)
PO Box 933
Manila
Philippines

PATRICK LAVELLE
Laboratoire d'Ecologie des Sols Tropicaux
Centre ORSTOM
93143-Bondy Cedex
France

ROD LEFROY
Department of Agronomy and Soil Science
University of New England
Armidale
NSW 2351
Australia

JOHN McCOLL
Division of Ecosystem Sciences
Department of Environmental Science
Policy and Management
108 Hilgard Hall
University of California
Berkeley
CA 94720, USA

PARAMU MAFONGOYA
SADC-ICRAF Agroforestry Project
c/o Department of Research and Specialist
Services
PO Box CY594
Causeway
Harare
Zimbabwe

JAKOB MAGID
Department of Agricultural Sciences
Royal Veterinary and Agricultural University
Thorvaldsensvej 40
DK 1871 FC
Frederiksberg
Denmark

HÅKAN MARSTORP
Department of Soil Sciences
Swedish University of Agricultural Sciences
Box 7014
S-750 07 Uppsala
Sweden

ROEL MERCKX
Laboratory of Soil Fertility and Soil Biology
Faculty of Applied Agricultural Sciences
KU Leuven
K. Mercierlaan 92
3001 Leuvan/Heverlee
Belgium

DARYL MOORHEAD
Department of Biological Sciences
Texas Tech University
Lubbock
TX 79409, USA

PHILIP MORRIS
Institute of Grassland and Environmental
Research
Plas Gogerddan
Aberystwyth
Dyfed, UK

TORSTEN MUELLER
Department of Agricultural Sciences
Royal Veterinary and Agricultural University
Thorvaldsensvej 40
DK 1871 FC
Fredericksberg
Denmark

ROBERT MYERS
IBSRAM
PO Box 9-109
Bangkhan
Bangkok 10900
Thailand

P.K. NAIR
University of Florida
Department of Forestry
118 Nesins-Ziegler Hall
Gainesville
FL 32611, USA

NIELS NIELSEN
Department of Agricultural Sciences
Royal Veterinary and Agricultural University
Thorvaldsensvej 40
DK 1871 FC
Fredericksberg
Denmark

MEINE VAN NOORDWIJK
ICRAF
PO Box 161
Bogor 1600
Indonesia

MALCOLM OADES
CRC for Soil and Land Management
Glen Osmond
SA 5064
Australia
and Department of Soil Science
Waite Agricultural Research Institute
The University of Adelaide
Glen Osmond
SA 5064
Australia

CHERYL PALM
Tropical Soil Biology and Fertility
Programme
PO Box 30592
Nairobi
Kenya

KEITH PAUSTIAN
Natural Resources Ecology Laboratory
Colorado State University
Fort Collins
CO 80523, USA

HENRY PHOMBEYA
Agroforestry Commodity Team
Chitedze Agricultural Research Station
PO Box 158
Lilongwe
Malawi

MARK ROBBINS
Institute of Grassland and Environmental
Research
Plas Gogerddan
Aberystwyth
Dyfed, UK

A. PHILIP ROWLAND
Institute of Terrestrial Ecology
Merlewood Research Station
Grange-Over-Sands
Cumbria LA11 6JU, UK

NTERANYA SANGINGA
Soil Microbiology
IITA Nigeria
c/o Lambourn and Co.
26 Dingwall Road
Croydon CR9 3EE, UK

TARA SEWELL
Fuel Science Program
Pennsylvania State University
University Park
PA 16802, USA

B. SINGH
Department of Agronomy and Soil Science
University of New England
Armidale
NSW 2351
Australia

ROBERT SINSABAUGH
Biology Department
University of Toledo
Toledo
OH 43606, USA

JAN SKJEMSTAD
Division of Soils
CSIRO
Glen Osmond
SA 5064
Australia
and CRC for Soil and Land Management
Glen Osmond
SA 5064
Australia

SEGLINDE SNAPP
The Rockefeller Foundation
PO Box 30721
Lilongwe 3
Malawi

MIKE SWIFT
Tropical Soil Biology and Fertility (TSBF)
c/o UNESCO ROSTA
UN Complex Gigiri
PO Box 30592
Nairobi
Kenya

GUANGLOG TIAN
International Institute of Tropical
Agriculture (IITA)
PMB 5320
Ibadan
Nigeria
c/o L.W. Lambourn & Co.
26 Dingwall Road
Croydon CR9 3EE, UK

A. TILL
Department of Agronomy and Soil Science
University of New England
Armidale
NSW 2351
Australia

BERNHARD VANLAUWE
Soil Microbiology
IITA Nigeria
c/o Lambourn and Co.
26 Dingwall Road
Croydon CR9 3EE, UK

PATMA VITYAKON
Department of Soil Science
Khon Kaen University
Khon Kaen
Thailand

CHRISTINE WACHENDORF
Ecosystem Research Center
University of Kiel
Germany

JOHN WAID
School of Microbiology
La Trobe University
Bundoora
Victoria 3083
Australia

DAVID WARDLE
AgResearch
Ruakura Agricultural Reseach Centre
Private Bag 3123
Hamilton
New Zealand

ANDY WHITMORE
DLO Research Institute for Agrobiological
and Soil Fertility Research
PO Box 129
NL-9750 AC Haren
The Netherlands

Preface

Biological management of nutrient supply to plants is intrinsically more complex than provision of nutrients as inorganic fertilizers. We need a predictive understanding of the factors that determine decomposability of plant litters (i.e. their nature), whether nutrients released are retained or lost from the system, whether rates of decomposition can be manipulated to improve nutrient use efficiency, and how the various fractions of plant residues translate into pools of organic matter in soil. Only then can predictive models for nutrient release, plant uptake and soil organic matter dynamics be truly tested and validated.

We had discussed the idea that a detailed discussion of contemporary ideas on the characterization and manipulation of plant quality and especially its role in soil organic matter formation and nutrient cycling was warranted for some time. At the International Society of Soil Science meeting in Acapluco, Mexico in 1994 we decided to take the plunge and organize a conference on which this book is based. Within two days we had arranged an advisory group, decided on dates and title and circulated a first announcement. The format of the meeting largely followed that of the earlier 'Beyond the Biomass' meeting in that many of the invited speakers were scientists from related fields of research to provide an opportunity for cross-fertilization of ideas. Other spoken papers were selected 'blind' from submitted abstracts in that names and affiliations were removed from the abstracts before we saw them and the choice of papers for oral presentation was based on a desire to select a wide range of novel contributions spanning the subject area. This resulted in an interesting mix of speakers including several who were still conducting their doctoral research. The range of selected contributions embraced managed and natural systems in a wide range of climates and ecosystems.

The international symposium 'Driven by Nature: Plant Litter Quality and Decomposition' was held between 17 and 20 September 1995 at Wye College, University of London, in Kent, UK. The meeting was attended by nearly 200 delegates from 34 countries and comprised eight keynote papers, 22 oral presentations and 103 posters, interspersed with discussions. The chapters in this book are largely based on oral presentations made at the meeting. The contributions range from a comprehensive historical overview, specific biochemical and faunal mediated processes through to agronomic and socioeconomic aspects of managing biological inputs. A number of chapters on simulation modelling examine how far current knowledge of the role of litter quality has been integrated into general concepts of decomposition and soil organic matter turnover.

From discussions at the meeting it emerged that comparison of results between different groups or laboratories was impeded by incompatabilty of results, both due to the different analytical methods used to assess plant litter quality and differences in the range of parameters selected for measurement. This problem was addressed in a specific workshop at the end of the conference and an attempt has been made to set guidelines for description of plant litter quality (see Palm and Rowland, Chapter 28). We conclude these proceedings with a brief summary which highlights some priorities for future research.

The symposium was organized with the support of the British Society of Soil Science (BSSS), the International Society of Soil Science (ISSS) and Tropical Soil Biology and Fertility Programme (TSBF). We acknowledge sponsorship from the Swiss Development Cooperation which allowed participation of several delegates from developing countries. We wish to thank Jo Anderson, Patrick Lavelle, Roel Merckx, Bob Myers, Meine van Noordwijk, Cheryl Palm, Mike Swift, Keith Syers, Richard Thomas, Segundo Urquiaga and Andy Whitmore for their advice and support. We are particularly grateful to Susan Simpson for her excellent assistance in organization of the conference (which she renamed 'Driven to Distraction') and the many in-house helpers who made sure that the symposium ran smoothly. We also thank the independent referees who so promptly and conscientiously reviewed the manuscripts and helped to make this book more than just a rough summary of results. We have chosen publication as a book, rather than in a journal, to communicate the output of the symposium in the hope that contributors would be speculative and explore new ideas for the benefit of future research.

Georg Cadisch and Ken Giller
Wye College, University of London
February 1996

Part I

Review: Any Progress?

1   Plant Litter Quality and Decomposition: An Historical Overview

O.W. Heal[1], J.M. Anderson[2] and M.J. Swift[3]

[1] *Institute of Ecology and Resource Management, University of Edinburgh, EH9 3JU, UK;*
[2] *Department of Biological Sciences, University of Exeter, Exeter EX4 4PS, UK;*
[3] *Tropical Soil Biology and Fertility, UN Complex Gigiri, Nairobi, Kenya*

Background

The influence of the quality of litter on its subsequent rate of decomposition and its influence on soil fertility has been recognized since the early stages of agriculture. As a research topic it goes back at least as far as the establishment of the long-term experiments at Rothamsted by Lawes and Gilbert 150 years ago, fuelled by the controversy with Liebig (1840) on the origin of nitrogen required by agricultural crops (Johnston, 1994). The pioneering studies by Lawes (1861) on the use of farmyard manures for the maintenance of soil fertility and Muller (1887) on the influence of soil and tree species on the development of humus forms, led to the development of management practices based on simple empirical indices of litter quality such as the C-to-N ratio. The expansion of decomposition research in the 1920s and 1930s was characterized by a wide variety of descriptive and experimental field and laboratory studies in a wide variety of disciplines (e.g Waksman, 1924; Campbell, 1929; Jensen, 1929; Bornebusch, 1930; Findlay, 1934; Linderstrom-Lang and Duspiva, 1935; Broadfoot and Pierre, 1939 and many others).

The emergence of more mechanistic understanding of decomposition can be related to a number of seminal papers in the period 1940 to 1970. A strengthened conceptual framework and a powerful analytical tool was provided by the definition of the decomposition constant, k, by Jenny, Gessel and Bingham (1949) and its later development by Olson (1963). This was linked to the development of ecosystem theory, with explicit recognition of the integral role of the decomposition subsystem for the maintenance of ecosystem functioning (Olson, 1963; Rodin and Basilevic, 1967; Odum, 1969). The mechanistic explanation of k, describing how the decay rates of the individual substrates which comprise litter, i.e. litter quality, combine to determine the overall decomposition rate, was provided by Minderman (1968). These theoretical developments, accompanied by the introduction of the litter bag technique (Bocock and Gilbert, 1957) and other more sophisticated methods (e.g. Mayaudon and Simonart, 1959), led to a wealth of field studies describing decomposition of litters in different ecosystems.

The International Biological Programme in the 1970s provided further impetus for research on decomposition and soil biology in relation to ecosystem structure and function, particularly in the Biome Programmes (Coupland, 1979; Breymeyer and Van Dyne, 1980; Bliss *et al.*, 1981; Reichle, 1981). These IBP studies and experience provided extensive quantitative information on the geographical variation in the control of decomposition by the environment, resource quality and organisms. This information was an important base which was developed into a formalized paradigm of decomposition processes in terrestrial ecosystems (Swift *et al.*, 1979).

What have been the developments in decomposition research in the 20 years since *Decomposition in Terrestrial Ecosystems* was written and what are the key questions that remain? About 1000 papers dealing directly with litter decomposition have been published in refereed journals since 1980. The total of relevant publications is probably an order of magnitude greater when conference proceedings or book chapters and topics such as specific substrates, enzymes, biochemical pathways, mineralization, trace gas emission, and decomposer organisms are considered. This consistent research effort into decomposition has been maintained through continued interest in ecosystem processes and soil biology, the availability of widely applicable techniques, plus new technical developments. Increased interest in the management of nutrients and organic matter in relation to sustainability of agriculture, agroforestry and forestry, plus concerns over the impacts of agricultural pollution, provided an applied focus and funding. New challenges, with increased requirements for prediction and scaling-up, have been provided by issues of the potential impacts of climate change and the contribution of decomposition to greenhouse gas emissions.

Thus decomposition research in general, and litter quality studies in particular, seems to have been subject to evolution rather than revolution. The older literature contains much that is familiar and in many cases, is still applicable. Accumulation of detail, formalization of concepts and use of more sensitive techniques have characterized progress. Progress has not been uniform however, as exemplified by the dramatic technical developments in molecular biology. While the techniques are increasingly used to probe for specific soil bacteria, such as nitrifiers and methanotrophs, their potential for investigating the structure of decomposer communities, particularly fungal communities, is only beginning to be explored (Egger, 1992; Ritz *et al.*, 1994). Similarly, the wealth of descriptive studies on soil organic matter over the last 60 years (Kononova, 1966), using the classical analysis of humus fractions, has

not been paralleled by the same rate of development in mechanistic understanding of soil organic matter stabilization (SOM) and turnover. This situation has changed dramatically over the last decade or so with the development of new physical and chemical fractionation techniques and the understanding of organic matter transformations provided by gas chromatography mass spectrometry (GCMS), pyrolysis mass spectrometry (PyMS) and non-invasive techniques such as nuclear magnetic resonance (NMR) and stable isotope analysis.

Given the large and wide range of research, the task of reviewing the last 20 or so years of research on litter quality and decomposition is obviously selective and subjective. As a framework we use the central concepts defined by Swift *et al.* (1979). The basic module (Fig. 1.1a) represents any organic resource which is changed from state R_1 to R_2 over time (t_1 to t_2). The module is repeated in the form of a cascade, a recurrent stepwise process in which the products of decomposition of one resource become the initial resources for subsequent modules in the cascade (Fig. 1.1b). The rate of change from R_1 to R_2 is regulated by a combination of three interacting groups of factors, the physicochemical environment (P) and the quality of the resource (Q) acting through their regulation of the decomposer community (O). Decomposition of any resource is the result of three component processes; catabolism (K), i.e. chemical changes such as mineralization giving rise to inorganic forms, and the synthesis of decomposer tissues and humus; comminution (C), by which there is a physical reduction in particle size and often selective redistribution of chemically unchanged litter; and leaching (L), which causes transport down the profile or removal from the system of labile resources in either changed or unchanged form. The module which we developed is sufficiently general to be applicable to all forms of decomposing organic matter in any terrestrial environment. The relative importance of the three regulating factors (O, P, Q) varies in a predictable way within and between sites, resulting in a hier-

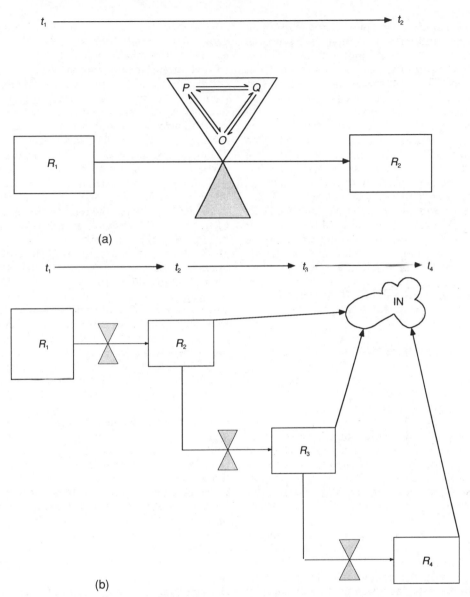

Fig. 1.1. The central concepts of decomposition of organic matter. (a) The regulation of decomposition by three interacting factors, the physico-chemical environment (*P*) and resource quality (*Q*) acting through decomposer organisms (*O*). (b) The decomposition cascade, with the module represented in (a) recurring as the primary resource is decomposed and redistributed through comminution, catabolism and leaching (from Swift *et al.*, 1979).

archical system of controls (Lavelle *et al.*, 1993).

It is the simple, logical structure of the module which seems to be its strength, providing a robust framework within which the relative emphasis of particular studies can be defined. The papers which follow in this book provide details of many facets of the subject of litter quality (*Q*). This introductory paper provides an 'hors d'oeuvre'. We identify some of the recent research trends which we consider have improved understanding of how litter quality influences subsequent decomposition processes and how this information can be applied to current and future management and policy issues. First, we review various aspects by which initial litter quality (Q) has been defined in relation to its subsequent decomposition. Following this extended section, current thinking on the production and fate of subsequent resources (the cascade) and the extrinsic physico-chemical factors (P), particularly climate change, which influence decomposition of litters are considered. Then, we move to the practicalities of agricultural management of litter Q and to the integration of the decomposition subsystem into the ecosystem. Organisms (O) are not considered separately; they are pervasive and their structure and function has recently been well reviewed in Ritz *et al.* (1994).

Definition of Litter Quality

The main drive in research has been to find ways of predicting the rate of litter decomposition and even more importantly, the rate of nutrient release, from the chemical composition of the resource. A variety of predictive equations has been proposed, mainly using various ratios of carbon, nitrogen, lignin and polyphenols (see Mafongoya *et al.*, Myers *et al.* and Vanlauwe *et al.*, Chapters 13, 17 and 12, of this volume for detailed review).

C-to-N remains a significantly important feature of such formulae but a number of lines of research have demonstrated important interactions with other factors. The C-to-N ratio was well established by the 1920s as a general index of the quality of litter (Waksman, 1924; Waksman and Tenney, 1928) and, according to Jensen (1929), it goes back as far as 1916. It still has widespread use. However, it is now generally accepted that to understand the mechanisms regulating processes of decomposition in a wide spectrum of resource types, the form of the carbon in plant cells as an energy source, the concentration of other nutrients, and the composition of various secondary plant

compounds ('modifiers'), can all be significant. Thus, whilst the C-to-N ratio is accepted as a general index of quality, the relative importance of the different chemical and physical components in different resources and their interaction is a matter of considerable debate and research effort.

Resource quality of plant materials is broadly defined by the major life forms (annuals, deciduous and evergreen perennials, woody plants etc.) and the environmental conditions under which they grow (Grime, 1988). The different ecological strategies of plants in the relative allocation of carbon and nitrogen to growth and production or plant protection compounds also have obvious implications for resource quality of the tissues. Deciduous tree leaf litter generally shows an inverse correlation between nitrogen concentrations and the concentrations of lignin and polyphenols. In evergreen conifers, nitrogen limitation is also associated with greater needle retention time on the trees, higher concentrations of lignin and secondary defence compounds, and hence lower litter resource quality (Flanagan and Van Cleve, 1983). Miller *et al.* (1979) showed that these feedbacks between N immobilization in litter and plant responses to N limitation could arrest the growth of a pine plantation on soil of low inherent fertility. Evergreen rainforests not only show a wider diversity of species than temperate forests, but also greater independent variation in the concentrations of nutrients, lignin and polyphenols in different species growing on the same soil type (Swift and Anderson, 1989). The relationships between these litter resource quality characteristics and decomposition rates are largely empirical and there is still poor understanding of the interactions of the biochemical constituents of litter at the cellular level where microbial decomposition takes place. However, Berendse *et al.* (1987) expressed explicit relationships between litter composition, mass loss and microbial growth in a theoretical model. The model considers the relative importance of lignin, carbohydrates masked by lignin, and free carbohydrates in relation to nitrogen. A negative relationship of mass loss to lignin concentration is predicted and a positive relationship to initial nitrogen concentration when N is limiting microbial growth. The model realistically simulates various field data for forest litters, but indicates the limitations of available chemical, as well as microbial information.

Nitrogen is one of the commonest factors limiting litter decomposition as it determines the growth and turnover of the microbial biomass mineralizing the organic C. Theoretically the optimum ratio of C-to-N for microbial growth is about 25, but fungi and bacteria can decompose resources with far higher ratios. Dead plant materials may contain between about 5% and 0.1% N and so C-to-N ratios ranging from 20 to 500 provide a general framework for considering the potential rates of litter decomposition. Only animal and microbial tissues with high protein content have C-to-N ratios below this range. There is, however, considerable 'noise' in the predictive value of C-to-N ratios because of differences in the biodegradability of constituent C and N compounds and the regulatory effects of other resource quality characteristics. These variables form a hierarchical series of controls along the C-to-N continuum. Materials with C-to-N ratios of <20 decompose rapidly, often with the release of ammonia, because nitrogenous compounds are metabolized as C sources. Plant materials with ratios between 25 and 75 include green leaves (both natural inputs and green manures) which have different biochemical composition to leaf litter, with high available C-to-N and P contents because they have not undergone the usual processes of senescence. Such resources usually decompose very rapidly but the N mineralization potential of some legume tissues may be affected by the complexing of protein by polyphenols when cells lyse, and the N-to-polyphenol ratio may be a better indicator of their quality. Cereal and legume straws and litters from annual crops usually contain less than 10–15% lignin and hence C-to-N ratios of 50 to 100 are reasonable predictors of decomposition rates since the higher ratios mainly reflect lower N concentrations in tissues rather than changes in the form of C. Over the same C-to-N range, however, litters from woody perennials contain increasing concentrations of lignin which is not only refractory to microbial enzymes but can also mask cell wall polysaccharides from attack. Hence, above C-to-N of about 75–100, the lignin-to-N ratio may be a better indicator of C availability to microorganisms. At higher C-to-N ratios characteristic of woody tissues, microbial attack is affected by the massive structure of crystalline cellulose, extensive lignification of tissues and the presence of modifiers such as condensed tannins and terpenes.

In the context of definition of litter quality in relation to decomposition, we now consider four aspects of particular note that have emerged: (i) changes over time in the variables which control decomposition; (ii) variation in litter quality within a species growing under different site conditions; (iii) the importance of variation in the biochemical composition of the lignin and other polymers; (iv) roots and root exudates.

Changes over time in the variables which control decomposition

Although the negative exponential model is a useful descriptor to define the rate of decomposition, it has long been recognized that the fractional loss rate declines with time (Minderman, 1968; Jenkinson, 1977; Melillo et al., 1989; Andren et al., 1990). In particular, the often ignored decline in respiration rate of litter over time reflects the decline in the quality of the remaining substrate. Reasons for this decline include the successive loss of more readily decomposed substrates, leaving the more resistant fractions, and possibly the formation of resistant polymeric compounds as described by Minderman (1968). These sequential patterns of resource utilization can result in a shift in the relative importance of different quality variables controlling decomposition and nutrient mineralization. The recognition that the factors which control decomposition change with time is one of the features justifying the cascade model.

The change in controls was clearly described by Berg and Staaf (1980). They showed a shift from nutrient control in the early stages of decomposition of *Pinus sylvestris* needles, to the dominance of lignin as the controlling factor in later stages (Fig. 1.2). Similarly, in reviewing nitrogen release from legumes in tropical agroforestry systems, Palm (1995) concluded that the polyphenol-to-nitrogen ratio may serve as a short-term index for green manures, while (lignin + polyphenol)-to-nitrogen provided an index of longer-term release for more woody and naturally senescent material. The general situation was well summarized by Fog (1988) who wrote:

> During the last decade, several papers have questioned the ability of the C-to-N ratio to predict C and N mineralization. Herman *et al.* (1977) were probably the first to realize that N had different effects on carbohydrate decomposition and lignin decomposition. C mineralization is much better predicted by a combination of the C-to-N ratio and

the lignin-to-carbohydrate ratio than the C-to-N ratio alone (Herman *et al.*, 1977; McClaugherty and Berg, 1987). The switch from N immobilization to N mineralization does not occur at a specific C-to-N ratio (Bosatta and Staaf, 1982; Berg and McClaugherty, 1987). In a mini-review Berg, (1986) concluded that the decomposition of forest litter is mainly governed by the rate of lignin decomposition, and that this rate, in turn, is increased by a high cellulose content and decreased by a high nitrogen content. The evidence from the present review is that this is the case not only for forest litter, but also for wood, straw and many other substances

The consequence of changes in the composition of a litter during decomposition is that correlations between the rate of mass or nutrient loss and simple expressions of the initial composition of the litter will have limitations. The correlations subsume a variety of interactions and the later the stage of decomposition the less important are the initial properties of the litter. As discussed later, the use of litter bags

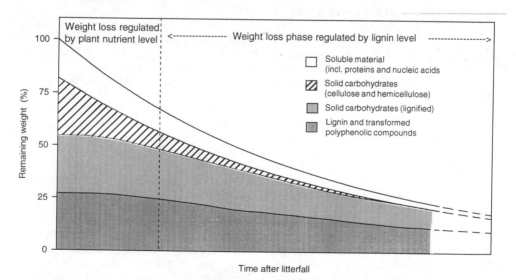

Fig. 1.2. Diagramatic representation of the decomposition of components of *Pinus sylvestris* needle litter. The initial phase of mass loss is regulated by available energy sources and nutrients, while the later phase is regulated by lignin concentration (adapted from Berg and Staaf, 1980).

tends to focus attention on the earlier stages of decomposition.

Changes are not restricted to chemistry. For example Gallardo and Merino (1993) in a study of leaf litter from different Mediterranean shrubs and trees, distinguished an initial leaching phase in which leaf toughness and toughness-to-P concentration of the original litter provided the best predictors of mass loss. In contrast, the cutin-to-N ratio or cutin concentration were the best predictors in the post-leaching phase. Similarly, microbial, particularly fungal, successions on decomposing litters reflect changes in litter composition, as do fauna with recognition of phases in palatability and interaction with microflora (e.g. Scheu and Wolters, 1991; Van Wensem *et al.*, 1993).

A main conclusion is that, while indices of decomposition based on initial litter quality provide a broad 'explanation' of decomposition, they are only a first approximation. Because of changes in the factors controlling decomposition over long time intervals and the progressive effect of extrinsic factors, the limitations in the predictive value of simple measures of initial composition must be recognized.

Variation in litter quality within a species growing under different site conditions

In studies on litter quality, there is increasing recognition that variations in site conditions can have considerable effect on the chemistry of litter generated by different individuals of the same plant species (Berg and Tamm, 1991). Sanger *et al.* (1996) have shown that *Pinus sylvestris* needles vary in carbohydrate composition and lignin polymerization in relation to nutrient status and pH of different soils. Similarly, Vitousek *et al.* (1994) found that when the tropical tree *Metrosideros polymorpha* was grown on dry Hawaiian lava flows, its litter decomposed twice as quickly as litter from the same species grown on wet sites, when both were measured in the same site. They concluded that the higher substrate quality from dry sites could be due to trade-offs among nutrient-use effi-

ciency, water-use efficiency and carbon gain by the plant when growing under different conditions. These modern examples confirm and amplify results from earlier studies (e.g. Handley, 1954; Davies *et al.*, 1964).

Research into herbivore and pathogen defence compounds provides an insight into causes of intraspecific variation in litter quality. Increased defence compounds tend to be generated when metabolic resources are present in excess of growth demands. For example, in sunny conditions with limiting nutrients, carbon will be relatively in excess, and carbon-based defences such as tannins and terpenoids will increase. The converse occurs under shaded conditions, and similar patterns are seen in nitrogen-based defence and nitrogen availability (Bryant *et al.*, 1983; Coley, 1987). Defence compounds generated in the live plant often, but not always, carry over into the litter. Thus 'mobile' defence compounds (e.g. alkaloids and monoterpenes) have a life-span of only a few hours or days in the plant, contrasting with the 'immobile' defences (e.g. tannins and lignins) which are expensive to construct, increase with leaf lifetime, and are not reclaimed during leaf senescence (Coley, 1987; Fig 1.3).

The importance of variation in the biochemical composition of the lignin and other polymers

Proximate analysis of organic fractions played an important part in the early research on litter quality. However, the tedious and inaccurate methods obscured the varying biochemical composition, despite the evidence from microbiology of the varying resistance to decomposition and distinctive inhibitory capabilities of different component compounds. The advent of more sophisticated techniques, particularly the many variants of mass spectroscopy, now allow the rapid and sensitive characterization of organic molecules, giving the capability to follow their degradation and synthesis (Petersson and Fry, 1987; Elliott and Cambardella, 1991; Mulder *et al.*, 1991; van Breemen, 1992).

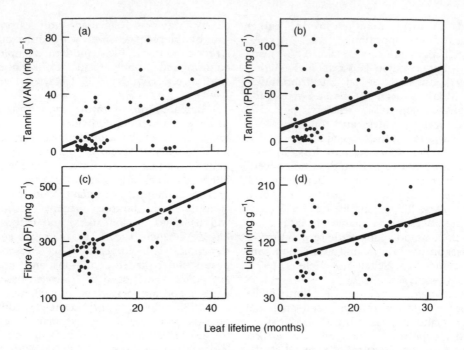

Fig. 1.3. Concentrations of condensed tannins (VAN=vanillin, PRO=proanthocyanins), fibre (ADF=acid detergent fibre) and lignin as a function of leaf lifetime for Panamanian tree species (from Coley, 1987).

Such enhanced characterization, especially when combined with C and N isotopes, has opened a veritable 'Pandora's box', with the danger of disappearing into ever finer detail with analysis of each litter providing a new variant and another publication. For example lignin in the walls of sclerenchyma and tracheids is formed from three precursor phenylpropanoid alcohols (coumaryl, coniferyl and sinapyl; see Hammel, Chapter 2, this volume). The relative ratios of these moieties within the lignin polymer are species dependent and result in differing decomposition characteristics (Hedges and Mann, 1979; Ander *et al.*, 1984; Brett and Waldron, 1990; Sanger *et al.*, 1996).

The strength of the new technologies lies in their ability to track and quantify pathways of decomposition, to identify synthesis of new organic fractions, and to define the differences in chemical composition of litters which do not conform to the general paradigm. Refinement of biochemical interactions between specific molecu-

lar configurations and the responsive microbial enzyme systems is justifiable in terms of better understanding. Decomposition is a process of equivalent status to photosynthesis and we need to understand it just as fully. However, when the aim is to develop practical and predictive procedures, we need to ask what level of refinement is necessary.

Roots and root exudates

There has been a wealth of rhizosphere and mycorrhizal research which has highlighted the critical role played by microorganisms in the transfer of nutrients from decomposing litters to plants. In contrast, roots, rhizodeposits and exudates have attracted little decomposition research effort, especially compared with plant inputs to soil from above ground. This is despite the importance of roots as the plant–soil interface and their importance in carbon flow: 16–33% of the total carbon assimilated by plants is released directly into soil contributing 30–60% of the organic pool in soil (e.g. Boone, 1994). The limited research on root inputs is particularly disappointing in relation to arable systems where root residues may be the only organic inputs and are major contributors to SOM replenishment.

The emphasis on root exudation and rhizodeposition reflects the importance of these substrates to rhizosphere microorganisms and the associated effects on plant nutrient availability. The advent of C isotope methods has significantly improved knowledge of decay rates and allowed discrimination between root and microbial respiration (Cheng et al., 1994; Swinnen et al., 1995). In general, the decomposition of root products conforms with the general pattern of above ground plant parts, being determined by the chemical composition. Results from different methods have been increasingly reconciled (Andren et al., 1990; Swinnen et al., 1995). The influence of nutrients, particularly N, has been widely recognized in most agricultural soils, related to the readily decomposable C source and the capacity of microorganisms to store excess C. However, whilst carbohydrates rather than lignins and defence compounds tend to dominate in non-woody species, the reverse may occur in roots of trees and shrubs. For example spruce and bilberry roots had higher C-to-N ratio, and contained more lignin and phenolic polymers than their needles and leaves (Gallet and Lebreton, 1995).

The four aspects of litter quality discussed above, concentrated on a conventional approach of analysis of physico-chemical composition (Q) of the litter. They repeatedly show the influence of plant growth characteristics in determining Q and, because of the considerable analytical effort, individual research projects are based on a limited number of litter types, usually fewer than 20. An unconventional approach has been adopted by Cornelissen and colleagues (UCPE, 1995; Cornelissen, 1996). As part of their Integrated Screening Programme (ISP), designed to define plant structural and functional characteristics in the context of the Strategy Theory of Grime (1988), the

weight loss of more than 100 species of leaf, needle and shoot litter was measured under standard field conditions. Weight loss was related to a variety of plant characteristics such as growth habit and strategy, evergreen versus deciduous habit, autumn coloration and evolutionary advancement. Litter chemistry was not defined, but there is clear evidence of the adaptive strategy of tissue defence, related, for example, to the length of life of the leaf. Thus, deciduous species with long life span have well developed defence mechanisms. They tend to conserve resources by withdrawing chlorophyll, leaving lignins and tannins which provide the brown or yellow-brown autumn colour of slow decomposing leaves (Fig.1.4).

Another characteristic discovered was that species that are successful in environmentally stressed habitats have relatively slow decomposing leaves; those from disturbed and productive habitats tend to have a relatively high decay rate. Although the correlations between decomposition rate and growth characteristics may provide only general predictive capability, they are easy to make, provide a logical explanation, and can be effective where expected changes in the composition of plant associations are not specific (e.g. responses to climate change).

The Resource Cascade

The decay continuum of plant litter to soil organic matter described by Melillo et al. (1989) represents the spine of the resource cascade (Fig. 1.1b). They defined two stages in the decay process, based mainly on a long-term study of Pinus resinosa. The early stage was characterized by the relatively rapid loss of cellulose associated with maximum cellulase activity and nitrogen immobilization. In the second stage, the much slower rate of mass loss was dominated by decomposition of lignocellulose (acid soluble sugars plus acid insoluble C compounds). The second phase was associated with nitrogen mobilization. The distinction between the two stages occurred when the 'Lignocellulose Index' reached

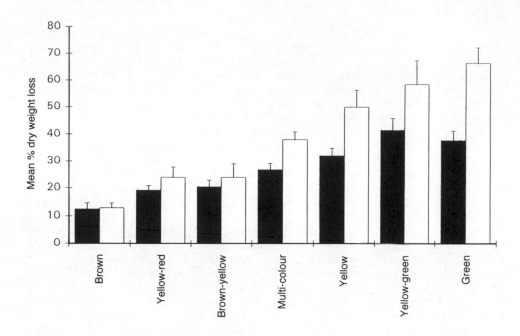

Fig. 1.4. Rates of mass loss of leaf and needle litters in relation to their colour at litter fall, based on measurement of 125 litter types decomposed for 8 weeks under standardized field conditions, with two mesh sizes, 0.3 mm (■) and 5.0 mm (□) (from Cornelissen, 1996).

about 0.7 (the LCI is the ratio of lignin-to-(lignin + cellulose)) and about 15–30% of the original litter remained. It was hypothesized that various resources, regardless of their initial chemical composition, were reduced to a lowest common denominator of chemical quality – an LCI of 0.7–0.8 – after which the composition changed little and decay was determined by environmental conditions. Both carbon and nitrogen dynamics are broadly described by the two-phase model, and Melillo *et al.* (1989) emphasized the need for concentrated research on the controls of the decay of the material late in the continuum.

The resource continuum of Melillo *et al.* (1989), while distinguishing two major stages in the resource cascade, concentrates on the residual mass of material. However, during decomposition parts of the resource are redistributed, interacting with the biochemical and physical environment in soils in both space and time (Aber and Melillo, 1982; Berg and Ekbohm, 1983). The redistribution of material highlights a methodological limitation in many studies of decomposition – the use of litter bag and similar containment systems. These methods focus on the material remaining which is made up of the more refractory primary plant materials and microbial products. They measure only net change and ignore the fate of material leaving the container, material which is usually more labile and often of higher nutrient content and may be more degradable than the residue. Microclimatic effects of litter bags, and the use of single species of litter rather than natural mixtures, are further limitations in field use. Containment methods, both in the field and laboratory, also tend to exclude fauna and vegetation; a feature which can facilitate controlled experiments to test the effect of presence and absence of selected organisms, but can lead to spurious results where exclusion is inadvertent. Litter bags are attractive in their simplicity and convenience and have made a great contribution to decomposition research. However, the widespread use of containment methods may have given us a biased and limited view of decomposition processes. As long as the limitations are recognized, litter bags still have an important role to play in the research armoury, especially in programmes requiring extensive observations, and when supplemented with more sensitive measures such as respiration and mineralization, and combined with isotope labelling.

Dependence on litter bags as a tool for understanding the course of decomposition has probably introduced conceptual artefacts because of focus on the residual resource rather than on the fate of labile constituents lost from the bags, or on material entering the bags. The resource cascade (Fig.1.1b) highlights the dispersion of material during decomposition. Transport of particulates by fauna, translocation by fungi and leaching by water, bring different organic fractions into environments different from that of the parent resource. The process of decomposition in soil can be viewed not so much as a time series, as shown by litter bag studies, but as a series of modules, coupled by inputs and outputs of carbon, nutrients and modifiers, and regulated by O and P. Four examples illustrate the importance of following the fate of fractions through the cascade.

1. The complex patterns of monomeric (flavonoids, phenolic acids) and polymeric (lignins, tannins) phenolic compounds generated from spruce and bilberry litters within mountain forest ecosystem profiles show the combination of transport and *in situ* production and decomposition that can be detected with sophisticated analysis (Gallet and Lebreton, 1995).
2. Inorganic soil constituents such as clays, silt and metal-sesquioxides can adsorb fresh organic substrates, i.e. those which have been generated close to or moved into contact with mineral fractions. It has been estimated that over 50% of soil organic matter is so associated (Greenland, 1965; Anderson et al., 1981). The resulting occlusion of organic matter, especially in silt microaggregates, is a major cause of its slow decomposition (e.g. Tisdall and Oades, 1982; Christensen, 1987; Elliott

and Cambardella, 1991; Sagger *et al.*, 1994) and is open to manipulation as seen in the case of conventional and no-tillage systems (Beare *et al.*, 1994). Transport by water and soil fauna, particularly earthworms, is probably an important mechanism facilitating contact between organic and inorganic fractions (Anderson, 1988a). Thus Zech (1991) quoted a laboratory experiment in which the rate of beech litter decomposition was increased by 32% through the action of lumbricid worms in the early stages of decay but decreased by 68% in the later stages. He explained the initial increase as a result of comminution and aeration stimulating microbial activity. The later decline was accounted for by the formation of organomineral complexes which protected the organic matter from microbial attack.
3. Zech (1991) also described an increase in dissolved organic carbon (DOC) mobilized in the O and A horizons, from 150 kg ha^{-1} in healthy spruce stands to 380 kg ha^{-1} in degraded stands. About 85–95% of the DOC was subsequently absorbed and mineralized in the B horizon and only 6–35 kg ha^{-1} year^{-1} was leached from the system. He suggested that the mineral soils act as a chromatographic system with a greater preference for hydrophobic substances. Hydrophilic substances tend to remain in solution and are more liable to microbial attack and leaching. Thus considerable amounts of organic C may be redistributed, selectively fractionated and mineralized within the soil profile as part of the resource cascade.
4. A fourth example emphasizes the need for further understanding of the mechanisms and significance of the resource cascade for litter decomposition in different systems. It is generally assumed that litter comminution by invertebrates enhances microbial decomposition by increasing the surface area exposed to fungal and bacterial attack. Many studies have also shown that the incorporation of leaf litter into mineral soils by earthworms accelerates rates of litter decomposition though the mechanisms, over and above stabilization of the microclimate, have not been clearly demonstrated. Anderson (1988b), how-

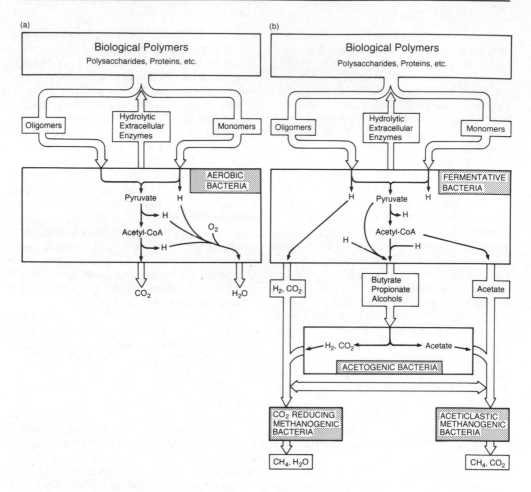

Fig. 1.5. Biochemical pathways of degradation of polymers under (a) aerobic and (b) anaerobic conditions (P. Westermann, personal communication).

ever, showed that although comminution of oak leaf litter by soil fauna enhanced N mineralization in lysimeters containing a moder woodland soil, the total mass of organic matter in the lysimeters was similar in treatments with and without fauna. In contrast to situations where litter incorporated into base-rich mineral soil significantly alters the chemical environment for decomposition, the biochemical conditions in oak litter and the acid surface organic matter horizons may have been rather similar at Anderson's site. Hence the chemical resource quality characteristics of the oak litter dominated decomposition rates irrespective of the change in physical state and location in the profile.

The resource cascade, as illustrated in the four examples above, is a complex expansion of the description provided by the simple decay curves and k values. The underlying biochemical and metabolic pathways for the various organic molecules can be readily defined under laboratory conditions, as in Fig. 1.5. Under field conditions, the spatial and physico-chemical heterogeneity of the soil, plus variations in the biological community, are superimposed on and interact with the sequence of biochemical transformations. The individual organic (and inorganic) components of the initial litter clearly interact selectively with the soil environment in space and time. The initial C-to-N and other ratios can provide a broad indication of the decomposition potential. In order to predict and manipulate the realised decomposition a more sophisticated understanding of the interactions which occur in the resource cascade is necessary. This is a subject that continues to demand collaboration between biologist, biochemist and soil scientist.

Extrinsic Factors (P) Influencing Litter Quality and Decay

The physico-chemical environment (P), particularly soil conditions and climate, causes the main variation in quality of litter input through selection of plant species and control of moisture and nutrient conditions. In recent years, decomposition researchers have been challenged to determine the changes that will occur in response to climate change. Much has been written on climate change effects and our ability to predict response to temperature and moisture variation is relatively well developed (Scharpenseel et al., 1990; Anderson, 1991, 1992). Despite this, some interesting questions remain to be quantified. For example, the effect of varying temperature response relationships of decomposers in different environments; the extent to which pattern of substrate utilization will shift with temperature; and the ways in which the delicate balance between aerobic and anaerobic processes will result in quantitative shifts between CO_2 and CH_4 emission. In contrast, our assessment of changes in litter Q that are likely to occur and the consequences for decay has been based on much weaker knowledge, especially when it comes to quantifying feedback effects both on the plant community and on the release of greenhouse gases. Two topics in particular, the effects of increased CO_2 and UV-B, have generated a new wave of research into litter Q.

Enhanced atmospheric CO_2 is widely held to increase photosynthesis, but with uncertain compensatory uptake of nutrients. The result tends to be a reduction in the nutrient concentration in plant tissues

with, for example, an increase in C-to-N ratio. There is evidence of considerable variation in plant species response (e.g. C_3 versus C_4 plants) and, within a species, variation in nutrient availability; a fall in tissue nutrient concentration is likely to be greatest in nutrient limited conditions (see also Arp et al., Chapter 15, this volume). An additional response by the plant may be to allocate excess photosynthate to secondary compounds (lignin) or to root exudates. The latter response could interact with mechanisms such as mycorrhizas to increase nutrient uptake. The net effect is a wide variation in recorded plant response. Cotrufo (1995) in a literature review found recorded response of N concentration in plant tissues ranging from a reduction to 40% of that in ambient CO_2, to an increase to nearly 160%, with a mean of a reduction to 87% . Thus, despite the positive response of mycorrhizas, N_2 fixers and rhizosphere microorganisms to enhanced CO_2 (Thomas et al., 1991; Zak et al., 1993; Norby, 1994; O'Neill, 1994), there is likely to be a general decline in litter Q in response to increased CO_2. The question of whether the decrease in litter Q is an artefact of short-term experiments seems to be answered by the observation of reduced Q in plants grown in environments where there has been long-term, natural enhancement of CO_2 (Korner and Miglietta, 1994).

The effect on decomposition of the CO_2 induced changes in litter Q, with increased lignin and reduced N concentrations are reasonably predictable (Fig. 1.6). It seems that our basic understanding of litter Q and decomposition is adequate to predict the general response. Quantitative prediction of nutrient release and feedback effects on vegetation are much less certain. One of the earliest studies on decomposition of CO_2 enhanced litter, that of Couteaux et al. (1991), also provided a warning on experimental technique and an insight into the influence of community structure on rates of decomposition. They showed that litter of sweet chestnut grown under enhanced CO_2 had a slower decay rate than control litter. The surprising result was that the reduction in decay rate

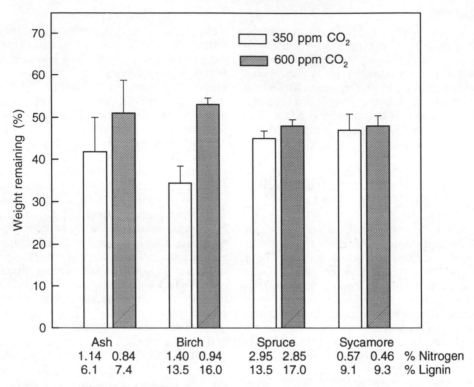

Fig. 1.6. Decomposition of litter of four tree species grown under ambient (350 ppm) and enhanced (600 ppm) levels of atmospheric CO_2. Birch and spruce litters were incubated in microcosms for 155 days, ash for 170 and sycamore for 243 days. Initial concentrations of nitrogen and lignin are shown (adapted from Cotrufo *et al.*, 1994).

was influenced by the complexity of the faunal community; an increase in faunal diversity reduced and even eliminated the impact of lower Q. The result may be an artefact of a short term experiment or an indication that increased biodiversity of soil functional groups tends to dampen responses to environmental change (Andren *et al.*, 1995; Janzen *et al.*, 1995; Verhoef and Brussard, 1990; Heal *et al.*, 1996)

The impact of increased UV-B on litter Q and decomposition presents an analogous situation. The increasing levels of UV-B at higher latitudes can cause changes in plant biochemistry, with increased pigmentation and secondary compounds. Detailed evidence is currently limited, but increased tannins and decreased α-cellulose in leaf litter of *Vaccinium* spp. have been recorded with experimentally increased UV-B (Gehrke *et*

al., 1995; Johanson *et al.*, 1995). Decay rate of litters from treated plants showed slight reductions in mass loss, respiration and nutrient mineralization. There was also evidence of a small direct effect of UV-B in reducing decomposition, and of changes in fungal community structure. Since direct UV exposure will only affect superficial litters, the main effects are anticipated to be through alteration in litter quality. However, the combination of indirect and direct effects could cause a reduction in decomposition rate of the order of 5–10% at higher latitudes and potentially a feedback effect through reduced nutrient availability in nutrient limited environments.

In both the CO_2 and UV-B examples quoted, there are problems in reconciling experimental evidence from process and population studies. Yet this is necessary if we are to answer such key questions as –

what degree of C accumulation will result from the gradual increases in CO_2 and UV-B, where, and over what time scales? There is some evidence that decomposer populations are already responding to climate change. Data on net CO_2 release indicate that tundra ecosystems in Alaska have moved from being net sinks of about 20–120 gC m^{-2} $year^{-1}$ to being net sources of 50–280 gC m^{-2} $year^{-1}$ (Oechel et al., 1993). Such a shift in CO_2 release presumably results from increased rates of decomposition induced by a combination of climate, CO_2 and UV-B, acting on increased production or stored C, or from a shift from anaerobic to aerobic activity (CH_4 emission was not taken into account).

If such shifts in C balance are occurring in response to changes in the extrinsic environment (P), then what are the changes in decomposition processes and populations? Are existing techniques sufficiently sensitive to detect predicted changes and to discriminate between the various factors controlling decomposition? What changes in nutrient distribution and dynamics have resulted from this shift in decomposition? These questions challenge the creativity of researchers in decomposition to predict from their theoretical and practical knowledge, to design critical experiments and observations, and to determine the relative importance of contributory processes. Arctic ecosystems will be the first to experience the impacts of increased greenhouse gases. They provide a real-world test-bed which can assist in resolving questions, including those on litter quality, that are likely to take longer to emerge at lower latitudes.

Management of Varying Litter Quality

Selman Waksman and co-workers stimulated interest in the use of organic inputs in a series of studies on the decomposition of crop residues. They developed theories linking the proximate composition, rate of decay and the release of nutrients which were the precursor for some of the current theories described earlier. How is the knowledge of litter quality and its effect on decomposition being translated into management practice? One particular application was in the development of the concept of 'synchrony', first outlined by Swift et al. (1980), and translated into the TSBF Programme (Tropical Soil Biology and Fertility Programme), initiated in 1984. The principle of 'synchrony' is – the release of nutrients from above-ground inputs and roots can be synchronized with plant growth demands (Myers et al., 1994).

This basic statement of the synchrony hypothesis was to stimulate research on decomposition processes in tropical agricultural (and forestry) systems with the target of improving the efficiency of utilization of organic materials as a source of nutrients with or without interaction with inorganic fertilizers. In the words of one of the principal scientists involved 'I want to be able to use a handful of organic matter with as much precision as I can a handful of compound fertilizer' (P.A. Sanchez, personal communication).

Initial evidence for the potential for synchrony was drawn from comparative observations on nutrient cycling in fallow and arable field plots in the humid tropics (Swift et al., 1981; Anderson and Swift, 1983). In experimental plots of cowpea, 63% of the N and 48% of the P was released from decomposing leaves within 3 weeks of the start of the rainy season, significantly in advance of crop plant growth. The fate of this material was not tracked in the experiments, but in the acid and coarse structured topsoils cultivated over the majority of the humid tropics, the nutrients released may be rapidly leached below the rooting depth of the young plants. In contrast, in a neighbouring bush fallow, while P was released at approximately the same rate from decomposing leaf litter, it took 15 weeks for release of the same proportion of N. Further, whereas there was a very narrow range of decay rates in the above-ground organic inputs in the crop field, the range in the fallow was much broader. From these initial observations it was postulated that if the dynamics of the organic system in an agricultural field could be brought closer to those in a natural system, then improvements in the efficiency of nutrient cycling could be achieved. Further, manipulation of organic inputs comprising a wide range of resource qualities could influence different parts of the nutrient cycle (Myers *et al.*, 1994).

In the last decade, there has been a substantial body of work on organic matter management in cropping systems, particularly but not exclusively in the tropics. These studies are reviewed elsewhere (Myers *et al.*, 1994; Palm, 1995; Myers *et al.*, chapter 17, this volume) but a few comments from the TSBF experience throughout the tropics illustrate some of the continuing issues, most of which have wider application.

1. Synchrony should always be seen in relative rather than absolute terms; a means of improving the efficiency of transfer of nutrients from organic input to crop, rather than to reduce nutrient losses to levels comparable with natural systems. In many cases the efficiency of transfer of nutrients is low, and often at levels similar to those for inorganic fertilizers in the same environment, that is 10–20% (Palm, 1988; Van der Meersch *et al.*, 1993; Myers *et al.*, 1994; Giller and Cadisch, 1995). The majority of information in tropical agriculture and agroforestry is from relatively high Q litters characterized by a rapid release pattern. The low Q litters seem to share the same low efficiency as rice straw, that is only 2–4% (Sisworo *et al.*, 1990). However, the essential test is whether a significant improvement can be achieved, within or between seasons.

2. Although resource quality has been the major focus of attention as a management tool, it is clearly only one of the factors determining nutrient release pattern. Placement, e.g. surface mulch or incorporation, may be just as important a means of manipulating nutrient availability, that is manipulation of the physical environment of the decomposing organic matter.

3. The value of organic inputs is by no means confined to their nutrient supplying power. Their influence on surface and soil temperature and moisture, on soil struc-

ture and ionic balance can be significant; something which needs to be recognized both experimentally and in terms of practical recommendations.

4. All residues seem to have the potential for significant residual effects via the transfer to soil organic matter or retention as undecomposed litter from year to year. The value of low Q litters may be underestimated in this respect. Thus Palm (1995) concluded that although N gain by the crop in the first year from low Q litter (high lignin and polyphenol) may be only 12% compared with 20% from high Q residues, loss in volatilization reduced from 20% to 5%, and 80% rather than 60% was retained as mulch or soil organic matter.

5. One of the components of the synchrony concept is that combinations of litters of varying Q can be used to regulate the timing (and position) of nutrient availability. Evidence to establish guidelines for such manipulation is still limited (Myers *et al.*, 1994; Palm 1995; Handayanto *et al.*, Chapter 14, this volume). In general, there is surprisingly little information on the conditions under which interaction between litters occurs (e.g. transfer of nutrients from high to low Q litters so reducing loss or damping the availability curve, but possibly increasing the decay rate of the low Q material – and possibly altering subsequent SOM-Q). An alternative is for little or no interaction between residues of different quality, the resultant nutrient availability curve being a mean of the component curves.

Within the TSBF programme research on soil fauna, particularly earthworms and termites, has shown the potential for management to maintain the important contribution which they play in decomposition, nutrient cycling, soil structure and crop productivity (Lavelle *et al.*, 1994). Management of microflora, other than through inoculation of *Rhizobium*, has shown little progress. However, it is clear from temperate studies that the transfer of nutrients by mycorrhizas to plants can be closely linked to the litter quality, particularly polyphenol content, with the potential to reduce nutrient leaching and denitrifica-tion in nutrient poor soils (Bending and Read, 1995; Northrup *et al.*, 1995).

An overwhelming conclusion is that, while the synchrony concept is philosophi-cally appealing and is very effective in focusing research, it has proved complex and resistant to clear management pre-scription. The research stimulated by the concept has identified the limitations in using the initial composition of the litter as a predictor of the longer term release of nutrients and the need to improve our understanding of the links between litter quality and SOM formation. These limita-tions do not negate the value of the syn-chrony concept; they argue for more sensitive understanding and caution against simplistic application.

Integration of Q into Ecosystem Models

A recurrent, underlying thread in this review has been the need to place the effects of litter Q into a broader context. To understand the effects of litter Q it is necessary to examine their interactions with O and P, to understand the plant physiology and the feedback effects within the system. In other words, to take an ecosystem view. Three developments in the last decade or so have been important in this respect.

1. A move from measurement of particu-lar species of litter in particular sites, towards much more extensive studies has proved difficult, but rewarding. The com-parative studies of Scots pine needle decomposition along climatic gradients initiated by Bjorn Berg have been seminal. This approach has complemented the syn-thesis of existing data to quantify, at a geo-graphical scale, the factors regulating decomposition (Meentemeyer and Berg, 1986). The comparative approach, using litters of both standard and variable Q, has been adopted in Europe in the CORE and DECO studies (Anderson *et al.*, 1992; Jansson and Reurslag, 1992), in North America by the US Long-Term Intersite

Decomposition Experiment Team (LIDET, 1995) and in the Canadian Intersite Decomposition Experiment (Trofymow *et al.*, 1995), and in the trop-ics by TSBF. This type of extensive, sys-tematic study over a range of climatic and soil conditions can quantify the relative influence of different factors and improve the basis for extrapolation of decomposi-tion data to large spatial scales. The diffi-culties of such coordinated, collaborative research are, however, not trivial, but are an important step for the research com-munity. The combination of such field data with explicit theoretical models is a powerful tool, as demonstrated by Berendse *et al.* (1987).

2. Representation of decomposition processes into ecosystem models has improved significantly and the CENTURY model in particular has had widespread application. In this, surface and root litters are separated in structural and metabolic fractions related to their decay rates. Sub-sequent decomposition generates a series of soil carbon pools with different decay potentials (microbial, active, slow, passive and leached) (Parton *et al.*, 1987). However, one of the problems here is to relate these to measurable and biochemi-cally understandable entities. Despite this, CENTURY has proved capable of simu-lating the decomposition and crop growth dynamics in tropical agriculture; systems for which it was not originally designed (Parton *et al.*, 1994).

The general principles by which decomposition is simulated in most models are very similar, with dependence on basic relationships between mass loss constants (k) and C-to-N, temperature and moisture. The influence of soil conditions is increas-ingly being expressed in terms of soil tex-ture or pH to enhance the ability to extrapolate over a range of sites. Mass loss is the usual measure and the relationships between it and nutrient release and avail-ability are usually assumed to be linear. This simplistic assumption works at a general level but may have severe limita-tions as demands for model precision increase and does not assist in developing our understanding of nutrient dynamics.

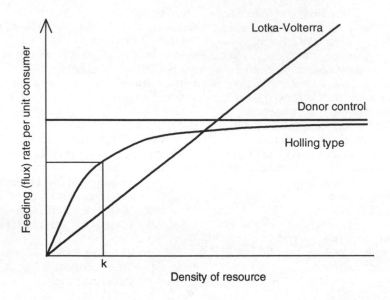

Fig. 1.7. Relationships between feeding rate and resource density for different population control theories as a basis for explicit representation of population dynamics into decomposition models (from Zheng, 1993).

Representation of the dynamics of organism populations and community structure in ecosystem models is still in its infancy – and is unnecessary for many purposes. However, to improve our basic understanding of soil biology and to contribute to the debate on biodiversity, it is important to stretch our expertise. Recent technical developments providing information on community function are encouraging (e.g. Zak *et al.*, 1994) and at least two lines of modelling are promising.

First, analysis of the trophic structure of the soil community provides a conceptual framework of relationships (Moore and Hunt, 1988; Moore and de Ruiter, 1991). This can accommodate information from detailed quantitative studies (e.g. Allen-Morley and Coleman, 1989; Clarholm, 1989; Couteaux *et al.*, 1991) and comparative studies of management treatments (Hendrix *et al.*, 1986). Expression of omnivore and other non-specific relationships, as well as non-trophic relationships, remain a major challenge.

Second, more explicit representations of the population dynamics which drive decomposition are being developed

(Sinsabaugh and Moorhead, 1994). For example, Bengtsson *et al.* (1996) explore the explicit relationships between soil food web dynamics and a general theory of carbon and nitrogen dynamics developed by Bosatta and Ågren (1991). Fundamentally the model is based on the theory that relationships in detritus food webs are donor-controlled, with consumer density having no influence on the rate of resource removal (Pimm, 1982; Begon *et al.*, 1986). The rationale for using donor-control as the theoretical basis may be somewhat pragmatic but other options, as described by Zheng (1993) (Fig. 1.7), create other problems. The aim to incorporate substrate quality, substrate modification and resource competition is laudable (Bengtsson *et al.*, 1996).

These two approaches have limited representation of the changes in community composition during decomposition and the variation in communities between resources of different quality. A possible basis might be the logical extension of the plant growth strategy theory of Grime (1988). The general life history characteristics of the succession of decomposers can

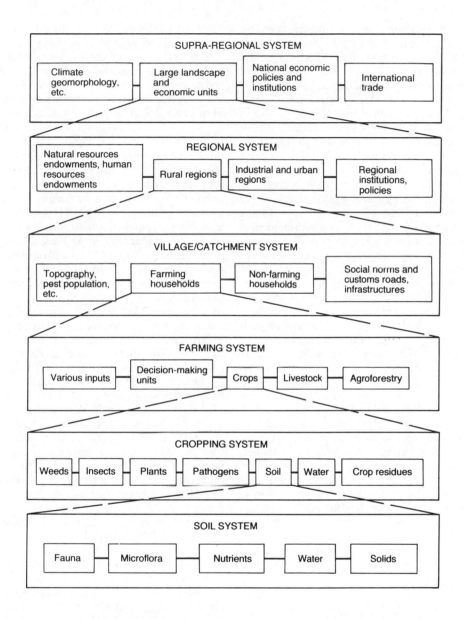

Fig. 1.8. A hierarchical model in which soil biological processes are placed in the context of agricultural systems (from Izac, 1994).

be related to the growth strategies, and hence biochemistry, of the plants from which the litter was derived (Pugh, 1980; Cooke and Rayner, 1984; Heal and Ineson, 1984; Swift and Heal, 1986; Andren *et al.*, 1995). There is considerable circumstantial evidence supporting the hypothesis that there are three primary ecological strategies, corresponding to habitat conditions defined by resource availability. While the hypothesis is usually applied to plants, it is also appropriate to the selection of different strategies by the microbial and faunal communities in response to variations in litter quality and to the change in the availability of resources during decomposition. The strategy concept provides considerable opportunity for prediction of response within the decomposer community (Grime and Anderson, 1986).

3. A third development takes a more expansionist view. So far we have taken the conservative view of the ecosystem, but that is not adequate. The concept of sustainability is a modern development to ensure the inclusion of *Homo sapiens* in the older perception of the ecosystem. Elliott *et al.* (1994) emphasized that 'humans plus the environment are taken as the ecosystem for the purposes of management'. Within TSBF, research on synchrony of decomposition with crop demand is integrated with farm management and the farm system. This recognizes that the production and decomposition of organic resources is only one of the potential resources to be considered in the decision-making process. Account must also be taken of social and economic factors in the farm, village, region and supra-regional systems (Izac, 1994; Swift *et al.*, 1994) (Fig. 1.8). The approach makes considerable demands on interdisciplinary cooperation, including interaction between researcher, farmer and policy maker. This level of cooperation and interaction is necessary if the research on decomposition is to be relevant, applicable and translated into action. It is an approach which is of general application and, despite sophisticated communications and organization, it is particularly needed in developed countries.

Conclusions

Research development tends to contain phases of data accumulation, often stimulated by technical innovation and leading to the emergence of new theory. Research on litter quality and decomposition has been through a very productive phase which has generated a wealth of descriptive data on the rates of decomposition of specific litters in specific sites, the organisms involved in the processes, and the relative importance of various controlling factors. The development of techniques in analysis of organic chemistry and in the use of both stable and radioisotopes have made important contributions to the characterization of litter composition and to quantifying the dynamics of decomposition. These methods have also had significant impacts on the understanding of the characteristics and dynamics of soil organic matter, a topic that is largely outside the scope of the present discussion on litter quality. In contrast, the rapidly developing techniques of molecular biology have, as yet, had relatively little impact on our understanding of the organisms involved in dynamics of decomposition. In addition, the application of measurements of microbial biomass has not been as productive as expected by its advocates. Research on microbial biomass has certainly focused attention on the organisms as a nutrient pool in its own right with distinctive turnover characteristics, but it has not significantly enhanced our understanding of the dynamics and function of soil microflora. The potential of the suite of methods now available has yet to be realized (Coleman *et al.*, 1994).

The developments in methodology and in the detailed understanding of decomposition will undoubtedly be reflected in the proceedings of this meeting. But to what extent has the increased detail and volume of information been converted into improvements in synthesis and theory? There have been a number of significant and important trends; refinements rather than major new theoretical insights. Some examples cited earlier are: (i) collaboration between researchers has provided geographically and environmentally extensive information on the factors controlling decomposition; (ii) increasingly explicit definition of decomposition theory and representation of litter and soil organic matter pools in ecosystem models to explore nutrient and trace gas release; (iii) recognition of the potential application of plant 'Strategy Theory' to decomposer populations and processes; (iv) the gradual development of concepts of food-web and other functional relationships which have helped to structure and quantify the inter-actions between organisms involved in the decomposer community; (v) the attempts to translate general concepts of decom-position into a practical context through the 'Synchrony' hypothesis.

While there have been some develop-ments in synthesis and theory over the last two decades, these do not seem to match the intensive efforts and wealth of results produced by researchers in decomposition over the same period. Are there other general patterns, concepts, models and theories which will emerge in the chapters that follow? The tasks are not trivial, but they are critical to the issues that have to be addressed over the next decade. Three major issues have emerged which demand the application of existing understanding and new research on decomposition at local, national, regional and global levels. The issues are – sustainability, biodiversity, and climate change. What are the chal-lenges for decomposition research? Some of the questions to be addressed are:

1. Sustainability. What are the changes in management of decomposition and nutri-ent mobilization which can maintain or enhance soil fertility, particularly where soils are vulnerable to degradation or there are significant losses from the system to other environments? By what methods can decomposition be managed to increase the efficiency of use of inorganic fertilizers, especially in soils of low fertility and areas of environmental sensitivity? How can we improve the integration of organic matter management into farming systems?

2. Biodiversity. Given the great diversity of soil organisms and complexity of food webs, are there particular ecosystems or environmental conditions in which reduc-tions in genetic or species diversity cause significant changes in decomposition processes? To what extent do decomposer organisms adapt to reductions in diversity by increasing their functional niche? Are there general relationships between species diversity and decomposer function and are there definable levels of diversity at which decomposer processes change significantly? Are there particular species or functional

groups for which reduction or extinction of populations causes a disproportionate change in rate of decomposition?

3. Climate change. To what extent are decomposition processes resistant to climate changes through the genetic, species and functional diversity of the biota? What are the most significant quan-titative feedback effects on plant growth and greenhouse gas emissions that will result from direct and indirect effects of change in climatic factors? To what extent will the differential dispersal abilities of decomposer organisms influence the change in distribution or characteristics of ecosystems? What are the most effective methods of decomposition management by which carbon sequestration and trace gas emissions can be controlled?

These and related questions require much more predictive application of decomposition information and its integra-tion with other disciplines. An expansion from site specific studies, exploration of the logical consequences of results, plus rigorous testing of predictions against independent observations, are all trends which should be emerging from the papers at this meeting.

Two apparently divergent pathways may be distinguishable and both are neces-sary in addressing these major issues. The most obvious pathway leads to an increas-ingly detailed and mechanistic understand-ing of decomposition processes and populations. This reductionist approach aims to quantify our understanding of decomposition at a level comparable to that of the other main biological process – photosynthesis. The second pathway leads to the production of tools for management and policy. To do this, the science of decomposition has to embrace and inte-grate with the constructs and information from other disciplines, including those of the social and economic sciences. It has to determine the properties of the decompos-er system which are relevant to the level of resolution being considered. It also requires presentation of conclusions at a level of generalization which, whilst rigor-ously derived and scientifically defensible,

can be translated into practicable recommendations. Both pathways involve the
further development of models, but the
detail of the mechanistic models required
to express the processes is quite distinct
from the management or policy related
models. The latter types, while based on
knowledge derived from process mechanisms, require much more succinct expression, little elaboration of detail, but must
have the ability to express geographical
and temporal variability in terms that are
relevant to the wider user audience.

The cascade model presented by Swift
et al., (1979) gave a highly mechanistic view
of the decomposition process. Not surprisingly the volume of work in the succeeding
17 years has revealed a wealth of detail that
demonstrates a much greater degree of variation in pattern, process and population
than the simple model predicts. The
developments in decomposition research
seem to have been evolutionary rather than
revolutionary. The major challenges which
now face the science will have to be

addressed largely through the use of existing
tools, but will require major developments
in thought and in application.

Finally, a word on communication.
Probably the most widely read publication
on litter quality this year was Charlesworth
(1995). She distilled the essence of management of Q, and placed it in the context
of O and P, in a single page, describing the
process of one of the oldest practices in
sustainable resource use – composting
(Frontispiece). We have a lot to learn
about synthesis and communication!

Acknowledgements

It is a particular pleasure to thank
Cheryl Palm for her input to this chapter
and to all our colleagues in TSBF from
around the world for their stimulus.
Thanks also to John Beckett, Librarian of
ITE Merlewood for help with literature,
and to the reviewers for their constructive
comments.

References

Aber, J. and Melillo, J. (1982) Nitrogen mineralization in decaying hardwood leaf litter as a
    function of initial nitrogen and lignin content. *Canadian Journal of Botany* 60,
    2263–2269.
Allen-Morley, C.R. and Coleman, D.C. (1989) Resilience of soil biota in various food webs
    to freezing perturbations. *Ecology* 70, 1127–1141.
Ander, P., Eriksson, K.E. and Hui-Sheng, Y. (1984) Metabolism of lignin derived aromatic
    acids by wood-rotting fungi. *Journal of General Microbiology* 130, 63–68.
Anderson, J.M. (1988a) Fauna mediated transport processes in soil. *Agriculture, Ecosystems
    and Environment* 24, 5–19.
Anderson, J.M. (1988b) Spatiotemporal effects of invertebrates on soil processes. *Biology and
    Fertility of Soils* 6, 216–227.
Anderson, J.M. (1991) The effect of climate change on decomposition in grassland and conif-
    erous forests. *Ecological Applications* 1, 326–347.
Anderson, J.M. (1992) Responses of soils to climate change. *Advances in Ecological Research*
    22, 163–210.
Anderson, J.M. and Swift, M.J. (1983) Decomposition in tropical forests. In: Sutton, S.C.,
    Whitmore, T.C. and Chadwick, A.C. (eds) *Tropical Rain Forest; Ecology and
    Management*. Blackwell, Oxford, pp. 287–310.
Anderson, D.W., Saggar, S., Bettany, J.R. and Stewart, J.W.B. (1981) Particle size fractions
    and their use in studies of soil organic matter: I. The nature and distribution of forms of
    carbon, nitrogen and sulphur. *Soil Science Society of America Journal* 45, 767–772.
Anderson, J.M., Beese, F., Berg, B., Bolger, T., Couteaux, M.M., Henderson, R., Ineson, P.,
    McCarthy, P., Palka, L., Raubuch, M., Splatt, P., Verhoef, H.A. and Willison, T. (1992)
    Mechanisms of nutrient turnover in the soil compartment of forests. In: Teller, A.,
    Mathy, P. and Jeffers, J.N.R. (eds) *Responses of Forest Ecosystems to Environmental Change*.
    Elsevier, London, pp. 342–350.

Andren, O., Bengtsson, J. and Clarholm, M. (1995) Biodiversity and species redundancy among litter decomposers. In: Collins, H.P., Robertson, G.P. and Klug, M.J. (eds) *The Significance and Regulation of Soil Biodiversity*. Kluwer Academic Press, Dordrecht, pp. 141–151.

Andren, O., Lindberg, T., Bostrom, U., Clarholm, M., Hansson, A.-C., Johansson, G., Lagerlof, J., Paustian, K., Persson, J., Pettersson, R., Schnurer, J., Sohlenius, B. and Wivstad, M. (1990) Organic carbon and nitrogen flows. In: Andren, O., Lindberg, T., Paustian, K. and Rosswall, T. (eds) *Ecology of Arable Lands – Organisms, Carbon and Nitrogen Cycling. Ecological Bulletins (Copenhagen)* 40, 85–126.

Beare, M.H., Cabrera, M.L., Hendrix, P.E. and Coleman, D.C. (1994) Aggregate-protected and unprotected organic matter pools in conventional- and no-tillage soils. *Soil Science Society of America Journal* 58, 787–795.

Begon, M., Harper, J.L. and Townsend, C.R. (1986) *Ecology: Individuals, Populations and Communities*. Blackwell, Oxford, 876 pp.

Bending, G.D. and Read, D.J. (1995) The structure and function of the vegetative mycelium of ectomycorrhizal plants. V. Foraging behaviour and translocation of nutrients from litter. *New Phytologist* 130, 401–409.

Bengtsson, J., Zheng, D.W., Agren, G.I. and Persson, T. (1996) Food webs in soil: an interface between population and ecosystem ecology. In: Jones, C.G. and Lawton, J. (eds) *Linking Species and Ecosystems*. Chapman and Hall, London (in press).

Berendse, F., Berg, B. and Bosatta, E. (1987) The effect of lignin and nitrogen on the decomposition of litter in nutrient-poor ecosystems: a theoretical approach. *Canadian Journal of Botany* 65, 1116–1120.

Berg, B. (1986) Nutrient release from litter and humus in coniferous forest soils – a mini review. *Scandinavian Journal of Forest Research* 1, 359–369.

Berg, B. and Ekbohm, G. (1983) Nitrogen immobilisation in decomposing needle litter at variable carbon:nitrogen ratios. *Ecology* 64, 63–67.

Berg, B. and McClaugherty, C. (1987) Nitrogen release from litter in relation to the disappearance of lignin. *Biogeochemistry* 4, 219–224.

Berg, B. and Staaf, H. (1980) Decomposition rate and chemical changes of Scots pine needle litter. II. Influence of chemical composition. In: Persson, T. (ed.) *Structure and Function of Northern Coniferous Forests – An Ecosystem Study. Ecological Bulletins (Stockholm)* 32, 373–390.

Berg, B. and Tamm, C.O. (1991) Decomposition and nutrient dynamics of litter in long-term optimum nutrition experiments *Scandinavian Journal of Forest Research* 6, 305–321.

Bliss, L.C., Heal, O.W. and Moore, J.J. (1981) *Tundra Ecosystems: A Comparative Analysis*. Cambridge University Press, Cambridge, 813 pp.

Bocock, K.I.. and Gilbert, O.J.W. (1957) The disappearance of leaf litter under different woodland conditions *Plant and Soil* 9, 179–185.

Boone, R.D. (1994) Light fraction soil organic matter: origin and contribution to net nitrogen mineralization. *Soil Biology and Biochemistry* 26, 1159–1168.

Bornebusch, C.H. (1930) The fauna of forest soil. *Forstlige Forsoksvaesen i Danmark* 11, 1–224.

Bosatta, E. and Agren, G.I. (1991) Dynamics of carbon and nitrogen in the soil: a generic theory. *American Naturalist* 138, 227–245.

Bosatta, E. and Staaf, H. (1982) The control of nitrogen turnover in forest litter. *Oikos* 39, 143–154.

Brett, C. and Waldron, K. (1990) *Physiology and Biochemistry of Plant Cell Walls*. Unwin Hyman, London, 194 pp.

Breymeyer, A. and van Dyne, G.M. (eds) (1980) *Grasslands: Systems Analysis and Management*. Cambridge University Press, Cambridge, 950 pp.

Broadfoot, W.M. and Pierre, W.H. (1939) Forest soil studies. I. Relation of rate of decomposition of tree leaves to their acid-base balance and other chemical properties. *Soil Science* 48, 329–348.

Bryant, J.P., Chapin III, F.S. and Klein, D.R. (1983) Carbon/nutrient balance of boreal plants in relation to vertebrate herbivory. *Oikos* 40, 358–368.

Campbell, W.G. (1929) The chemical aspect of the destruction of oakwood by powder-post and death-watch beetles, *Lyctus* spp. and *Xestobium* sp. *Biochemical Journal* 23, 1290–1293.

Charlesworth, K. (1995) Life, the universe & (almost) everything! No. 240. Muck & Magic. *New Scientist* 146, No.1978, 53.

Cheng, W., Coleman, D.C., Carroll, C.R. and Hoffman, C.A. (1994) Investigating short-term carbon flows in the rhizosphere of different plant species using isotopic trapping. *Agronomy Journal* 86, 782–788.

Christensen, B.T. (1987) Decomposability of organic matter in particle size fractions from field soil with straw incorporation. *Soil Biology and Biochemistry* 19, 429–435.

Clarholm, M. (1989) Effects of plant–bacterial–amoebal interactions on plant uptake of nitrogen under field conditions. *Biology and Fertility of Soils* 8, 373–378.

Coleman, D.C., Dighton, J., Ritz, K. and Giller, K.E. (1994) Perspectives on the compositional and functional analysis of soil communities. In: Ritz, K., Dighton, J. and Giller, K.E. (eds) *Beyond the Biomass*. Wiley, Chichester, pp. 261–271.

Coley, P.D. (1987) Interspecific variation in plant anti-herbivore properties: the role of habitat quality and rate of disturbance. *The New Phytologist* 106 (Suppl.), 251–263.

Cooke, R.C. and Raynor, A.D.M. (1984) *The Ecology of Saprophytic Fungi*. Longman, London, 415 pp.

Cornelissen, J.H.C. (1996) An experimental comparison of leaf decomposition rates in a wide range of temperate plant species and types. *Journal of Ecology* (in press).

Cotrufo, M.F. (1995) Effects of enriched atmospheric concentration of carbon dioxide on tree litter decomposition. PhD Thesis, University of Lancaster, Lancaster, UK.

Cotrufo, M.F., Ineson, P. and Rowland, A.P. (1994) Decomposition of tree leaf litters grown under elevated CO_2: effect of litter quality. *Plant and Soil* 163, 121–130.

Coupland, R.T. (ed.) (1979) *Grassland Ecosystems of the World: Analysis of Grasslands and their Uses*. Cambridge University Press, Cambridge, 401 pp..

Couteaux, M.M., Mousseau, M., Celerier, M.-L. and Bottner, P. (1991) Increased atmospheric CO_2 and litter quality: decomposition of sweet chestnut leaf litter with animal food webs of different complexities. *Oikos* 61, 54–64.

Davies, R.I., Coulson, C.B. and Lewis, D.A. (1964) Polyphenols in plant, humus and soil. IV. Factors leading to increase in biosynthesis of polyphenol in leaves and their relationship to mull and mor formation. *Journal of Soil Science* 15, 310–318.

Egger, K.N. (1992) Analysis of fungal population structure using molecular techniques. In: Carroll, G.C. and Wicklow, D.T. (eds) *The Fungal Community: Its Organisation and Role in the Ecosystem*. Marcel Dekker, New York, pp. 193–208.

Elliott, E.T. and Cambardella, C.A. (1991) Physical separation of soil organic matter. *Agriculture, Ecosystems and Environment* 34, 407–419.

Elliott, E.T., Janzen, H.H., Campbell, C.A., Cole, C.V. and Myers, R.J.K. (1994) Principles of ecosystem analysis and their application to integrated nutrient management and assessment of sustainability. In: Wood, R.C. and Dumanski, J. (eds) *Sustainable Land Management for the 21st Century*. Marcel Dekker, New York, pp. 35–57.

Findlay, W.P.K. (1934) Studies in the physiology of wood-decay fungi. 1. The effects of nitrogen content upon the rate of decay. *Annals of Botany* 46, 109–117.

Flanagan, P.W. and van Cleve, K. (1983) Nutrient cycling in relation to decomposition and organic matter quality in taiga ecosystems. *Canadian Journal of Forest Research* 13, 795–817.

Fog, K. (1988) The effect of added nitrogen on the rate of decomposition of organic matter. *Biological Reviews* 63, 433–462.

Gallardo, A. and Merino, J. (1993) Leaf decomposition in two Mediterranean ecosystems of southwest Spain: influence of substrate quality. *Ecology* 74, 152–161.

Gallet, C. and Lebreton, P. (1995) Evolution of phenolic patterns in plants and associated litters and humus of a mountain forest ecosystem. *Soil Biology and Biochemistry* 27, 157–166.

Gehrke, C., Johanson, U., Callaghan, T.V., Chadwick, D. and Robinson, C.H. (1995) The impact of enhanced ultraviolet-B radiation on litter quality and decomposition processes in *Vaccinium* leaves from the Subarctic. *Oikos* 72, 213–222.

Giller, K.E. and Cadisch, G. (1995) Future benefits from biological nitrogen fixation – an ecological approach to agriculture. *Plant and Soil* 174, 255–277.

Greenland, D.J. (1965) Interaction between clays and organic compounds in soils. II. Adsorption of soil organic compounds and its effect on soil properties. *Soil Fertility* 28, 521–532.

Grime, J.P. (1988) The C-S-R model of primary plant growth strategies – origins, implications and tests. In: Gottleib, L.D. and Jain, S.K. (eds) *Plant Evolutionary Biology*. Chapman and Hall, London, pp. 371–393.

Grime, J.P. and Anderson, J.M. (1986) Environmental controls over organism activity: Introduction. In: Van Cleve, K., Chapin III, F.S., Flanagan, P.W., Viereck, L.A. and Dyrness, C.T. (eds) *Forest Ecosystems in the Alaskan Taiga*. Springer, New York, pp. 89–95.

Handley, W.R.C. (1954) Mull and mor formation in relation to forest soils. *Bulletin of the Forestry Commission* 23, 1–115.

Heal, O.W. and Ineson, P. (1984) Carbon and energy flow in terrestrial ecosystems: Relevance to microflora. In: Klug, M.J. and Reddy, C.A. (eds) *Current Perspectives in Microbial Ecology*. American Society for Microbiology, Washington, pp. 394–404.

Heal, O.W., Struwe, S. and Kjoller, A. (1996) Diversity of soil biota and ecosystem function. In: Walker, B. and Steffen, W. (eds) *Global Change in Terrestrial Ecosystems*. IGBP Vol. 1. Cambridge University Press, Cambridge (in press).

Hedges, J.I. and Mann, D.C. (1979) The characterisation of plant tissues by their lignin oxidation products. *Geochemica et Cosmochemica Acta* 43, 1803–1807.

Hendrix, P.F., Parnalee, R.W., Crossley, D.A., Coleman, D.C., Odum, E.P. and Groffman, P.M. (1986) Detritus food webs in conventional and non-tillage agroecosystems. *BioScience* 36, 374–380.

Herman, W.A., McGill, W.B. and Dormaar, J.F. (1977) Effects of initial chemical composition on decomposition of roots of three grass species. *Canadian Journal of Soil Science* 57, 205–215.

Izac, A.M.N. (1994) Ecological-economic assessment of soil management practices for sustainable land use in tropical countries. In: Greenland, D.J. and Szabolcs, I. (eds) *Soil Resilience and Sustainable Land Use*. CAB International, Wallingford, pp. 77–96.

Jansson, P-E. and Reurslag, A. (1992) Climatic influence on litter decomposition: Methods and some results of a NW-European transect. In: Teller, A., Mathy, P. and Jeffers, J.N.R. (eds) *Response of Forest Ecosystems to Environmental Changes*. Elsevier, London, pp. 351–358.

Janzen, R.A., Dormaar, J.F. and McGill, W.B. (1995) A community-level concept of controls on decomposition processes: decomposition of barley straw by *Phanerochaete chrysosporium* or *Phlebia radiata* in pure and mixed culture. *Soil Biology and Biochemistry* 27, 173–180.

Jenkinson, D.S. (1977) Studies on the decomposition of plant material in soil. V. The effects of plant cover and soil type on the loss of carbon from ^{14}C labelled ryegrass decomposing under field conditions. *Journal of Soil Science* 28, 209–213.

Jenny, H., Gessel, S.P. and Bingham, F.T (1949) Comparative study of decomposition rates in temperate and tropical regions. *Soil Science* 68, 419–432.

Jensen, H.L. (1929) On the influence of carbon:nitrogen ratios of organic material on the mineralisation of nitrogen. *Journal of Agricultural Science* 19, 71–82.

Johanson, U., Gehrke, C., Bjorn, L.O., Callaghan, T.V. and Sonesson, M. (1995) The effects of enhanced UV-B radiation on a Subarctic heath ecosystem. *Ambio* 24, 106–111.

Johnston, A.E. (1994) The Rothamsted Classical Experiments. In: Leigh, R.A. and Johnston, A.E. (eds) *Long-term Experiments in Agricultural and Ecological Sciences*. CAB International, Wallingford, pp. 9–37

Kononova, M.M. (1966) *Soil Organic Matter*, 2nd edn. Pergamon Press, Oxford.

Korner, C. and Miglietta, F. (1994) Long term effects of naturally enriched CO_2 on Mediterranean grassland and forest trees. *Oecologia* 99, 343–351.

Lavelle, P., Blanchart, E., Martin, A. and Martin, S. (1993) A hierarchical model for decomposition in terrestrial ecosystems: Application to soils of the humid tropics. *Biotropica* 25, 130–150.

Lavelle, P., Dangerfield, M., Fragoso, C., Eschenbrenner, V., Lopez-Hernandez, D., Pashanasi, B. and Brussard, L. (1994) The relationship between soil macrofauna and tropical soil fertility. In: Woomer, P.L. and Swift, M.J. (eds) *The Biological Management of Tropical Soil Fertility*. Wiley, Chichester, pp. 137–170.

Lawes, J. (1861) On the application of different manures to different crops and on their proper distribution on the farm. Private publication cited by Dyke, G.V. (1993) *John Lawes of Rothamsted: Pioneer of Science Farming and Industry*. Hoos Press, Harpenden, UK.

Liebig, J. von, (1840) Organic Chemistry and its Application to Agriculture and Physiology. Cited in Dyke, G.V. (1993) *John Lawes of Rothamsted: Pioneer of Science Farming and Industry*. Hoos Press, Harpenden, UK.

Linderstrom-Lang, K. and Duspiva, F. (1935) Beiträge zur enzymatischen Histochemie. XVI. Die Verdauung von Keratin durch die Larten der Kleidermovte. *Hoppe-Seyler's Zeitschrift für Physiologische Chemie* 237, 131–158.

Long-Term Intersite Decomposition Experiment Team (LIDET) (1995) *Meeting the Challenge of Long-Term, Broad-Scale Ecological Experiments*. Publication No. 19, LTER Network Office. Seattle, Washington, 23 pp.

Mayaudon, J. and Simonart, P. (1959) Etude de la decomposition de la materiere organique dans le sol au moyen de carbon radioactif. III. Decomposition des substances solubles dialysables des proteines et des humicelluloses *Plant and Soil* 11, 170–175.

McClaugherty, C. and Berg, B. (1987) Cellulose, lignin and nitrogen concentrations as rate regulating factors in late stages of forest litter decomposition. *Pedobiologia* 30, 101–112.

Meentemeyer, V. and Berg, B. (1986) Regional variation in rate of mass loss of *Pinus sylvestris* needle litter in Swedish pine forests as influenced by climate and litter quality. *Scandinavian Journal of Forest Research* 1, 167–180.

Mellillo, J.M., Aber, J., Linkens, A.E., Ricca, A., Fry, B. and Nadelhoffer, K. (1989) Carbon and nitrogen dynamics along the decay continuum: plant litter to soil organic matter. In: Clarholm, M. and Bergstrom, L. (eds) *Ecology of Arable Lands*. Kluwer Academic Press, Dordrecht, pp. 53–62.

Miller, H.G., Cooper, J.M., Miller, J.D. and Pauline, O.J.L. (1979) Nutrient cycles in pine and their adaptation to poor soils. *Canadian Journal of Forest Research* 9, 19–26.

Minderman, G. (1968) Addition, decomposition and accumulation of organic matter in forests. *Journal of Ecology* 56, 355–362.

Moore, J.C. and de Ruiter, P.C. (1991) Temporal and spatial heterogeneity of trophic interactions within below-ground food webs. *Agriculture, Ecosystems and Environment* 34, 371–397.

Moore, J.C. and Hunt, H.W. (1988) Resource compartmentation and the stability of real ecosystems. *Nature* 333, 261–263.

Mulder, M.M., Pureveen, J.B.M., Boon, J.J. and Martinez, A.T. (1991) An analytical pyrolysis mass spectrometric study of *Eucryphia cordifolia* wood decayed by white-rot and brown-rot fungi. *Journal of Analytical and Applied Pyrolysis* 19, 175–193.

Muller, P.E. (1887) *Studien über die natürlichen Humusformen und deren Einwirkungen auf Vegetation und Boden*. Springer, Berlin.

Myers, R.J.K., Palm, C.A., Cuevas, E., Gunatilleke, I.U.N. and Brossard, M. (1994) The synchronisation of nutrient mineralisation and plant nutrient demand. In: Woomer, P.L. and Swift, M.J. (eds) *The Biological Management of Tropical Soil Fertility*. Wiley, Chichester, pp. 81–116.

Norby, R.J. (1994) Issues and perspectives for investigating root response to elevated atmospheric carbon dioxide. *Plant and Soil* 165, 9–20.

Northrup, R.R., Yu Zengshou, Dahlgren, R.A. and Vogt, K.A. (1995) Polyphenol control of nitrogen release from pine litter. *Nature* 377, 227–229.

Odum, E.P. (1969) The strategy of ecosystem development. *Science* 164, 262–270.

Oechel, W.C., Hastings, S.J., Vourlitis, G., Jenkins, M., Riechers, G. and Grulke, N. (1993) Recent changes of Arctic tundra ecosystems from a net carbon dioxide sink to a source. *Nature* 361, 520–523.

Olson, J.S. (1963) Energy storage and the balance of producers and decomposers in ecological systems. *Ecology* 44, 322–331.

O'Neill, E.G. (1994) Responses of soil biota to elevated atmospheric carbon dioxide. *Plant and Soil* 165, 55–65.

Palm, C.A. (1988) Mulch quality and nitrogen dynamics in alley cropping systems in the Peruvian Amazon. PhD thesis, North Carolina State University, Raleigh, NC, USA.

Palm, C.A. (1995) Contribution of agroforestry trees to nutrient requirements of intercropped plants. *Agroforestry Systems* 30, 105–124.

Parton, W.J., Schimel, D.S., Cole, C.V. and Ojima, D.S. (1987) Analysis of factors controlling soil organic matter levels in Great Plains grasslands. *Soil Science Society of America* 51, 1173–1179.

Parton, W.J., Woomer, P.L. and Martin, A. (1994) Modelling soil organic matter dynamics and plant productivity in tropical ecosystems. In: Woomer, P.L. and Swift, M.J. (eds) *The Biological Management of Tropical Soil Fertility*. Wiley, Chichester, pp. 171–188.

Peterson, B.J. and Fry, B. (1987) Stable isotopes in ecosystem studies. *Annual Review of Ecology and Systematics* 18, 293–320.

Pimm, S.L. (1982) *Food Webs*. Chapman and Hall, London, 219 pp.

Pugh, G.J.F. (1980) Strategies in fungal ecology. *Transactions of the British Mycological Society* 75, 1–14.

Reichle, D.E. (ed) (1981) *Dynamic Properties of Forest Ecosystems*. Cambridge University Press, Cambridge, 683 pp.

Ritz, K., Dighton, J. and Giller, K.E. (1994) *Beyond the Biomass*. Wiley, Chichester, 275 pp.

Rodin, L.E. and Basilevic, N.I. (1967) *Production and Mineral Cycling in Terrestrial Vegetation*. Oliver and Boyd, Edinburgh, 288 pp.

Saggar, S., Tate, K.R., Feltham, C.W., Childs, C.W. and Parshotam, A. (1994) Carbon turnover in a range of allophanic soils amended with ^{14}C labelled glucose. *Soil Biology and Biochemistry* 26, 1263–1271.

Sanger, L.J., Cox, P., Splatt, P., Whelan, M.J. and Anderson, J.M. (1996) The variation in quality and decomposition potential of *Pinus sylvestris* litter from sites with different soil characteristics: lignin and phenylpropanoid signatures. *Soil Biology and Biochemistry* (in press).

Scharpenseel H.W., Schomaker, M. and Ayoub, A. (eds) (1990) *Soils on a Warmer Earth*. Elsevier, Amsterdam, 274 pp.

Scheu, S. and Wolters, V. (1991) The influence of fragmentation and bioturbation on the decomposition of C-14 labelled beech leaf litter. *Soil Biology and Biochemistry* 23, 1029–1034.

Sinsabaugh, R.L. and Moorhead, D.L. (1994) Resource allocation to extracellular enzyme production: a model for nitrogen and phosphorus control of litter decomposition. *Soil Biology and Biochemistry* 26, 1305–1311.

Sisworo, W.H., Mitrosuhardjo, M.M. Rasjid., H. and Myers, R.J.K. (1990) The relative roles of N fixation, fertilizer, crop residues and soil in supplying N in multiple cropping systems in a humid, tropical upland cropping system. *Agrivita* 15, 69–75.

Swift, M.J. and Anderson, J.M. (1989) Decomposition as an ecosystem process. In: Werger, M.A. (ed.) *Ecosystems of the World, Vol 13B Tropical Rain Forests*. Elsevier, London, pp. 547–569.

Swift, M.J. and Heal, O.W. (1986) Theoretical considerations of microbial succession and growth strategies: intellectual exercise or practical necessity? In: Jensen,V., Kjoeller, A. and Sorensen, L.H. (eds) *Microbial Communities in Soil*. Elsevier, Amsterdam, pp. 115–131.

Swift, M.J., Heal, O.W. and Anderson, J.M. (1979) *Decomposition in Terrestrial Ecosystems*. Blackwell Scientific Publications, Oxford, 372 pp.

Swift, M.J., Cook, A.G. and Perfect, T.J. (1980) The effects of changing agricultural practice on the biology of a forest soil in the sub-humid tropics. 2. Decomposition. In: Furtado, J.I. (ed.) *Tropical Ecology and Development*. International Society of Tropical Ecology, Kuala Lumpur, pp. 341–348.

Swift, M.J., Russel-Smith, A. and Perfect, T.J. (1981) Decomposition and mineral nutrient dynamics of plant litter in a regenerating bush-fallow in the sub-humid tropics. *Journal of Ecology* 69, 981–995.

Swift, M.J., Bohren, L., Carter, S.E., Izac, A.M. and Woomer, P.L. (1994) Biological management of tropical soils: integrating process research and farm practice. In: Woomer, P.L. and Swift, M.J. (eds) *The Biological Management of Tropical Soil Fertility*. Wiley, Chichester, pp. 209–227.

Swinnen, J., van Veen, J.A. and Merckx, R. (1995) Root decay and turnover of rhizodeposits in field-grown winter wheat and spring barley estimated by ^{14}C pulse-labelling. *Soil Biology and Biochemistry* 27, 211–218.

Thomas, R.B., Richter, D.D., Ye, H., Heine, P.R. and Strain, B.R. (1991) Nitrogen dynamics and growth of seedlings of an N-fixing tree (*Gliricidia sepium* (Jacq. Walp.)) exposed to elevated atmospheric carbon dioxide. *Oecologia* 8, 415–421.

Tisdall, J.M. and Oades, J.M. (1982) Organic matter and water-stable aggregates in soils. *Soil Science* 33, 141–163.

Trofymow, J.A., Preston, C.M. and Prescott, C.E. (1995) Litter quality and its potential effect on decay rates of materials from Canadian forests. *Water, Air and Soil Pollution* 82, 215–226.

UCPE (1995) *Report of the Unit of Comparative Plant Ecology 1992–1994*. UCPE, Sheffield, 39 pp.

Van Breeman, N. (ed.) (1992) *Decomposition and Accumulation of Organic Matter in Terrestrial Ecosystems: Research Priorities and Approaches*. Ecosystems Research Report 1. Commission of the European Communities, Brussels, 105 pp.

Van der Meersch, M.K., Merkx, R. and Mulongoy, K. (1993) Evolution of plant biomass and nutrient content in relation to soil fertility changes in two alley cropping systems. In: Mulongoy, K. and Merkx, R. (eds) *Soil Organic Matter Dynamics and Sustainability of Tropical Agriculture*. Wiley, Chichester, pp. 143–154.

Van Wensem, J., Verhoef, H.H. and Vanstraalen, N.M. (1993) Litter degradation stage as a factor for isopod interaction with mineralization processes. *Soil Biology and Biochemistry* 25, 1175–1183.

Verhoef, H.A. and Brussard, L. (1990) Decomposition and nitrogen mineralisation in natural and agroecosystems: the contribution of soil animals. *Biogeochemistry* 11, 175–211.

Vitousek, P.M., Turner, D.R., Parton, W.J. and Sanford, R.L. (1994) Litter decomposition on the Mauna-Loa environmental matrix, Hawaii: Patterns, mechanisms and models. *Ecology* 75, 418–429.

Waksman, S.A. (1924) Influence of microorganisms upon the carbon–nitrogen ratio in soil. *Journal of Agricultural Science* 14, 555–562.

Waksman, S.A. and Tenney, F.G. (1928) Composition of natural organic materials and their decomposition in the soil, III. The influence of nature of plant upon the rapidity of its decomposition. *Soil Science* 26, 155–171.

Zak, D.R., Pregitzer, K.S., Curtis, P., Teeri, J.A., Fogel, R. and Randlett, D.L. (1993) Elevated atmospheric CO_2 and feedback between carbon and nitrogen cycles in forested ecosystems. *Plant and Soil* 151, 105–117.

Zak, J.C., Willig, M.R., Moorhead, D. and Wildmen, H.G. (1994) Functional diversity of microbial communities: a quantitative approach. *Soil Biology and Biochemistry* 26, 1101–1108.

Zech, W. (1991) Litter decomposition and humification in forest soils. In: Van Breeman, N. (ed.) *Decomposition and Accumulation of Organic Matter in Terrestrial Ecosystems: Research Prioities and Approaches*. Ecosystems Research Report 1, Commission of the European Communities, Brussels, pp. 46–51.

Zheng, D.W. (1993) *Influence of Soil Food Web Structure on Decomposition in Terrestrial Ecosystems: Theoretical Investigations of Donor-controlled Linear Decomposition Systems*. Swedish University of Agricultural Sciences, Report 61, Uppsala, 33 pp.

Part II

Pathways and Processes in Litter Deomposition

---

2   Fungal Degradation of Lignin

K.E. Hammel

Institute for Microbial and Biochemical Technology, Forest Products Laboratory, Forest Service, US Department of Agriculture, Madison, WI 53705, USA

Importance of Lignin Biodegradation

Of all naturally produced organic chemicals, lignin is probably the most recalcitrant. This is consistent with its biological functions, which are to give vascular plants the rigidity they need to stand upright and to protect their structural polysaccharides (cellulose and hemicelluloses) from attack by other organisms. Lignin is the most abundant aromatic compound on earth, and is second only to cellulose in its contribution to living terrestrial biomass (Crawford, 1981). When vascular plants die or drop litter, lignified organic carbon is incorporated into the top layer of the soil. This recalcitrant material has to be broken down and recycled by microorganisms to maintain the earth's carbon cycle. Were this not so, all carbon would eventually be irreversibly sequestered as lignocellulose.

Lignin biodegradation has diverse effects on soil quality. The microbial degradation of litter results in the formation of humus, and ligninolysis probably facilitates this process by promoting the release of aromatic humus precursors from the litter. These precursors include incompletely degraded lignin, flavanoids, terpenes, lignans, condensed tannins, and suberins (Hudson, 1986). Undegraded lignocellulose, e.g. in the form of straw, has a deleterious effect on soil fertility because decomposing (as opposed to already decomposed) lignocellulose supports high populations of microorganisms that may produce phytotoxic metabolites. High microbial populations in undecomposed litter also compete with crop plants for soil nitrogen and other nutrients (Lynch and Harper, 1985). By breaking down the most refractory component of litter, ligninolysis thus contributes to the removal of conditions that inhibit crop productivity.

Conditions that disfavour the biological breakdown of lignocellulose lead to soils with pronounced accumulations of litter. For example, the soils of coniferous forests in the northwest United States may contain 50 years of accumulated litterfall, because the low pH of the litter and the lack of summer rainfall inhibit microbial activity. In mature forests of this type, woody material such as dead trunks and branches can constitute 50–60% of the litter. By contrast, the soils under broadleaf forests in the eastern United States accumulate only a few years' worth of litter, and soils in some tropical rain forests accumulate virtually none, because conditions are more favourable for decomposition (Spurr and Barnes, 1980). Warm temperature, high moisture content, high oxygen availability, and high palatability of the litter to microorganisms all favour decomposition. The more highly lignified litter is, the less digestible it is, and the more its decomposition depends on the unique organisms that can degrade lignocellulose.

Ecology of Fungal Lignocellulose Degradation

The organisms principally responsible for lignocellulose degradation are aerobic filamentous fungi, and the most rapid degraders in this group are Basidiomycetes (Kirk and Farrell, 1987). The ability to degrade lignocellulose efficiently is thought to be associated with a mycelial growth habit which allows the fungus to transport scarce nutrients, e.g. nitrogen and iron, over a distance into the nutrient-poor lignocellulosic substrate that constitutes its carbon source. It is curious in this regard that Actinomycetes (i.e. bacteria with a mycelial growth habit) have not evolved the capacity to degrade lignocellulose efficiently. It is possible that they have the ability to modify lignin somewhat, but no evidence has accumulated to show that they can degrade it (Kirk and Farrell, 1987).

Fungal wood decay

The course of fungal lignocellulose degradation is most readily observable in intact dead wood which, despite its complex ultrastructure, is actually the simplest and best characterized form of litter. In wood, three distinct types of fungal decay can be distinguished: white rot, brown rot, and soft rot (Eriksson et al., 1990).

White rot fungi are the most abundant degraders of wood in nature. Their strategy is to decompose the lignin in wood so that they can gain access to the cellulose and hemicelluloses that are embedded in the lignin matrix. Under optimal conditions, the rates at which white rot fungi mineralize lignin rival their rates of polysaccharide degradation. Basidiomycetes and xylariaceous Ascomycetes that cause white rot are the organisms principally responsible for wood decay in hardwood forests and in tropical forest ecosystems, and also play a prominent role in temperate coniferous forests (Eriksson et al., 1990; Blanchette, 1991; Dix and Webster, 1995). Several hundred species in numerous taxa have been described. Communities of white rot fungi and associated organisms in forest ecosystems have been described in extensive studies by British researchers (Rayner and Boddy, 1988; Boddy, 1992; Dix and Webster, 1995).

The brown rot fungi comprise a relatively small group of Basidiomycetes that decay the cellulose in wood preferentially. They do not degrade the lignin extensively, although they modify it by demethylating it. Brown rot fungi thus stand out as an exception to the usually valid observation that lignocellulose must be delignified first if organisms are to gain access to plant cell wall polysaccharides. The biochemical system that enables brown rot fungi to circumvent the lignin while degrading the cellulose and hemicelluloses in wood has not been characterized. Although these fungi secrete cellulases and hemicellulases, the enzymes are too large to penetrate the cell wall matrix in wood, and it is evident that other degradative systems must participate as well. Brown rot fungi make a large contribution to wood decay, especially in coniferous forests (Dix and Webster, 1995), and the residual modified lignin they leave behind is an important humus precursor (Hudson, 1986) . They deserve much more research attention, but have been difficult to study because they do not exhibit full degradative activity on defined media *in vitro* (Eriksson et al., 1990).

The soft rot fungi are Ascomycetes and Deuteromycetes that decay water-saturated (but not totally anaerobic) wood, as well as wood prone to fluctuating moisture regimes. Soft rot fungi are slower and less aggressive decayers than white and brown rot fungi, and are probably less important degraders in a quantitative sense. They attack the polysaccharide component of wood preferentially, but appear to have some ability to decompose lignin (Dix and Webster, 1995). Soft rot fungi have received little research attention, and their degradative mechanisms remain unknown.

Fungal leaf litter decay

The processes by which fungi degrade leaf litter, as opposed to woody litter, are poorly understood. In some cases, leaves are colonized shortly after they fall by Basidiomycetes. For example, *Marasmius androsaceus* is an early colonizer and degrader of pine needles, a relatively recalcitrant and long-lived form of leaf litter (Hudson, 1986; Dix and Webster, 1995). Older analyses indicate that conifer needles contain significant levels of lignin (Theander, 1978), but it remains to be shown whether the Basidiomycetes that are early colonizers of leaf litter are ligninolytic.

In most cases, leaf litter decomposition is more complex, involving a succession of biodegradative activities that precede attack by lignocellulose degraders (Hudson, 1986; Dix and Webster, 1995). The process typically begins with colonization by bacteria, Ascomycetes, and imperfect fungi that consume the least recalcitrant components present, e.g. sugars, starch, and low molecular weight extractives. The cellulose present in non-lignified leaf tissues is then attacked by some of these organisms, but there is no evidence that lignin is degraded during this early stage of decay. Subsequently, the remaining lignified litter is modified by fauna such as earthworms, millipedes, slugs, and termites, which macerate lignocellulose mechanically in a process that releases some digestible cellulose. Bacteria and fungi in the guts of these invertebrates then assist in the breakdown of this cellulose, but they do not degrade the lignin component appreciably. Instead, this mechanically modified lignocellulose is released relatively unchanged and becomes part of the soil organic matter. Fragmentation by animals significantly accelerates the degradation rate of the tougher types of litter such as tree leaves, but probably plays a lesser role in the degradation of soft herbaceous litter (Dix and Webster, 1995). Finally, the modified but still lignified litter is colonized by Basidiomycetes that degrade it further.

It is generally assumed that basidiomycete degraders of non-woody litter are ligninolytic, i.e. that they are more like white rotters than brown rotters, but so far little research has been done to confirm this view. The commercial edible mushroom *Agaricus bisporus* is the one litter decomposer whose degradative mechanisms have received some research attention. It degrades both cellulose and lignin, the former more rapidly (Wood and Leatham, 1983; Durrant *et al.*, 1991), and contains ligninolytic enzymes (Bonnen *et al.*, 1994). If *A. bisporus* is typical of other litter-decomposing Basidiomycetes, it is probably correct to infer that fungal ligninolysis is a significant process in non-woody litter. However, it remains unclear to what extent ligninolysis in litter plays the essential role that it does in wood by exposing trapped cellulose to fungal attack. Leaf litter is already finely milled by the time most Basidiomycetes colonize it, and certainly contains bioavailable cellulose, as shown by the fact that non-ligninolytic fungi can deplete cellulose during the composting of litter (Dix and Webster, 1995).

Lignin Structure

It is evident from the preceding discussion that fungal ligninolysis is an important component of the process by which some types of litter are degraded. Other, basically non-ligninolytic mechanisms, e.g. those of the brown rot type, may be equally important, but our current understanding of litter decomposition is still so fragmentary that attack of the white rot type is the only degradative component we can discuss at a mechanistic level. To do this, we first need to understand how the structure of lignin makes it the recalcitrant material that it is.

Lignin is formed in vascular plant cell walls by the oxidative coupling of several related phenylpropanoid precursors: coniferyl alcohol, sinapyl alcohol, and *p*-hydroxycinnamyl alcohol (see also Bavage *et al.*, Chapter 16, this volume). Peroxidases or laccases in the plant cell wall oxidize these monomers by one electron, yielding transient resonance-stabilized phenoxy radicals that then polymerize in a variety of configurations. The possible ways that the precursors can couple can be portrayed on paper simply by drawing the conventional resonance forms of the phenoxy radicals, and then by linking the most important of these in various pairwise combinations. This subject has been extensively reviewed (Adler, 1977; Higuchi, 1990), and it will suffice here simply to say that lignin consists primarily of the inter-monomer linkages shown in Fig. 2.1, and that the arylglycerol-β-aryl ether structure circled in the figure is quantitatively the most important of these, constituting over 50% of the polymer. Lignin is covalently associated with hemicelluloses in the cell wall via numerous types of linkage. Among the most important are ether bonds between the benzylic carbon of lignin and the carbohydrate moiety, ester bonds between the benzylic carbon of lignin and uronic acid residues, and lignin–glycosidic bonds. In graminaceous plants, hydroxycinnamic acid residues are frequent in the lignin, and are attached to hemicelluloses via ester linkages. The matrix of lignin and hemicellulose encrusts and protects the cellulose of the plant cell wall (Jeffries, 1990).

Fungi that degrade lignin are faced with several problems. Since the polymer is

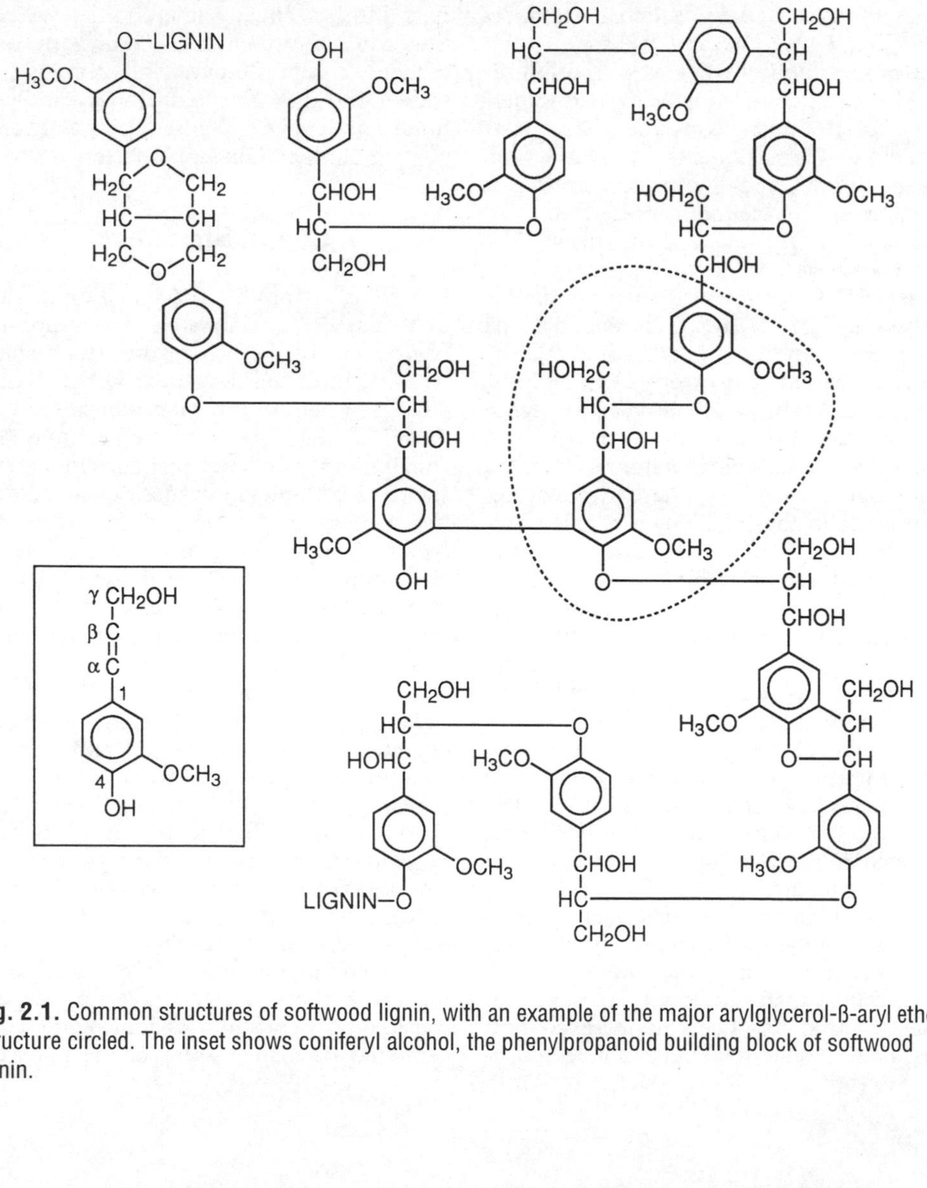

Fig. 2.1. Common structures of softwood lignin, with an example of the major arylglycerol-β-aryl ether structure circled. The inset shows coniferyl alcohol, the phenylpropanoid building block of softwood lignin.

extremely large and highly branched, ligninolytic mechanisms must be extracellular. Since it is interconnected by stable ether and carbon–carbon bonds, these mechanisms must be oxidative rather than hydrolytic. Since lignin consists of a mixture of stereoirregular units, fungal ligninolytic agents have to be much less specific than typical biological catalysts. Finally, the fact that lignin is insoluble in water limits its bioavailability to ligninolytic systems and dictates that ligninolysis is a slow process.

Measurement of Ligninolysis

The delignification of a solid lignocellulosic substrate is often assessed by the simple procedure of removing its low molecular weight components by extraction, weighing the leftover woody residue, degrading the remaining polysaccharide component with strong acid, and then reweighing the leftover insoluble lignin, which is chemically modified and referred to as Klason lignin. Klason lignin determinations are relatively simple to perform and can be useful if the investigator is confident that interfering substances are not present (Theander and Westerlund, 1993). For example, convincing data showing that certain xylariaceous Ascomycetes delignify wood have been obtained with the Klason procedure (Nilsson et al., 1989). However, this method is subject to errors if it is used on plant tissues that contain other high molecular weight components that are not removed in the initial extraction and acid treatment. Interfering substances of this type may include proteins and tannins (Theander and Westerlund, 1993). For example, Klason lignin analyses of dried pine needles indicate that they contain up to 30% lignin by weight (Theander, 1978), but this value exceeds that found even in many woods and is probably too high. On the other hand, the Klason procedure tends to underestimate the amount of lignin in annual plants, because some of the polymer is acid-soluble and consequently lost during the hydrolysis of polysaccharides.

The fungal delignification of woody material can be monitored by electron microscopy (Blanchette, 1991) or light microscopy with selective staining (Srebotnik and Messner, 1994), although these procedures are relatively complex and only semiquantitative. Microscopy is, of course, useful only when the substrate being investigated still contains lignified cell walls, but these techniques might be used with advantage to assess ligninolysis in relatively intact twigs or conifer needles.

Several chemical procedures have also been introduced for the estimation of lignin content. For example, pulverized wood samples can be treated with acetyl bromide in acetic acid, and the absorbance of the resulting solution at 280 nm can be compared with the absorbances obtained from known lignin standards. Methods of this type are subject to interference from other components, but can be useful for the comparision of closely related lignocellulosic samples (Theander and Westerlund, 1993).

A less direct but very flexible approach to the study of fungal ligninolysis is to assess not whether the growth substrate itself is being delignified, but rather whether the fungus degrades a simpler target molecule whose breakdown indicates that ligninolytic systems must be functioning. Model substrates of this type can be infiltrated into the organism's natural lignocellulosic growth medium, e.g. wood or litter, or they can be used as probes in defined liquid growth media (Kirk et al., 1975, 1978; Srebotnik et al., 1994). The most frequently used probes are ^{14}C-labelled synthetic lignins, which can be prepared by polymerizing ^{14}C-labelled p-hydroxycinnamyl alcohols (e.g. labelled coniferyl alcohol, see inset to Fig. 2.1) with horseradish peroxidase (Kirk and Brunow, 1988). The advantages of this approach are that synthetic lignins contain the same intermonomer structures that natural ones do (although they differ considerably in the relative frequency of each substructure type), and that they provide a simple and foolproof assay for ligninolysis: ^{14}CO$_2$ produced during degradation of the radiolabelled polymer can simply be trapped in alkali and determined by scintillation counting. The principal disadvantage of the method is that it is expensive and requires facilities for radiochemical organic synthesis: the necessary ^{14}C-labelled *p*-hydroxycinnamyl alcohols have to be prepared in the laboratory from simpler commercially available labelled precursors. It is also necessary to ensure that the synthetic lignins used are too large to be taken up intracellularly by the organism under investigation – this is generally done by subjecting the synthetic polymer to gel permeation chromatography before it is used and retaining only material with a molecular weight greater than about 1500. Because of their utility, radiolabelled synthetic lignins and newer polymeric lignin model compounds that represent the major substructure in lignin (Kawai *et al.*, 1995) are finding increasing use in studies of fungal ligninolysis.

Fungal Ligninolytic Mechanisms

Ligninolytic fungi are not able to use lignin as their sole source of energy and carbon. Instead, they depend on the more digestible polysaccharides in lignocellulosic substrates, and the primary function of ligninolysis is to expose these polysaccharides so that they can be cleaved by fungal cellulases and hemicellulases. In most fungi that have been examined, ligninolysis occurs during secondary metabolism, i.e. under nutrient limitation. With this approach, the fungus avoids synthesizing and secreting metabolically expensive ligninolytic agents when substrates more accessible than lignocellulose are present. The limiting nutrient for fungal growth in most woods and soils is probably nitrogen, and most laboratory studies of ligninolytic fungi have been done in nitrogen-limited culture media (Kirk and Farrell, 1987). However, a few ligninolytic fungi, e.g. some species of *Bjerkandera*, are ligninolytic even when sufficient nitrogen is present (Kaal *et al.*, 1993).

Given the chemical recalcitrance of lignin (Adler, 1977), it is evident that white rot fungi must employ unusual mechanisms to degrade it. Research has characterized several of these mechanisms in some detail, and has shown that they all display one fundamental similarity: they depend on the generation of lignin free radicals which, because of their chemical instability, subsequently undergo a variety of spontaneous cleavage reactions.

Lignin peroxidases

Lignin peroxidases (LiPs) were the first ligninolytic enzymes to be discovered (Glenn *et al.*, 1983; Tien and Kirk, 1983). They occur in some frequently studied white rot fungi, e.g. *Phanerochaete chrysosoporium*, *Trametes versicolor* and *Bjerkandera* sp. (Kirk and Farrell, 1987; Kaal *et al.*, 1993; Orth *et al.*, 1993) but are evidently absent in others, e.g. *Dichomitus squalens*, *Ceriporiopsis subvermispora* and *Pleurotus ostreatus* (Périé and Gold, 1991; Kerem *et al.*, 1992; Orth *et al.*, 1993; Rüttimann-Johnson *et al.*, 1993). LiPs resemble other peroxidases such as the classical, extensively studied enzyme from horseradish, in that they contain ferric heme and operate via a typical peroxidase catalytic cycle (Kirk and Farrell, 1987; Gold *et al.*, 1989). That is, LiP is oxidized by H_2O_2 to a two-electron deficient intermediate, which returns to its resting state by performing two one-electron oxidations of donor substrates. However, LiPs are more powerful oxidants than typical peroxidases are, and consequently oxidize not only the usual peroxidase substrates such as phenols and anilines, but also a variety of non-phenolic lignin structures and other aromatic ethers that resemble the basic structural unit of lignin (Kersten *et al.*, 1990). The simplest aromatic substrates for LiP are methoxylated benzenes and benzyl alcohols, which have been used extensively by enzymologists to study LiP reaction mechanisms. The H_2O_2-dependent oxidation of veratryl alcohol (3,4-dimethoxybenzyl alcohol) to veratraldehyde is the basis for the standard assay used to detect LiP in fungal cultures (Kirk *et al.*, 1990).

The LiP-catalysed oxidation of a lignin substructure begins with the abstraction of one electron from the donor substrate's aromatic ring, and the resulting

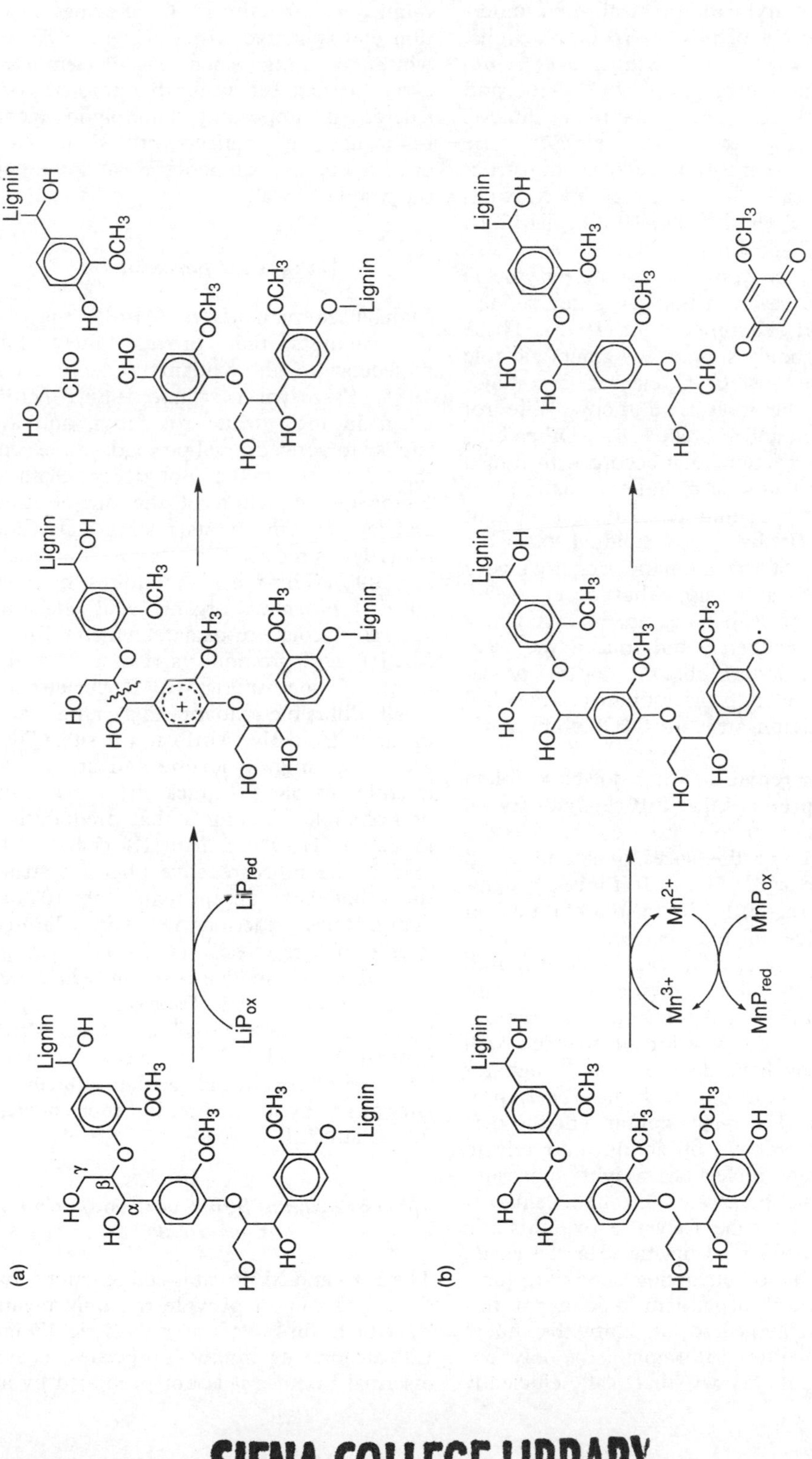

Fig. 2.2. (a) Cleavage of a recalcitrant internal non-phenolic arylglycerol-ß-aryl ether lignin structure by oxidized lignin peroxidase. (b) Cleavage of a reactive terminal phenolic arylglycerol-ß-aryl ether structure by oxidized manganese peroxidase.

species, an aryl cation radical, then undergoes a variety of postenzymatic reactions. (Kersten *et al.*, 1985; Schoemaker *et al.*, 1985; Hammel *et al.*, 1986; Kirk and Farrell, 1987). For example, dimeric model compounds that represent the major arylglycerol-β-aryl ether lignin structure undergo C_α-C_β cleavage upon oxidation by LiP (Kirk *et al.*, 1986) (Fig. 2.2). Synthetic polymeric lignins are also cleaved at this position by the enzyme *in vitro*, in a reaction that gives net depolymerization (Hammel *et al.*, 1993). These results strongly support a ligninolytic role for LiP, because C_α-C_β cleavage is a major route for ligninolysis in many white rot fungi (Kirk and Farrell, 1987). Other LiP-catalysed reactions that accord with fungal ligninolysis *in vivo* include aromatic ether cleavage at C_β and C_α-oxidation without cleavage. It has been pointed out that ionization of the aromatic ring to give a cation radical is also what occurs when lignin model substrates are analysed in a mass spectrometer, and indeed the fragmentation pattern obtained by this procedure is similar to that obtained when LiP acts on lignin structures (Dolphin *et al.*, 1987).

There remains an unresolved problem with the proposal that LiP catalyses fungal ligninolysis: LiP, like other enzymes, is too large to enter the pores in sound wood (Srebotnik *et al.*, 1988). If it initiates ligninolysis directly, LiP must therefore act at the surface of the secondary cell wall. Fungal attack of this type is indeed found, but electron microscopic observations also indicate that white rot fungi can remove lignin from the interior of the cell wall before they have degraded it enough for enzymes to penetrate. It has been proposed that LiP might circumvent the permeability problem by acting indirectly to oxidize low molecular weight substrates that could penetrate the lignocellulosic matrix and act themselves as oxidants at a distance from the enzyme (Harvey *et al.*, 1986), but no convincing candidate for a diffusible LiP-dependent oxidant of this type has emerged so far. Notwithstanding these difficulties, LiP remains the only fungal oxidant known that can efficiently mimic, *in vitro*, the C_α-C_β cleavage reaction that is characteristic of ligninolysis by white rot fungi such as *Phanerochaete chrysosporium*. LiP must therefore be considered an important ligninolytic agent, but it may act in concert with other, smaller oxidants that can penetrate and open up the wood cell wall.

Manganese peroxidases

Manganese peroxidases (MnPs) may be the catalysts that provide these low molecular weight oxidants (Glenn *et al.*, 1986; Paszczynski *et al.*, 1986). MnPs occur in most white rot fungi, and are similar to conventional peroxidases, except that Mn(II) is the obligatory electron donor for reduction of the one-electron deficient enzyme to its resting state, and Mn(III) is produced as a result (Wariishi *et al.*, 1992). This reaction requires the presence of bidentate organic acid chelators such as glycolate or oxalate, which stabilize Mn(III) and promote its release from the enzyme. The resulting Mn(III) chelates are small, diffusible oxidants that can act at a distance from the MnP active site. They are not strongly oxidizing and are consequently unable to attack the recalcitrant non-phenolic structures that predominate in lignin. However, Mn(III) chelates do oxidize the more reactive phenolic structures that make up approximately 10% of lignin. These reactions result in a limited degree of ligninolysis via C_α-aryl cleavage and other degradative reactions (Fig. 2.2) (Wariishi *et al.*, 1991; Tuor *et al.*, 1992). It is an interesting possibility that MnP-generated Mn(III) might cleave phenolic lignin structures in this fashion to facilitate later attack by the bulkier but more powerful oxidant LiP.

Co-oxidation of lignin via production of oxyradicals

The LiP- and MnP-catalysed reactions just described cannot provide the only means by which fungi cleave polymeric lignin. LiP, despite its unique properties, is not essential because it is not produced by all white rot fungi during ligninolysis. MnP-generated Mn(III) cannot be wholly responsible because white rot fungi that lack LiP are nevertheless able to degrade the non-phenolic lignin structures that resist attack by chelated Mn(III) (Srebotnik et al., 1994). Other ligninolytic mechanisms must therefore exist.

Recent work indicates that the production of diffusible oxyradicals by MnP may supply one such mechanism. In the presence of Mn(II), MnP promotes the peroxidation of unsaturated lipids, generating transient lipoxyradical intermediates that are known to act as potent oxidants of other molecules (Moen and Hammel, 1994). The MnP/lipid peroxidation system, unlike MnP alone, oxidizes and cleaves non-phenolic lignin model compounds. It also depolymerizes both non-phenolic and phenolic synthetic lignins, which strongly supports a ligninolytic role for this system in vivo (Bao et al., 1994). Although lipid peroxidation has previously been implicated in a variety of biological processes, e.g. aging and carcinogenesis (Halliwell and Gutteridge, 1989), we believe this is the first evidence that microorganisms may use it as a biodegradative tool.

Laccases

Laccases are blue copper oxidases that catalyse the one-electron oxidation of phenolics and other electron-rich substrates. Most ligninolytic fungi produce laccases, P. chrysosporium being a notable exception. Laccases contain multiple copper atoms which are reduced as the substrates are oxidized. After four electrons have been received by a laccase molecule, the laccase reduces molecular oxygen to water, returning to the native state. The action of laccase on lignin resembles that of Mn(III) chelates, in that phenolic units are oxidized to phenoxy radicals, which can lead to degradation of some structures (Kawai et al., 1988). In the presence of certain artificial auxiliary substrates, the effect of laccase can be enhanced so that it oxidizes non-phenolic compounds that otherwise would not be attacked, but it is not yet known whether natural versions of

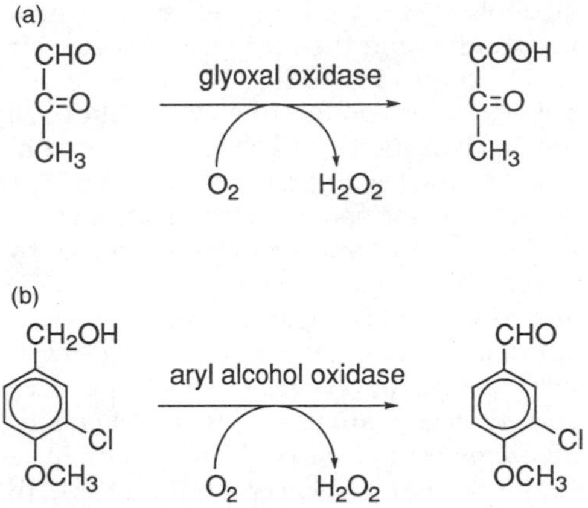

Fig. 2.3. Production of extracellular H_2O_2 by (a) the glyoxal oxidase of *Phanerochaete chrysosporium* and (b) the aryl alcohol oxidase of *Bjerkandera* sp.

such auxiliary substrates function in vivo in lignin biodegradation (Bourbonnais and Paice, 1992), and indeed, the actual role of laccase has yet to be fully clarified.

Peroxide-producing enzymes

To support the oxidative turnover of the LiPs and MnPs responsible for ligninolysis, white rot fungi require sources of extracellular H_2O_2. This need is met by extracellular oxidases that reduce molecular oxygen to H_2O_2 with the concomitant oxidation of a cosubstrate (Fig. 2.3). One such enzyme, found in P. chrysosporium and many other white rot fungi, is glyoxal oxidase (GLOX) (Kersten, 1990). GLOX accepts a variety of 1–3 carbon aldehydes as electron donors. Some GLOX substrates, e.g. glyoxal and methylglyoxal, are natural extracellular metabolites of P. chrysosporium (Kersten, 1990). Another substrate for the enzyme, glycolaldehyde, is released as a cleavage product when the major arylglycerol-β-aryl ether structure of lignin is oxidized by LiP (Hammel et al., 1994).

Aryl alcohol oxidases (AAOs) provide another route for H_2O_2 production in some white rot fungi. In certain LiP-producing species of *Bjerkandera*, chlorinated anisyl alcohols are secreted as extracellular metabolites and then reduced by a specific AAO to produce H_2O_2 (de Jong et al., 1994). It is noteworthy that, although many alkoxybenzyl alcohols are LiP substrates, chloroanisyl alcohols are not. The use of a chlorinated benzyl alcohol as an AAO substrate thus provides a strategy by which the fungus separates its ligninolytic and H_2O_2-generating pathways. A different approach is employed by some LiP-negative species of *Pleurotus*, which produce and oxidize a mixture of benzyl alcohols, including anisyl alcohol, to maintain a supply of H_2O_2 (Guillén et al., 1992). In yet other fungi, intracellular sugar oxidases might be involved in H_2O_2 generation (Kirk and Farrell, 1987).

Detection of ligninolytic enzymes in complex substrates

Once a fungus has been shown to degrade lignin in experiments with radiolabelled synthetic lignins, the question arises as to which ligninolytic enzymes the organism is expressing. If the degradation experiments have been done in defined liquid media, standard assays for LiP, MnP, laccase, and the various H_2O_2-producing oxidases can be done with little difficulty. However defined growth media that elicit the full expression of ligninolysis have not been developed for many fungi. Therefore, in experiments with previously uninvestigated fungi that grow on litter, it is more pertinent to ask what ligninolytic enzymes are expressed in the natural growth substrate. This remains a difficult question because many peroxidases are easily inactivated by phenols or other inhibitors that occur in lignocellulosic substrates, and it is consequently difficult to assay these important ligninolytic enzymes reliably in solid state cultures (Datta et al., 1991). Investigators must therefore turn to indirect methods for the detection of ligninolytic enzymes. One of these is to infiltrate a high molecular weight lignin model compound into the lignocellulosic substrate, and then to determine by subsequent product analysis whether the fungus cleaves it in the same way that purified LiP does (Kawai et al., 1995). Another approach currently under development, and useful when the gene for the enzyme of interest has been sequenced, is to isolate fungal RNA from the substrate and use reverse transcription/polymerase chain reaction techniques to determine whether the gene for the enzyme is being expressed (Gold and Alic, 1993; Lamar et al., 1995). These new research tools should help to alleviate our severe lack of knowledge about degradative mechanisms in litter-decomposing fungi.

References

Adler, E. (1977) Lignin chemistry. Past, present and future. *Wood Science and Technology* 11, 169–218.

Bao, W., Fukushima, Y., Jensen, K.A., Jr, Moen, M.A. and Hammel, K.E. (1994) Oxidative degradation of non-phenolic lignin during lipid peroxidation by fungal manganese peroxidase. *FEBS Letters* 354, 297–300.

Blanchette, R.A. (1991) Delignification by wood-decay fungi. *Annual Review of Phytopathology* 29, 381–398.

Boddy, L. (1992) Development and function of fungal communities in decomposing wood. In: Carroll, D.T. and Wicklow, D.T. (eds) *The Fungal Community*. Marcel Dekker, Inc., New York, pp. 749–782.

Bonnen, A.M., Anton, L.H. and Orth, A.B. (1994) Lignin-degrading enzymes of the commercial button mushroom, *Agaricus bisporus*. *Applied and Environmental Microbiology* 60, 960–965.

Bourbonnais, R. and Paice, M.G. (1992) Demethylation and delignification of kraft pulp by *Trametes versicolor* laccase in the presence of 2,2'-azinobis-(3-ethylbenzthiazoline-6-sulphonate). *Applied Microbiology and Biotechnology* 36, 823–827.

Crawford, R.L. (1981) *Lignin Biodegradation and Transformation*. John Wiley and Sons, New York, 154 pp.

Datta, A., Bettermann, A. and Kirk, T.K. (1991) Identification of a specific manganese per-oxidase among ligninolytic enzymes secreted by *Phanerochaete chrysosporium* during wood decay. *Applied and Environmental Microbiology* 57, 1453–1460.

De Jong, E., Cazemier, A.E., Field, J.A. and de Bont, J.A.M. (1994) Physiological role of chlorinated aryl alcohols biosynthesized de novo by the white rot fungus *Bjerkandera* sp. strain BOS55. *Applied and Environmental Microbiology* 60, 271–277.

Dix, N.J. and Webster, J. (1995) *Fungal Ecology*. Chapman and Hall, London, 549 pp.

Dolphin, D., Nakano, T., Maione, T.E., Kirk, T.K. and Farrell, R. (1987) Synthetic model ligninases. In: Odier, E. (ed.) *Lignin Enzymic and Microbial Degradation*. Institute National de la Recherche Agronomique, Paris, pp. 157–162.

Durrant, A.J., Wood, D.A. and Cain, R.B. (1991) Lignocellulose biodegradation by *Agaricus bisporus* during solid substrate fermentation. *Journal of General Microbiology* 137, 751–755.

Eriksson, K-E.L., Blanchette, R.A. and Ander, P. (1990) *Microbial and Enzymatic Degradation of Wood and Wood Components*. Springer-Verlag, Berlin, 407 pp.

Glenn, J.K., Morgan, M.A., Mayfield, M.B., Kuwahara, M. and Gold, M.H. (1983) An extracellular H_2O_2-requiring enzyme preparation involved in lignin biodegradation by the white-rot basidiomycete *Phanerochaete chrysosporium*. *Biochemical and Biophysical Research Communications* 114, 1077–1083.

Glenn, J.K., Akileswaran, L. and Gold, M.H. (1986) Mn(II) oxidation is the principal func-tion of the extracellular Mn-peroxidase from *Phanerochaete chrysosporium*. *Archives of Biochemistry and Biophysics* 251, 688–696.

Gold, M.H. and Alic, M. (1993) Molecular biology of the lignin-degrading basidiomycete *Phanerochaete chrysosporium*. *Microbiological Reviews* 57, 605–622.

Gold, M.H., Wariishi, H. and Valli, K. (1989) Extracellular peroxidases involved in lignin degradation by the white rot basidiomycete *Phanerochaete chrysosporium*. *American Chemical Society Symposium Series* 389, 127–140.

Guillén, F., Martínez, A.T. and Martínez, M.J. (1992) Substrate specificity and properties of the aryl-alcohol oxidase from the ligninolytic fungus *Pleurotus eryngii*. *European Journal of Biochemistry* 209, 603–611.

Halliwell, B. and Gutteridge, J.M.C. (1989) *Free Radicals in Biology and Medicine*, 2nd edn. Clarendon Press, Oxford, 543 pp.

Hammel, K.E., Kalyanaraman, B. and Kirk, T.K. (1986) Substrate free radicals are interme-diates in ligninase catalysis. *Proceedings of the National Academy of Sciences (USA)* 83, 3708–3712.

Hammel, K.E., Jensen, K.A., Jr, Mozuch, M.D., Landucci, L.L., Tien, M. and Pease, E.A. (1993) Ligninolysis by a purified lignin peroxidase. *Journal of Biological Chemistry* 268, 12274–12281.

Hammel, K.E., Mozuch, M.D., Jensen, K.A., Jr. and Kersten, P.J. (1994) H_2O_2 recycling during oxidation of the arylglycerol β-aryl ether lignin structure by lignin peroxidase and glyoxal oxidase. *Biochemistry* 33, 13349–13354.

Harvey, P.J., Schoemaker, H.E. and Palmer, J.M. (1986) Veratryl alcohol as a mediator and the role of radical cations in lignin degradation by *Phanerochaete chrysosporium*. *FEBS Letters* 195, 242–246.

Higuchi, T. (1990) Lignin biochemistry: biosynthesis and biodegradation. *Wood Science and Technology* 24, 23–63.

Hudson, H.J. (1986) *Fungal Biology*, Cambridge University Press. Cambridge, 298 pp.

Jeffries, T.W. (1990) Biodegradation of lignin-carbohydrate complexes. *Biodegradation* 1, 163–176.

Kaal, E.E.J., de Jong, E. and Field, J.A. (1993) Stimulation of ligninolytic peroxidase activity by nitrogen nutrients in the white rot fungus *Bjerkandera* sp. strain BOS55. *Applied and Environmental Microbiology* 59, 4031–4036.

Kawai, S., Umezawa, T. and Higuchi, T. (1988) Degradation mechanisms of phenolic β-1 lignin substructure model compounds by laccase of *Coriolus versicolor*. *Archives of Biochemistry and Biophysics* 262, 99–110.

Kawai, S., Jensen, K.A., Jr, Bao, W. and Hammel, K.E. (1995) New polymeric model substrates for the study of microbial ligninolysis. *Applied and Environmental Microbiology* 61, 3407–3414.

Kerem, Z., Friesem, D. and Hadar, Y. (1992) Lignocellulose degradation during solid-state fermentation: *Pleurotus ostreatus* vs. *Phanerochaete chrysosporium*. *Applied and Environmental Microbiology* 58, 1121–1127.

Kersten, P.J. (1990) Glyoxal oxidase of *Phanerochaete chrysosporium*: its characterization and activation by lignin peroxidase. *Proceedings of the National Academy of Sciences (USA)* 87, 2936–2940.

Kersten, P.J., Tien, M., Kalyanaraman, B. and Kirk, T.K. (1985) The ligninase of *Phanerochaete chrysosporium* generates cation radicals from methoxybenzenes. *Journal of Biological Chemistry* 260, 2609–2612.

Kersten, P.J., Kalyanaraman, B., Hammel, K.E., Reinhammar, B. and Kirk, T.K. (1990) Comparison of lignin peroxidase, horseradish peroxidase and laccase in the oxidation of methoxybenzenes. *The Biochemical Journal* 268, 475–480.

Kirk, T.K. and Brunow, G. (1988) Synthetic ^{14}C-labeled lignins. *Methods in Enzymology* 161, 65–73.

Kirk, T.K. and Farrell, R.L. (1987) Enzymatic 'combustion': the microbial degradation of lignin. *Annual Review of Microbiology* 41, 465–505.

Kirk, T.K., Connors, W.J., Bleam, R.D., Hackett, W.F. and Zeikus, J.G. (1975) Preparation and microbial decomposition of synthetic [14C]lignins. *Proceedings of the National Academy of Sciences (USA)* 72, 2515–2519.

Kirk, T.K., Schultz, E., Connors, W.J., Lorenz, L.F. and Zeikus, J.G. (1978) Influence of culture parameters on lignin metabolism by *Phanerochaete chrysosporium*. *Archives of Microbiology* 117, 277–285.

Kirk, T.K., Tien, M., Kersten, P.J., Mozuch, M.D. and Kalyanaraman, B. (1986) Ligninase of *Phanerochaete chrysosporium*. Mechanism of its degradation of the non-phenolic aryl-glycerol β-aryl ether substructure of lignin. *The Biochemical Journal* 236, 279–287.

Kirk, T.K., Tien, M., Kersten, P.J., Kalyanaraman, B., Hammel, K.E. and Farrell, R.L. (1990) Lignin peroxidase from fungi: *Phanerochaete chrysosporium*. *Methods in Enzymology* 188, 159–171.

Lamar, R.T., Schoenike, B., Vanden Wymelenberg, A., Stewart, P., Dietrich, D.M. and Cullen, D. (1995) Quantitation of fungal mRNAs in complex substrates by reverse transcription PCR and its application to *Phanerochaete chrysosporium*-colonized soil. *Applied and Environmental Microbiology* 61, 2122–2126.

Lynch, J.M. and Harper, S.H.T. (1985) The microbial upgrading of straw for agricultural use. *Philosophical Transactions of the Royal Society (London)* B 310, 221–226.

Moen, M.A. and Hammel, K.E. (1994) Lipid peroxidation by the manganese peroxidase of *Phanerochaete chrysosporium* is the basis for phenanthrene oxidation by the intact fungus. *Applied and Environmental Microbiology* 60, 1956–1961.

Nilsson, T., Daniel, G., Kirk, T.K. and Obst, J.R. (1989) Chemistry and microscopy of wood decay by some higher ascomycetes. *Holzforschung* 43, 11–18.

Orth, A.B., Royse, D.J. and Tien, M. (1993) Ubiquity of lignin-degrading peroxidases among various wood-degrading fungi. *Applied and Environmental Microbiology* 59, 4017–4023.

Paszczynski, A., Huynh, V.-B. and Crawford, R.L. (1986) Comparison of ligninase-I and peroxidase-M2 from the white-rot fungus *Phanerochaete chrysosporium*. *Archives of Biochemistry and Biophysics* 244, 750–765.

Périé, F.H. and Gold, M.H. (1991) Manganese regulation of manganese peroxidase expression and lignin degradation by the white rot fungus *Dichomitus squalens*. *Applied and Environmental Microbiology* 57, 2240–2245.

Rayner, A.D.M. and Boddy, L. (1988) *Fungal Decomposition of Wood. Its Biology and Ecology*. John Wiley, Chichester, 587 pp.

Rüttimann-Johnson, C., Salas, L., Vicuña, R. and Kirk, T.K. (1993) Extracellular enzyme production and synthetic lignin mineralization by *Ceriporiopsis subvermispora*. *Applied and Environmental Microbiology* 59, 1792–1797.

Schoemaker, H.E., Harvey, P.J., Bowen, R.M. and Palmer, J.M. (1985) On the mechanism of enzymatic lignin breakdown. *FEBS Letters* 183, 7–12

Spurr, S.H. and Barnes, B.V. (1980) *Forest Ecology*, 3rd edn. John Wiley & Sons, New York, 687 pp.

Srebotnik, E. and Messner, K. (1994) A simple method that uses differential staining and light microscopy to assess the selectivity of wood delignification by white rot fungi. *Applied and Environmental Microbiology* 60, 1383–1386.

Srebotnik, E., Messner, K. and Foisner, R. (1988) Penetrability of white rot-degraded pine wood by the lignin peroxidase of *Phanerochaete chrysosporium*. *Applied and Environmental Microbiology* 54, 2608–2614.

Srebotnik, E., Jensen, K.A., Jr and Hammel, K.E. (1994) Fungal degradation of recalcitrant nonphenolic lignin structures without lignin peroxidase. *Proceedings of the National Academy of Sciences (USA)* 91, 12794–12797.

Theander, O. (1978) Leaf litter of some forest trees. Chemical composition and microbiological activity. *TAPPI Journal* 61, 69–72.

Theander, O. and Westerlund, E. (1993) Quantitative analysis of cell wall components. In: Jung, H.G., Buxton, D.R., Hatfield, R.D. and Ralph, J. (eds) *Forage Cell Wall Structure and Digestibility*. American Society of Agronomy, Crop Science Society of America, and Soil Science Society of America, Madison, Wisconsin, 794 pp.

Tien, M. and Kirk, T.K. (1983) Lignin-degrading enzyme from the hymenomycete *Phanerochaete chrysosporium* Burds. *Science* 221, 661–663.

Tuor, U., Wariishi, H., Schoemaker, H.E. and Gold, M.H. (1992) Oxidation of phenolic aryl-glycerol β-aryl ether lignin model compounds by manganese peroxidase from *Phanerochaete chrysosporium*: oxidative cleavage of an α-carbonyl model compound. *Biochemistry* 31, 4986–4995.

Wariishi, H., Valli, K. and Gold, M.H. (1991) *In vitro* depolymerization of lignin by manganese peroxidase of *Phanerochaete chrysosporium*. *Biochemical and Biophysical Research Communications* 176, 269–275.

Wariishi, H., Valli, K. and Gold, M.H. (1992) Manganese(II) oxidation by manganese peroxidase from the basidiomycete *Phanerochaete chrysosporium*. Kinetic mechanism and role of chelators. *Journal of Biological Chemistry* 267, 23688–23695.

Wood, D.A. and Leatham, G.F. (1983) Lignocellulose degradation during the life cycle of *Agaricus bisporus*. *FEMS Microbiology Letters* 20, 421–424.

3 Plant Degradation by Ruminants: Parallels with Litter Decomposition in Soils

A. Chesson

Rowett Research Institute, Bucksburn, Aberdeen AB21 9SB, UK

Introduction: The Rumen Environment

All mammals lack the capacity to degrade the structural polysaccharides that form the cell walls of higher plants and, with the important exception of starch, the storage polysaccharides laid down by plants. Herbivores (and omnivores) are dependent for this function on the action of micro-organisms contained and nurtured within their digestive tracts, using the end products of microbial fermentation in support of their own nutrition. A variety of strategies have evolved for maximizing the extent of cell wall degradation, usually by extending the contact time between the gut microflora and cell wall substrate. Microbial fermentation in simple-stomached (monogastric, non-ruminant) animals takes place in the lower digestive tract after removal of simple sugars and starch in the stomach and small intestine, either in the colon itself (e.g. humans) or, additionally, in well-developed caeca (e.g. the horse). Cell wall digestion and nitrogen recovery may be further enhanced by coprophagy. This is an arrangement also adopted by termites and other insects living on plant detritus (Varma *et al.*, 1994).

Ruminants are the most extensively adapted of all animals for the utilization of plant material with a high cell wall content. They are able to retain large amounts of ingested plant material in contact with a high concentration of microorganisms for longer periods than most other herbivores.

This is achieved in a complex stomach consisting of three functionally distinct parts; the reticulo-rumen, the omasum and the abomasum (Fig. 3.1). The reticulo-rumen itself, where the bulk of fermentation occurs, is a large sac fed directly by the oesophagus. Thus in the ruminant animal, unlike the non-ruminant, all ingested foods are first attacked by the rumen microflora. Only after an extended period of microbial digestion is residual food material and part of the microflora directed through the omasum to the abomasum or true stomach. Here host peptic secretions begin the process of the breakdown of microbial and other proteins which have survived rumen fermentation.

The reticulo-rumen

A crucial factor in the evolution of a true symbiotic relationship between the ruminant host and its microbial population was the anaerobic nature of the environment provided by the host. In the absence of molecular oxygen the flora could only partially reduce the carbohydrate substrate consumed by the animal to volatile fatty acids (VFA). The host, however, could make use of its aerobic capacity to further oxidize the VFA to CO_2 and water. There is no doubt that the host is, in energy terms, the major beneficiary of this arrangement since the approximate yield of ATP from the anaerobic dissimilation of hexose is 5 mol ATP per mol of hexose, compared with approximately 38 mol ATP

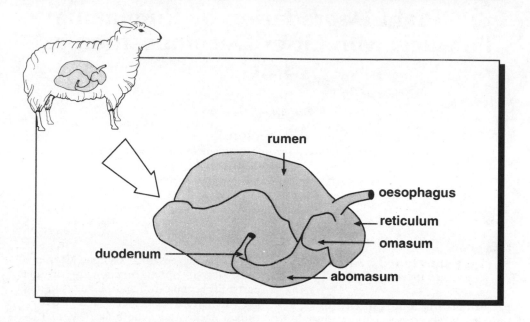

Fig. 3.1. The ruminant digestive tract. Only the modified stomach, consisting of the reticulo-rumen, omasum and abomasum, which is unique to the ruminant is shown. From the duodenum onwards the basic features of the digestive tract are common to all mammals.

yielded under aerobic conditions. In return, the host provides the microflora with a constant temperature, an adequate supply of nitrogen in the form of ammonia released at the rumen wall from urea diffused back into the rumen and a copious supply of buffered saliva which maintains the pH at near neutrality.

The stability provided by the host and the throughput of substrate allows concentrations of obligatory anaerobic bacteria to be maintained at around 10^{11} viable cells ml^{-1} rumen liquor. In addition the rumen contains a mixed population of entodiniomorph and holotrich protozoa (approximately 10^6 ml^{-1}) and an apparently unique group of obligatory anaerobic chytrid fungi (approximately 10^5 thallus forming units ml^{-1}) whose taxonomy is imperfectly known (Trinci *et al.*, 1994). Although only a few organisms among this varied flora are able to hydrolyse cell wall polysaccharides, attempts to establish a stable, functioning cell wall degrading population of bacteria in gnotobiotic lambs required the introduction of over 100 different strains (Fonty *et al.*, 1983) pointing to the com-

plexity of the rumen ecosystem.

The rumen is essentially an aquatic environment in which the bulk of ingested plant material occurs as a partially submerged raft in the upper layer of the rumen liquor. However the rumen is far from static. While the sheep rumen contains some 8–10 l of liquid and the rumen of a dairy cow approximately 70 l, the regular contractions of the rumen wall ensure that some 10,000 l of liquid or more flow through the raft each day. From the viewpoint of the resident flora the rumen has much in common with a flowing stream and presents the microorganisms with the problem of obtaining control over their substrate and avoiding washout. The cellulolytic bacteria do this by forming a close association with the ingested plant particles; the pleomorphic *Fibrobacter succinogenes* altering its cell outline to fit that of the substrate and the *Ruminococus* spp. (*R. albus* and *R. flavefaciens*) binding less closely via extensions to their glycocalyx (Weimer, 1992; Pell and Schofield, 1993). Similarly the rumen fungi colonize by means of motile zoospores (occasionally detached plantlets), and the rhizoids produced on germination penetrate the plant fragment growing between adjacent walls and, on occasions penetrating through the cell wall by an appressorium-type mechanism (Ho *et al.*, 1988; Trinci *et al.*, 1994). The protozoa do not colonize the larger plant fragments but engulf small plant particles, bacteria and starch granules (see Williams and Coleman, 1992).

A proportion of the solid digesta in the rumen is regularly regurgitated, mixed with fresh saliva and chewed during rumination. This is the mechanism for particle comminution and the breaking open of intact cells. Depending on the proportion of cell wall in the diet, the animal may be involved in this important process for up to 8 hours a day. Microbial degradation of walls may increase fragility and reduce particle density, but does not directly reduce particle size. Particle size and density are the major determinants of residence time in the rumen (Ullyatt *et al.*, 1986). In animals consuming poor quality hay and cereal straws with a high lignified cell wall content, the mean residence time in the rumen is in the region of 72-96 h. At this rate of throughput the animal can barely sustain itself and is said to be fed close to maintenance. In contrast, in a high-yielding dairy cow fed high quality hay and concentrate at two to three times maintenance, the mean residence time of the cell wall fraction is probably closer to 24 h.

The omasum

The entrance to the omasum lies close to that of the oesophagus and is fed mainly by the liquid first squeezed from the food bolus during rumination rather than directly from the reticulo-rumen. This part of the stomach complex has a series of leaf-like internal structures giving it a large surface area relative to its volume. As this implies, it is one of the principal sites for the absorption of water, VFA and dissolved minerals. In addition it acts as a sieve allowing the passage from the rumen of smaller plant fragments suspended in

water but impeding and retaining the larger particles. The control allowed by the selective release of smaller particles from the rumen ensures that contact time with feed particles is maximized without compromising the throughput of digestible material. This enables the animal to gain from plant material with a high cell wall content sufficient nutrients in excess of those needed for maintenance to lay down lean tissue, and to produce milk or fibre.

The abomasum and beyond

The abomasum is equivalent to the stomach of monogastrics and represents the start of a gastrointestinal tract which differs little in form or function from other mammals. In theory, all storage polysaccharide and soluble sugars have been removed by rumen fermentation from any plant materials entering this part of the tract leaving only the more recalcitrant part of the plant. In practice, and depending on the nature of the diet, a proportion of starch and fermentable cell wall may escape rumen degradation to be attacked by host amylolytic secretions and the hindgut flora. In addition, a substantial proportion of the small particles passing the omasum are viable and dead bacteria and these form the main source of protein for the animal. Amino acids are released from bacterial protein by endogenous secretions typical of all mammals, beginning with pepsin in the abomasum, and are absorbed in the upper parts of the small intestine. This ensures that the nitrogen required to support the growth of the microflora is not lost to the host as occurs in monogastric species that do not practise coprophagy.

Differences between gut and soil habitats

The rumen ecosystem differs from soil habitats in a number of important respects, foremost of which is the time scale of degradation. The retention time for plant cell walls within the rumen rarely exceeds 96 h. The rate of colonization and degradation during this period exceeds that commonly encountered in or on soil and makes comparisons with litter degradation difficult. However, the 40-60% loss of dry matter from cereal straws normally recorded in the rumen (Tuah *et al.*, 1986) compared with an annual loss of approximately 60% from straw ploughed under (Smith and Peckenpaugh, 1986: Wessen and Berg, 1986) attests to the efficiency of the rumen process. Such a short time scale does not allow for microbial succession and ingested plant material is simultaneously colonized by the rumen fungi and bacteria. Other factors that contrast with most soil and litter ecosystems and need to be considered in making comparisons include:

- the controlled and near constant environmental conditions maintained in the rumen;
- the general absence of limiting factors such as the supply of nitrogen and phosphorus in the rumen environment;
- the highly anaerobic (<0.005 mol oxygen, Hillman *et al.*, 1985) nature of the rumen environment and the absence of degradation processes dependent on molecular oxygen;
- the high numbers of viable microorganisms adapted to rumen conditions (i.e. tolerance to VFA, length of life-cycle) and their inability to survive elsewhere;
- the aqueous nature of the rumen environment;
- the animals preference for the consumption of green plant material with readily degradable cell walls and, given the opportunity, the ability of the grazing or browsing animal to select the most degradable material.

Provided that these differences are recognized, then some parallels can be drawn between the degradation of plant detritus and cell wall degradation in the rumen. Both are microbial processes involving a common substrate and requiring enzymes with the same specificity to achieve breakdown. Both processes are aided by comminution by animals and both are subject to the same limitations on the rate and, sometimes, the extent of degradation.

The factors that operate to limit or impede cell wall degradation arise from the organization of the plant material, the size and composition of the microbial population and the activities of the soil fauna or host animal. In general, it can be argued that plant-derived factors determine the *maximum rate* of degradation and, in the rumen and in anoxic soils at least, also the *maximum extent* of degradation. The other factors that may influence the degradation process intervene to reduce the actual rate and extent of degradation to below the maximum set by the plant.

The limitations imposed by the plant on the microbial degradation of cell walls operate at several levels of organization ranging from the microscopic to the molecular. These are summarized in Table 3.1 and further considered below.

Influence of Plant Anatomy on Degradation Processes

The integrity of the plant material has a marked effect on colonization in all environments. The cuticle, which covers all aerial parts of the plant, impedes access to underlying tissues and microorganisms can penetrate only through the stomata and lenticels of intact tissue or where the cuticle has been damaged. In the animal this is achieved by an initial mastication which often extensively disrupts the ingested material providing access to the

Table 3.1. Factors influencing cell wall degradation at various levels of organization of plants used for animal feeding.

Level of organization	Dimension	Effect
Plant anatomy	10^{-2}–10^{-5} m	Governs accessibility of microorganisms to substrate
Cell wall architecture	10^{-6}–10^{-9} m	Governs accessibility of enzymes to substrate
Polymer chemistry	10^{-9}–10^{-10} m	Governs enzyme–substrate interactions

luminal surface of broken cells and to broken edges of cell walls. Once the disrupted plant material enters the rumen and is exposed to the very high concentration of microorganisms present, colonization is rapid and dry matter loss from cell walls can usually be detected within an hour of consumption (Chesson *et al.*, 1986). In contrast, fallen plant litter is deposited in a far more intact form than is supplied to an animal, physical disruption is a much slower process and the number of viable organisms available for immediate colonization generally is far smaller.

All plant materials consist of a heterogeneous population of cell types contributing to different tissues and plant parts, each with walls which differ in dimension and chemical composition. Compositional differences are greatest in green plants where metabolically active cells with readily degradable primary walls exist alongside dead cells with extensively secondary-thickened and lignified walls

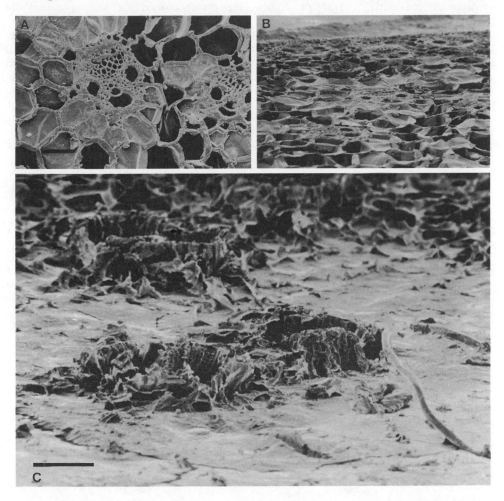

Fig. 3.2. Scanning electron micrographs of maize stem before and after digestion. (a) A 100 µm section shown in transverse section. (b) The same section tilted through 30°. (c) A tilted 100 µm section after 24 h digestion by rumen microorganisms. All of the inner parenchyma, most of the outer parenchyma and the phloem have been degraded leaving the xylem tissues of the vascular bundles and the sclerenchyma standing proud of the surface. Bar = 100 µm.

which are inherently resistant to degradation. In older annual plants or in senescent material where most or all cells have undergone lignification, compositional differences are often much reduced and may be evident only in the polyphenolic fraction of the cell walls. The consequences of this heterogeneity are evident in any microscopic observation of plant tissues undergoing digestion (Fig. 3.2). Primary cell types, such as leaf mesophyll or phloem, are clearly degraded in preference to the lignified xylem and sclerenchyma with thinner walled cells like parenchyma and epidermis occupying an intermediate position (Akin, 1979, 1989). These differences in the degradation characteristics and the contribution made by individual cell types to the overall degradation of the plant are often not appreciated when quantitative experiments are made with whole plant samples.

Although working with wall preparations from individual cells can provide useful information, in such preparations all cells are equally accessible to cell wall degrading microorganisms and therefore equally available for attack. This is evidently not the case *in vivo* (Grabber and Jung, 1991; Chesson, 1993). Cells which are inherently degradable may be slow to be colonized or remain undegraded either because they remain entire or because they are located in a position in which access by microorganisms is prevented by other cells. Epidermal cells are often highly degradable in isolation but little attacked *in situ* because they are sandwiched between an inert cuticle and resistant sclerenchyma. Similarly many vessel cells may remain free from colonization as long as the bundle sheath remains entire (Akin, 1988; Wilson, 1993). At present, methods for assessing the importance of anatomical barriers to colonization and degradation remain descriptive. However recent developments in two- and three-dimensional image analysis coupled with the use of confocal microscopy allow the extent of degradation of the wall of each individual cell to be measured separately (Travis *et al.*, 1993).

Cell Wall Structure and Degradation

Architecture of the primary wall

The most commonly encountered form of primary cell walls from higher plants,

Table 3.2. The polymer composition of primary and secondary-thickened Type I and Type II cell walls. The values given are calculated from experimental data obtained from isolated cells types extracted from ryegrass leaf, cocksfoot stem and a brassica species (kale). The data is intended only to provide a guide to composition since the actual composition will vary with age and phylogenic origin. In addition, a number of assumptions have been made in assigning sugar units to polymer types which may not hold for all plant materials.

| | Cell wall organic matter (g kg^{-1}) | | | |
| | Type I | | Type II | |
Polymer	Primary	Lignified	Primary	Lignified
Total protein	110	64	83	18
Total phenolics	41	117	20	73
Total carbohydrate	849	829	897	909
Cellulose	386	352	484	467
Mixed-linked glucan	–	–	33	26
Xyloglucan	76	22	127	31
Arabinoxylan	22	276	153	327
Arabinogalactan	131	79	50	17
Rhamnogalacturonan	234	100	50	41

referred to as a Type I wall by Carpita and Gibeaut (1993), consists essentially of a dual network of polysaccharide polymers (Table 3.2). This is admirably presented in diagrammatic form by Carpita and Gibeaut (1993). The first network, representing some 65% of the mass of the wall, is formed from the cellulose microfibrils cross-linked by chains of xyloglucan. Regions of each xyloglucan bind strongly to two or more microfibrils and interlock with other xyloglucan polymers. The second network, consisting primarily of pectic polysaccharide and representing some 20% of the wall, intertwines between the cellulose–xyloglucan network. The two networks remain discrete and tend to be physically associated rather than chemically bonded. Thus the bulk of the pectic network can be extracted from the wall by chelating agents at room temperatures; conditions under which covalent bonds would remain intact. Further evidence for duality of the polysaccharide structure is provided by the response of tomato and other cells adapted in culture to the herbicide 2,6-dichlorobenzonitrile (dichlobenil;), a potent inhibitor of cellulose biosynthesis. Although the cellulose–xyloglucan network is entirely absent from the walls of these cells, the pectic network remains and is sufficient to allow the cell to develop and function normally (Shedletzky et al., 1992).

In addition to polysaccharide, the wall contains a substantial number of proteins which may account for a further 10% of wall dry matter. The function of relatively few of the several hundred proteins found within the primary wall is known. Many have enzymatic activity and are concerned with the remodelling of the growing wall while others have structural role. Once the period of cell wall extension is complete and the cell has reached its maximum dimensions, one protein in particular increases in amount and seems to have a role in locking the wall into its final shape. This is the hydroxyproline-rich glycoprotein extensin (Lamport, 1986) which occurs as a rod-like structure and may, via isodityrosine cross units (Cooper et al., 1987) contribute a third network to the mature primary wall.

Members of the Poaceae (Gramineae) and a few closely related genera possess a primary cell wall that differs in many respects from the Type I model but is still based on a network structure (Table 3.2). In this type of wall the proportions of xyloglucan, pectic material and protein are much reduced. Xyloglucan polymers are shorter and closely associate with a single cellulose microfibril for their entire length and their role in cross-linking between cellulose microfibrils is replaced by arabinoxylan. The cellulose (xyloglucan)–arabinoxylan network represents approximately 80% of Type II cell walls. The pectic network appears structurally similar to that found in Type I cell walls but accounts for only 4-6% of wall dry matter (Carpita, 1989; Chesson et al., 1995a). Structural proteins of Type II walls are threonine-rich glycoproteins with sequences which show some homology with the hydroxproline-rich extensin (Kieliszewski et al., 1990). However, they are found in smaller amounts than extensin and probably play a secondary role to phenolic cross-links in fixing the shape of the mature cell.

Secondary thickening and lignification

During the process of secondary thickening additional cellulose and xylan is laid down on the luminal side of both Type I and Type II primary cell walls. The cellulose microfibrils are often deposited in a more organized and commonly orientated form than is typical of the primary wall and, in extensively thickened walls, may form more than one layer with a change in orientation between layers. The xylan laid down is a more linear polymer than the arabinoxylan typical of the Type II primary wall. Extended regions of the xylan chain are free from carbohydrate substituents and carry only occasional acetyl groups on C2 or C3 of individual xylopyranosyl units. Where carbohydrate substitution occurs the majority of side chains consist of single units of α-linked glucuronic acid or its 4-O-methyl derivative. For this reason the xylan found in secondary wall layers is commonly referred to as a glucuronoxylan. In annual plants there is little evidence of the presence of other polysaccharides occurring in other than trace amounts and the secondary wall layer(s) can be thought of as a single poly-saccharide network.

With few exceptions, secondary thickening is accompanied by lignification, a process that occurs in the primary wall as the polysaccharides forming the secondary wall layers are being laid down (Scobbie *et al.*, 1993). Thereafter the newly-formed secondary wall layer(s) become progres-sively lignified. The nature of the lignin polymer formed appears to differ both with time and cell type and, as a result, there appears to be heterogeneity in lignin struc-ture both between and within cell walls (Monties, 1989; Terashima *et al.*, 1993). In woody angiosperms the initial stages of lig-nification appear to involve the deposition of a polymer rich in 4-hydroxyphenyl-propanoid (H) units with equal or lesser amounts of guaiacyl (G; 4-hydroxy, 3-methoxyphenylpropanoid) units. As lignifi-cation spreads to the earliest-formed secondary wall layer, the concentration of H units decreases to be replaced by increas-ing amounts of syringyl (S; 3,5-dimethoxy-4-hydroxyphenylpropanoid) units. In the later stages of lignification both G and S units are deposited in ratios which depend on the particular cell type examined. There is also indirect evidence obtained by observing the incorporation of tritiated pre-cursors that the earliest formed lignin has a higher degree of carbon to carbon bonding than is found in lignin laid down during the latter stages of maturation (He and Terashima, 1989) although this may simply reflect the greater proportion of S units pre-sent in which 5-5' bonding is denied.

The process of lignin biogenesis in Type II cell walls appears to follow the general angiosperm pattern (He and Terashima, 1991) although there is one notable difference: the high concentration of the bifunctional phenolic acids, *p*-coumaric and ferulic acids, found asso-ciated with lignin. Approximately half of the ferulate present is both ester bonded to wall polysaccharide through its acid function and ether bonded to other phenylpropanoid units through the phenolic hydroxyl thus forming a bridging unit between lignin and carbohydrate (Lam *et al.*, 1992, 1994). This structure is consistent with the view that ferulate, which is inserted onto xylan by the action of a postulated feruloyl transferase (Myton and Fry, 1994) before export of the poly-saccharide through the plasmalemma and into the wall, acts as an initiation site for lignification (Ralph *et al.*, 1995). The deposition of *p*-coumarate tends to paral-lel the formation of lignin and it appears to serve a function within the plant which is different to that of ferulate. Although a small amount of *p*-coumaric acid may be detected ester-linked to carbohydrate at the earlier stages of Type II wall matura-tion (Mueller-Harvey *et al.*, 1986) it is not generally associated with the primary wall. The bulk of that deposited in the wall is thought to be ester-linked to the γ-carbon of phenylpropanoid units (Ralph *et al.*, 1994) resulting in chain termination.

Ferulic acid also occurs as an ester-linked substituent of pectic poly-saccharides in some Type I cell walls (Fry, 1982) and may also act as an initiation site during lignin deposition in the primary wall forming ferulate bridges between pec-tic material and lignin. However, plants with Type I cell walls clearly lack the mechanism for inserting ferulic acid into xylans and, in its absence, such bridge structures cannot be formed in the secon-dary wall of these plants. Instead the lignin is more directly associated with carbohydrate through the formation of benzyl ether and other alkali-stable bonds (Watanabe *et al.*, 1989). This difference is evident when the alkali-solubility of lignin is examined. In plants with Type II cell walls in which much of the lignin is bonded to cell wall carbohydrate through ester-linked ferulate, some 50–90% of the total lignin can be solubilized with 1 M NaOH at room temperature compared with 0–10% of the total lignin in plants with Type I cell walls (Chesson, 1983).

Architecture and the physical properties of cell walls

The three-dimensional lattice structure created by the polysaccharide network of the primary and secondary walls imposes physical characteristics which have clear consequences for wall degradability. Foremost among these is the pore structure, defined essentially by the spacing between individual polysaccharide chains, but further modified by the other wall components. The visualization of the polysaccharide network in cell walls of onion, a monocotyledonous plant with Type I walls, by a freeze-fracture electron microscopic technique enabled direct measurement of pore size to be made (McCann *et al.*, 1990). Pore diameters of approximately 10 nm were measured in the intact wall, increasing to 20 nm after extraction of the pectic network. These values were approximately twice those of previous measurements made using the solute exclusion technique (Stone and Scallan, 1968) with dextrans and polyethylene glycol probes of known diameter. Results obtained by solute exclusion with a variety of vegetation suggested that effective pore diameters of 3.5–5.5 nm are more commonly encountered (Carpita *et al.*, 1979; Gogarten, 1988; Flournoy *et al.*, 1993). However, the electron micrographs were made with uniform primary cells from a storage organ from which all bound water had been removed. Values would be expected to be larger than those obtained from hydrated samples and from walls with a load bearing function.

Although solute exclusion and gold probe methods can provide a rough guide to the mean size of pores likely to be encountered, they are insufficiently sensitive to distinguish between populations of

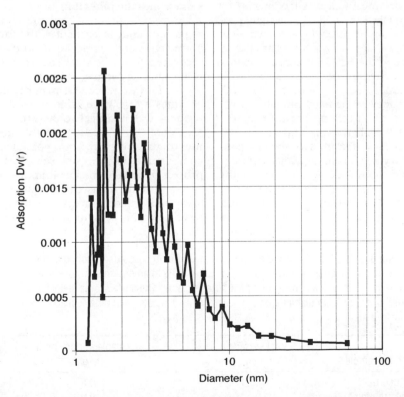

Fig. 3.3. The pore regime of maize stem measured by gas adsorption. Results are expressed as the relationship between pore diameter and Dv(r), the volume based pore size distribution, which provides a better estimate of relative pore numbers.

pores or to measure the relative frequency at which they occur. Physical chemistry methods, notably gas adsorption techniques, are able to discriminate between pores of the diameters expected (Fig. 3.3). The pore regime shown in Fig. 3.3 for cell walls of a mature maize internode determined by gas adsorption is typical of most vegetative cell walls. Pore diameters between 2 and 8 nm are typical with the relative numbers weighted towards pores of smaller diameter. The occurrence of several discrete populations of pores and the similarity between pore regimes in different plants reflects the prevalent spacings between wall polymers and the degree of commonality of structure that exists between plants of different phylogenic origin.

The limited porosity of the plant cell wall has important implications for the process of digestion. Pores of <4–5 nm diameter are too small to allow ingress by even relatively low molecular weight proteins of the order of 25 kDa. Enzymes of greater molecular weight are restricted to the surface of the cell. In practice, polysaccharidase activities are rarely detected in rumen liquor because most cell wall degrading enzymes remain intimately associated with the attacking organism or the surface of the plant wall. This adaptation appears to have the dual benefit of conserving enzymes in an aqueous environment and physically orientating the enzyme for maximum effect. The major rumen activities appear to exist as multi-enzyme complexes of one million daltons or more (Lamed et al., 1987; Huang and

Forsberg, 1990; Wilson and Wood, 1992) and to show some similarity with the cellulosomal structures postulated for the compost bacterium, Clostridium thermocellum (Béguin et al., 1992; Béguin and Aubert, 1994). However, a scaffold protein responsible for locking the various enzymes together has yet to be identified in the complexes produced by rumen bacteria. One apparent difference between Clostridium and the rumen cellulolytic bacteria is the identification of a number of genes coding for bifunctional enzymes in the rumen bacteria. These have two catalytic centres able to attack simultaneously two similar or different cell wall linkages (Flint et al., 1993). Any enzyme complex restricted to the surface of the wall would be faced with a variety of glycosidic and other linkages. Bifunctional enzymes may represent an efficient way of packaging all of the activities needed to erode the wall surface into the minimum space.

Enzymes produced by soil-inhabiting organisms and, in particular, the fungi are more typically extracellular and readily recovered from the litter/soil environment (Deng and Tabatabai, 1994; Gander et al., 1994). However, the vast majority of these enzymes are still too large to easily penetrate the wall and freely diffuse through the wall pores. As a result selective degradation of wall polymers is not possible and all components of the wall are lost by a process of surface erosion at approximately the same rate. Thus the available surface area and regiochemistry of the surface layers are the keys to understanding the degradative process.

Table 3.3. The total surface area ($m^2 g^{-1}$) measured by gas adsorption, and the calculated maximum surface area ($m^2 g^{-1}$) available to an enzyme of molecular weight >20 kDa of a number of typical forage plants. The surface area associated with pores of 3 nm diameter or less was excluded in calculating the available surface area.

Plant sample	Total surface area ($m^2 g^{-1}$)	Available surface area ($m^2 g^{-1}$)
Lucerne	2.1	0.8
Maize stem	2.4	0.9
Grass (Timothy)	4.9	2.6
Wheat straw	3.2	1.7

Available surface area and the importance of comminution

Nitrogen gas used for gas adsorption/desorption measurements is able to freely enter intact cells and the micro- and mesopores of the individual cell walls. The measure of surface area obtained from the volume of gas adsorbed is thus a measure of *total surface area* and is not affected by the integrity of the tissue or the porosity of the wall. The total surface area of most plant material is lower than might be expected and typically is of the order 2–8 $m^2 g^{-1}$ (Table 3.3). Although, as might be expected, surface area is lost during degradation, the surface area per unit weight of residue at any given time in the process stays constant or even increases slightly. This pattern of change is consistent with cell walls whose pore structure is homogeneous but which are losing mass by surface erosion. The small increase in area per unit mass can be explained by the small increase in the diameter of the cell lumen as the inner surface of the cell wall is eroded. However the total surface area available to nitrogen is substantially greater than that available to an enzyme. The total surface area is better calculated discounting the area associated with pores below an arbitrarily chosen diameter (Table 3.3).

Provided microbial numbers are not a limiting factor, which is rarely the case in the rumen, and given that degradation is a wholly superficial process, then the total surface area will define a maximum possible rate of degradation and the true available surface area, the real rate. In the laboratory it is possible to fully disrupt cell wall preparations and to ensure that the total and available surface areas are essentially the same. When this is done it can be shown that the rate of degradation is a constant independent of cell type or age (Lopez *et al.*, 1993). However, in practice, cells remain entire through much of the degradative process and their luminal surfaces are not available to attacking microorganisms and their enzymes. A principal aim of feed processing is to disrupt feed structure and increase the available surface to increase the rate of digestion of the feed.

This, in turn, reduces the energy and time the host animal needs to spend on rumination. It is evident that a parallel role is played by the microfauna found in litter and soils and many reports have documented the increased availability of nutrients or rate of litter breakdown in their presence (Scheu and Wolters, 1991; Tian *et al.*, Chapter 9, this volume).

Polymer Chemistry

Cell wall polysaccharides

The fine structure of isolated cell wall polysaccharides can have a marked effect on their breakdown by enzymes *in vitro*. The extent of degradation usually is inversely related to the degree of substitution of the backbone and presence of appropriate debranching enzymes have been shown on numerous occasions to enhance the action of the enzymes attacking the main chain (Biely, 1985). However, there is little evidence to show that such constraints act *in vivo* in the rumen. The rumen microflora is well adapted to the nature of the plant feed consumed by the host and is able to produce the necessary debranching activities in tandem with enzymes active against the main chain. In the absence of lignin, or after chemical delignification, cell walls are fully degradable. The thin wall of primary cells such as leaf mesophyll, for example, disappear within 6–8 h of ingestion (Chesson *et al.*, 1986)

The fine structure of cellulose has often been cited as a constraint to digestion by gut microorganisms. The evidence for this is slight and seems largely based on an extrapolation from the kinetics of degradation of modified cellulose or cellulose from cotton and wood of varying degrees of crystallinity. Crystallinity (crystallinity index, CI) is a measure of order within the molecule determined by IR spectroscopy, X-ray diffraction (Chesson, 1983) or, more recently, by ^{13}C NMR (Cyr *et al.*, 1990). In fact, the CI of forage cellulose is of a low degree of order overall compared with cotton and the cellulose found in most woods. Regions of higher order do exist within forage cellulose, although it is not known whether these are associated with the walls of specific cell types, specific regions of the wall or are more generally distributed. There is, however, no evidence to suggest that these more highly ordered regions of the molecule demonstrate an increased resistance to degradation by gut microorganisms since the CI of forage cellulose does not increase during digestion *in vivo* (Beveridge and Richards, 1975). In fact *in vitro* evidence suggests that crystallinity is less of a factor than the surface area of the fibril accessible to bacteria and their enzymes. Samples of cellulose having similar gross surface areas but different CI were degraded at similar rates by mixed rumen microorganisms (Weimer *et al.*, 1990). This observation is wholly consistent with a superficial attack by a bacterial enzyme complex ordered for maximum effect in which chain ends exposed by the action of endo-β-glucanases are attacked by other glycosidases in the complex before the glycosidic linkage reforms. The enzymes responsible for cellulose degradation produced by the aerobic fungi predominant in litter and the upper layers of the soils, which exist in a less ordered state than the cellulosomal-like complexes of the bacteria, are more likely to be affected by the CI of the cellulose substrate.

Phenylpropanoids and the extent of degradation

Lignin is the overriding factor determining the digestibility of the cell wall component of plant feed by ruminants and other herbivores. A broad inverse correlation exists between cell wall degradability and lignin content although the variation found within any one group of closely related plants also suggests that the nature of the lignin–carbohydrate association can be as important as the total amount of lignin present.

Lignin has long been used as an inert marker to calculate cell wall digestibility on the assumption that it is not degraded within the strongly anaerobic environment provided by the rumen. Although considerable amounts of lignin are released during cell wall degradation and can be detected as soluble lignin–carbohydrate complexes (LCC) in the rumen liquor (Conchie *et al.*, 1988), all such phenyl-propanoid-rich material is eventually precipitated by the more acid environment of the abomasum. Some minor modifications to lignin structure probably do occur in the rumen (demethoxylation, metabolism of phenolic acids) and there is an apparent potential for the anaerobic cleavage of 8-O-4 aryl ether bonds (Chen *et al.*, 1985). However, such changes are evidently small and virtually all of the lignin ingested can be recovered in the faeces.

Lignin acts to protect the structural polysaccharide to which it is covalently linked. The most probable mechanism for this is steric hindrance but the hydrophobicity of the complex also may mitigate against hydrolysis. Thus LCC exposed at the surface of the cell wall are inherently resistant to microbial attack and either are not degraded or are degraded at a rate substantially lower than any surrounding polysaccharide free from any association with lignin (Fig. 3.4). Despite a tendency for most LCC to be undermined by the breakdown of surrounding polysaccharide and released from the wall, with time, LCC accumulate at the wall surface at the expense of non-associated polysaccharide. This changes significantly the regiochemistry of the surface layer slowing and eventually preventing further degradation from occurring. The rate at which this inert LCC surface develops is a product of the initial concentration of lignin in the wall. It is fastest in the most extensively lignified tissues which, as a consequence, show the least loss of dry matter. As a consequence, the overall lignin content of the plant material consumed effectively determines the maximum possible extent of degradation under rumen conditions.

The constraint imposed by lignin is not absolute even within the rumen. Artificially extended holding times (>120 h) or the reintroduction of 'fully' digested material will always lead to a small additional loss. Digestibility in the

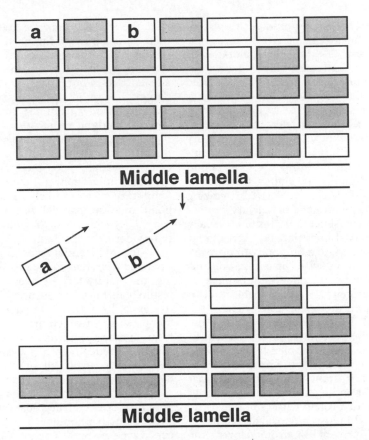

Fig. 3.4. A diagrammatic representation of changes at the surface of a lignified cell wall seen in cross-section undergoing microbial attack in the rumen. Attack is from the luminal side of the wall towards the middle lamella. The open boxes represent regions of polysaccharide closely associated with lignin and the darker boxes free polysaccharide. At the start of digestion the surface (top layer) consists of both lignified and non-lignified polysaccharide. As digestion proceeds free polysaccharide is preferentially degraded leaving a more extensively lignified surface which inhibits further digestion. Some lignin-carbohydrate complex (a and b) is undermined by the degradation of surrounding polysaccharide and released, contributing to the soluble lignin–carbohydrate complex found in rumen liquor.

rumen is probably a poor guide to the loss of biomass in soils because of the longer time span even in the absence of lignolytic activities. In practice, in litter and soils, colonization by a lignolytic flora will ensure that there is no limit imposed by the substrate on the eventual extent of breakdown although other environmental factors may, and often do, intervene. None the less, the inhibitory effects of lignin remain evident and lignin (or lignin + condensed tannin) content is strongly correlated with nitrogen release in many soils (Vallis and Jones, 1973; Thomas and Asakawa, 1993).

Condensed tannin and nitrogen availability

Condensed tannins, rather confusingly referred to as polyphenols in the context of litter degradation, are commonly encountered in feed legumes and in particularly high concentration in the leaves of some tree legumes used as browse or chopped for more general feed use (Dzowela *et al.*, 1995; Perez-Maldonado *et al.*, 1995). Animals that commonly browse tannin-rich material have adapted to resist some of the deleterious effects shown in grazing species. In browsing animals such as deer, an element of protection is offered by a proline-rich salivary protein which has a high affinity for tannins and which complexes the tannin before feed enters the rumen (Robbins *et al.*, 1987; D'Mello, 1992). In grazing species, in the absence of salivary protection, the presence of tannin usually reduces palatability, intake, dry matter digestibility and, in common with nitrogen mineralization in soils, nitrogen availability. The ability of condensed tannins to complex protein is very variable and is dependent on the phylogenic origin of the tannin, pH, cationic environment and the nature of the protein being complexed. In general maximum complex formation will occur around the pI of the protein since at this pH electrostatic repulsion is minimal allowing bonding between carbonyl groups on the protein with tannin hydroxyl groups. Ribulose biphosphate carboxylase (leaf protein 1) from clover, for example, has a pI of 5.5 but will complex over a far wider range in the rumen. Much less is known about tannins from tropical forage or tree legumes and their interaction with dietary protein than is known about tannins from temperate forage legumes (Perez-Maldonado *et al.*, 1995).

Enhanced Degradation

Manipulation of cell wall degradation can take two forms, manipulating the process or modifying the substrate. In the former case the goal is to match as closely as possible the rate and/or extent of degradation with the maximum possible values set by the plant and in the latter case to reset the maximum values. Feed processing (milling, pelleting, extrusion, micronizing) can and is used to disrupt the cellular integrity of the plant material and maxi-

mize the available surface area. However, there is a concomitant need to preserve particles of sufficient size to ensure that retention time within the rumen is adequate to maintain the extent of degradation. Too small a particle size will result in early outflow from the rumen and, at best, a hindgut rather than rumen fermentation. Feeding strategies seek to match the supply of nitrogen with the energy available to the microflora to maintain microbial numbers and avoid temporary excesses likely to contribute to pollution. The parallel with the soil concept of synchrony with plant demand is self-evident.

More substantial gains in the rate and extent of degradation are possible by altering the nature of the plant. Immediate results can be achieved using a variety of oxidative, hydrothermal and alkaline treatments, some of which are intended as on-farm processes while others require purpose built plant (Chesson *et al.*, 1995b). Essentially all of the treatments disrupt the LCC and reduce the lignin content of the plant, thereby increasing the potential extent of degradation (Chesson, 1993). Such treatments are effective but expensive in both energy and financial terms and have found favour only when feed resources are scarce or when the treatments are directly or indirectly subsidized.

Biological treatments for feeds

Harnessing the capacity of lignolytic fungi to reduce lignin content and improve cell wall degradability as a feed pre-treatment has obvious parallels with litter decomposition. As a concept it is anything but new. *Palo-podrido*, the highly digestible decayed wood of several hardwood trees has been used for centuries as animal feed in Southern Chile (Zadrazil *et al.*, 1982). Aside from this natural phenomenon, two fundamental problems have prevented biological pretreatments becoming a commercially viable proposition:

● the need to avoid or reduce competition from faster growing organisms;
● the inevitable loss of carbohydrate and

digestible dry matter that accompanies lignin breakdown.

Attempts to resolve the problem of competition have led to a divergence of effort with the use of bulk sterilization and purpose designed solid substrate fermenters offered as one solution (Reid, 1989; Zadrazil *et al.*, 1990a). This approach may be coupled with edible fruiting body production, usually *Plerotus* spp., improving the economics of an otherwise expensive treatment (Zadrazil *et al.*, 1990b). The alternative, composting under conditions which selectively favour the introduced lignolytic strain, is seen as a low-cost process applicable to non-industrialized countries and requiring the minimum of capital outlay. *Coprinus fimetarius* is often the organism of choice since it will, unlike the majority of competing organisms, tolerate the high pH condition produced by ammonia (urea) treatment of rice or wheat straw substrates. The fungus also captures some of the excess ammonia-N which is lost to the atmosphere during urea-treatment increasing the protein content and improving the amino acid balance of the treated product (Singh *et al.*, 1992).

Breakdown of lignin is energetically unfavourable to microorganisms and appears an act of necessity rather than choice. As a result, other substrates, when available, are attacked in preference and lignin degradation is invariably accompanied by the loss of cell wall carbohydrate. It is relatively rare for the rate of lignin breakdown to exceed that of carbohydrate and so treatments with most fungi tend to decrease the digestibility of the treated product. Even when fungal strains do exhibit some selectivity for lignin and produce a more digestible product, often the gains made are offset by overall loss of digestible dry matter which occurs during fungal growth. Recombinant technology, which has been extensively applied to *Phanerochaete chrysosporium* (Gold and Alic, 1993) and, to a lesser extent, to other lignolytic strains, may offer solutions to this problem. In the meantime, attempts to apply cell-free enzymes to delignify crop residues have proved singularly unsuccessful (Khazaal *et al.*, 1993).

There is little doubt that, in the longer term, molecular biology will be used to alter the lignin content and nutritive value of forage plants, providing a more permanent solution to improving poor quality feed materials. Antisense RNA technology has already been used to down-regulate enzymes in the phenylpropanoid pathway with consequences for digestibility (Halpin *et al.*, 1994; Bavage *et al.*, Chapter 16, this volume) and such methods are now being applied to elite crop lines. Residues from transgenic plants would be expected to degrade faster in agricultural soils than conventional plants.

The Rumen – An Optimum Habitat for Cell Wall Decomposition

Fluctuations in a wide range of ecological factors including temperature, rainfall and soil moisture, the availability of nitrogen, phosphorus and other limiting nutrients, and the biological activity of both the flora and fauna, affect the decomposition of plant detritus in litter and soils. In many respects the rumen represents an ideal habitat in which most of these variables are controlled and adjusted for maximum effect. The rate of cell wall degradation under these conditions, which is measured in hours, can give some indication of the severe constraints to degradation that operate in soils and litter where degradation may be measured in weeks. However, the rumen has evolved to avoid, as far as possible, the problem of lignin and related compounds. Consumption is restricted to plant material in which the lignin content is less than approximately 15% of cell wall dry matter and often is well below 10%. At this level much of the lignin occurs in discrete units and can be solubilized if not degraded. Above 15% lignin is found in larger interconnected structures and is not amenable to solubilization. As a result, the extent of cell degradation is substantially reduced.

When making comparisons it is important to recognize that the rumen is a specialized ecosystem. The rumen flora are unable to survive outside the digestive tract and soil borne organisms fail to flourish in the rumen. The rumen is an optimum environment for plant degradation, but only for selected substrates. Herbivores are unable to decompose the more woody detritus that makes up a substantial proportion of plant litter.

Acknowledgements

The author thanks Sandra Murison (Rowett Research Institute) for providing the SEM shown in Fig. 3.2 and Peter Gardner (Rowett Research Institute) for the gas adsorption data shown in Fig. 3.3 and Table 3.3. Financial support for this work was provided by the Agriculture, Environment and Fisheries Department of the Scottish Office.

References

Akin, D.E. (1979) Microscopic evaluation of forage digestion by rumen microorganisms. A review. *Journal of Animal Science* 48, 701–710.

Akin, D.E. (1988) Biological structure of lignocellulose and its degradation in the rumen. *Animal Feed Science and Technology* 21, 295–310.

Akin, D.E. (1989) Histological and physical factors affecting digestibility of forages. *Agronomy Journal* 81, 17–25.

Béguin, P. and Aubert, J.P. (1994) The biological degradation of cellulose. *FEMS Microbiology Reviews* 13, 25–58.

Béguin, P., Millet, J. and Aubert, J.P. (1992) Cellulose degradation by *Clostridium thermocellum:* from manure to molecular biology. *FEMS Microbiology Letters* 100, 523–528.

Beveridge, R.J. and Richards, G.N. (1975) Investigation of the digestion of cell-wall components of spear grass and of cotton cellulose by viscometry and by X-ray diffraction. *Carbohydrate Research* 43, 163–172.

Biely, P. (1985) Microbial xylanolytic systems. *Trends in Biotechnology* 3, 286–290.

Carpita, N.C. (1989) Pectic polysaccharides of maize coleoptiles and proso millet cells in liquid culture. *Phytochemistry* 28, 121–125.

Carpita, N.C. and Gibeaut, D.M. (1993) Structural models of primary cell walls in flowering plants: consistency of molecular structure with the physical properties of the walls during growth. *The Plant Journal* 3, 1–30.

Carpita, N.C., Sabularse, D., Montezinos, D. and Delmer, D.P. (1979) Determination of the pore size of cell walls of living plant cells. *Science* 205, 144–1147.

Chen, W., Supanwong, K., Ohmiya, K., Shimizu, S. and Kawakami, H. (1985) Anaerobic degradation of veratrylglycerol-β-guaiacyl ether and guaiacoxyacetic acid by mixed rumen bacteria. *Applied and Environmental Microbiology* 50, 1451–1456.

Chesson, A. (1983) Effects of sodium hydroxide on cereal straws in relation to the enhanced degradation of structural polysaccharides by rumen microorganisms. *Journal of the Science of Food and Agriculture* 32, 745–758.

Chesson, A. (1993) Mechanistic models of forage cell wall degradation. In: Jung, H.G., Buxton, D.R., Hatfield, R.D. and Ralph, J. (eds) *Forage Cell Wall Structure and Digestibility*. American Society of Agronomy, Inc., Crop Science Society of America, Inc., Soil Science Society of America, Inc., Madison, Wisconsin, USA, pp. 347–376.

Chesson, A., Stewart, C.S., Dalgarno, K. and King, T.P. (1986) Degradation of isolated grass mesophyll, epidermis and fibre cell walls in the rumen and by cellulolytic rumen bacteria in axenic culture. *Journal of Applied Bacteriology* 60, 327–336.

Chesson, A., Gordon, A.H. and Scobbie, L. (1995a) Pectic polysaccharides of mesophyll cell walls of perennial ryegrass leaves. *Phytochemistry* 38, 579–583.

Chesson, A., Forsberg, C.W. and Grenet, G. (1995b) Improving the digestion of plant cell walls and fibrous feeds. In: Journet, M., Grenet, E., Farce, M.-H., Theriez, M. and Demarquilly, C. (eds) *Recent Developments in the Nutrition of Herbivores*. INRA Editions, Paris, pp. 249–277.

Conchie, J., Hay, A.J. and Lomax, J.A. (1988) Soluble lignin–carbohydrate complexes from sheep rumen fluid. Their composition and structural features. *Carbohydrate Research* 177, 127–151.

Cooper, J.B., Chen, J.A., van Holst, G.J. and Varner, J.E. (1987) Hydroxyproline-rich glycoproteins of plant cell walls. *Trends in Biochemistry* 12, 24–27.

Cyr, N., Elofson, R.M. and Mathison, G.W. (1990) Determination of crystallinity of carbohydrates by ^{13}C cross polarization/magic angle spinning NMR with application to the nutritive value of forages. *Canadian Journal of Animal Science* 70, 695–701.

Deng, S.P. and Tabatabai, M.A. (1994) Cellulase activity of soils. *Soil Biology and Biochemistry* 26, 1347–1354.

D'Mello, J.P.F. (1992) Chemical constraints to the use of tropical legumes in animal nutrition. *Animal Feed Science and Technology* 38, 237–261.

Dzowela, B.H., Hove, L., Topps, J.H. and Mafongoya, P.L. (1995) Nutritional and antinutritional characters and rumen degradability of dry matter and nitrogen from some multipurpose tree species with potential for agroforestry in Zimbabwe. *Animal Feed Science and Technology* 55, 207–214.

Flint, H.J., Martin, J., McPherson, C.A., Daniel, A.S. and Zhang, J.-X. (1993) A bifunctional enzyme, with separate xylanase and β-(1,3-1,4)-glucanase domains, encoded by the xynD gene of *Ruminococcus flavefaciens*. *Journal of Bacteriology* 175, 2943–2951.

Flournoy, D.S., Paul, J.A., Kirk, K. and Highley, T.L. (1993) Changes in the size and volume of pores in sweetgum wood during simultaneous rot by *Phanerochaete chrysosporium* Burds. *Holzforschung* 47, 297–301.

Fonty, G., Gouet, P.L., Jouany, J.P. and Semaud, J. (1983) Ecological factors determining establishment of cellulolytic bacteria and protozoa in the rumens of meroxenic lambs. *Journal of General Microbiology* 129, 213–223.

Fry, S.C. (1982) Phenolic components of the primary cell wall. Feruloylated oligosaccharides of *D*-galactose and *L*-arabinose from spinach polysaccharide. *Biochemical Journal* 203, 493–504.

Gander, L.K., Hendricks, C.W. and Doyle, J.D. (1994) Interferences, limitations and improvement in the extraction and assessment of cellulase activity in soil. *Soil Biology and Biochemistry* 26, 65–73.

Gogarten, J.P. (1988) Physical properties of the cell wall of photoautotrophic suspension cells from *Chenopodium rubrum* L. *Planta* 174, 333–339.

Gold, M.H. and Alic, M. (1993) Molecular biology of the lignin-degrading basidiomycete *Phanerocheate chrysosporium*. *Microbiology Reviews* 57, 605–622.

Grabber, J.H. and Jung, G.A. (1991) *In vitro* disappearance of carbohydrates, phenolic acids, and lignin from parenchyma and sclerenchyma cell walls isolated from cocksfoot. *Journal of the Science of Food and Agriculture* 57, 315–323.

Halpin, C., Knight, M.E., Foxon, G.A., Campbell, M.M., Boudet, A.M., Boon, J.J., Chabbert, B., Tolier, M.-T. and Schuch, W. (1994) Manipulation of lignin quality by downregulation of cinnamyl alcohol dehydrogenase. *The Plant Journal* 6, 339–350.

He, L. and Terashima, N. (1989) Formation and structure of lignin in monocotyledons I. Selective labeling of the structural units of lignin in rice plant (*Oryza sativa*) with ^{3}H and visualization of their distribution in the tisue by microautoradiography. *Mokuzai Gakkaishi* 35, 116–122.

He, L. and Terashima, N. (1991) Formation and structure of lignin in monocotyledons IV. deposition process and structural diversity of the lignin in the cell wall of sugarcane and rice plant studied by ultraviolet microspectroscopy. *Holzforschung* 45, 191–198.

Hillman, K., Lloyd, D. and Williams, A.G. (1985) Use of a portable quadrupole mass spectrometer for the measurement of dissolved gas concentrations in the ovine rumen liquor in situ. *Current Microbiology* 12, 335–340.

Ho, Y.W., Abdullah, N. and Jalaludin, S. (1988) Penetrating structures of anaerobic rumen fungi in cattle and swamp buffalo. *Journal of General Microbiology* 134, 177–181.

Huang, L. and Forsberg, C.W. (1990) Cellulose digestion and cellulase regulation and distribution in *Fibrobacter succinogenes* subsp. *succinogenes*. *Applied and Environmental*

Microbiology 56, 1221–1228.

Khazaal, K.A., Owen, E., Dodson, A.P., Palmer, J. and Harvey, P. (1993) Treatment of barley straw with ligninase: effect of activity and fate of the enzyme shortly after being added to straw. *Animal Feed Science and Technology* 41, 15–27.

Kieliszewski, M.J., Leykam, J.F. and Lamport, D.T.A. (1990) Structure of the threonine-rich extensin from *Zea mays*. *Plant Physiology* 92, 316–326.

Lam, T.B.T., Iiyama, K. and Stone, B.A. (1992) Cinnamic acid bridges between cell wall polymers in wheat and phalaris internodes. *Phytochemistry* 31, 1179–1183.

Lam, T.B.T., Iiyama, K. and Stone, B.A. (1994) An approach to the estimation of ferulic acid bridges in unfractionated cell walls of wheat internodes. *Phytochemistry* 37, 327–333.

Lamed R.J., Naimark J., Morgenstern E. and Bayer E.A. (1987) Specialised cell surface structures in cellulolytic bacteria. *Journal of Bacteriology* 169, 3792–3800.

Lamport, D.T.A. (1986) The primary cell wall: a new model. In: Young, R.A. and Rowell, R.M. (eds) *Cellulose: Structure, Modification and Hydrolysis*. John Wiley, New York, pp. 77–90.

Lopez, S., Murison, S.D., Travis, A.J. and Chesson, A. (1993) Degradability of parenchyma and sclerenchyma cell walls isolated at different developmental stages from a newly extended maize internode. *Acta Botanica Neerlandica* 42, 165–174.

McCann, M.C., Wells, B. and Roberts, K. (1990) Direct visualization of cross-links in the primary plant cell wall. *Journal of Cell Science* 96, 323–334.

Monties, B. (1989) Molecular structure and biochemical properties of lignins in relation to possible self-organisation of lignin networks. *Annales des Sciences Forestieres (Paris)* 46(Suppl.), 848s–855s.

Mueller-Harvey, I., Hartley, R.D., Harris, P.J. and Curzon, E.H. (1986) Linkage of *p*-coumaroyl and feruloyl groups to cell-wall polysaccharides of barley straw. *Carbohydrate Research* 104, 121–138.

Myton, K.E. and Fry, S.C. (1994) Intraprotoplasmic feruloylation of arabinoxylans in *Festuca arundinacea* cell cultures. *Planta* 193, 326–330.

Pell, A.N. and Schofield, P. (1993) Microbial adhesion and degradation of plant cell walls. In: Jung, H.G., Buxton, D.R., Hatfield R.D. and Ralph, J. (eds) *Forage Cell Wall Structure and Digestibility*. American Society of Agronomy, Inc., Crop Science Society of America, Inc., Soil Science Society of America, Inc., Madison, Wisconsin, USA, pp. 397–423.

Perez-Maldonado, R.A., Norton, B.W. and Kerven, G.L. (1995) Factors affecting *in vitro* formation of tannin–protein complexes. *Journal of the Science of Food and Agriculture* 69, 291–298.

Ralph, J., Hatfield, R.D., Quideau, S., Helm, R.F., Grabber, J.H. and Jung, H.J.G. (1994) Pathway of *p*-coumaric acid incorporation into maize lignin as revealed by NMR. *Journal of the American Chemical Society* 116, 9448–9456.

Ralph, J., Grabber, J.H. and Hatfield, R.D. (1995) Lignin-ferulate cross-links in grasses: active incorporation of ferulate polysaccharide esters into ryegrass lignins. *Carbohydrate Research* 275, 167–178.

Reid, I.D. (1989) Solid state fermentation for biological delignification. *Enzyme and Microbial Technology* 11, 786–803.

Robbins, C.T., Mole, S., Hagerman, A.E. and Hanley, T.R. (1987) Role of tannins in defending plants against ruminants: reduction in dry matter digestion. *Ecology*, 68, 1606–1615.

Scheu, S. and Wolters, V. (1991) Influence of fragmentation and bioturbation on the decomposition of ^{14}C-labelled beech leaf litter. *Soil Biology and Biochemistry* 23, 1029–1034.

Scobbie, L., Russell, W., Provan, G.J. and Chesson, A. (1993) The newly extended maize internode: a model for the study of secondary cell wall formation and consequences for digestibility. *Journal of the Science of Food and Agriculture* 61, 217–225.

Shedletzky, E., Shmuel, M., Trainin, T., Kalman, S. and Delmer, D. (1992) Cell wall structure in cells adapted to growth on the cellulose-synthesis inhibitor 2,6–dichlorobenzonitrile (DCB): a comparison between two dicotyledonous plants and a graminaceous monocot. *Plant Physiology* 100, 120–130.

Singh, K., Singh, G.P. and Gupta, B.N. (1992) Biochemical studies of *Coprinus fimetarius* inoculated straws. *Indian Journal of Microbiology* 32, 473–477.

Smith, J.H. and Peckenpaugh, R.E. (1986) Straw decomposition in irrigated soil. Comparison of twenty-three cereal straws. *Soil Science Society of America Journal* 50, 928–932.

Stone, J.E. and Scallan, A.M. (1968) A structural model for the cell wall of water-swollen wood pulp fibres based on their accessibility to macromolecules. *Cellulose Chemistry and Technology* 2, 343–358.

Terashima, N., Fukushima, K., He, L.-F. and Takabe, K. (1993) Comprehensive model of the lignified plant cell wall. In: Jung, H.G., Buxton, D.R., Hatfield, R.D. and Ralph, J. (eds) *Forage Cell Wall Structure and Digestibility*. American Society of Agronomy, Inc., Crop Science Society of America, Inc., Soil Science Society of America, Inc., Madison, Wisconsin, USA, pp. 247–270.

Thomas, R.J. and Asakawa, N.M. (1993) Decomposition of leaf litter from tropical forage grasses and legumes. *Soil Biology and Biochemistry* 25, 1351–1361.

Travis, A.J., Murison, S.D., Chesson, A. and Perry, P. (1995) Quantitative assessment of cell wall degradability in forages and crop residues using 3–D microscopy. *Food Structure* (in press).

Trinci, A.P.J., Davies, D.R., Gull, K., Lawrence, M.I., Neilsen, B.B., Rickers, A. and Theodorou, M.K. (1994) Anaerobic fungi of herbivorous animals. *Mycological Research* 93, 129–152.

Tuah, A.K., Lufadeju, E. and Orskov, E.R. (1986) Rumen degradation of straw. 1. Untreated and ammmonia-treated barley, oat and wheat straw varieties and triticale straw. *Animal Production* 43, 261–269.

Ulyatt, M.J., Dellow, D.W., John, A., Reid, C.S.W. and Waghorn, G.C. (1986) Contribution of chewing during eating and rumination to the clearance of digesta from the ruminoreticulum. In: Milligan, L.P., Grovum, W.L. and Dobson, A. (eds) *Control of Digestion and Metabolism in Ruminants*. Prentice Hall, Englewood Cliffs, New Jersey, USA. pp. 498–514.

Vallis, I. and Jones, R.J. (1973) Net mineralization of nitrogen in leaves and leaf litter of *Desmodium intortum* and *Phaseolus atropurpureus* mixed with soil. *Soil Biology and Biochemistry* 5, 391–398.

Varma, A., Kolli, B.K., Paul, J., Saxena, S. and Konig, H. (1994) Lignocellulose degradation by microorganisms from termite hills and termite guts: a survey on the present state of art. *FEMS Microbiology Reviews* 15, 9–28.

Watanabe, T.J., Ohnishi, J., Yamasaki, Y., Kaizu, S. and Koshijima, T. (1989) Binding-site analysis of the ether linkage between lignin and hemicelluloses in lignin-carbohydrate complexes by DDQ oxidation. *Agricultural and Biological Chemistry* 53, 2233–2252.

Weimer, P.J. (1992) Cellulose degradation by ruminal microorganisms. *Critical Reviews in Biotechnology* 12, 189–223.

Weimer, P.J., Lopez-Guisa, J.M. and French, A.D. (1990) Effect of cellulose fine structure on kinetics of its digestion by mixed ruminal microorganisms *in vitro*. *Applied and Environmental Microbiology* 56, 2421–2429.

Wessen, B. and Berg, B. (1986) Long-term decomposition of barley straw. Chemical changes and growth of fungal mycelium. *Soil Biology and Biochemistry* 18, 55–59.

Williams, A.G. and Coleman, G.S. (1992) *The Rumen Protozoa*. Springer-Verlag, New York, pp. 441.

Wilson, J.R. (1993) Organization of forage plant tissues. In: Jung, H.G., Buxton, D.R., Hatfield, R.D. and Ralph, J. (eds) *Forage Cell Wall Structure and Digestibility*. American Society of America, Inc., Madison, Wisonsin, USA, pp. 1–32.

Wilson, C.A. and Wood, T.M. (1992) Studies on the cellulase of the rumen anaerobic fungus *Neocallimastix frontalis*, with special reference to the capacity of the enzyme to degrade crystalline cellulose. *Enzyme and Microbial Technology* 14, 258–264.

Zadrazil, F., Grinbergs, J and Gonzalez, A. (1982) 'Palo podrido' – decomposed wood which was used as feed. *European Journal of Applied Microbiology and Biotechnology* 15, 167–171.

Zadrazil, F., Janssen, H., Diedrichs, M. and Schuchardt, F. (1990a) Pilot-scale reactor for solid-state fermentation of lignicellulosics with higher fungi: production of feed, chemical feedstocks and substrates suitable for biofilters. In: Coughlan, M.P. and Amaral Collaco, M.T. (eds) *Advances in Biological Treatments of Lignocellulosic Materials*. Elsevier Applied Science, London pp. 31–41.

Zadrazil, F., Diedrichs, M., Janssen, H., Schuchardt, F. and Park J.S. (1990b) Large scale solid-state fermentation of cereal straw with *Pleurotus* spp. In: Coughlan, M.P. and Amaral Collaco, M.T. (eds) *Advances in Biological Treatments of Lignocellulosic Materials*. Elsevier Applied Science, London, pp. 43–58.

4 Role of Phenolic Secondary Metabolites in Plants and their Degradation in Nature

J.B. Harborne

Department of Botany, University of Reading, Whiteknights, Reading RG6 6AS, UK

Introduction

Unlike most other classes of secondary constituents (e.g. alkaloids, essential oils, etc.), the plant polyphenols are distinctive in their universal occurrence in vascular plants. All plants have a characteristic 'phenolic' fraction. Part is bound to the cell wall as lignin or as ferulic acid esterified to the hemicellulose, but much is present in water-soluble form, in glycosidic combination. There are many different structures; over 8000 plant products have phenolic substitution. Only a few of these are widely distributed: simple phenolic acids like 4-hydroxybenzoic; phenylpropanoids such as caffeic and ferulic acids; and flavonoids such as kaempferol and quercetin.

Plant polyphenols fall into two groups: those of low molecular weight; and those which are oligomeric or polymeric, namely the tannins (Table 4.1). The high molecular weight condensed tannins have a wide distribution, occurring in the leaves of all ferns and gymnosperms and of about half the families (the woody members) of angiosperms. By contrast the hydrolysable tannins, which are based biosynthetically on gallic acid, have a more restricted occurrence, being found only in the dicotyledons, in some 15 of the 40 orders. Plant polyphenols can occur throughout the plant including flower, fruit, seed, stem and root. However the most constant occurrence is in the leaf and the phenolic profile of the leaf will provide a good indication of the kind of phenolics that are likely to be present in other plant organs. Most plant surveys for phenolics have concentrated on leaf tissue, because of its accessibility not only with living plants but also with herbarium specimens. Leaf phenolic profiles do vary according to physiological status and age, but patterns are relatively constant throughout the life of the leaf.

Phenolic constituents have a variety of roles in the life of the plant. The anthocyanins, for example, are major contributors to plant colours in the pink, red, mauve and blue range. The colourless flavonoids have been implicated as ultraviolet protective agents in leaves, some flavonol glycosides present in pollen have a role in reproduction while flavones in root exudates are signal molecules for nodulation in the legume–*Rhizobium* symbiosis. Other leaf phenolics and especially the tannins are probably valuable for protecting plants from herbivory or microbial infection, as will be described below. Economically, plant polyphenols are important because they make major contributions to the taste, flavour and colour of our food and drink. The flavour and taste of tea for example is related to the fact that the tea leaf contains up to 30% of its dry weight as catechin-based polyphenol. Phenolics are also involved as active principles in a number of medicinal plants.

Table 4.1. The major classes of phenolics in plants.

Number of carbon atoms	Basic skeleton	Class	Example
6	C_6	Simple phenols	Catechol, hydroquinone
		Benzoquinones	2,6-Dimethoxybenzoquinone
7	C_6—C_1	Phenolic acids	p-Hydroxybenzoic, salicylic
8	C_6—C_2	Acetophenones	3-Acetyl-6-methoxybenzaldehyde
		Phenylacetic acids	p-Hydroxyphenylacetic
9	C_6—C_3	Hydroxycinnamic acids	Caffeic, ferulic
		Phenylpropenes	Myristicin, eugenol
		Coumarins	Umbelliferone, aesculetin
10	C_6—C_4	Naphthoquinones	Juglone, plumbagin
13	C_6—C_1—C_6	Xanthones	Mangiferin
14	C_6—C_2—C_6	Stilbenes	Lunularic acid
		Anthraquinones	Emodin
15	C_6—C_3—C_6	Flavonoids	Quercetin, malvin
		Isoflavonoids	Genistein
18	$(C_6$—$C_3)_2$	Lignans	Podophyllotoxin
30	$(C_6$—C_3—$C_6)_2$	Biflavonoids	Amentoflavone
n	$(C_6$—$C_3)_n$	Lignins	–
	$(C_6)_n$	Catechol melanins	–
	$(C_6$—C_3—$C_6)_n$	Flavolans (condensed tannins)	Procyanidin
	$(C_6$—$C_1)_n$	Gallotannins	Geraniin

Anthocyanins

These phenolic pigments provide pink to blue colours in flowers and fruits and have a well-defined role in attracting pollinators to the flowers for cross-pollination and animals to fruits for seed dispersal. There is natural selection for pollinator flower colour preferences so that blue (delphinidin-based) flowers predominate in temperate floras and scarlet (pelargonidin-based) flowers are common in tropical floras. Anthocyanin pigments are also regularly present in leaves either as a 'young leaf' flush, as permanent coloration (e.g. copper beech) or as autumnal red coloration. The purpose of leaf coloration is more obscure than that of flower or fruit and there is still some uncertainty why these pigments should appear in vegetative tissues. Their occurrence in part may be related to environmental factors and plants under stress (e.g. nitrogen deficiency) may develop anthocyanin coloration in leaves and stems. The production of a 'flush' of anthocyanin coloration in juvenile leaves may be a device to avoid insect predation. This is based on the fact that many phytophagous insects are geared to green leaves as a food source, so that red leaves are not recognized for this purpose. In support of this view it is known that aphids and butterflies avoid feeding on red cabbage varieties, although these plants are just as favourable as green forms for reproduction and survival (van Emden, 1989).

Anthocyanins have relatively unstable chromophores, which are easily attacked by hydroxyl ions at the acid pH of the cell sap causing decolorization and ring opening. They are necessarily stabilized *in vivo* by intermolecular copigmentation with flavones, by intramolecular copigmentation with the aromatic acyl substituents that may be present attached through sugar to the chromophore, or by molecular stacking in the case of O-methylated pigments.

Anthocyanins of flowers generally have a short half-life, coinciding with the relatively fugitive nature of floral tissues. Thus delphinidin glycosides in the plant *Petunia hybrida* corollas are reported to have a half-life of 25–31 h based on tracer studies (Steiner, 1971). Similarly, examination of anthocyanin synthesis in *Pisum sativum* flowers has demonstrated the presence of a very active system of synthesis and turnover. The catabolism of anthocyanins may involve the removal of the sugar from the 3-hydroxyl, which immediately causes discoloration and an enzyme called anthocyanase has been described which carries out this step. Alternatively, anthocyanins with catechol moieties (e.g. cyanidin and delphinidin glycosides) may undergo phenolase oxidation, again leading to loss of colour and disruption of the aromatic groups (Harborne, 1980).

Leaf anthocyanins may be turned over more slowly than floral anthocyanins but are likely to be metabolically active. The flush coloration of juvenile leaves is known to disappear relatively rapidly as the leaf develops. Leaves of species with permanent leaf colour probably lose most of their anthocyanin during senescence. It seems unlikely in general that much anthocyanin will remain in the leaf after it has died. However, it is always possible that anthocyan colour may become fixed in 'plant tissue' by covalent binding to cell wall polysaccharide. There are rare examples of flowers (e.g. *Limonium*) which on drying retain their coloration for many years.

Internal Leaf Flavonoids

The most convincing function prescribed to the abundant phenolics within the leaf is that of ultraviolet (UV) protection. Evidence, which is still very incomplete, is based on observations that the synthesis of many of these phenolics is induced by light, especially by UV light, and some of them are located specifically in the upper epidermal cells of the leaves. They also absorb strongly in the 280–320 nm (UV-B) region. This is true of the flavones and flavonols but also of related phenylpropanoids, such as chlorogenic acid (λ max 320 nm). Other roles are less clear cut. They can be toxic to some phytophagous insects (e.g. *Helicoverpa*) but the majority of such species readily metabolize and detoxify dietary phenolics; a few Lepidoptera such as the marbled white butterfly even sequester and store dietarily-derived flavonoids in body and wing. Also, these glycosidic constituents show relatively low antifungal activity and do not appear obviously to be protective against microbial infection.

How far these colourless flavonoids are turned over during the life of the leaf is not clear. An experiment with chickpea (*Cicer arietinum*) leaves suggested a half-life for the kaempferol and quercetin glycosides present of between 7 and 12 days. On the other hand, an experiment with *Cucurbita maxima* seedlings suggested relatively short half-lives (36–48 h) for the biosynthetically intermediate flavonol glycosides (monoside, bioside) but no turnover of the end products (triosides) (Luckner, 1990).

Leaf flavonoids do not appear to be leached out by rainfall since model allelopathic experiments with grass leaves failed to remove most of the flavonoids present. Only simple phenolic acids (e.g. vanillic, ferulic) appeared in quantity in the leachates (Chou, 1989). During leaching, hydrolysis of ester linkages must occur since free acids not present in the living leaves were detected in the leachate. Leaf flavonoids do appear to be relatively stable. They are readily detected in herbarium leaf samples of 100 years old or more. They have also been detected in fossilized leaves (Harborne and Turner, 1984). Recent investigation in these laboratories of dead leaves of several representative temperate tree species showed that according to the HPLC profiles the flavonols of the living leaves are relatively unchanged in the dead leaves. It seems possible therefore that many of the internal phenolics will survive in leaf litter for considerable periods.

The pathways of degradation of flavonoids are reasonably well documented from *in vitro* experiments. Quercetin, for example is converted to gossypetin by *Pseudomonas* and then broken down to protocatechuic and keto acids. In fungi, quercetin is converted to a depside and then to phloroglucinol carboxylic acid, while in plants it is oxidized first to a 2,3-diol and then broken into its constituent parts. The dihydrochalcone phloridzin, the major phenolic of the apple tree, is metabolized either by direct cleavage between the two aromatic rings or by hydroxylation in the 3-position followed by further phenolase oxidation to the *ortho*-quinone (Luckner, 1990). It is interesting to note here that phloridzin and some of its phenolic breakdown products are reported to occur in soil under apple trees and to survive long enough in that soil to cause allelopathic effects on subsequent planting of apple seedlings (Rice, 1984).

External Leaf Phenolics

Phenolics have been recognized as occurring externally on the upper leaf surface, usually admixed with leaf wax, in many plants. Such compounds have been uncovered partly during investigations of constitutive antifungal agents in plants and partly during surveys of leaf surfaces for lipophilic *O*-methylated flavonoids. In both cases, the compounds are obtained by a brief 30 s dip of the leaf into either methanol or dichloromethane; this avoids extracting any of the internal constituents. All such compounds are therefore lipophilic and this is borne out by their chemical structures (presence of *O*-methylation or isopentenyl substitution), although they retain some degree of water solubility. It is clear that they must have some solubility in the aqueous film of moisture present in the phyllosphere in order to have an inhibitory effect on the germination of fungal spores on leaf surfaces.

Some indication of the range of phenolic structures that have been encountered as antifungal agents on plant surfaces are given in Table 4.2. Structures listed include flavanones (sakuranctin), simple alkyl phenols, flavonol methyl ethers (quercetin 7,3'-dimethyl ether) and isoflavones (luteone). In most cases, there are circumstantial data supporting an antifungal role (see Grayer and Harborne, 1994). How far these phenolics undergo metabolism and turnover is not clear. However, our own unpublished experiments on the acylated kaempferol glycoside of *Quercus* leaf and the 4-coumaric acid methyl ester of *Betula pendula* show that these phenols are readily leached from the surface by standing the leaves in water overnight. This suggests that they are likely to suffer leaching during heavy rain and they have to be continually synthesized within the leaf in order to maintain reasonable levels at the leaf surface. It may also

Table 4.2. Antifungal phenolics obtained from plant surfaces.

Phenolic	Source
Kaempferol 3-(2,3-diacetyl-4,6-di-*p*-coumaryl)glucoside	*Quercus cerris* leaf surface
4-Coumaric acid ester	*Betula pendula* leaf surface
6-Isopentenylnaringenin	*Humulus lupulus* resin
5-Pentadecylresorcinol	*Mangifera indica* fruit peel
Chrysin dimethyl ether	*Helichrysum nitens* leaf surface
Quercetin 7,3'-dimethyl ether	*Wedelia biflora* leaf surface
Sakuranetin	*Ribes nigrum* leaf surface glands
Luteone	*Lupinus albus* leaf surface
Pinocembrin	*Populus deltoides* leaf glands

For references, see Grayer and Harborne (1994), except that the first two examples are from unpublished results of the author.

seem likely that such surface constituents will be lost from the leaf during leaf death and will therefore not be present in leaf litter.

Leaf Tannins

Tannins, especially the condensed flavolans, occur very widely in plants and may be present in leaves in concentrations of 5% dry weight or more. They represent a barrier to herbivory through their intense astringent effects on mammalian taste. Tannins are antinutritional in farm animals, when present in certain fodder crops, although the cause of these effects is still uncertain. It may be linked to inhibition of digestive enzymes, formation of less-digestible complexes with dietary protein or inhibition of microbial flora.

The majority of experiments carried out in the last decade confirm the view that tannins are significant feeding barriers to grazing mammals not only due to anti-nutritional effects but also because of their direct toxicity. The toxic effects of condensed tannins in unadapted mammals are now well established. Weaning hamsters that are treated with a diet containing 4% dry weight of sorghum tannin suffer severe weight loss and then perish within 3–21 days (Butler et al., 1986). Likewise, hydrolysable tannins have been shown to be responsible for poisoning in farm animals eating plants containing high levels. The tannin β-punicalagin, present in leaves of the bushy tree *Terminalia oblongata,* produces what is known as yellow-wood poisoning, causing liver damage in cattle and sheep feeding on this plant. Also the poisoning of farm animals by the buds, leaves, branches and fruits of *Quercus* species, a worldwide phenomenon, is almost certainly due to the hydrolysable tannins present. It has been suggested that the toxic action is due to the release of free phenols produced during detoxification *in vivo.* Indeed, phenol *p*-cresol, catechol and pyrogallol have all been detected in the blood of poisoned cattle. Likewise, co-evolutionary adaptation in mammals to high tannin diets is

well established from the experiments of Butler *et al.* (1986). Such adaptation involves the increased synthesis of a series of unique proline-rich proteins in the parotid glands. These salivary proteins have a high affinity for condensed tannins and remove them by binding to them at early stage in the digestive process. While this adaptation is well pronounced in some mammals (rats, rabbits and hares) it is weak or absent from others (e.g. sheep, cows).

What is most astonishing about animals adapted to feeding on tannin rich diets is that they can distinguish the quality of the tannin and prefer one plant species over another because of differences in the tannins present. This is true of snowshoe hares in Alaska, who show a threefold preference for leaves of bitter brush, *Purshia tridentata,* over those of blackbrush, *Coleogne ramossima.* This difference in feeding behaviour appears to be due entirely to chemical variation in the procyanidins present. In blackbrush, they are based entirely on epicatechin units, while in bitter brush they are based on a one to one ratio of catechin and epicatechin units (Clausen *et al.,* 1990). Perhaps the former polymer has a lower affinity for the salivary gland proteins in the snowshoe hare than the latter material.

Relatively little is known in detail of the metabolism and turnover of plant tannins, partly because of the difficulties of accurately measuring the quantities present in the leaf at any given time. Seasonal changes certainly occur and there is possibly an increase to a maximum in mid-season and a decrease as the leaf ages (Baldwin *et al.,* 1987; Iason *et al.,* 1995; Mafongoya, Chapter 13, this volume). Our own unpublished experiments with four *Quercus* species, in which condensed and hydrolysable tannins were measured separately against accurate standards, confirmed this trend in both first and second flush leaves. This may reflect an increase in bound or non-extractable tannin with time and there have been suggestions of an increase in polymerization and binding to cell wall with increasing leaf maturity. It could also reflect active turnover of tannin, as has been suggested by Wood *et al.* (1994) from experiments on Nepalese fodder trees. Only labelling experiments would indicate unambiguously whether tannins are turned over or not during normal leaf growth.

Condensed tannins are not particularly stable when isolated in pure form, undergoing slow aerial oxidation and giving rise to coloured by-products. They would therefore appear to be stabilized *in vivo* within the leaf by loose association with other leaf constituents. They are readily detected in herbarium specimens of some age, although the concentrations may be reduced. How stable they remain after leaf fall and whether they remain in an extractable form is not certain. They may well be subject to leaching and there is evidence in the case of leaf prunings of *Calliandra* of a loss of polyphenols (including tannin) after standing for 5 days in water (Handayanto *et al.*, 1995).

Leaf Phenols

While attention has concentrated on plant tannins as major defensive agents against herbivory, the role of associated low molecular weight phenolics has been somewhat neglected. It is possible that they may act synergistically with the tannins in providing plant defence. However, there is also good evidence that they are capable on their own of providing feeding deterrents. Recent research (Table 4.3) has shown that simple phenols of *Populus* and *Salix* are antifeedants to a range of different animals. How far these results apply to other low molecular weight plant phenolics and other herbivores is not yet clear, but other grazing birds besides the ruffed grouse are known to be sensitive to dietary phenolics (Harborne, 1993).

Another role that has been proposed for such phenols is as allelopathic agents (Rice, 1984). In particular, the cinnamic acid ferulic has been implicated as such an agent in several plants, producing an inhibitory effect on the germination of seed of competing species. It occurs for example in the foliage of the shrub *Arctostaphylos glandulosa* in bound form and has been detected in the free state in leaf leachates and in the soil underneath the shrub (Harborne, 1993).

One other simple phenol that has received much attention as an allelopathic agent is juglone (5-hydroxynaphthoquinone). This occurs in the walnut tree *Juglans nigra* in bound form as the 4-glucoside of 1,4,5-trihydroxynaphthalene and this undergoes hydrolysis and oxidation to the free toxin juglone during its release from the walnut tissues. Its effectiveness in preventing germination of seeds of competing plant species is exemplified by its

Table 4.3. Deterrent phenolics of woody plants.

Deterrent phenolic	Antifeedant activity	Plant source
Salicortin, tremulacin	Gypsy moth *Lymantria dispar*	*Populus* spp.
Tremulacin	Swallowtail *Papilio glauca*	*P. tremuloides*
Salicortin	Willow beetle *Phratora volgatissina*	*Salix viminalis*
Salicortin (and its hydrolysis products)	Snowshoe hare *Lepus americanus*	*Populus balsamifera*
Coniferyl benzoate	Ruffed grouse *Bonasa umbellus*	*P. tremuloides*
Platyphylloside	Mountain hare *Lepus timidus*	*Populus* spp.

For references, see Harborne (1993).

ability at a concentration of 0.002% to inhibit germination of lettuce seed. Measurements of juglone concentration in the soil under walnut trees show that relatively large amounts (3.6–4.0 µg g^{-1} soil) can be recovered in the top layers (0–8 cm deep), while lesser amounts (1 µg g^{-1} soil) are still detectable to a depth of 1.8 m. How long juglone persists in the soil under walnut trees is still unknown. A soil bacterium named *Pseudomonas* J1 has been identified which will degrade juglone but it is not known whether it is sufficiently widespread to be effective in turning over the amounts of juglone that accumulate under walnut trees (Williamson and Weidenhamer, 1990).

Conclusion

Plant phenolics have a number of interesting and varied ecological roles in the life of a plant. In particular, they appear to be important in protecting plants from herbivory and/or microbial infection, although the evidence for this is still largely circumstantial and requires much further experimental verification. In spite of current emphasis on function, our knowledge of phenolic metabolism and turnover within the plant is still fragmentary. Furthermore,

we know little about the precise mechanisms which allow the release of phenolics from living tissues, although there is good evidence that some low molecular weight phenols (but not the flavonoids) are leached by rainfall from the leaves of trees and shrubs. The fate of phenolics in the dying leaf and the precise nature of the phenolics in leaf litter still need further investigation. Considering the high concentrations of oxidizing enzymes, especially phenolases, in leaf tissues one might expect considerable degradation and destruction of phenolics during leaf fall. While such chemical changes undoubtedly take place, it does appear that a significant concentration of the phenolic fraction, some of it bound to protein or cell wall, remains within the leaf for some years. For example, phenolic contents of up to 1.8 g^{-1} kg (measured as tannic acid equivalents) have been recorded for pine litter in Northern California (Northup *et al.*, 1995). Here the total phenolic content was determined by a simple colour reaction and the nature of the compounds present has not been determined. Certainly, the dynamic aspects of phenolic degradation in leaf litter require much further experimental investigation (using for example HPLC monitoring) before we know exactly what changes occur.

References

Baldwin, I.T., Schultz, J.C. and Ward, D. (1987) Patterns and sources of leaf tannin variation in yellow birch and sugar maple. *Journal of Chemical Ecology* 13, 1069–1078.

Butler, L.G., Rogler, J.C., Mehansho, H. and Carlson, D.M. (1986) Dietary effects of tannins. In: Cody, V., Middleton, E. and Harborne, J.B. (eds) *Plant Flavonoids in Biology and Medicine*. Alan Liss, New York, pp. 141–158.

Chou, C.H. (1989) Comparative phytotoxic nature of leachate from four subtropical grasses. *Journal of Chemical Ecology* 15, 2149–2159.

Clausen, T.P., Provenza, F.D., Burritt, E.A., Reichardt, P.B. and Bryant, J.P. (1990) Ecological implications of condensed tannin structure: a case study. *Journal of Chemical Ecology* 16, 2381–2392.

Grayer, R. and Harborne, J.B. (1994) A survey of antifungal compounds from higher plants, 1982–1993. *Phytochemistry* 37, 19–42.

Handayanto, E., Cadisch, G. and Giller, K.E. (1995) Manipulation of quality and mineralisation of tropical legume tree prunings by varying nitrogen supply. *Plant and Soil* 176, 149–160.

Harborne, J.B. (1980) Plant phenolics. In: Bell, E.A. and Charlwood, B.V. (eds) *Encyclopedia of Plant Physiology*, New Series, Volume 8. Springer-Verlag, Berlin, pp. 329–402.

Harborne, J.B. (1993) *Introduction to Ecological Biochemistry*, 4th edn. Academic Press, London, 318 pp.

Harborne, J.B. and Turner, B.L. (1984) *Plant Chemosystematics*. Academic Press, London, 562 pp.

Iason, G.R., Hodgson, J. and Barry, T.N. (1995) Variation in condensed tannin concentration of a temperate grass in relation to season and reproductive development. *Journal of Chemical Ecology* 21, 1103–1112.

Luckner, M. (1990) *Secondary Metabolism in Microorganisms, Plants and Animals*, 3rd edn. Springer-Verlag, Berlin, 563 pp.

Northup, R.R., Yu, Z., Dahlgren, R.A. and Vogt, K.A. (1995) Polyphenol control of nitrogen release from pine litter. *Nature* 377, 227–229.

Rice, E.L. (1984) *Allelopathy*, 2nd edn. Academic Press, New York, 422 pp.

Steiner, A.M. (1971) Turnover of anthocyanin 3-monoglucosides in petals of *Petunia hybrida*. *Zeitschrift für Pflanzenphysiologie* 65, 210–222.

Van Emden, H.F. (1989) *Pest Control*, 2nd edn. Edward Arnold, London, 117 pp.

Williamson, G.B. and Weidenhamer, J.D. (1990) Bacterial degradation of juglone; evidence against allelopathy? *Journal of Chemical Ecology* 16, 1739–1742.

Wood, C.D., Tiwari, B.N., Plumb, V.E., Powell, C.J., Roberts, B.T., Sirimane, V.D.P., Rossiter, J.T. and Gill, M. (1994) Interspecies differences and variability with time of protein precipitation activity of extractable tannins, crude protein, ash, and dry matter content of leaves from 13 species of Nepalese fodder trees. *Journal of Chemical Ecology* 20, 3149–3162.

5 Decomposition Induced Changes in the Chemical Structure of Fallen Red Pine, White Spruce and Tamarack Logs

J.A. Baldock[1*], T. Sewell[2*] and P.G. Hatcher[2]

[1]Petawawa National Forestry Institute, Canadian Forest Service, Chalk River, Ontario, Canada K0J 1J0; [2]Fuel Science Program, Pennsylvania State University, University Park, PA 16802, USA

Introduction

Fallen logs can contribute significantly to carbon storage in forest ecosystems (Harmon *et al.*, 1986; Spies, 1988; Harmon *et al.*, 1990), can play an important role in nutrient cycling by influencing mineralization and immobilization processes (Sollins *et al.*, 1987; MacMillan, 1988), and provide a source of material for the formation of humus as decomposition processes proceed. Changes in the chemical structure of logs as the extent of decomposition increases have been studied using solid-state ^{13}C NMR (nuclear magnetic resonance) spectroscopy. Decomposition of Chilean ulmo wood (*Eucryphia cordifolia*) by a brown-rot fungus (unidentified) resulted in an almost complete removal of *O*-alkyl carbon and a concentration of aromatic carbon, while a white rot fungus (*Ganoderma australe*) selectively removed aromatic structures (Martinez *et al.*, 1991). Preston *et al.* (1990) noted that decomposition of Douglas fir (*Pseudotsuga menziesii*) and western hemlock (*Tsuga heterophylla*) logs was characterized by an initial loss of *O*-alkyl carbon and a concomitant increase in the concentration of aromatic carbon; however, for western red cedar (*Thuja plicata*) logs, little change in chemical composition was observed despite mass loss and physical collapse as the extent of decomposition increased.

Although solid-state ^{13}C NMR is well suited to determine the gross chemical changes as wood decomposes, the question arises as to whether the accumulation of aromatic structures during decomposition processes occurs as a result of preservation of unaltered or altered lignin molecules or the synthesis of other aromatic materials.

Ertle and Hedges (1985) and Hedges *et al.* (1988) showed that the ratio of the amount of more oxidized lignin monomers (carboxylic acids) to that of the corresponding aldehyde monomers released during a CuO oxidation procedure can provide an index of the extent of lignin degradation. This ratio has been referred to as the Ad:Al (acid:aldehyde) ratio and was shown to increase as the extent of fungal degradation of wood increased (Hedges *et al.*, 1988). Using a tetramethylammonium hydroxide (TMAH) thermochemolysis procedure which breaks the lignin molecule down into its methylated monomer derivatives, Hatcher and Minard (1995) also noted an increase in the Ad:Al ratio with increasing extent of decomposition of Douglas fir wood. Hatcher *et al.* (1995) compared Ad:Al results obtained using the TMAH procedure to CuO results obtained by deMontigny *et al.* (1993) for the same series of degraded wood samples. The Ad:Al ratios obtained using TMAH were linearly correlated to those obtained using CuO, and provided a more sensitive

* Present address: CSIRO Division of Soils, PMB 2, Glen Osmond, SA 5064, Australia.

indicator of the extent of lignin alteration due to the larger range of values obtained. The TMAH procedure also appeared to produce monomers which may reflect the state of preservation of the monomer side-chain carbons.

The objectives of this work were to quantify the chemical changes associated with the decomposition of small diameter logs (<20 cm) of three species (red pine, white spruce and tamarack) using solid-state ^{13}C NMR, and to investigate the degree of lignin alteration in the logs as the extent of decomposition increased using the TMAH thermochemolysis procedure.

Methodological Approach

Fallen logs in a red pine (*Pinus resinosa*), white spruce (*Picea glauca*), and a tamarack (*Larix laricina*) plantation located in the Petawawa Research Forest near Chalk River, Ontario (lat. 45° 57′, long. 77° 34′, mean annual temperature 5.0°C, mean annual precipitation 820 mm) were classified according to a system developed by Fogel *et al.* (unpublished report) and used by Sollins *et al.* (1987). Briefly, decomposition class 0 corresponded to wood which was living and undecomposed when sampled and in progressing from decomposition class 1 through to 5 the extent of decomposition increased. In the pine, spruce, and tamarack plantations, fallen logs exhibiting the characteristics of decomposition classes 1 to 3, 1 to 4, and 1 to 2, respectively, were identified. Wood disks, 5–10 cm in length and not exceeding 20 cm in diameter, were collected from living trees (decomposition class 0) and 10 different fallen logs from each decomposition class. Where bark was no longer attached to the log, it was collected from the interface between the log and forest litter. The wood disks and bark were dried at 40°C to constant mass. Bark was removed from the dried wood disks, where required, and sapwood was separated from heartwood using a bandsaw. All samples were ground to 40 mesh using a Wiley mill. Composite samples were prepared for the bark, sapwood and heartwood of each

decomposition class for each species by taking 1.0 g of the dried and ground material from each of the 10 replicate disks collected.

A conventional solid-state cross-polarization ^{13}C NMR spectrum was obtained for each composite sample using a Chemagnetics Inc. CMC-100 spectrometer and the following experimental conditions: spectrometer frequency of 25.2 MHz, contact time of 1 ms, pulse delay time of 1 s, a spectral width of 14 kHz, and a spinning speed of approximately 3.5 kHz. Preliminary experiments were performed to ensure that the contact time and pulse delay time used were adequate. Chemical shift values were set using hexamethylbenzene and referenced back to TMS (tetramethylsilane). The spectra acquired for the various decomposition classes of red pine sapwood are presented in Fig. 5.1. Alkyl carbon found in methyl, methylene and methine structures are seen in 10–45 ppm region. Peaks with chemical shift values in the vicinity of 66, 73, 86 and 105 ppm are derived from carbohydrate structures, principally cellulose and hemicellulose in undecomposed wood, although amine carbon in proteins, the oxygenated carbon of the methoxyl groups and propyl side chains in lignin structures, and other types of carbon may also appear in this region. Aromatic carbon appears over the region of 110–140 ppm unless it is oxygenated, in which case it can be seen in the 140–165 ppm region. Signals derived from carbonyl and amide carbon occur over the 165–225 ppm region; however, on the basis of the acquired spectra, integration of the carbonyl region was confined to 165–200 ppm. In solid-state ^{13}C NMR, the signal intensity observed for a given chemical shift value, when expressed as a function of the total spectral intensity acquired, represents the proportion of that particular type of carbon found in the sample. Since similar chemical structures have similar chemical shift values, the proportion of the total spectral intensity observed in each of the following chemical shift ranges was determined: 10–45, 45–110, 110–165, and 165–200 ppm. These chemical shift ranges were given the

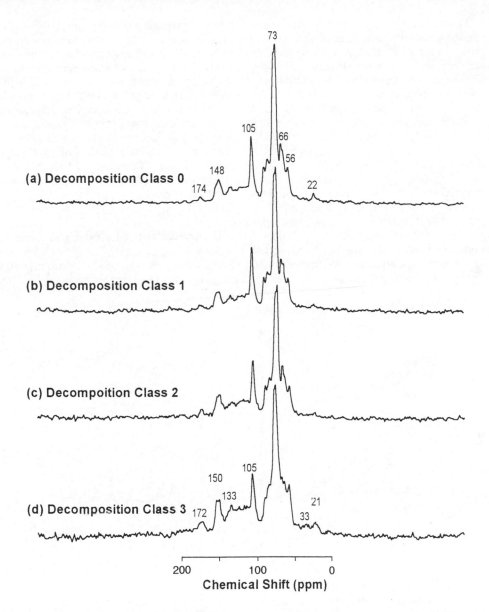

Fig. 5.1. Solid-state ^{13}C NMR spectra acquired for sapwood collected from red pine logs exhibiting the characteristics of decomposition classes 0 to 3.

general labels alkyl, *O*-alkyl, aryl and carbonyl, respectively. These labels are indicative of the dominant forms of C thought to be present in each region. However, because of overlap between adjacent regions and the presence of additional types of C, the proportions of each type of C are approximate.

The TMAH thermochemolysis analyses were also performed on the composite samples. Approximately 0.8 mg of composite sample was placed in a 2 ml glass ampoule with 100 µl of a 25% TMAH in methanol solution. The methanol was evaporated to dryness under vacuum. The ampoules were flame sealed under vacuum and heated at 250°C for 30 min. After cooling, the ampoules were opened and reaction products were removed by washing all inner surfaces with dichloromethane (CH_2Cl_2). The dichloromethane solution was concentrated to approximately 100 µl and 1 µl was injected splitless onto a J&W DB5 30 m capillary column (i.d. 0.25 mm, phase thickness 0.12 µm) fitted in a Carlo Erba 500 gas chromatograph interfaced to a Kratos MS80 mass spectrometry system. After 25 s, the split valve was opened. The column was heated from an initial temperature of 60°C to 150°C at a rate of 15°C min^{-1}, and then from 150°C to a final temperature of 280°C at a rate of 4°C min^{-1}. Total ion current was used to produce the chromatograms, and mass spectra, obtained at a scan rate of 0.6 s decade^{-1} of mass with

a 0.2 s magnet settling time added, were used to identify the compounds associated with each peak in the chromatograms. Examples of the chromatograms obtained for undecomposed and decomposed white spruce sapwood are presented in Fig. 5.2. Ad:Al ratios were calculated by expressing the total ion current associated with the acid peak in the chromatogram as a function of that for the aldehyde peak. For each composite sample, the entire TMAH thermochemolysis procedure was repeated from two to nine times.

Degradation of Red Pine Logs

Integration of the ^{13}C NMR spectra acquired for the components of the red

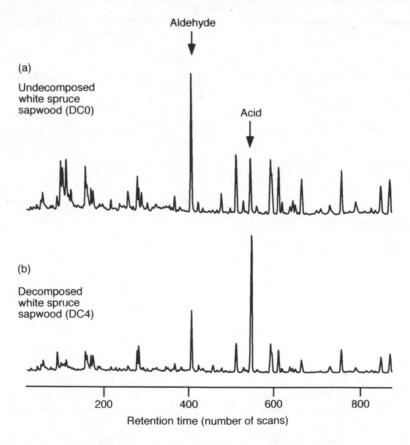

Fig. 5.2. Total ion current chromatograms of the TMAH thermochemolysis products of (a) fresh undecomposed white spruce sapwood (decomposition class 0 = DC0) and (b) decomposed white spruce sapwood (decomposition class 4 = DC4).

Table 5.1. Chemical composition of bark, sapwood, and heartwood of red pine, white spruce, and tamarack logs decomposed to different extents as revealed by solid-state ^{13}C NMR spectroscopy.

Species	Component	Decomposition class	Percentage of total ^{13}C NMR spectral area			
			Carbonyl	Aryl	O-Alkyl	Alkyl
Red pine	Bark	0	3	27	61	9
		1	4	23	61	12
		2	4	23	61	12
	Sapwood	0	2	19	68	11
		1	2	20	73	4
		2	3	23	71	3
		3	6	26	62	6
	Heartwood	0	2	18	74	5
		1	2	17	74	7
		2	2	18	72	8
		3	2	19	72	7
White spruce	Bark	0	3	24	58	16
		1	5	25	50	20
		2	3	17	62	18
		3	5	24	56	15
		4	4	18	60	18
	Sapwood	0	4	19	71	6
		1	3	19	73	5
		2	2	17	74	7
		3	3	27	65	5
		4	2	25	67	6
	Heartwood	0	5	21	68	5
		1	3	18	76	3
		2	2	19	72	7
		3	2	23	72	3
		4	3	29	62	6
Tamarack	Bark	0	5	24	54	18
		1	5	29	52	14
		2	6	28	44	22
	Sapwood	0	6	21	70	4
		1	2	18	75	5
		2	3	22	70	5
	Heartwood	0	4	20	73	3
		1	2	18	76	5
		2	2	19	78	2

pine logs indicated that signals derived from O-alkyl and aryl carbon were dominant (Table 5.1). Signal intensity was also observed in the alkyl and carbonyl regions; however, changes induced as the decomposition class of the logs increased were confined to the contents of O-alkyl and aryl C. For bark, a constant increase in O-alkyl C and decrease in aryl C was noted as the extent of decomposition increased indicating a preferential loss of aromatic structures, possibly due to a leaching of soluble phenolic compounds such as tannins, or to a preferential degradation of

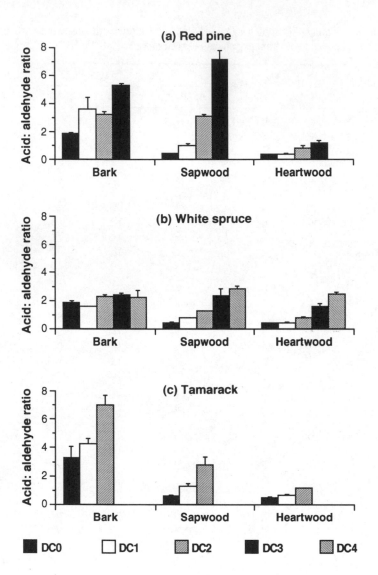

Fig. 5.3. Acid:aldehyde ratios obtained for bark, sapwood and heartwood of (a) red pine, (b)white spruce and (c) tamarack logs exhibiting different extents of decomposition. The extent of decomposition increases in progressing from decomposition classes 0 to 4. Error bars represent the standard deviation of the mean acid:aldehyde values.

lignin. The loss of *O*-alkyl C and accumulation of aryl C observed for sapwood was consistent with a preferential degradation of carbohydrate structures (e.g. cellulose and hemicellulose) and an associated concentration of lignin. The lack of any significant changes in the composition of heartwood, as the decomposition class of the logs increased, suggested that microbial colonization and decomposition of the heartwood was limited. Such a limitation may have resulted from an inability of the decomposer community to fully colonize the logs in the time provided and/or, as indicated by Eriksson *et al.* (1990), by the presence of substances which limit microbial activity.

An increase in the Ad:Al ratios of red pine bark and sapwood was noted in progressing from undecomposed to decomposition class 3 logs, indicating an increase in the extent of lignin degradation (Fig 5.3a). The accumulation of aryl structures noted in the ^{13}C NMR spectra of sapwood was thus expected to result from an accumulation of altered lignin molecules and not from a selective preservation of unaltered lignin. The minor change in Ad:Al ratio observed for the heartwood was consistent with the lack of significant chemical changes noted in the ^{13}C NMR data.

Degradation of White Spruce Logs

As observed for the red pine logs the majority of ^{13}C NMR signal intensity was observed in the O-alkyl and aryl spectral regions. The chemical changes noted in progressing from fresh undecomposed white spruce logs through to the logs of decomposition class 4 were more variable than those observed for the red pine logs (Table 5.1); however, the general trends appeared to be similar. Little change was noted in the composition of bark as extent of decomposition increased and a loss of O-alkyl C and an accumulation of aryl C were noted for sapwood and heartwood. The chemical changes noted for the white spruce heartwood were of a similar magnitude to that noted for the sapwood, indicating that in contrast to red pine, the heartwood of white spruce logs appeared to be as susceptible to microbial attack as the sapwood. The chemical changes noted for the sapwood and heartwood are consistent with a preferential utilization of carbohydrate structures over lignin during the decomposition process.

In contrast to red pine bark, little change in the Ad:Al ratio was noted as extent of decomposition increased for white spruce bark, suggesting that the extent of lignin alteration remained constant as the logs were degraded. For sapwood and heartwood, the progressive increase in Ad:Al values noted as decomposition continued, indicated that the accumulation of aryl structures, noted in the ^{13}C NMR spectra, resulted from an accumulation of altered lignin. The similar Ad:Al values obtained for the sapwood and heartwood also supported the ^{13}C NMR results suggesting that white spruce heartwood was as susceptible to decomposition as sapwood.

Degradation of Tamarack Logs

In the tamarack plantation, no logs could be found which exhibited the characteristics associated with decomposition class 3 or greater, presumably because of the young age of the plantation (<25 years). As a result, the extent of decomposition of the tamarack logs was more limited than for the red pine and spruce logs. The ^{13}C NMR spectra were again dominated by O-alkyl and aryl carbon, but a significant amount of alkyl C was observed in the bark. In progressing from fresh undecomposed logs to decomposition class 2 logs, little change in the chemical composition of sapwood and heartwood was evident; however, a reduction in O-alkyl C and minor increases in aryl and alkyl C in bark were noted.

Increases in the Ad:Al values were observed for the bark and sapwood as decomposition proceeded. Given that only minor changes were noted in the NMR spectra of these materials, it would appear that the lignin alteration began before the onset of observable differences in NMR spectra. The lack of a significant increase in the Ad:Al values of heartwood compared with sapwood suggested that microbial colonization of the heartwood was limited as was observed for the red pine logs.

Conclusions

Decomposition induced changes in the chemical composition of the materials included in this study varied with both wood component and species; however, with only one exception (red pine bark), the general pattern of decay involved a loss of O-alkyl (carbohydrate) C and an accumulation of aryl C contained in altered lignin structures. In the red pine and tamarack logs, decomposition of the heartwood lagged behind that of the sapwood, whereas the heartwood and sapwood of white spruce appeared equally susceptible to decomposition. Given that the white spruce logs were of an equivalent or larger diameter than the red pine and tamarack logs and all logs had similar exposure times, it is unlikely that this resistance to decomposition was a function of heartwood placement within the log and the length of time of exposure to decomposition. It is suggested that the resistance of red pine and tamarack heartwood to decomposition resulted from the presence of compounds capable of restricting microbial colonization.

The information obtained from solid-state ^{13}C NMR and TMAH thermochemolysis analyses in this study was complementary and the combined use of these techniques provided a more detailed assessment of the changes which occurred during the decomposition of fallen logs than could be obtained using either technique alone. Changes in the Ad:Al ratio appeared to be a sensitive index of the extent of lignin degradation and variations in this ratio were often observed before significant changes in ^{13}C NMR spectra became apparent. Such an observation is not entirely surprising since alterations in the composition of the lignin propyl side chains could occur without influencing ^{13}C NMR spectra because signals from side chain carbons are hidden under O-alkyl signals of carbohydrates.

References

deMontigny, L.E., Preston, C.M., Hatcher, P.G. and Kogel-Knaber, I. (1993) Comparison of humic horizons from two ecosystem phases on northern Vancouver Island using ^{13}C CPMAS NMR spectroscopy and CuO oxidation. *Canadian Journal of Soil Science* 73, 9–25.

Eriksson, K.E.L., Blanchette, R.A. and Ander, P. (1990) *Microbial and Enzymatic Degradation of Wood and Wood Components.* Springer-Verlag, New York, 407 pp.

Ertle, J.R. and Hedges, J.I. (1985) Sources of sedimentary humic substances: vascular plant debris. *Geochimica Cosmochimica Acta* 49, 2097–2107.

Harmon, M.E., Ferrell, W.K. and Franklin, J.F. (1990) Effects on carbon storage of conversion of old-growth forests to young forests. *Science* 247, 699–702.

Harmon, M.E., Franklin, J.F., Swanson, F.J. Sollins, P., Gregory, S.V., Lattin, J.D., Anderson, N.H., Cline, S.P., Aumen, N.G., Sedell, J.R., Lienkeenper, G.W., Cromach, K, Jr and Cummins, K.W. (1986) Ecology of coarse woody debris in temperate ecosystems. *Advances in Ecological Research* 15, 133–302.

Hatcher, P.G. and Minard, R.D. (1995) Comment on the origin of benzenecarboxylic acids in pyrolysis methylation studies. *Organic Geochemistry* 23, 991–994.

Hatcher, P.G., Nanny, M.A., Minard, R.D., Dibble, S.D. and Carson, D.M. (1995) Comparison of two thermochemolytic methods for the analysis of lignin in decomposing wood: The CuO oxidation method and the method of thermochemolysis with tetramethylammonium hydroxide (TMAH). *Organic Geochemistry* 23, 881–888.

Hedges, J.I., Blanchette, R.A., Weliky, K. and Devol, A.H. (1988) Effects of fungal degradation on the CuO oxidation products of lignin: A controlled laboratory study. *Geochimica Cosmochimica Acta* 52, 2717–2726.

MacMillan, P.C. (1988) Decomposition of coarse woody debris in an old-growth Indiana forest. *Canadian Journal of Forest Research* 18, 1353–1362.

Martínez, A.T., González, A.E., Valmaseda, M., Dale, B.E., Lambregts, M.J. and Haw, J.F. (1991) Solid-state NMR studies of lignin and plant polysaccharide degradation by fungi. *Holzforschung* 45, 49–54.

Preston, C.M., Sollins, P. and Sayer, B.G. (1990) Changes in organic components for fallen logs in old-growth Douglas-fir forests monitored by ^{13}C nuclear magnetic spectroscopy. *Canadian Journal of Forest Research* 20, 1382–1391.

Sollins, P., Cline, S.P., Verhoeven, R., Sachs, D. and Spycher, G. (1987) Patterns of log decay in old-growth Douglas-fir forests. *Canadian Journal of Forest Research* 17, 1585–1595.

Spies, T.A., Franklin, J.F. and Thomas, T.B. (1988) Coarse woody debris in Douglas-fir forests of western Oregon and Washington. *Ecology* 69, 1689–1702.

6 Solid-state NMR Investigations of Organic Transformations during the Decomposition of Plant Material in Soil

D.W. Hopkins[1] and J.A. Chudek[2]

[1]Department of Biological Sciences and
[2]Department of Chemistry, University of Dundee, Dundee DD1 4HN, UK

Introduction

Understanding the biogeochemical events that occur when plant residues decompose in soil is restricted by the inability to follow the transformations of organic compounds without the complications and compromises of extracting organic matter from soil. Solid-state ^{13}C nuclear magnetic resonance (NMR) spectroscopy offers an opportunity for *in situ* characterization of soil organic matter (SOM), i.e. in whole soil samples, at the level of functional groups (Wilson, 1987; Skjemstad *et al.*, Chapter 20, this volume) although severe quantification problems mean that NMR spectra are best interpreted comparatively (Kinchesh *et al.*, 1995a). Many solid-state ^{13}C NMR spectra have been recorded for whole soils and organic-rich fractions from soils using the naturally abundant ^{13}C to provide semi-quantitative snap-shots of the C distribution between different functional groups. The application of NMR to mineral soils is, however, restricted by the Fe, Cu and Mn contents of soil, the paramagnetic properties of which interfere with the NMR signals, and by the low natural abundance of [13]C (Skjemstad *et al.*, Chapter 20, this volume).

We have attempted to overcome some of the problems of investigating the transformations of SOM in soils by incorporating ^{13}C-enriched grass material into soil, following the subsequent redistribution of ^{13}C and determining the rates at which the NMR signals assigned to particular functional groups change. This approach offers the opportunity to follow the component processes of decomposition against the unlabelled SOM background. We have compared the forms of ^{13}C which accumulate during the decomposition of grass material with that which accumulates during microbial metabolism of ^{13}C-glucose in soil. We have also related the rates of decomposition of SOM in size fractions from grassland soils to their chemical compositions, as revealed by solid-state ^{13}C NMR.

Previous studies have shown that ^{13}C from enriched glucose added to soil is detected in a range of functional groups in the SOM, of which *O*-alkyl-C associated with the larger (250-2000 μm) and less-dense fractions and alkyl-C associated with the clay-sized fraction were some of the more significant (Baldock *et al.*, 1989, 1991). However, pure organic compounds are not the natural substrates of soil microorganisms and it was our objective to investigate the decomposition of more complex and realistic natural substrates, such as plant material. Six months after the incorporation of ^{13}C from enriched wheat straw into acid upland peat, the ^{13}C was present primarily as *O*-alkyl-C in larger size fractions (500–1000 μm) and both as methyl-C and alkyl-C and as *O*-alkyl-C in the smaller size fractions (5–50 μm) (Benzing-Purdie *et al.*, 1992).

Transformations during decomposition

[13]C-enriched *Lolium perenne* leaf material (12 atom% [13]C; C-to-N = 20; 39% by weight cellulose; 29% by weight hemicellulose; 6% by weight lignin, Allen *et al.*, 1974) was added to a sandy-loam textured podzolic soil (pH 7.0 (water), 3.4% total C) at the rate of 40 mg plant material g^{-1} soil. The progress of plant material decomposition was followed by both weekly determinations of CO_2 evolution rate (measured by gas chromatography) and determination of ^{13}C remaining in the soil when samples were taken at 0, 14, 28, 56, 112 and 224 days (measured by mass spectrometry). The soil was maintained at 22°C and the moisture content periodically adjusted to 50% water-holding capacity.

After 224 days, 48% of the added ^{13}C had been lost. This loss can be accounted for as CO_2, since the additional CO_2 evolved between 0 and 224 days from the amended soils over that from the unamended soil was equivalent to about 50% of the added material. There was good correlation between the measurements of ^{13}C remaining and the cumulative CO_2 lost ($R^2 = 0.92$, $n = 12$), but the ^{13}C data are likely to be more reliable because calculation of the CO_2 lost involved extrapolation from rates over 24 h to an estimate of the amount evolved over 7 days and then summing the weekly measurements. The overall decline in ^{13}C followed first-order kinetics with a rate constant (k) of 0.0030 day^{-1} ($R^2 = 0.79$, $n = 12$), which corresponds to 333 days mean residence time (MRT) (i.e. k^{-1}). The overall decay was, of course, the result of several component processes with different rate constants (Jenkinson and Rayner, 1977) and it is almost certainly for this reason that the first-order function accounted for less than 80% of the variation in ^{13}C remaining with time.

Cross polarization magic angle spinning ^{13}C NMR spectra were recorded for samples of the whole soil, after drying and grinding (<100 μm), using a Bruker AM 300 MHz/WB FT NMR spectrometer with 500 μs contact time, 1 s relaxation delay, 6.8 μs [13]C 90° pulse and 4 kHz MAS in Zirconia rotors with Kel-F caps. The rotors contained 0.4 g soil on each occasion and a relatively small number of scans (2040) were recorded for each sample so that soil-only controls gave spectra with no significant signals above the noise.

For the spectra collected at the early stages of decomposition, there were clear NMR signals at 10–45 ppm, 45–60 ppm (present as a shoulder on the larger signal centred at 75 ppm), 60–90 ppm, 90–110 ppm and 160-200 ppm (Fig. 6.1). The chemical assignments and probable classes of organic compounds for these and other chemical shift ranges are shown in Table 6.1. Some of the signals in the spectra are spinning side bands, with those at 140 and 200 ppm, either side of the signal centred around 170 ppm, in the day 28 spectrum in Fig. 1, being the more obvious. These features have not been used in the characterization of the ^{13}C distribution.

The resonance at 10–45 ppm (methyl- and alkyl-C) probably arose mainly from aliphatic lipids, fatty acids and waxes (Oades *et al.*, 1987; Wilson, 1987) with a smaller contribution from hemicelluloses (Kolodziejski *et al.*, 1982) and possibly acetyl substituents in lignin (Nordén and Berg, 1990). There was a small signal in the aromatic region (110–160 ppm) which may be attributed to either aromatic-C in lignin or a spinning side band, however, the relatively low lignin content of the grass material suggests that acetyl substituents from lignin made only a relatively small contribution to the 10–45 ppm signal. The small signal at 45–60 ppm is likely to have been due to methoxyl-C and/or alkyl-amino-C (Wilson, 1987). Lignin is a significant source of methoxyl-C, but as with acetyl substituents of lignin in the signal at 10–45 ppm, methoxyl-C was probably only a relatively small component of the 45–60 ppm shoulder. The signal at 60–90 ppm (*O*-alkyl-C) probably arose from oxygenated C in carbohydrates. The C2, C3 and C5 carbons of cellulose would have given signals between 70 and 75 ppm, the C4 at about 87 ppm and the C6 at about 65 ppm (Wilson, 1987). In

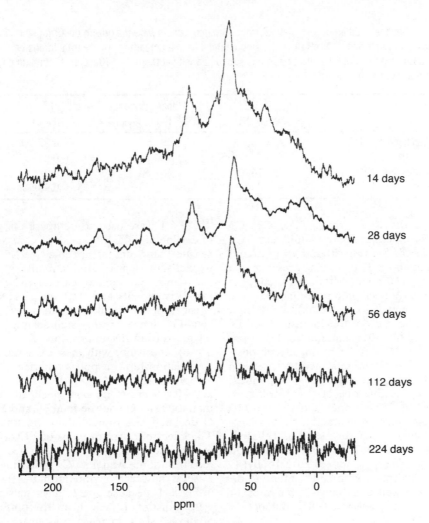

Fig. 6.1. CP MAS ^{13}C NMR spectra for podzolic soil plus ^{13}C-enriched *Lolium perenne* (40 mg g^{-1} soil, 12 atom%) incubated for different periods of time recorded using a Bruker AM 300 MHz/WB FT NMR spectrometer (500 µs contact time, 1 s relaxation delay, 6.8 µs ^{13}C 90° pulse and 4 kHz MAS) over 2040 scans.

Table 6.1. Chemical shift ranges, chemical assignments and classes of compounds (Baldock *et al.*, 1991).

Shift range (ppm)	Type of C	Class of compounds
10–45	Methyl- and alkyl-C	Lipids, waxes and aliphatic hydrocarbons
45–60	Methoxyl- and alkyl-amino-C	Lignin substituents, amino acids and amino sugars
60–90	*O*-alkyl-C	Carbohydrates
90–110	Acetal- and ketal-C	Carbohydrates
110–160	Aromatic-C	Phenyl-propylene sub-units of lignin
160–200	Carbonyl-C	Organic acids and amino acids/peptides

Table 6.2. Decline in different forms of ^{13}C from enriched *Lolium perenne* during decomposition in podzolic soil as detected by NMR. A_t is the signal intensity as a percentage of the total intensity (0–200 ppm) at time = 0, A_0 is the percentage signal intensity at the outset (0 days), t = time in days and k is the rate constant in day^{-1}.

Assignment	$A_t = A_0 e^{-kt}$	Mean residence time (days)	R^2 ($n = 6$)
Total-C, as seen by NMR	$A_t = 93 e^{-0.017t}$	59	0.97
Methyl- and alkyl-C	$A_t = 17 e^{-0.014t}$	71	0.86
Alkyl-amino- and methoxyl-C	$A_t = 14 e^{-0.018t}$	56	0.97
O-alkyl-C	$A_t = 30 e^{-0.020t}$	50	0.98

addition, the ring carbons (C1, C2, C3 and C5) of hemicelluloses would also have given signals between 70 and 75 ppm and the C4 at about 87 ppm. The presence of a signal 90–110 ppm (acetal- and ketal-C) is likely to have been produced mainly by the C1 (dioxygenated-C) of cellulose (Wilson, 1987). The ratio of the intensities of the signals at 60–90 ppm and 90–110 ppm was relatively constant during decomposition which indicates that the C responsible was in the same class of compounds. Furthermore, this type of ^{13}C distribution between the 60–90 ppm and the 90–110 ppm resonances is characteristic of plant-derived polysaccharides, such as cellulose (Oades *et al.*, 1987). Signals at 160–200 ppm are usually attributed to carbonyl-C groups which may include amide-C in peptides as well as carboxyl-C in a range of organic acids (Wilson, 1987; Baldock *et al.*, 1991).

After 224 days all the signals had been lost from the NMR spectra recorded under the particular conditions employed for the spectra in Fig. 6.1. However, the total signal intensity (0–200 ppm) was significantly correlated with the amount of ^{13}C remaining in the soil at different times between 0 and 224 days ($R^2 = 0.92$, $n = 6$) and the decline in total intensity followed first-order kinetics (Table 6.2). This is a somewhat fortuitous observation because the relationship between signal intensity in CP MAS ^{13}C NMR spectra and ^{13}C is not the same for all types of ^{13}C. As decomposition proceeded, the intensities of the particular signals declined in all cases. The spectral integrals for some of the shift ranges in

Table 6.1 were used to compare the rates of decline of different types of ^{13}C in the *L. perenne* amendment (Table 6.2). The weaker signals, attributed to methoxyl- and alkyl-amino-C and to carbonyl-C, were lost between 56 and 112 days. The intensities of the O-alkyl-C and acetal- and ketal-C signals were significantly correlated during decomposition ($R^2 = 0.98$, $n = 6$), consistent with their close molecular relationship in polysaccharides. The carbohydrate-C (O-alkyl-, acetal- and ketal-C) from *L. perenne* declined faster than the total C and the methyl- and alkyl-C declined less rapidly than the total C. The effect of different types of ^{13}C declining at different rates would have been ^{13}C redistribution. Major ^{13}C redistribution was not, however, obvious from the spectra in Fig. 6.1, because by 224 days the signals were indistinguishable from the noise. At 56 days, a slight increase in intensity of the alkyl-C signal centred at about 40 ppm was detected relative to the other signals. This would account for the slower rate of decline of the 10–45 ppm signal intensity (Table 6.2) and which is presumed to be the result of accumulation of microbial metabolites.

Although the k values for the declines in signal intensities (Table 6.2) provide a means of comparing the dynamics of the different forms of C, they are clearly substantial over-estimates of the true decomposition rates, because the NMR data imply that there was no ^{13}C left in the soil while the ^{13}C content had only declined to 52% of that at the outset. Also, they cannot be regarded as truly quantitative

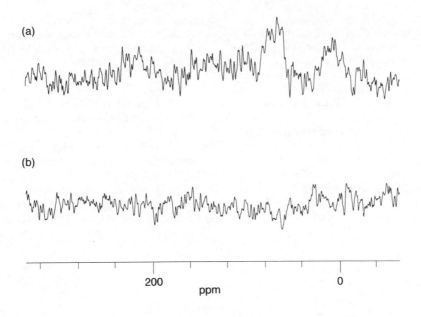

Fig. 6.2. CP MAS ^{13}C NMR for whole soil (a) and the heavy (>1.13 g cm^{-3}) fraction (b), separated from the podzolic soil by centrifugation on a colloidal silica medium, after incubation with 40 mg ^{13}C-enriched (12 atom%) g^{-1} soil *Lolium perenne* for 224 days. Spectra were recorded using a Chemagnetics CMX LITE 300 MHz NMR spectrometer (500 μs contact time, 1 s relaxation delay, 4.0 μs ^{13}C 90° pulse and 4 kHz MAS) over 1600 and 2852 scans for a and b, respectively.

because of the potential presence of spinning sidebands beneath the signals. There are a number of possible reasons for the loss of ^{13}C signal:

1. Decline in ^{13}C content during decomposition and associated ^{13}C redistribution into a wide range of metabolites as a result of microbial activity had led to all forms of ^{13}C being below the detection threshold of the NMR spectrometer under the operating parameters employed.
2. During decomposition, the plant material and its decomposition products had become associated with paramagnetic species in the soil, particularly Fe, which act to mask NMR signals.

Re-examination of the whole soil and the heavy fraction (density >1.13 g cm^{-3}, separated by ultrasonic dispersion and centrifugation on colloidal silica medium) samples taken at 224 days with a more sensitive Chemagnetics CMX LITE 300 MHz NMR spectrometer gave the spectra in Fig. 6.2. These spectra are the differences between the spectra for the ^{13}C-amended whole soil or heavy fraction and the spectra of the corresponding unamended samples recorded with 500 μs contact time, 1 s relaxation delay, 4.0 μs ^{13}C 90° pulse and 4 kHz MAS in Zirconia rotors with Kel-F caps for 1600 scans in the case of the whole soils and 2852 scans for the heavy fractions. Part of the increased sensitivity was because the rotors for this instrument are larger and contained 0.7 g soil when packed. The spectrum for the whole soil clearly shows the residual *O*-alkyl-C from carbohydrates, although the corresponding acetal- and ketal-C signal is less clear (Fig. 6.2). There is also clear evidence of an increase in the intensity of the methyl-and alkyl-C, relative to that of the *O*-alkyl-C, compared with that seen in Fig. 6.1. This relative increase in methyl- and alkyl-C is consistent with the slower rate of decline previously seen and the fact that the first-order function accounted for only 86% of the variation between alkyl- and methyl-C signal intensity with time compared with 97% for the other types of ^{13}C (Table 6.2). The NMR spectrum of the heavy fraction had no interpretable signals (Fig. 6.2). This indicates that virtually all the ^{13}C was present in the light fraction, however, the small mass of the corresponding light fraction prevented NMR spectra from being recorded. It is also clear that the alkyl- and methyl-C which had accumulated in the whole soil during decomposition was not preferentially associated with the heavy fraction. This C was almost certainly,

therefore, in the light fraction, possibly as microbial biomass and metabolites associated with undecomposed or only partially decomposed plant residues.

Synthesis of Microbial Metabolites

The hypothesis that microbial metabolism of carbohydrates led to accumulation of methyl-C and alkyl-C was investigated by determining the shorter-term (within 14 days) fate of uniformly-enriched ^{13}C-glucose added to an organic soil (Fig. 6.3).

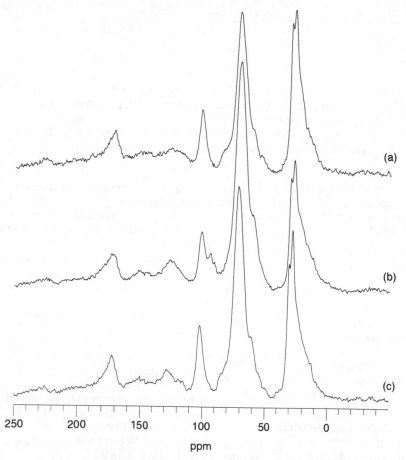

Fig. 6.3. CP MAS ^{13}C NMR spectra for stagnohumic gley soil (a), soil plus 34 mg uniformly-enriched ^{13}C-glucose (99 atom%) g^{-1} soil after both 0 (b) and 14 days (c) incubation at 22°C. Spectra were recorded using a Chemagnetics CMX LITE 300 MHz NMR spectrometer (500 μs contact time, 1 s relaxation delay, 4.0 μs ^{13}C 90° pulse and 4 kHz MAS) over 2048 scans in each case.

The soil used was from the organic horizon (49% total C) of a stagnohumic gley soil, which as a result of a large lime application (20 t $CaCO_3$ ha^{-1}) 13 years previously was near neutrality (pH 6.8) and had a large, active microbial community (Isabella and Hopkins, 1994). A total of 34 mg uniformly-labelled ^{13}C-glucose (99 atom%) was added per g soil (this concentration was only about double the minimum saturating concentration of glucose, which was 16 mg g^{-1} soil (L. Hawaleschka, J.A. Chudek and D.W. Hopkins, unpublished data). The soil was incubated at 22°C, dried and ground before CP MAS ^{13}C spectra were recorded with the Chemagnetics NMR spectrometer (parameters as above except that the rotors contained only 0.2 g soil because of the lower soil density) for 2048 scans. A soil amenable to NMR due to low Fe content was used in this experiment so that the addition of ^{13}C-glucose could be sufficiently small to avoid the soil microbial community being saturated and to prevent the NMR signals from extracellular ^{13}C-glucose dominating the spectra over the whole period of the incubation, yet sufficiently large for the ^{13}C retained in the soil after respiratory loss as CO_2 to be detected against the SOM background (L. Hawaleschka, J.A. Chudek and D.W. Hopkins, unpublished data).

Considering the spectra in Fig. 6.3, at the outset the intensity of O-alkyl-C (60–90 ppm) signal was considerably enhanced by ^{13}C-glucose addition compared with the unamended soil and a small but clear signal can be seen at 95 ppm in the acetal- and ketal-C region, which is due to the C1 of glucose. After 14 days, the O-alkyl-C signal had declined significantly, although not to the same relative intensity as before ^{13}C-glucose addition indicating that there was some residual ^{13}C-glucose. After 14 days there was a marked increase in the intensity of the 10–45 ppm signal, with a particularly sharp signal at 27 ppm characteristic of alkyl-C. This ^{13}C could be newly-synthesized microbial metabolites or biomass, although the apparent absence of accumulated amide-^{13}C (160–200 ppm) and alkyl-amino-^{13}C (45–60 ppm) tends to suggest that the biomass had not become heavily labelled.

Chemical Composition and Decomposition

The decomposition rates of SOM in particle size fractions (>2000 μm and >63 μm) isolated from two clay-loam textured stagnogley soils, in one of which SOM had accumulated under highly acidic (pH = 4.3) conditions (soil 1 in Table 6.3) and the other of which had little SOM accumulation and nearer neutral pH (5.8) (soil 2 in Table 6.3) (Hopkins *et al.*, 1993). The decomposition rates of these fractions (Table 6.3) were determined by incubating them in a common test soil for 80 days at 25°C and fitting the CO_2 evolution data, obtained using alkali traps (Hopkins *et al.*, 1993), to first-order decay functions. Isolating the particles from the soils would have disrupted most physical protection of the SOM allowing the influence of chemical composition of the SOM alone to be related to decomposition rate. The larger size fractions

Table 6.3. Properties of some particle size fractions from two grassland stagnogley soils. Soil 1 fractions were obtained from an acid soil in which organic matter was accumulating and soil 2 fractions were obtained from near-neutral soil (Hopkins *et al.*, 1993).

Soil	Size fraction (μm)	C content (mgC g^{-1} soil)	Rate constant, k (day^{-1})	R^2 (n = 6)
1	>2000	130	−0.0024	0.98
	>63	91	−0.00036	0.96
2	>2000	70	−0.0014	0.98
	>63	40	−0.00025	0.94

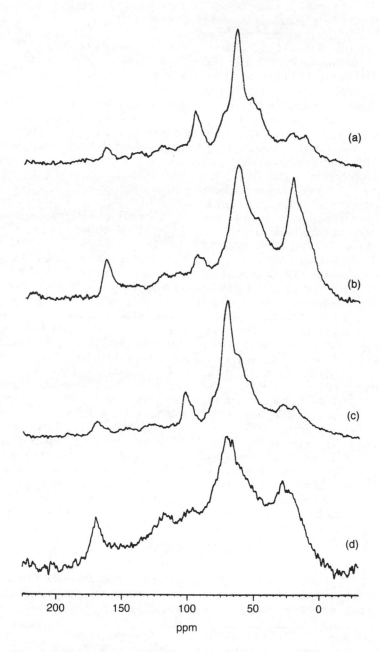

Fig. 6.4. CP MAS ^{13}C NMR spectra for the >2000 μm (a) and >63 μm (b) size fractions from acidic stagnogley soil (1, in Table 6.3) and the >2000 μm (c) and <63 μm (d) size fractions from a near-neutral stagnogley soil (2, in Table 6.3). The spectra were recorded using a Bruker AM 300 MHz/WB FT NMR spectrometer (500 μs contact time, 1 s relaxation delay, 6.8 μs ^{13}C 90° pulse and 4 kHz MAS) over 3100, 1060, 2000 and 8658 scans for samples A, B, C and D, respectively.

(>2000 μm) from both soils decomposed at different rates, but the decomposition rates of these fractions were substantially greater than the smaller fractions (>63 μm). The CP MAS ^{13}C NMR spectra were recorded with a Bruker AM 300 MHz/WB FT NMR spectrometer (parameters as above) over 3100 scans for the >2000 μm fraction (a), 1060 scans for the <63 μm fraction from soil 1, 2000 scans for the >2000 μm fraction (c), and 8658 scans for the <63 μm fraction (d) from soil 2 (Table 6.3; Fig. 6.4). The main differences in the chemical composition, as seen by NMR, were between the small and the large fractions, rather than between the fractions of the same size from the different soils (Fig. 6.4). Decomposition was most rapid for the fractions with large O-alkyl-, acetal-and ketal-C signals consistent with a high carbohydrate content and relatively slow for the fractions with large alkyl- and methyl-C signals. Such observations are completely consistent with the increase in alkyl- and methyl-C during decomposition (Figs 6.2 and 6.3) and with the decline in the particle size with which organic matter is associated as decomposition advances (Christensen, 1987; Oades et al., 1987).

Conclusions

These investigations have allowed us to compare the rates at which different forms of ^{13}C from enriched L. perenne material decline in soil during decomposition. The data indicate that carbohydrate-C declined most rapidly and that the main products arising during the decomposition and loss of approximately 50% of the added C were characterized by methyl- and alkyl-C. Under field conditions loss of about 50% of C from a L. perenne amendment would be expected by about 6 months, and the absolute rate of loss would decline with time (Jenkinson, 1965). Considerable redistribution of C would be likely as the proportion of more recalcitrant compounds increases, and our data indicate

that the main type of C which had accumulated in slowly degrading soil fractions was methyl- and alkyl-C.

Our investigations have focused on the early phases of decomposition of plant material. From solid-state NMR investigations of soils subject to longer-term (many decades) treatments, it has been suggested that carbohydrate-C is the most active fraction in SOM, since it contributes the most intense NMR signal for soils in which SOM accumulates and declines relative to other forms of C when a change in management results in a decline in total SOM (Kinchesh et al., 1995b). NMR analyses of soils from a moorland site, where severe fires separated by 19 years have removed virtually all the SOM, and from the adjacent unburned site have provided complementary data (D.W. Hopkins, C.R. Goldspink and J.A. Chudek, unpublished data). At this site, O-alkyl-C accumulated fastest over the first 18 years after burning consistent with plant litter accumulation and over the subsequent 19 years the proportion of alkyl- and methyl-C increased at approximately twice the rate of all other forms of C.

The next stage in the work presented here is to follow the transformations of ^{13}C through the forms of C that appear to have accumulated. This will have to be done with soil samples increasingly hostile to NMR because of the declining ^{13}C content due to decomposition, but for an improved understanding of the longer-term transformations of organic matter derived from plant litter it will be necessary to follow the ^{13}C into the next generation of accumulating molecules.

Acknowledgements

This work was supported by the UK Biotechnology and Biological Sciences Research Council and the University of Dundee's research initiative fund. We are grateful to D. Barraclough, E.A. Webster, B.L. Isabella and R.L. MacKay.

References

Allen, S.E., Grimshaw, H.M., Parkinson, J.A. and Quarmby, C. (1974) *Chemical Analysis of Ecological Materials*. Blackwell Scientific Publications, Oxford, 252 pp.

Baldock, J.A., Oades, J.M., Vassallo, A.M. and Wilson, M.A. (1989) Incorporation of uniformly labelled ^{13}C-glucose into the organic fraction of soil. Carbon balance and CP/MAS ^{13}C NMR measurements. *Australian Journal of Soil Research* 27, 725–746.

Baldock, J.A., Currie, G.J. and Oades, J.M. (1991) Organic matter as seen by solid-state ^{13}C nuclear magnetic resonance spectroscopy and pyrolysis tandem mass spectrometry. In: Wilson, W.S. (ed.) *Advances in Soil Organic Matter Research*. Royal Society of Chemistry, Cambridge, pp. 45–60.

Benzing-Purdie, L., Cheshire, M.V., Williams, B.L., Ratcliffe, C.I., Ripmeester, J.A. and Goodman, B.A. (1992) Interactions between peat and sodium acetate, ammonium sulphate, urea or wheat straw during incubation studied by ^{13}C and ^{15}N NMR spectroscopy. *Journal of Soil Science* 43, 113–125.

Christensen, B.T. (1987) Decomposability of organic matter in particle size fractions from field soils with straw incorporation. *Soil Biology and Biochemistry* 19, 429–435.

Hopkins, D.W., Chudek, J.A. and Shiel, R.S. (1993) Chemical characterization and decomposition of organic matter from two contrasting grassland soil profiles. *Journal of Soil Science* 44, 147–157.

Isabella, B.L. and Hopkins, D.W. (1994) Nitrogen transformations in a peaty soil improved for pastoral agriculture. *Soil Use and Management* 10, 107–111.

Jenkinson, D.S. (1965) Studies on the decomposition of plant material in soil. I. Losses of carbon from ^{14}C labelled ryegrass incubated with soil in the field. *Journal of Soil Science* 16, 104–115.

Jenkinson, D.S. and Rayner, J.H. (1977) The turnover of soil organic matter in some of the Rothamsted classical experiments. *Soil Science* 123, 298–305.

Kinchesh, P., Powlson, D.S. and Randall, E.W. (1995a) ^{13}C NMR studies of organic matter in whole soils: I. Quantitation possibilities. *European Journal of Soil Science* 46, 125–138.

Kinchesh, P., Powlson, D.S. and Randall, E.W. (1995b) ^{13}C NMR studies of organic matter in whole soils: II. A case study of some Rothamsted soil. *European Journal of Soil Science* 46, 139–146.

Kolodziejski, W., Frye, J.S. and Maciel, G.E. (1982) Carbon-13 nuclear magnetic resonance spectrometry with cross-polarization and magic-angle spinning for analysis of lodgepole pine wood. *Analytical Chemistry* 54, 1419–1424.

Nordén, B. and Berg, B. (1990) A non-destructive method (solid-state ^{13}C NMR) for determining organic chemical components of decomposing litter. *Soil Biology and Biochemistry* 22, 271–275.

Oades, J.M., Vassallo, A.M., Waters, A.G. and Wilson, M.A. (1987) Characterization of organic matter in particle size and density fractions from a red-brown earth by solid-state ^{13}C NMR. *Australian Journal of Soil Research* 25, 71–82.

Wilson, M.A. (1987) *NMR Techniques and Applications in Geochemistry and Soil Chemistry*. Pergamon Press, Oxford, 353 pp.

7 Kinetically Defined Litter Fractions Based on Respiration Measurements

H. Marstorp

Department of Soil Sciences, Swedish University of Agricultural Sciences, Box 7014, S-750 07 Uppsala, Sweden

Plant litter quality is often determined in terms of its chemical components. Typically, the contents of water- or ethanol-soluble components of litters are used as measures of easily degradable material, while the fibre fractions are considered to be more difficult to degrade or to have decomposition-modifying properties. Changes in the chemical composition of litter as well as rates of product formation and nutrient release have been used to calculate degradation rates of chemical litter fractions, often expressed as first-order rate constants (e.g. Dendooven, 1990). In mathematical models chemical fractions are typically grouped together in pools according to their rate constants (e.g. van Veen *et al.*, 1985). As an alternative to defining fractions based on their chemical characteristics I propose that plant litter be separated into 'kinetically defined fractions' (KDF) based on direct measurements of the rates at which they are utilized by microorganisms. In this chapter a method is presented in which respiration measurements on decomposing litter are used to determine such fractions. As outlined, the approach is mainly applicable to easily degradable litter components.

CO$_2$-Evolution and Decomposition

Heterotrophic soil microorganisms degrade litter and other organic materials to obtain energy for their growth and maintenance. During aerobic degradation the end products in the energy-yielding reactions are CO$_2$ and water. As a consequence, CO$_2$ evolution can be used as a measure of microbial activity and amounts decomposed, although part of the decomposed substrate is also used for microbial synthesis. Respiration is easy to measure, at least in laboratory experiments. Methods for measuring respiration have been recently reviewed by Zibilske (1994) and Alef (1995). In contrast to the litter-bag technique or other similar methods where mass loss and changes in chemical composition are monitored, respiration measurements do not give direct information about the type of litter substances decomposed. Nevertheless, decomposition rates and amounts decomposed can be used to characterize the litter components utilized by the decomposers. Innumerable investigations of this kind have been carried out with various types of plant materials and specific substances, such as glucose, under a variety of environmental conditions. However, only a few studies with plant materials have been conducted in which measurements were carried out frequently enough to allow determination of decomposition rates and sequences for the most easily degradable litter components. In one such study Reber and Schara (1971) were able to separate the decomposition of sugars, amino acids and phenols in straw leachates. Their measurements were made hourly, and the respiration rates showed several peaks, reflecting the different components.

With an apparatus for obtaining automatic CO_2 measurements, such as that described by Nordgren (1988), the time resolution (minutes to hours) necessary to follow the decomposition of even the most easily degradable litter components can be acquired. The decomposition of such substrates is closely related to microbial growth. After an initial lag decomposers will show exponential growth if provided with high enough substrate concentrations. Although developed for single-species batch cultures, a relationship for growth-associated product formation (Pirt, 1975; Stenström et al., 1991) can be applied to the respiration data. The product formed (in this case CO_2) in an infinitely small time interval dt is given by

$$dp = q_p \, x \, dt \qquad (7.1)$$

where p is the product concentration (CO_2), q_p is the specific rate of product formation and x the concentration of microbial biomass. During exponential growth, if q_p is constant the following relationship will be obtained:

$$\frac{dp}{dt} = q_p \, x_0 \, e^{\mu t} \qquad (7.2)$$

or in logarithmic form

$$\ln \frac{dp}{dt} = \mu t + \ln q_p \, x_0 \qquad (7.3)$$

where x_0 is the concentration of biomass at the start of growth, and μ is the specific growth rate. From equation (7.3) it follows that the specific growth rate can be estimated from the slope of the line when $\ln \frac{dp}{dt}$ is plotted against time (Fig. 7.1). The length of the lag phase can be calculated (Nordgren et al., 1988) as the period between substrate addition and the point in time at which the regression line of $\ln \frac{dp}{dt}$ against time equals ln for the initial rate of product formation after substrate addi-tion (Fig. 7.1). This definition of the lag phase is analogous to the lag phase based on growth curves (Pirt, 1975).

Lag phase lengths and specific growth rates calculated with this relationship have been shown to differ considerably between litter components in a given soil (Marstorp, 1996a). Consequently, the specific growth rate and lag phase length can be used to characterize the litter components undergoing decomposition. In soil with its diversity of organisms, the specific microbial growth rate estimated from respiration curves is a mean value representing the rates of all organisms growing on a particular substrate. However, non-growing organisms can also influence the estimated value of μ. Summing the constant respiration rate of non-growing organisms and that of exponentially growing organisms results in a somewhat lower μ value (Fig. 7.2). Moreover, an apparent lag results when initially only a fraction of the population grows at its maximum rate (Pirt, 1975). When the lag phase is determined on the basis of CO_2 production an apparent lag can be the result of CO_2 produced by non-growing, but respiring, organisms. Consequently, the lag phases measured can be apparent, at least in part (Fig. 7.2). The possibility that non-growing and growing organisms can simultaneously contribute to respiration has also been proposed very recently by Stenström and Stenberg (1995). They presented a method for separating the initial respiration rate after glucose addition (the SIR response, Anderson and Domsch, 1978) into two components, one from growing organisms and one from non-growing ones.

Determination of 'Kinetically Defined Fractions'

Methodological approach

Several respiration peaks can be detected with automatic CO_2 measurements during the early phase of plant litter decomposition (Figs 7.3 and 7.4). During a peak, the increase in respiration rate becomes exponential after an initial lag if the peak is large enough. This reflects the fact that the microorganisms grow exponentially above a certain concentration of the litter components used as substrate. Thus, a respiration peak represents a group of litter

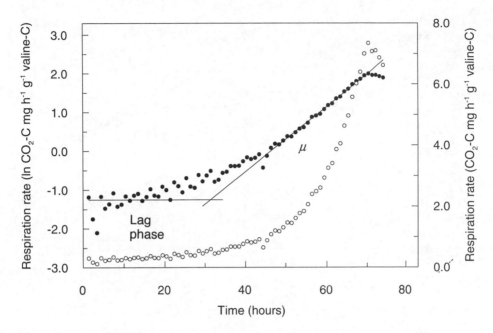

Fig. 7.1. Respiration rate (○ absolute; ● ln values) during decomposition of valine. Valine was added in an amount equivalent to 20 mg C jar[-1] (40 g soil d.w.). Specific growth rate (μ) and lag phase length are calculated as indicated.

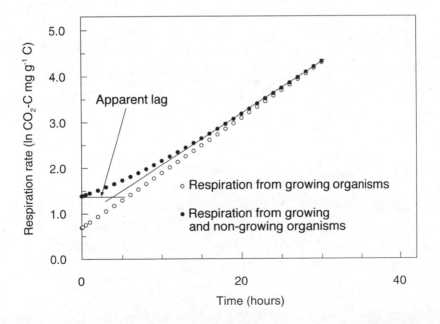

Fig. 7.2. Hypothetical curve showing the effect of respiration from non-growing microorganisms on the occurrence of an apparent lag phase and on the value of the estimated specific growth rate.

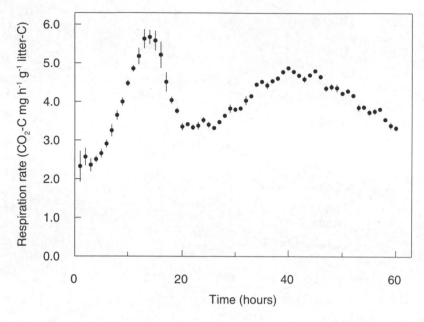

Fig. 7.3. Respiration rate during decomposition in soil of sugarbeet leaves. Bars indicate SEM ($n=4$). Leaves were added to each jar (40 g soil d.w.) in amounts equivalent to 40 mg total C.

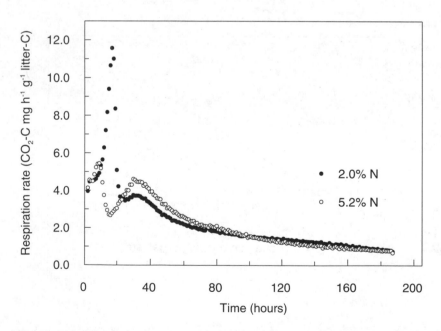

Fig. 7.4. Respiration rate during decomposition of ryegrass shoots of two different qualities showing two respiration peaks. Shoots were added to each jar (40 g soil d.w.) in amounts equivalent to 50 mg total C (redrawn from Marstorp, 1996a).

components, making up a 'kinetically defined fraction' (KDF), that is utilized during a single growth phase. During such a phase, the litter components may be used simultaneously or sequentially by a single microbial species or several species, but they still make up a common fraction. The litter components making up the KDF can then be identified by the following approach:

1. Comparisons of continuous respiration curves for whole-plant litter with corresponding curves for some of its chemical fractions, in order to roughly determine the nature of the components making up a kinetic fraction. Examples of such chemical fractions include ethanol and water extracts and the litter components remaining after extraction. If necessary, nutrients should be added to avoid nutrient-limited decomposition. These comparisons assume that respiration from the separate chemical fractions is additive. The validity of this assumption for extraction with water is supported by the fact that addition of an easily available substrate (glucose) has been shown not to influence the mineralization of non-water-soluble straw components (Cochran *et al.*, 1988). However, interactive effects have been reported after more powerful extraction procedures (Vanlauwe *et al.*, 1994) such as a hot extraction with neutral detergent solution according to Goering and van Soest (1970).

2. Comparisons of characteristics of microbial growth on whole-plant litter with those of growth on model substances, i.e. litter components in their pure form. If the decomposition is associated with exponential microbial growth, the model substances can be characterized by estimating lag phase lengths and specific microbial growth rates using a relationship for growth-associated product formation as described in the previous section. Model substances having lag phase lengths and specific growth rates that allow decomposition during a certain respiration peak are identified. Single litter components or a group of components making up a KDF and model substances must be present in

concentrations large enough to promote exponential microbial growth. The litter components must also be as available in the litter material as they are in their pure forms.

3. The contents of major components identified in steps 1 and 2 are then determined in litters of different qualities. Quantitative relationships between the contents of these components and the amounts of CO_2 respired during the different respiration peaks are subsequently examined. As a first approach it can be assumed that the mineralization of the different litter components is additive, i.e. no interactions occur. Comparisons of evolved amounts of CO_2 during a respiration peak with initial contents of the identified substances in the shoots require that overlapping respiration peaks can be separated and integrated. Proportions of the litter components used for synthesis and respiration might also vary depending on their nature (e.g. Linton and Stephenson, 1978; Dendooven, 1990) and concentration (Bremer and Kuikman, 1994). As a result, the amount of CO_2 evolved might not be proportional to the amount of C in identified litter components, although it might still be linearly related. An example of such a relationship is given in the next section (Fig. 7.5).

4. Detailed studies are needed to further elucidate the microbial utilization of litter components; for example, any interactions that can occur between identified litter components need to be determined. Such work requires discrete measurements of other variables, such as $^{14}CO_2$ and inorganic-N, at specific stages of decomposition which are identified through continuous respiration measurements.

An example

The approach outlined in the preceding section was used in a decomposition study with *Lolium multiflorum* shoots (Marstorp, 1996a). The quality of the shoot materials used was changed by varying the amount of N fertilizer added (six different rates) during cultivation. Two respiration peaks, representing two different kinetic fractions,

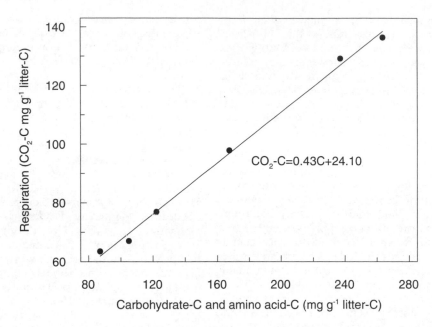

Fig. 7.5. Accumulated respiration during decomposition of ryegrass shoots of six different qualities as a function of soluble non-structural carbohydrates and free amino acids in the shoots. Ryegrass shoots were added to each jar (40 g soil d.w.) in amounts equivalent to 50 mg total C (redrawn from Marstorp, 1996a).

were identified during the first week of decomposition (Fig. 7.4). The first peak was absent when the litter was extracted with water before incubation, indicating that this KDF consisted of water-soluble litter components.

The duration of the first peak was compared with the length of lag phases obtained for some major water-soluble litter components such as sugars and amino acids found in the *L. multiflorum* shoots. The length of the lag phase for glucose, fructose and sucrose, respectively, fell within the first respiration peak of the *L. multiflorum* shoots. Specific growth rates on these substances were of the same magnitude as those obtained on the shoots during the first peak. The amino acids that had a lag phase encompassed by the first respiration peak accounted for 83–85% of the total free amino acid C content in the different shoots. N mineralization was observed for the shoots with the two highest N contents, whereas at lower N

contents net immobilization was observed, indicating that both N containing and non-N containing litter components were decomposed during the first peak.

The sum of amino acid-C and sugar-C was, however, lower than the amount of CO_2-C evolved during the first peak. Consequently, other substances must also have been mineralized. When the amount of fructan-C in the shoots was added to the sum of sugars and amino acids, a strong linear relationship ($R^2 = 0.996$) was obtained with the accumulated amount of CO_2-C respired from the start of the experiment to the end of the first respiration peak (Fig. 7.5). It was concluded that glucose, fructose, sucrose, free amino acids and probably also fructans made up a single kinetic fraction reflected in the first respiration peak

The second respiration peak of the *L. multiflorum* shoots consisted of both water-soluble and non-water soluble litter components, as revealed by comparisons of respiration curves obtained from extracted and non-extracted shoots. The amount of CO_2 evolved during the second peak also increased with increasing protein content, but the amounts respired were higher than the amount of protein-C. N immobilization was also observed at all N-levels during this period. Consequently, the components decomposed during the second respiration peak, i.e. the second KDF, must have consisted of proteins as well as nitrogen-free-soluble and non-soluble plant components.

It could not be determined whether sugars, amino acids and fructans were decomposed simultaneously or sequentially, although they were identified as a single KDF in the referred-to investigation discussed. This matter was studied in a supplementary investigation using some of the identified components, i.e. glucose, glutamic acid, alanine and valine, which were cross-wise ^{14}C-labelled (Marstorp, 1996b). The three amino acids were chosen because their lag phases differed in length. Since the decomposition process was followed continuously through respiration measurements, $^{14}CO_2$ could be sampled at well-defined moments such as during the lag- and exponential phases, at inflection points, etc. It was shown that glucose and the amino acids were mineralized simultaneously when combined. In each combination the component with the shortest lag phase was mineralized preferentially. In cases where the lag phases for two components were of equal length, the component supporting the highest specific growth rate was mineralized preferentially. These experiments also showed that interactions occurred between components that affected their decomposition rates. For example, glucose hastened the mineralization of the amino acids that had longer lag phases than glucose. As a result, interactions must be considered when identifying KDFs.

Advantages and Possibilities

The proposed method for delimiting KDFs and identifying the related litter

components can overcome many of the difficulties associated with chemical characterization. For example, the solubility of a litter component is not necessarily related to its decomposability. The experiment carried out by Marstorp (1996a) with *L. multiflorum* shoots showed that the boundaries for the water-soluble fraction did not coincide with the respiration peaks. The first respiration peak and a part of the second respiration peak were derived from water-soluble substances, but during the later part of the second peak non-water soluble substances were decomposed too. Consequently, water solubility was not an adequate measure of litter quality during the early phase of decomposition. Although not shown experimentally, it is not likely that the boundaries for an ethanol-soluble fraction would correspond with those of the first respiration peak since the fructans in the litter are only partly soluble in ethanol (Smith, 1973). Moreover, the chemical composition of a litter fraction based on solubility criteria is variable. For example, the nature and proportions of water-soluble litter components, such as sugars, amino acids and fructans, will vary with species, stage of development, fertilizer regime, etc. (Nowakowski, 1962; Goswami and Willcox, 1969; Smith, 1973). As a consequence, such fractions are poor predictors of litter decomposition rates or nutrient turnover. Like fractions based on solubility criteria, the composition of kinetic fractions is also variable. Still, in a KDF, litter components are always grouped together according to their rate of microbial utilization, irrespective of their chemical nature.

The grouping of litter components in KDFs according to their rate of microbial utilization may enable better predictions of nutrient turnover. For example, the sequence in which proteins and carbohydrates are utilized may influence the extent of N mineralization at low soil inorganic N levels. If the carbohydrates are decomposed before the proteins, N might limit microbial growth, resulting in a zero-order, non-growth related decomposition (Alexander and Scow, 1989). As a consequence, N immobilization would be lower

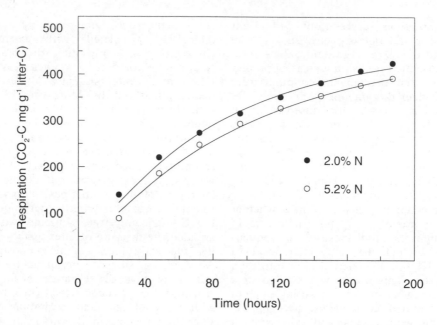

Fig. 7.6. First-order rate equation fitted to the data shown in Fig. 7.4 but summed on a 24-h basis. Ryegrass shoots of different quality were added to each jar (40 g soil d.w.) in amounts equivalent to 50 mg total C.

than in the case of a growth-linked decomposition.

The automatic CO_2 measurements and determination of KDFs give much more information than a conventional curve-fitting based on daily CO_2 values. As an illustration, in the earlier described experiments with *L. multiflorum* shoots (Marstorp, 1996a) two distinct KDFs, as well as the major litter components making up the fractions, were identified, especially those making up the first fraction. Two respiration curves from this experiment were summed on a 24 h basis (Fig. 7.6). The summed curves were then fitted to a first-order rate equation, and a good fit was obtained ($R^2=0.99$). Although the rate constants were slightly different for the two types of shoot material (0.32 and 0.26 day^{-1}), the sizes of estimated pools were the same. No information was obtained through this 'curve fitting exercise' that could relate the composition of the ryegrass shoots to their decomposition pattern. The good agreement obtained between a first-order rate equation and

measured values resulted from the fact that there were not enough data points to reveal the appropriate kinetics. This might be a common phenomenon, as pointed out by Alexander and Scow (1989). As a consequence, first-order-rate kinetics can be useful tools for making rough comparisons, but they are inadequate for more detailed studies of easily degradable litter components.

Interactions between litter components complicate the comparison of model substances and litter materials in terms of their specific growth rates and lag phase lengths. However, the proposed method for delimiting KDF can be used to elucidate the nature and extent of such interactions. Model experiments can be set up in which the decomposition of identified litter components is followed in more detail. Continuous respiration measurements allow microbial substrate utilization to be monitored with a high temporal resolution, thereby making it possible to also measure other variables at well-defined moments. Most variables commonly measured in soil microbial studies would be more informative if they were related to a well-defined moment during a growth period rather than to a certain time after substrate addition. Some efforts in this direction have been made in the area of sugar and amino acid decomposition (Marstorp, 1996b). Further development of this technique would enable concepts from pure and mixed culture utilization of multiple substrates (Harder and Dijkhuizen, 1982; Wanner and Egli, 1990; Egli *et al.*, 1993; Gottschal, 1993) to be applied in studies of litter decomposition in soil.

The method outlined for identifying and determining KDF based on microbial growth characteristics was developed mainly for the most readily available litter components. Most of the model substances used were simple, water-soluble substances. Although it can be difficult to use this approach on structural litter components, it might be extended to non-structural polymers, at least to water-soluble ones. Difficulties can arise in finding suitable model substances for those polymers that can vary in composition and structure, such as chain length and degree of branching, with stage of plant development, variety, species, etc. In such cases the polymers would have to be prepared from the actual plant litter. Nevertheless, the grouping of litter components in KDF through the use of continuous respiration measurements might provide valuable information on the microbial use of certain polymers. However, it may be necessary to apply more complex kinetics and to measure other variables reflecting degradation of the polymers.

Acknowledgement

This work was supported by the Swedish Council for Forestry and Agricultural Research.

References

Alef, K. (1995) Soil respiration. In: Alef, K. and Nannipieri, P. (eds) *Methods in Applied Soil Microbiology and Biochemistry*. Academic Press, London, pp. 214–222.

Alexander, M. and Scow, K.M. (1989) Kinetics of biodegradation in soil. In: Sawhney, B.L. and Brown, K. (eds) *Reactions and Movements of Organic Chemicals in Soil*. Soil Science Society of America and American Society of Agronomy, Madison, USA, pp. 243–269.

Anderson, J.P.E. and Domsch, K.H. (1978) A physiological method for the quantitative measurements of microbial biomass in soil. *Soil Biology and Biochemistry* 10, 215–221.

Bremer, E. and Kuikman, P. (1994) Microbial utilization of ^{14}C[U]glucose in soil is affected by the amount and timing of glucose additions. *Soil Biology and Biochemistry* 26, 511–517.

Cochran, V.L., Horton, K.A. and Cole, C.V. (1988) An estimation of microbial death rate and limitations of N and C during wheat straw decomposition. *Soil Biology and Biochemistry* 20, 293–298.

Dendooven, L. (1990) Nitrogen mineralization and nitrogen cycling. PhD thesis, The Catholic University of Leuven, Leuven, Belgium.

Egli, T., Lendenmann, U. and Snozzi, M. (1993) Kinetics of microbial growth with mixtures of carbon sources. *Antonie van Leeuwenhoek* 63, 289–298.

Goering, H.K. and Van Soest, P.J. (1970) Forage fiber analysis. *Agriculture Handbook 379*. United States Department of Agriculture, Washington DC, 20 pp.

Goswami, A.K. and Willcox, J.S. (1969) Effect of applying increasing levels of nitrogen to ryegrass, I. Composition of various nitrogenous fractions and free amino acids. *Journal of the Science of Food and Agriculture* 20, 592–595.

Gottschal, J.C. (1993) Growth kinetics and competition – some contemporary comments. *Antonie van Leeuwenhoek* 63, 299–313.

Harder, W. and Dijkhuizen, L. (1982) Strategies for mixed substrate utilization in microorganisms. *Philosophical Transactions of the Royal Society of London B* 297, 459–480.

Linton, J.D. and Stephenson, R.J. (1978) A preliminary study on growth yields in relation to the carbon and energy content of various organic growth substrates. *FEMS Microbiology Letters* 3, 95–98.

Marstorp, H. (1996a) Influence of soluble carbohydrates, free amino acids, and protein content on the decomposition of *Lolium multiflorum* shoots. *Biology and Fertility of Soils,* 21, 257–263.

Marstorp, H. (1996b) Interactions in the microbial use of soluble plant components in soil. *Biology and Fertility of Soils,* 22, 45–52.

Nordgren, A. (1988) Apparatus for the continuous, long-term monitoring of soil respiration rate in large number of samples. *Soil Biology and Biochemistry* 20, 955–957.

Nordgren, A., Bååth, E. and Söderström, B. (1988) Evaluation of soil respiration characteristics to assess heavy metal effects on soil microorganisms using glutamic acid as a substrate. *Soil Biology and Biochemistry* 20, 949–954.

Nowakowski, T.Z. (1962) Effects of nitrogen fertilizers on total nitrogen, soluble nitrogen and soluble carbohydrate contents of grass. *Journal of Agricultural Science, Cambridge* 59, 387–392.

Pirt, J.S. (1975) *Principles of Microbe and Cell Cultivation.* Blackwell Scientific Publications, Oxford, 274 pp.

Reber, H. and Schara, A. (1971) Degradation sequences in wheat straw extracts inoculated with soil suspensions. *Soil Biology and Biochemistry* 3, 381–383.

Smith, D. (1973) The nonstructural carbohydrates. In: Butler, G.W. and Bailey, R.W. (eds) *Chemistry and Biochemistry of Herbage,* Vol 1, Academic Press, London, pp. 105–155.

Stenström, J. and Stenberg, B. (1995) Division of substrate-induced respiration (SIR) into growing and nongrowing microorganisms. In: *Abstract, 7th International Symposium Microbial Ecology,* International Committee of Microbial Ecology, Santos, Brazil.

Stenström, J., Hansen, A. and Svensson, B. (1991) Kinetics of microbial growth-associated product formation. *Swedish Journal of Agriculture Research* 21, 55–62.

Van Veen, J.A., Ladd, J.N. and Amato, M. (1985) Turnover of carbon and nitrogen through the microbial biomass in a sandy loam and clay soil incubated with [^{14}C(U)]glucose and [^{15}N](NH$_4$)$_2$SO$_4$ under different moisture regimes. *Soil Biology and Biochemistry* 17, 747–756.

Vanlauwe, B., Dendooven, L. and Merckx, R. (1994). Residue fractionation and decomposition: The significance of the active fraction. *Plant and Soil* 158, 263–274.

Wanner, U. and Egli, T. (1990) Dynamics of microbial growth and cell composition in batch culture. *FEMS Microbial Reviews* 75, 19–44.

Zibilske, L.M. (1994) Carbon mineralization. In: *Methods of Soil Analysis, Part 2- Microbiological and Biochemical Properties.* Soil Science Society of America, Madison, pp. 836–863.

Part III

Foraging, Feeding and Feedbacks

8 Linkages between Soil Biota, Plant Litter Quality and Decomposition

D.A. Wardle[1] and P. Lavelle[2]

[1]*AgResearch, Ruakura Agricultural Research Centre, Private Bag 3123, Hamilton, New Zealand;* [2]*Laboratoire d'Ecologie des Sols Tropicaux, Centre ORSTOM, 93143-Bondy Cedex, France*

Introduction

In most terrestrial ecosystems, the majority of net primary production enters the decomposition subsystem as plant litter. The breakdown of this litter is determined by a range of factors which each operate at vastly different spatial and temporal scales. In the hierarchical model concept of Lavelle *et al.* (1993) it is proposed that processes which operate at large spatial scales constrain those which operate at smaller scales; climatic factors help determine abiotic soil characteristics which in turn help determine litter quality and ultimately the activity and composition of soil microbial and invertebrate communities. It is also important to note that there are numerous 'feedbacks' in this hierarchy, with those factors acting at smaller spatial scales also influencing those which operate at larger scales. Thus the effects of plant litter quality on the soil biota regulate the extent to which the biota in turn facilitates the decomposition of plant litter.

Although the microflora is the biotic component that has the enzyme complement most appropriate for the breakdown of plant litter (and is therefore responsible for most of the soil activity: Petersen and Luxton, 1982), it is the nature of biotic interaction between the microflora and soil fauna that is critical in regulating those processes. These interactions broadly occur at three levels of resolution: (i) 'microfood-webs', involving soil nema-todes, protozoa and their predators; (ii) 'litter-transforming systems' involving soil mesofauna and some macrofauna, in which interactions take place in purely organic structures like fragmented material and faecal pellets and (iii) 'ecosystem engineers' involving larger organisms, which interact with microorganisms in both the 'internal' and 'external' rumen, and which build organo-mineral physical structures that significantly alter the habitat for smaller organisms. Predation and competition are often important in regulating the components of the microfood-web system (Wardle and Yeates, 1993) while mutualistic interactions (between gut or soil microflora and soil fauna) become increasingly important for larger organisms, which rely on microbial enzymes to aid digestion of resources (Lavelle, 1994). These interactions are highly important in relation to litter quality: if an organism is regulated by competition (i.e. resource availability) then it is more likely to be regulated by litter quality (bottom-up effects) than if the dominant means of regulation is predation (top-down effects). Mutualistic relationships between microflora and soil fauna are directly related to litter quality, since these interactions are essentially adaptations to the otherwise recalcitrant nature of plant litter.

In this chapter we evaluate linkages between plant litter quality, soil biota and decomposition processes. Specifically we address both sides of the 'Driven by Nature' issue: (i) the effects of quality of litter input on components of the soil biota; and (ii) the effects of soil biota and their interactions (including adaptive strategies) on patterns of decomposition and nutrient release.

Microflora-Litter Linkages

The lowest trophic level to gain nutrition from plant litter, i.e. the primary saprophytes, consists mainly of two morphologically disjunct groups, the bacteria and the fungi. The mass of these components, or the 'microbial biomass', is very strongly related to resource quality. This is shown by the strong relationships which microbial biomass often demonstrates with soil and litter nitrogen status (Wardle, 1992), by the rapid response of microflora to readily-available carbon sources (Dighton, 1978), and by the negative relationship between the microflora and various compounds such as tannins and 'lignin' (Rayner and Boddy, 1988; Šlapokas and Granhall, 1991). Changes in the microbial biomass during decomposition are usually concomitant with changes in plant litter quality. It has been proposed, particularly for resistant substrates such as wood (Cooke and Rayner, 1984) and some litter types (Wardle, 1993) that as decomposition proceeds the microbial community changes from a disturbed (open) one, to a closed one where microbial competition is most likely to occur, and then to a stressed one (in litter at least this stress appears to emanate from declining pH and exhaustion of key nutrients). Usually microbial biomass is greatest at the closed stage, during which resource competition results in available resources being the most limiting factor (Wardle, 1993).

The species composition of litter fungi is particularly responsive to changes in litter quality. Garrett (1963) proposed a hypothesis of fungal succession in which sugar fungi were followed by those which could degrade cellulose and lignin as the litter quality changed. Although it is now recognized that the mechanisms of fungal succession are much more complex than this, it is still apparent that resource quality does play a role in determining fungal successional patterns. Further evidence for the role of litter quality in influencing fungal species composition emerges when different litter types are compared. For example, Widden (1986) determined that saprophytic fungal community structures strongly mirrored forest vegetation composition, and hence the types of litter input. Similarly Robinson et al. (1994) observed that leaves and internodes of wheat (differing in C-to-N ratio and physical structure) supported vastly different fungal communities. Part of the difference in fungal community structure between litter types is undoubtedly related to differences in the nature of interspecific relationships. Fungi often demonstrate intense competitive interactions, and the most realistic experiments investigating competitive ability (i.e. those on 'real' substrates, not agar) show that competitive balances between different fungal species can depend, quite significantly, on the nature of the resources present (Carreiro and Koske, 1992; Wardle et al., 1993).

Less is understood as to how bacteria respond to litter quality, although they show at least some relationship with changes in quality during decomposition (e.g. Cornejo et al., 1994). However, bacteria are usually associated with aqueous pores in litter and soil, rather than intimately associated with litter as are fungi, and there is both empirical and theoretical evidence that the nature of resources present is less likely to directly affect bacteria than fungi (Wardle, 1995); see below.

If plant litter quality impacts upon the soil microflora, and in particular the fungal component, then it is reasonable to expect that this will in turn determine decomposition rates (Lavelle et al., 1993). For relatively resistant litter types at least this appears to be the case; for example, Flanagan and van Cleve (1983) presented data which indicated that the total mycelial length in organic matter under four different tree species differed considerably, and that this mycelial length was in turn very strongly correlated ($r > 0.90$) with both N and P mineralization.

Interactions Involving Microfauna

Bacteria are predominantly consumed by bacterial-feeding nematodes (which engulf their prey whole) and protozoa, while fungi are fed upon by fungal-feeding nematodes (which pierce their prey with stylets and ingest their cytoplasmic contents). These microfauna are in turn consumed by top predatory nematodes. The microflora, microfauna, and their predators, comprise the 'microfood-web' system which consists of fungal-based and bacterial-based energy channels (Moore and Hunt, 1988), and in which components may be regulated by competition (through resource limitation, or 'bottom-up' effects) and predation (or 'top-down' effects). Feeding relationships are usually direct, and mutualistic interactions are rare between organisms in adjacent trophic levels.

Using data from an ongoing field experiment, Wardle and Yeates (1993) demonstrated that microbe-feeding nematodes generally did not increase in abundance even when microbial mass was greatly enhanced by resource addition. However, there was a notable increase in top-predatory nematodes, indicative of a tritrophic effect in which the intermediate trophic level (microbe-feeding nematodes) is regulated mainly by top-down effects. It is also apparent from this study, and other data (Wardle, 1995) that, at least in some situations, fungi are resource-limited and respond to resource addition (and are thus likely to benefit by enhanced resource quality) and generally are not strongly regulated by invertebrate grazing pressure, probably because they have the structural and chemical complexity to avoid grazing. Bacteria, meanwhile, that lack the mechanisms to escape grazing to the same degree, are regulated mainly by predation, and do not reach sufficient population sizes for resources to become limiting. Thus, bacteria can enter strong predator–prey relationships with microfauna (Stout, 1980). It would therefore appear that the two components of the microfood-web system which are the most resource-limited, and thus likely to respond directly to litter quality are the fungi and top predatory nematodes.

Further evidence of this is found in a subsequent study at the same site (Wardle et al., 1995), in which changes in micro-food-webs were followed in a sawdust mulch over a 3-year period. Over the first year all groups demonstrated predictable increases as the food web developed. However, even though the N and P content of the sawdust continued to increase over the subsequent 2 years, only fungi and (to a lesser extent) top predatory nematodes showed any response to this enhancement of resource quality; the other groups all instead entered predator–prey cycles with each other which induced substantial temporal variations in population size (Fig. 8.1). These observations also suggest that in the largely aquatic, bacterial-based compartment of this micro-food-web system predation becomes increasingly important at lower trophic levels, which is consistent with the hypothesis of Menge and Sutherland (1976), developed for aquatic systems. Meanwhile in the sessile, fungal-based compartment, competition and predation are important at alternate trophic levels, consistent with the hypothesis of Hairston et al. (1960), developed for sessile, terrestrial plant-dominated systems. Our findings of the relative importance of bottom-up and top-down effects in regulating certain components of the microfood-web system (principally nematode groups) are independently confirmed by modelling-based studies (De Ruiter et al., 1995).

We are not suggesting that populations of bacteria and microbe-feeding microfauna do not respond to resource addition and quality. In fact, bacterial-feeding microfauna can show extremely rapid increases with the addition of fresh substrate (Christensen et al., 1992) and respond differentially to the nature of organic matter input from different plant species (Griffiths et al., 1992; Yeates, 1987). However, increases of this sort may be transient and we believe that after resource addition, bacteria and their predators soon enter into predator–prey cycles (Stout, 1980; Wardle et al., 1995)

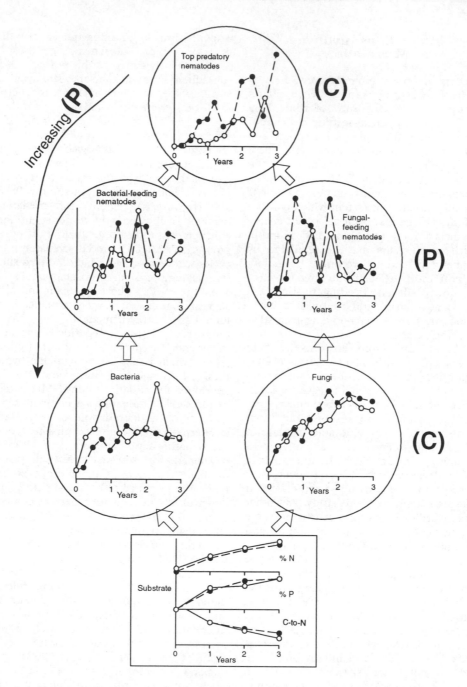

Fig. 8.1. Temporal dynamics of nematode populations, microbial mass and substrate quality in field-placed sawdust. Data presented are relative values, from two sites (represented by solid and dashed lines respectively). After the first year only the fungi show a statistically significant relationship with substrate quality. 'C' and 'P' indicate that competition (resource-limitation) and predation respectively are mainly responsible for regulating each group of organisms. Data derived from Wardle *et al.* (1995).

which are independent of all but relatively large changes in resource quality.

It is the nature of these predator–prey relationships involving bacteria and their grazers which cause them to be of fundamental importance in nutrient mineralization (Clarholm, 1985; Ingham *et al.*, 1985). Bacterial-feeding microfauna enhance nutrient release directly through excretion, and indirectly through maintaining bacteria in a logarithmic growth phase (Trofymow and Coleman, 1982). Griffiths (1994) showed that, over most credible C-to-N ratios of protozoa and bacterial prey, a significant proportion of N ingested by protozoa is likely to be excreted (i.e. 60% if bacteria and protozoa both have a C-to-N ratio of 5). Nutrient status of bacteria (indicative of resource quality) can be very important in determining the degree of mineralization; Darbyshire *et al.* (1994) determined that ammonium excretion by a soil ciliate was significantly enhanced by increasing the nitrogen content of its prey, and that phosphorus was only mineralized when protozoa were fed with bacteria with a high P content. This mineralization is very important in increasing plant growth (and implicitly litter quality), and it has been shown, at least in microcosm experiments, that microfaunal predation enables plant uptake of nitrogen which would otherwise be immobilized in bacterial cells (Kuikman and van Veen, 1989).

Interactions Involving Arthropods

The most intensively studied arthropods in the decomposer subsystem are the springtails and mites. Springtails, cryptostigmatid mites and astigmatid mites are often assumed to be fungivorous and/or saprophagous while many of the components of the Protostigmata and Mesostigmata are assumed to be predatory (see Petersen and Luxton, 1982; Vannier, 1985). However, these distinctions are blurred by omnivory (Wardle, 1995). Although less well understood, larger arthropods may be important both as saprophages (e.g. members of the Isopoda, Formicidae, Diplopoda and

Coleoptera) and predators (e.g. members of the Arachnomorpha, Chilopoda and Coleoptera).

Many mesofaunal species, in particular springtails, graze directly on fungal hyphae, although the extent of direct grazing may depend on season and resource quality (Anderson, 1977). Since litter quality determines the nature of the fungal community present (see earlier discussion) it is apparent that this will impact upon those organisms which feed on them. There is strong evidence that springtails selectively graze some fungal species in preference to others (e.g. Visser and Whittaker, 1977) and will actively seek out preferred fungal species over comparatively large distances (Bengtsson *et al.*, 1994). Earlier-colonizing fungi are often eaten in preference to those which colonize later, meaning that grazing of fungi may contribute in part to observed fungal successions (Frankland, 1992). The preference of springtails for earlier-successional fungi (Klironomas *et al.*, 1992) may explain why some springtails dominate earlier rather than later during the decomposition of litter (Wardle *et al.*, 1995). Grazing by springtails also has the potential to benefit fungal production, mainly through dispersal of propagules, removal of senescent hyphae (Visser, 1985) and induction of compensatory growth (Bengtsson *et al.*, 1993), although overgrazing usually proves to be detrimental (e.g. Hanlon and Anderson, 1980). Nutrient concentrations may regulate whether or not stimulatory effects on fungi occur. For example, Hanlon (1981a) found fungal activity to be enhanced by springtails under high substrate nutrient concentrations but reduced by springtails under lower ones; Teuben (1991) found the reverse effect. Therefore these effects do not work in easily predictable directions.

The 'true' saprophagous arthropods, which directly consume litter, fulfil the role of litter transformers. Because of their limited range of enzymes, they are dependent upon microflora for conditioning resources before digestion. This is particularly apparent in litter of very low quality such as wood (Setälä and Marshall, 1994), and fungal amelioration of wood quality through enzyme activities, improved carbon-to-nutrient ratios, and breakdown of inhibitory compounds appears necessary for subsequent colonization of saprophagous arthropods (Swift and Boddy, 1984). This faunal–microbial association can benefit the microflora considerably due to the fragmentation of litter and reduction of its structual complexity (Tajovský et al., 1992). The transformed litter, in the form of faecal pellets, enhances microbial activity and thus breakdown of the undigested material. This system serves as an 'external rumen', with arthropods reingesting their faecal material so as to reabsorb substrates made available following egestion, and thus optimizing their overall nutrient uptake efficiency. This adaptive strategy is of particular importance in relatively fresh litter with little microbial conditioning, and in nutrient-poor litter; arthropods prevented from reingesting faecal material in such situations can show considerable stress effects (Swift et al., 1979; Hassall and Rushton, 1982).

Although litter-transforming arthropods can induce short-term enhancement of microbial activity, the structures they create can also induce longer-term inhibition of microbial activity and decomposition, through increased compaction, reduced porosity, and less favourable moisture status (Hanlon and Anderson, 1980). This is particularly apparent when the faecal pellets are in the form of small, unstable aggregates (Lavelle, 1994). Reduction of pore size in faecal pellets may result in bacteria being enhanced at the expense of fungi (Hanlon, 1981b).

Litter quality also affects interactions between different faunal groups in the litter transformation and microfood-web systems. For example Tajovský et al. (1992) found that transformation of litter by millipedes caused population increases of both protozoa and nematodes. Interactive effects between faunal groups and litter quality are shown particularly strongly in the study of Coûteaux et al. (1991), in which chestnut leaves were grown in ambient and elevated CO_2, so as to produce leaf litter which differed in quality, principally C-to-N ratio. Addition of nematodes, springtails and isopods all influenced protozoan populations, but these effects were unpredictable and vastly different for the two litter types (Fig. 8.2). A similar pattern emerged in relation to the effects of litter-transformers on nematodes. Since most of these organisms (protozoa, springtails, isopods and most of the nematodes) were of comparable trophic levels (i.e. consumers of microflora and litter) these results suggest that complex competitive and mutualistic relationships exist between different faunal groups, and that changes in plant litter quality can strongly influence biotic interactions (and hence populations of soil organisms) but not in predictable directions.

Litter quality helps determine the ability of saprophagous arthropods to release nutrients. Seastadt (1984) found, upon surveying the literature on short-term experiments, that faunal effects on decomposition and mineralization of nutrients were very variable, and it is likely that this is a function of the nature of the resources present. Teuben (1991) suggested that fungivorous springtails and saprophagous isopods had a buffering effect, both in stimulating nutrient availability and microorganisms in low nutrient conditions, and inhibiting these properties in nutrient-rich conditions. Further data confirming this is presented by Blair et al. (1992) and van Wensem et al. (1993) (and to some extent by Coûteaux et al., 1991, see Fig. 8.2). However, this relationship is probably not universal since stimulation of microflora by mesofauna can also result in nutrient immobilization, reducing nutrient release rates (Seastadt and Crossley, 1980). Blair et al. (1992) concluded that the N-pool of decomposing litter is dynamic with simultaneous mineralization and immobilization occurring; it is the balance between these which will determine the effects that fauna have on decomposition and mineralization, and this is likely to be regulated by litter quality. The effects of arthropods on mineralization are known to significantly alter plant growth, particularly in controlled conditions. In a microcosm system, Setälä and Huhta (1991) found

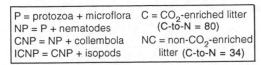

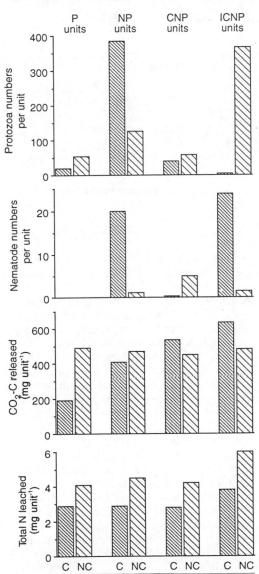

Fig. 8.2. Protozoa and nematode numbers, and C and N mineralization, in chestnut litter of two contrasting qualities, and in response to various faunal additions, after 24 weeks in experimental 'units'. Drawn from Tables 2 and 3 of Coûteaux *et al.* (1991).

that invertebrates (including arthropods) substantially enhanced both growth and nutrient content of *Betula pendula* seedlings; their results suggest evidence for a positive feedback whereby arthropod feeding activity may potentially improve the nutrient status of subsequent plant litter.

Predatory arthropods consume fauna in both the microfood-web and litter-transforming systems and, although the extent to which they are regulated by litter quality is not well known, any response they show to resource quality is likely to be indirect and related to lower trophic levels (i.e. similar to the trends discussed for top predatory nematodes). However, these organisms do have the potential to exert significant top-down effects, resulting in altered rates of litter breakdown. This is demonstrated in two studies (Santos *et al.*, 1981; Kajak *et al.*, 1993) in which manipulations of predators appear to induce a 'trophic cascade' (similar to those identified in aquatic systems by Carpenter *et al.*, 1988) resulting in an alteration of rates of plant litter mineralization (Fig. 8.3).

Interactions Involving Ecosystem Engineers

Soil invertebrates that can dig soil and produce organomineral structures have been called 'ecosystem engineers' (Stork and Eggleton, 1992; Jones *et al.*, 1994). They comprise the largest soil invertebrates and social insects. When present, they influence the existence of other organisms that are smaller and/or produce purely organic structures, i.e. the litter-transformers, components of 'microfood-webs' and the entire microfloral community (Lavelle, 1994). Ecosystem engineers develop efficient internal mutualistic relationships with microflora which allow them to digest otherwise resistant material. Termites, endogeic and anecic (but not epigeic) earthworms and ants are major components of this group. Some large diplopods, isopods, and even terrestrial crabs that are

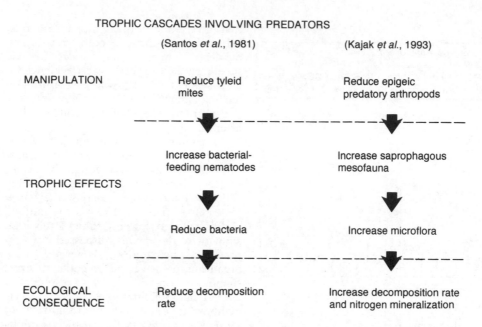

Fig. 8.3. Evidence from two studies for trophic cascades which regulate decomposition of plant litter.

capable of digging burrow systems may be additional components with significant impacts in some situations. These organisms have a very important role in creating habitats for other organisms, and thus serve as excellent examples of 'metabionts' (see Waid, Chapter 11, this volume).

The organisms in this group and their spheres of influence (i.e. the earthworm 'drilosphere' and termite 'termitosphere') are powerful regulators of microbial activity, and hence of decomposition (Lavelle and Gilot, 1994). They are able to digest a number of resistant compounds such as tannin-protein complexes and 'lignin', but they may also be inhibited by the accumulation of certain secondary compounds in litter. Therefore, although their geographic distribution is limited mainly by macroclimate, litter quality is also a relevant determinant of their distribution and activity. These organisms may in turn affect the quality of litter produced, although there is currently little evidence of this in the literature. Further, behavioural and physiological peculiarities of these organisms may profoundly affect the timing and location of decomposition processes.

The quality of litter directly affects its palatability. A number of experiments have shown clear choices by litter-feeding earthworms when given different types of litter (Satchell, 1967; Ferrière and Bouché, 1985). On the same basis termite groups have been classified depending on their preference for dead grass, leaf litter, wood, or 'humus'. In environments where tree species that produce litters of different qualities coexist, earthworm communities may demonstrate obvious patchy spatial distributions. In the Amazon forest of French Guyana, trees may produce litters with highly differing qualities (Charpentier *et al.*, 1995). *Dicorynia guianensis* produces litter with a very high content of polyphenolic complexes which sequester over 80% of litter nitrogen. In contrast, *Qualea* spp. produce litter with low levels of phenolic complexes, but high aluminium levels. The litter which accumulates under *Qualea* trees is favourable for endogeic earthworms which are found at the foot of these trees where they form clearly delineated patches in which litter decomposition is fast, while they are absent from the litter of *D. guianensis* (Fig. 8.4). The litter that accumulates at the base of *D. guianensis* trees is conducive for ingress by plant roots, and a thick root mat develops which

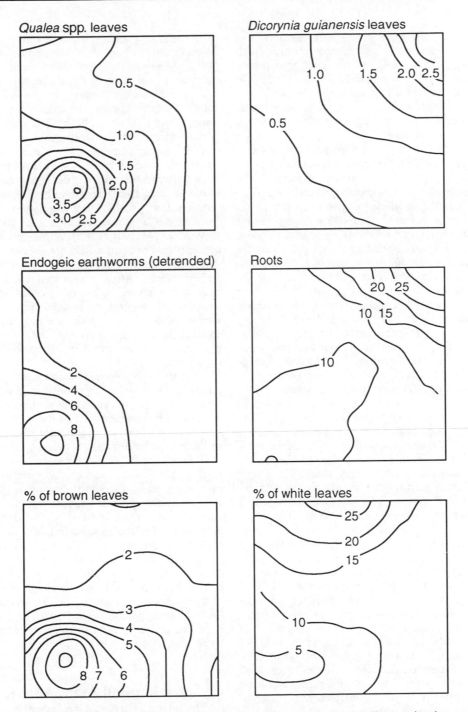

Fig. 8.4. Spatial distribution of leaves of *Qualea* spp. and *Dicorynia guianensis* in litter, endogeic earthworms, root abundance in soil, brown leaves and white leaves (colonized by white rot fungi) in a 20 × 20 m plot of tropical rainforest in French Guyana. Numbers on isoclines represent densities m^{-2}.

contains a high proportion of roots of a third tree species, *Eperua falcata*. Such 'single-tree' effects are not uncommon (Boettcher and Kalisz, 1991, 1992) and they may result in the formation of a mosaic of units with rather different decomposer communities and turnover rates of decomposing material.

Following litter deposition, fungal colonization may increase the palatability and digestibility of substrate material. For example, fungus-growing termites can only digest lignocellulosic material after it has been attacked by fungi. These fungi release an enzyme which adds its effects to those of enzymes produced by the termites themselves, allowing digestion of the substrate (Rouland *et al.*, 1990). Anecic earthworms are known to preferentially ingest litter after it has experienced a preliminary fungal attack (Cortez and Hameed, 1988), and some species may accumulate litter around the opening of their burrows forming 'middens' where a preliminary digestion of the 'external rumen' type is observed (Hamilton and Sillman, 1989). Such processes may occur over much larger scales during vegetation succession, e.g. when forests develop, mature, senesce, and develop again. In alpine forests of France, Bernier and Ponge (1993, 1994) found that earthworms which are almost absent during the growing and mature phases of coniferous forest development suddenly appear in units containing senescent and dead trees. At that stage, it seems that spruce litter that accumulates over decades in a moder type of humus suddenly becomes palatable to earthworms. In a few years, a high proportion of this 'free' organic matter is mixed with soil minerals by anecic earthworm activity, and a large flush of available nutrients occurs which is capable of sustaining rapid growth of young forest units. Again, the occurrence of forest patches of different ages, and hence with litter of different qualities, results in a patchy distribution of earthworm populations.

The activities of earthworms and termites may affect litter quality both directly (through improving plant nutrient supply) and indirectly (by influencing vegetation composition in their spheres of influence). As an example of a direct effect, Spain *et al.* (1992) observed in a pot experiment that inoculation of the endogeic earthworm *Millsonia anomala* caused an increase in the concentrations of root nitrogen and phosphorus, and shoot phosphorus of the African fodder grass *Panicum maximum*, as well as changes in the shoot : root ratio. This could be indicative of a positive feedback between root litter quality and earthworm activity. In relation to indirect effects, termites may greatly affect vegetation type through the large structures that they create (Spain and McIvor, 1988). Some termites accumulate nutrients and clay minerals into their mounds thus creating sites with high fertility status. After colony death (which may occur several years or decades after foundation) vegetation invades this soil. Composition of this plant community is specific, and in some East African savannas, these sites are preferentially colonized by shrubs.

Further, ecosystem engineers may affect decomposition processes by transferring litter into structures that they have created. This sometimes results in patchy distributions that are frequently observed in their populations. Sequestration of litter at different stages of decomposition in termite and earthworm structures can also be important in controlling decomposition. For example, coarse organic debris included in compact casts of the endogeic earthworm *Millsonia anomala* had a much slower decomposition rate than similar debris in control, non-aggregated soil (Martin, 1991). Therefore, one year after having been egested by the earthworm, soil had an 11% higher content of this fraction despite the partial digestion that had occurred during transit through the earthworm gut.

Above-ground Associations: Invasive Plants, Biodiversity and Herbivores

The relationship between plant litter quality and decomposition are ultimately large spatial scales, including those relating to the nature of the plant community present (c.f. Lavelle *et al.*, 1993). Grime (1979) proposed a theory in which plants growing in communities could be classified as according to three main strategies, namely 'ruderals' (adapted to disturbed sites), 'competitors', and 'stress-tolerators' (adapted to non-varying, harsh conditions). This is relevant in terms of litter quality; ruderals produce litter with a high nutrient status and few secondary metabolites while stress-tolerators produce nutrient-poor litter with high concentrations of secondary metabolites (Hobbie, 1992). The ecological consequences of strategy-induced effects can be demonstrated through two examples involving invasion by plants with vastly different strategies. The first example involves the invasive pasture weed *Carduus nutans* in grasslands dominated by *Lolium perenne* and *Trifolium repens* in the Waikato area of New Zealand (Wardle *et al.*, 1994). This species has a ruderal strategy, and establishes as randomly-located individual plants, usually as a result of local disturbance. Its litter has a high nitrogen content (3.5%) and decomposes extremely rapidly (half-life = 15 days). The litter is clearly favourable for soil biota: the microbial biomass, springtail populations and epigeic arthropod populations are, respectively 1.2, 3-and 1.5-fold greater in patches with *C. nutans* litter than in adjacent pasture. *C. nutans* thus induces random patches of temporarily enhanced biological activity on an otherwise relatively invariant landscape. The second example involves the invasive ericaceous dwarf shrub *Empetrum hermaphroditum* in the Swedish boreal forest. This plant is highly stress tolerant and invades late successional systems as randomly-located clones (Zackrisson *et al.*, 1995). It produces high concentrations of secondary metabolites, and is highly allelopathic, providing one of the few unequivocal examples of allelopathy in the literature (Nilsson, 1994). The decomposition rate of *E. hermaphroditium* litter in microcosms is around 77% of that of the other dominant ground-dwelling shrub present, *Vaccinium myrtillus* (Wardle, Nilsson and

Zackrisson, unpublished). Further, the decomposition rate of litter of various species placed in *E. hermaphroditum* humus is only 70% of that placed in *V. myrtillus* humus. This appears to reflect the negative effects of *E. hermaphroditum* on the soil biota: microbial biomass in *E. hermaphroditum* humus is only 60% of that in *V. myrtillus* humus, and soil macrofauna is almost completely absent. As a result, invasion by *E. hermaphroditum* results in thick humus layers, with long term effects on nutrient availability and the soil biota; for example, in humus profiles, layers with high pollen counts of *E. hermaphroditum* (indicating that the humus is of *E. hermaphroditum* origin) show suppressed microbial activity relative to layers with low *E. hermaphroditum* pollen even 1000 years following formation. Invasion of *E. hermaphroditum* is reversed by wildfire, and this is at least partially attributable to adsorption of inhibitory compounds by newly produced charcoal (Zackrisson *et al.*, 1996). Thus when the natural fire-cycle is interrupted, *E. hermaphroditum* has the potential to induce general decline of the boreal forest ecosystem.

An extension of these individual-species effects are multiple-species effects. There has been considerable recent interest in the effects of plant species richness on ecosystem function (e.g. Vitousek and Hooper, 1993), although the ideas have actually been expressed considerably earlier; Odum (1969) speculated if diversity could alter physical stability in the ecosystem, and whether species richness was 'a necessity for long life of the ecosystem'. These ideas are directly relevant to plant litter quality: the majority of net primary production enters the decomposition system as litter, and if such a hypothesis is correct then species richness of plant litter would be expected to have important ecological effects. Although litter diversity may be an important component of litter quality, this aspect has received little attention to date. This hypothesis can be tested using 'litter-mix' experiments, in which the effects of mixing litter of different species are compared with what would be expected based on the effects of the

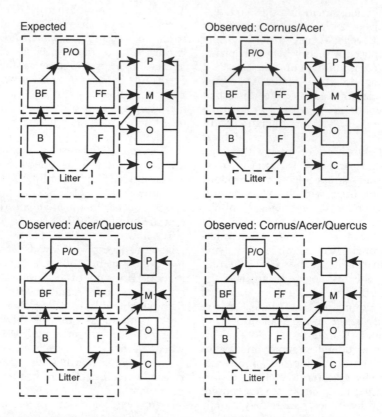

Fig. 8.5. Effects of litter diversity on microbial and faunal components of the detritus food-web in forest ecosystems. For each functional group the area of each rectangle represents the observed mass or population of that component in the two- or three-species litter mixes, relative to the expected values derived from litter monocultures, assuming effects of all litter-types are purely additive. Symbols: B = bacteria; F = fungi; BF = bacterial-feeding nematodes; FF = fungal-feeding nematodes; P/O = predatory/omnivorous nematodes; C = Collembola; O = oribatid mites; M = mesostigmatid mites; P = protostigmatid mites. Calculated from data presented by Blair *et al.* (1990).

components in monoculture. Chapman *et al.* (1988) observed, in considering litter in monoculture and mixed stands of four tree species, that mixture effects were often different from that expected based on the monocultures, but these effects were not predictable. Specifically, in a mixed *Pinus sylvestris – Picea abies* stand microbial activity, nutrient mobilization, and populations of springtails and earthworms were greater than expected, while in *P. abies – Alnus glutinosa* and *P. abies – Quercus petraea* mixes microbial activity and nutrient mobilization were less than expected. In their study, biodiversity of

litter input therefore altered ecosystem function but not necessarily always in the same direction. Blair *et al.* (1990) evaluated decomposition processes and components of the soil biota, in litterbags containing monocultures, and two- and three-species mixtures of *Acer rubrum*, *Cornus florida* and *Quercus prinus*. Decay rates of mixed litter did not differ from what was expected based on monoculture values, but nitrogen release in the mixtures was greater than expected in the initial phases of the study. These differences in mineralization appeared to be related to differences in the abundances of decomposer organisms. For example, fungivorous nematodes and mesofaunal groups were respectively generally greater and generally lesser in the mixtures than what would be expected based on the monocultures (Fig. 8.5). Results from Blair et al. (1990) indicate that litter biodiversity may affect different pools of soil organisms but not in easily predictable directions. Wardle et al. (1996) conducted a litterbag study in which mixtures of between two and eight species (together with appropriate monoculture litterbags) were randomly generated from a pool of 32 plant species (including grasses, herbs and trees), so as to evaluate whether decomposition, nitrogen mineralization and litter microbial biomass were affected mainly by species richness or mainly by the degree of dissimilarity of different litter types. Their results showed that species richness of litter may have either positive or negative effects on these properties, depending upon the types of plants considered. Further, while increasing diversity from one to two species had detectable effects, there was no consistent effect observed of increasing species richness from two to eight species, indicating that considerable species redundancy exists in relation to the ecological effects of litter diversity. There is also evidence that plant litter decomposition can be altered by plant species diversity in controlled experimental conditions (Wardle and Nicholson, 1996) although again, these effects do not work in predictable directions.

Finally, above-ground trophic diversity also impacts upon litter quality. Foliar herbivory can substantially alter litter quality and hence soil organisms. This can happen in two ways, with vastly different consequences. The first involves herbivores inducing a shift in the plant's chemistry and hence quality of litter return. Thus, herbivores have often been shown to stimulate decomposer organisms (e.g. Ingham and Detling, 1984; Seastadt et al., 1988). These effects are probably partly due to enhancement of nitrogen status of root litter produced by grazed plants (Seastadt et al., 1988), although grazing optimization of root productivity and herbivore-induced root mortality (and associated resource-input) may also contribute (Merrill et al., 1994). The second involves herbivores preferentially grazing some plant species, allowing others to dominate. The preferentially eaten plants are also those with superior litter quality, so the community becomes dominated by plants with poor litter quality. This effect is demonstrated by Pastor et al. (1988; 1993), who found that selective browsing by moose of aspen, poplar and ash (all with high litter quality) allowed spruce (with poor litter quality and a high cellulose content) to dominate. The net result is a lower soil microbial biomass, lower decomposition rate and less nitrogen mineralization, and ultimately retarded development of the soil profile.

Conclusions

Plant litter quality is of critical importance in regulating both the soil biota and the nature of soil biotic interactions, and these effects operate at three different scales of resolution; the 'microfood-web' system, 'litter-transformer' system, and the system involving 'ecosystem engineers'. These systems are constrained by the input of plant litter, which is in turn determined by the nature of the plant community. Although in any plant community there are usually several types of litter being returned to the soil, the effects of litter diversity on the soil biota and biotic processes has been largely unexplored; there is, however, evidence that litter diversity effects do not work in predictable directions, and that there may be considerable species redundancy. The importance of plant litter input in determining soil biological interactions is also apparent in studies which have considered the effects of spatial placement of plant litter (and consequently spatial heterogeniety), and those which have considered changes in components of the decomposer system (including humus dynamics: Bernier and Ponge, 1994) during vegetation succession. Ultimately the 'Driven by Nature' issue involves two components, biota, and the effects of biotic interactions on soil ecological processes, and an understanding of both components is essential in understanding the ecological effects of plant litter quality.

Acknowledgements

We thank Drs L. Greenfield, M.-C. Nilsson, J. Springett and G. Yeates for many constructive suggestions on earlier versions of the chapter.

References

Anderson, J.M. (1977) The organisation of soil animal communities. *Ecological Bulletins* 25, 15–23.

Bengtsson, G., Hedlund, K. and Rundgren, S. (1993) Patchiness and compensatory growth in a fungus-Collembola system. *Oecologia* 93, 295–302.

Bengtsson, G., Hedlund, K. and Rundgren, S. (1994) Food- and density-dispersal: evidence from a soil collembolan. *Journal of Animal Ecology* 63, 513–520.

Bernier, N. and Ponge, J.-F. (1993) Dynamique et stabilité des humus au cours du cycle sylvogénétique d'une pessière d'altitude. *Comptes-Rendus des Séances de l' Académie des Sciences de Paris, Serie 3, Sciences de la Vie* 316, 647–651.

Bernier, N. and Ponge, J.F. (1994) Humus forms during the sylvogenic cycle in a mountain spruce forest. *Soil Biology and Biochemistry* 26, 183–220.

Blair, J.M., Parmelee, R.W. and Beare, M.H. (1990) Decay rates, nitrogen fluxes and decomposer communities in single and mixed-species foliar litter. *Ecology* 71, 1976–1985.

Blair, J.M., Crossley, D.A. and Callaham, L.C. (1992) Effects of litter quality and microarthropods on N-dynamics and retention of exogenous ^{15}N in decomposing litter. *Biology and Fertility of Soils* 12, 241–252.

Boettcher, S.E. and Kalisz, P.J. (1991) Single-tree influence of earthworms in forest soils in eastern Kentucky. *Soil Science Society of America Journal* 55, 862–865.

Boettcher, S.E. and Kalisz, P.J. (1992) Single tree influences on soil properties in the mountains of eastern Kentucky. *Ecology* 71, 1365–1372.

Carpenter, S.R., Kitchell, J.E. and Hodgson, J.R. (1988) Cascading trophic interactions and lake productivity. *BioScience* 35, 634–639.

Carreiro, M.M. and Koske, R.E. (1992) The effect of temperature and substratum on competition among three species of forest litter microfungi. *Mycological Research* 96, 19–24.

Chapman, K., Whittaker, J.B. and Heal, O.W. (1988) Metabolic and faunal activity in litters of tree mixtures compared with pure stands. *Agriculture, Ecosystems and Environment* 24, 33–40.

Charpentier, F., Grandval, A., LeRoy-Guillaume, C., Rossi, J.P. and Lavelle, P. (1995) Single-tree effects on soil fauna communities in a tropical forest of French Guyana. In: *Driven by Nature: Plant Litter Quality and Decomposition – Abstracts.* Wye College, University of London, p. 47.

Christensen, S., Griffiths, B.S., Ekelund, F. and Rønn, R. (1992) Huge increase in bacterivores on freshly killed barley roots. *FEMS Microbiology Ecology* 86, 303–310.

Clarholm, M. (1985) Possible roles for roots, bacteria, protozoa and fungi in supplying nitrogen to plants. In: Fitter, A.H. (ed.) *Ecological Interactions in Soil.* Blackwell Scientific Publications, Oxford, pp. 355–365.

Cooke, R.C. and Rayner, A.D.M. (1984) *The Ecology of Saprophytic Fungi.* Longman, London, 415 pp.

Cornejo, F.H., Varela, A. and Wright, S.J. (1994) Tropical forest litter decomposition under seasonal drought: nutrient release, fungi and bacteria. *Oikos* 70, 183–190.

Cortez, J. and Hameed, R. (1988) Effects de la maturation des litières de ray-grass (*Lolium perenne* L.) dans le sol sur leur consommation et leur assimilation par *Lumbricus terrestris* L. *Revue d' Ecologie et de Biologie du Sol* 25, 397–412.

Coûteaux, M.-M., Mousseau, M., Célérier M.-L. and Bottner, P. (1991) Increased atmospheric CO_2 and litter quality: decomposition of sweet chestnut leaf litter with animal food webs of different complexities. *Oikos* 61, 54–64.

Darbyshire, J.F., Davidson, M.S., Chapman, S.J. and Ritchie, S. (1994) Excretion of nitrogen and phosphorus by the soil ciliate *Colpoda steinii* when fed upon the bacteria *Arthrobacter* sp. *Soil Biology and Biochemistry* 26, 1193–1199.

De Ruiter, P.C., Neutel, A.-M. and Moore, J.C. (1995) Energetics, patterns of interaction strengths, and stability in real ecosystems. *Science* 269, 1257–1260.

Dighton, J. (1978) Effects of synthetic lime aphid honeydew on populations of soil organisms. *Soil Biology and Biochemistry* 10, 369–376.

Ferrière, G. and Bouché, M. (1985) Première mesure écophysiologique d'un débit dans un animal endogé: le débit d'azote de *Nicodrilus longus* (Ude) (Lumbricidae, Oligochaeta) dans la prairie de Citeaux. *Comptes Rendus de l'Académie des Sciences de Paris* 301, 789–794.

Flanagan, P.W. and van Cleve, K. (1983) Nutrient cycling in relation to decomposition and organic-matter quality in taiga ecosystems. *Canadian Journal of Forest Research* 13, 795–817.

Frankland, J.C. (1992) Mechanisms in fungal succession. In: Carroll, G.C. and Wicklow, D.T. (eds), *The Fungal Community - Its Organisation and Role in the Ecosystem*, 2nd edn. Marcel Dekker, New York, pp. 383–401.

Garrett, S.D. (1963) *Soil Fungi and Soil Fertility*. Pergamon Press, New York, 165 pp.

Griffiths, B.S. (1994) Soil nutrient flow. In: Darbyshire, J.F. (ed.) *Soil Protozoa*. CAB International, Wallingford, pp. 65–91.

Griffiths, B.S., Welschen, R., Arendonk, J.J.C.M. and Lambers, H. (1992) The effect of nitrate-nitrogen on bacteria and the bacterial-feeding fauna in the rhizosphere of different grass species. *Oecologia* 91, 253–259.

Grime, J.P. (1979) *Plant Strategies and Vegetation Processes*. Wiley, Chichester, 222 pp.

Hairston, N.G., Smith, F.E. and Slobodkin, L.B. (1960) Community structure, population control and competition. *American Naturalist* 94, 421–425.

Hamilton, W.E. and Sillman, D.Y. (1989) Effects of earthworm middens on the distribution of soil microarthropods. *Biology and Fertility of Soils* 8, 279–284.

Hanlon, R.D.G. (1981a) Influence of grazing by Collembola on the activity of senescent fungal colonies grown on different nutrient concentrations. *Oikos* 36, 362–367.

Hanlon, R.D.G. (1981b) Some factors influencing microbial growth on soil animal faeces. II. Bacterial and fungal growth on soil animal faeces. *Pedobiologia* 21, 264–270.

Hanlon, R.D.G. and Anderson, J.M. (1980) The influence of macroarthropods feeding activities on microflora in decomposing oak leaves. *Soil Biology and Biochemistry* 12, 255–261.

Hassall, M. and Rushton, S.P. (1982) The role of coprophagy in the feeding strategies of terrestrial arthropods. *Oecologia* 53, 374–381.

Hobbie, S.E. (1992) Effects of plant species on nutrient cycling. *Trends in Ecology and Evolution* 7, 336–339.

Ingham, R.E. and Detling, J.K. (1984) Plant-herbivore interactions in a North American mixed grass prairie. III. Soil nematode populations and root biomass on *Cynomys lubovicionus* colonies and adjacent uncolonised areas. *Oecologia* 63, 307–313.

Ingham, R.E., Trofymow, J.A., Ingham, E.R. and Coleman, D.C. (1985) Interactions of bacteria, fungi and their nematode grazers on nutrient cycling and plant growth. *Ecological Monographs* 55, 119–140.

Jones, C.G., Lawton, J.H. and Shachak, M. (1994) Organisms as ecosystem engineers. *Oikos* 69, 373–386.

Kajak, A., Chmielewski, K., Kaczmarek, M. and Rembialkowska, E. (1993) Experimental studies on the effects of epigeic predators on matter decomposition processes in managed peat grasslands. *Polish Ecological Studies* 17, 289–310.

Klironomas, J.N., Widden, P. and Deslandes, I. (1992) Feeding preferences of the collembola *Folsomia candida* in relation to microfungal successions of decaying litter. *Soil Biology and Biochemistry* 24, 685–692.

Kuikman, P.J. and van Veen, J.A. (1989) The impact of protozoa on the availability of bacterial nitrogen to plants. *Biology and Fertility of Soils* 8, 13–18.

Lavelle, P. (1994) Faunal activities and soil processes: adaptive strategies that determine ecosystem function. In: *XV ISSS Congress Proceedings, Vol. 1 : Introductory Conferences.* Acapulco, Mexico, pp. 189–220.

Lavelle, P. and Gilot, C. (1994) Priming effects of macroorganisms on microflora: a key process of soil function? In: Ritz, K., Dighton, J. and Giller, K.E. (eds) *Beyond the Biomass - Compositional and Functional Analysis of Soil Microbial Communities.* Wiley, Chichester, pp. 173–180.

Lavelle, P., Blanchart, E., Martin, A., Martin, S., Spain, A.V., Toutain, F., Barois, I. and Schaefer, R. (1993) A hierarchical model for decomposition in terrestrial ecosystems: application to soils of the humid tropics. *Biotropica* 25, 130–150.

Martin, A. (1991) Short- and long-term effects of the endogenic earthworm *Millsonia anomala* (Omodeo) (Megascolecidae, Oligochaeta) of tropical savannas, on soil organic matter. *Biology and Fertility of Soils* 11, 234–238.

Menge, B.A. and Sutherland, J.P. (1976) Species diversity gradients: synthesis of the roles of predation, competition and spatial heterogeneity. *American Naturalist* 110, 351–369.

Merrill, E.U., Stanton, N.L. and Hak, J.C. (1994) Responses of bluebunch wheatgrass, Idaho fescue, and nematodes to ungulate grazing in Yellowstone National Park. *Oikos* 69, 231–240.

Moore, J.C. and Hunt, H.W. (1988) Resource compartmentation and the stabilty of real ecosystems. *Nature* 333, 261–263.

Nilsson, M.-C. (1994) Separation of allelopathy and resource competition by the boreal dwarf shrub *Empetrum hermaphroditum* Hagerup. *Oecologia* 98, 1–7.

Odum, E.P. (1969) The strategy of ecosystem development. *Science* 164, 262–270.

Pastor, J., Naiman, R.J., Dewey, B. and McInnes, P. (1988) Moose, microbes and the boreal forest. *Bioscience* 38, 770–777.

Pastor, J., Dewey, B., Naiman, R.J., McInnes, P.F. and Cohan, Y. (1993) Moose browsing and soil fertility in the boreal forests of Isle Royal National Park. *Ecology* 74, 467–480.

Petersen, H. and Luxton, M. (1982) A comparative analysis of soil fauna populations and their role in soil decomposition processes. *Oikos* 39, 287–388.

Rayner, A.D.M. and Boddy, L. (1988) Fungal communities in the decay of wood. *Advances in Microbial Ecology* 10, 115–166.

Robinson, C.H., Dighton, J., Frankland, J.C. and Roberts, J.D. (1994) Fungal communities on decaying wheat straw of different resource qualities. *Soil Biology and Biochemistry* 26, 1053–1058.

Rouland, C., Brauman, A., Keleke, S., Labat, M., Mora, P. and Renoux, J. (1990) Endosymbiosis and exosymbiosis in the fungus-growing termites. In: Lésel, R. (ed.) *Microbiology in Poikilotherms.* Elsevier, Amsterdam, pp. 79–82.

Santos, P.F., Phillips, J. and Whitford, W.G. (1981) The role of mites and nematodes in early stages of buried litter decomposition in a desert. *Ecology* 62, 664–669.

Satchell, J.E. (1967) Lumbricidae. In: Burges, N.A. and Raw, F. (eds) *Soil Biology.* Academic Press, London, pp. 259–322.

Seastadt, T.R. (1984) The role of microarthropods in decomposition and mineralisation processes. *Annual Review of Entomology* 29, 25–46.

Seastadt, T.R. and Crossley, J.R. (1980) Effects of microarthropods on the seasonal dynamics of nutrients in forest litter. *Soil Biology and Biochemistry* 12, 337–342.

Seastadt, T.R., Ramundo, R.A. and Hayes, D.C. (1988) Maximisation of densities of soil animals by foliage herbivory: empirical evidence, graphical and conceptual models. *Oikos* 51, 243–248.

Setälä, H. and Huhta, V. (1991) Soil fauna increase *Betula pendula* growth: laboratory experiments with coniferous forest floor. *Ecology* 72, 665–671.

Setälä, H. and Marshall, V.G. (1994) Stumps as habitats for Collembola during succession from clear-cuts to old-growth Douglas-fir forests. *Pedobiologia* 38, 307–326.

Šlapokas, T. and Granhall, U. (1991) Decomposition of willow-leaf litter in a short-rotation forest in relation to fungal colonisation and palatability to earthworms. *Biology and*

Fertility of Soils 10, 241–248.

Spain, A.V. and McIvor, J.G. (1988) The nature of herbaceous vegetation associated with termitaria in north-eastern Australia. *Journal of Ecology* 76, 181–191.

Spain, A.V., Lavelle, P. and Mariotti, A. (1992) Stimulation of plant growth by tropical earthworms. *Soil Biology and Biochemistry* 24, 1629–1634.

Stork, N.E. and Eggleton, P. (1992) Invertebrates as determinants and indicators of soil quality. *American Journal of Alternative Agriculture* 7, 38–47.

Stout, J.D. (1980) The role of protozoa in nutrient cycling and energy flow. *Advances in Microbial Ecology* 4, 1–50.

Swift, M.J. and Boddy, L. (1984) Animal-microbial interactions in wood decomposition. In: Anderson, J.M., Rayner, A.D.M. and Walton, D.H.W. (eds) *Invertebrate-Microbial Interactions*. Cambridge University Press, Cambridge, pp. 89–131.

Swift, M.J., Heal, O.W. and Anderson, J.M. (1979) *Decomposition in Terrestrial Ecosystems*. University of California Press, Berkeley 372 pp.

Tajovský, K., Šantrucková, H., Hánêl, L., Balík, V. and Lukešová, A. (1992) Decomposition of faecal pellets of the millipede *Glomeris hexasticha* (Diplopoda) in forest soil. *Pedobiologia* 36, 146–158.

Teuben, A. (1991) Nutrient availability and interactions between soil arthropods and microorganisms during decomposition of coniferous litter: a mesocosm study. *Biology and Fertility of Soils* 10, 256–266.

Trofymow, J.A. and Coleman, D.C. (1982) The role of bacterivorous and fungivorous nematodes in cellulose and chitin decomposition in the context of a root/rhizosphere/soil conceptual model. In: Freckman, D.A. (ed) *Nematodes in Soil Ecosystems*. University of Texas Press, Austin, pp. 117–138.

Vannier, G. (1985) Modes d'exploitation et partage des ressources alimentaires dans le systéme saprophage por les microarthropodes du sol. *Bulletin d'Ecologie* 16, 9–18.

Van Wensem, J., Verhoef, H.A. and van Straalen, N.M. (1993) Litter degradation stage as a prime factor for isopod interactions with mineralisation processes. *Soil Biology and Biochemistry* 25, 1175–1183.

Visser, S. (1985) Role of the soil invertebrates in determining the composition of soil microbial communities. In: Fitter, A.H. (ed.) *Ecological Interactions in Soil*. Blackwell Scientific Publications, Oxford, pp. 297–317.

Visser, S. and Whittaker, J.B. (1977) Feeding preferences for certain fungi by *Onichiurus subtenuis* (Collembola). *Oikos* 28, 320–325.

Vitousek, P.M. and Hooper, D.U. (1993) Biological diversity and terrestrial ecosystem biogeochemistry. In: Schulze, E.-D. and Mooney, H.A. (eds) *Biodiversity and Ecosystem Function*. Springer-Verlag, Berlin, pp. 3–14.

Wardle, D.A. (1992) A comparative assessment of factors which influence microbial biomass carbon and nitrogen levels in soils. *Biological Reviews* 67, 321–358.

Wardle, D.A. (1993) Changes in the microbial biomass and metabolic quotient during leaf litter succession in some New Zealand forest and scrubland ecosystems. *Functional Ecology* 7, 346–355.

Wardle, D.A. (1995) Impact of disturbance on detritus food-webs in agroecosystems of contrasting tillage and weed management practices. *Advances in Ecological Research* 26, 105–185.

Wardle, D.A. and Nicholson, K.S. (1996) Synergistic effects of grassland plant species on the soil microbial biomass and activity: implications for ecosystem-level effects of enriched plant diversity. *Functional Ecology* 10 (in press).

Wardle, D.A. and Yeates, G.W. (1993) The dual importance of competition and predation as regulatory forces in terrestrial ecosystems: evidence from decomposer food webs. *Oecologia* 93, 303–306.

Wardle, D.A., Nicholson, K.S., Ahmed, M. and Rahman, A. (1994) Interference effects of the invasive plant *Carduus nutans* L. against the nitrogen fixation ability of *Trifolium repens* L. *Plant and Soil* 163, 287–297.

Wardle, D.A., Parkinson, D. and Waller, J.E. (1993) Interspecific competitive interactions between pairs of fungal species in natural substrates. *Oecologia* 94, 165–172.

Wardle, D.A., Yeates, G.W., Watson, R.N. and Nicholson, K.S. (1995) Development of the decomposer food web, trophic relationships and ecosystem properties during a three-year primary succession of sawdust. *Oikos* 73, 155–166.

Wardle, D.A., Bonner, K.I. and Nicholson, K.S. (1996) Biodiversity and plant litter: experimental evidence which does not support the view that enhanced species richness improves ecosystem function. *Oikos* (in press).

Widden, P. (1986) Microfungal community structure from forest soils in southern Quebec, using discriminant function and factor analysis. *Canadian Journal of Botany* 64, 1402–1412.

Yeates, G.W. (1987) How plants affect nematodes. *Advances in Ecological Research* 17, 61–113.

Zackrisson, O., Nilsson, M.-C., Steijlen, I. and Hörnberg, G. (1995) Regeneration pulses and climate-vegetation interactions in non-pyrogenic boreal Scots pine stands. *Journal of Ecology* 83, 469–483.

Zackrisson, O., Nilsson, M.-C. and Wardle, D.A. (1996) Key ecological function of charcoal from wildfire in the Boreal forest plant-soil system. *Oikos* (in press).

9 Soil Fauna-mediated Decomposition of Plant Residues under Constrained Environmental and Residue Quality Conditions

G. Tian[1], L. Brussaard[2], B.T. Kang[1] and M.J. Swift[3]

[1]*International Institute of Tropical Agriculture (IITA), PMB 5320, Ibadan, Nigeria, c/o L.W. Lambourn & Co., 26 Dingwall Road, Croydon CR9 3EE, UK;* [2]*Department of Terrestrial Ecology and Nature Conservation, Wageningen Agricultural University, Bornsesteeg 69, 6708 PD Wageningen, The Netherlands;* [3]*Tropical Soil Biology and Fertility Programme (TSBF), c/o UNESCO ROSTA, UN Complex, Gigiri, PO Box 30592, Nairobi, Kenya*

Introduction

The decomposition of plant residues is influenced by resource quality, decomposer organisms and environmental conditions (Swift *et al.*, 1979). The effect of quality, as defined by chemical composition of the plant residues (C-to-N ratio, lignin and polyphenol contents), on decomposition has been well documented (e.g. Melillo *et al.*, 1982; Palm and Sanchez, 1990; Tian *et al.*, 1992b). Moisture and temperature have fundamental effects on plant residue decomposition as they control decomposer activity (Swift *et al.*, 1979; Tian *et al.*, 1993a).

Decomposer organisms consist of a complex community of soil biota including microflora and soil fauna. Fungi and bacteria are ultimately responsible for the biochemical processes in the decomposition of organic residues. Soil fauna enhance the biodegradation and humification of organic residues in several ways: (i) by comminuting organic residues and increasing the surface area for microbial activity; (ii) by producing enzymes which break down complex bio-molecules into simple compounds, and polymerize compounds to form humus; (iii) by improving the environment for microbial growth and interactions. For example, earthworms increase the decomposition of organic residues and the release of nutrients by comminuting plant residues, by incorporating the organic matter into the soil (Lavelle, 1988; Tian *et al.*, 1995), and by producing casts enriched with microflora and enzymes (Syers *et al.*, 1979; Mulongoy and Bedoret, 1989). Termites are known to be efficient in digesting cellulose and in some cases also lignified substances (Lee and Wood, 1971; Wood, 1988). Millipedes break down plant litter and mix it with mineral soil, which they ingest (Kevan, 1962; Tian *et al.*, 1995). By feeding on microflora, protozoa and nematodes can increase N mineralization (Kuikman and van Veen, 1989; Bouwman *et al.*, 1994), while predatory mites have a stabilizing effect on such interactions (Brussaard *et al.*, 1995).

Under normal conditions soil fauna and microflora form an integrated system for decomposition of organic residues. However, under environmental disturbance, such as conversion of natural to agricultural systems, and elevation of atmospheric CO_2, the integrated system of soil fauna and microflora in plant residue decomposition is altered. By reviewing the results of a recently completed study on plant residue decomposition and other published work with an emphasis on the soil organisms, we will address the hypothesis that soil invertebrate effects on the decomposition of plant residues are more significant as regulating factors such as soil moisture and residue quality become more constraining. In the hierarchy of determinants of the decomposition process as proposed by Lavelle *et al.* (1993), this means that, while the soil biota in general is negatively affected under constrained conditions, this affects the soil fauna ('macroorganisms' in the scheme of Lavelle *et al.*, 1993) less than soil microorganisms.

Soil fauna also affect the long-term release of nutrients by their influence on the humification of organic matter (e.g. Brussaard and Juma, 1996) and the build-up and maintenance of soil structure (e.g. Lee and Foster, 1991). But, these subjects are beyond the scope of the present chapter.

Soil Fauna-mediated Decomposition Associated with Conversion of Natural to Agricultural Systems in the Tropics

Conversion of natural to agricultural systems is known to cause drastic changes of the environment for the soil decomposer community. First, the buffer provided by a dense vegetation against fluctuations in microclimate is lost with land clearing, resulting in a harsher environment with large extremes in soil temperature and moisture conditions. Second, a change from natural vegetation to annual crops decreases the plant species richness, resulting in a reduction in diversity of food sources, hence, narrowing the spectrum of residue quality (see also Wardle and Lavelle, Chapter 8, this volume). These changes are more likely to have a larger effect on small, delicate organisms with a limited foraging range and cryptozoic

habits than on larger and more mobile animals with omnivorous or carnivorous feeding habits, except when the soil is tilled. We will not address soil tillage here.

Critchley *et al.* (1979) studied the soil microclimate and invertebrate soil fauna in cultivated and bush plots in the humid tropics. They observed that in the cultivated area at 10 cm depth the diurnal temperature ranged between 26 and 32°C, compared to the bush where the temperature at the same depth was almost constant at 25°C over the 4 days of measurement. The bush showed a consistently higher soil moisture content than the cultivated area throughout the year. Similar observations were reported by Lal and Cummings (1979). As a result, a lower activity of the majority of the surface soil fauna was observed in the cultivated compared with the bush area (Critchley *et al.*, 1979). For subterranean fauna, microarthropods were lower in number in cultivated land than bush. With a few exceptions, notably Prostigmata, populations of Acari and Collembola decreased with the conversion from bush to cultivated land (Table 9.1). Tian (unpublished data) similarly observed that the mean mite, springtail and total microarthropod numbers were significantly higher in plots with secondary growth than in continuously cropped plots.

We used litter bags to measure the microarthropod-mediated decomposition in three continuous maize/cassava cropping systems: one with *Leucaena leucocephala* as a hedgerow tree (LAC), one with *Pueraria phaseoloides* as a herbaceous legume relay crop (PRC) and one without hedgerow or herbaceous legumes (C). A bush-fallow with *Chromolaena odorata* after 3 (BR1), 15 (BR2), and 27 (BR3) months of fallow served for comparison. *Senna siamea* leaves were kept in surface-placed litter bags with two mesh sizes: 0.5 mm to allow access of microarthropods and 0.08 mm to exclude microarthropods. The microarthropod-mediated decomposition was expressed as the relative increase in the decomposition rate constants of *Senna* leaves in the 0.5 mm compared with the 0.08 mm mesh-size litter bags. As expected, the decomposition of leaf litter was

Table 9.1. Mean population densities of Acari and Collembola in a Alfisol for the 0–5 cm and 5–10 cm soil depth (number m^{-2}) (from Critchley *et al.*, 1979).

Soil depth	Bush plot	Cultivated plot
0–5 cm		
Acari		
Cryptostigmata	17,403	1,991
Prostigmata	8,477	9,839
Mesostigmata	3,557	1,121
Astigmata	1,372	708
Collembola		
Isotomidae	6,963	1,425
Entomobryidae	2,421	754
Onychiuridae	1,628	243
Sminthuridae	1,128	620
Poduridae	665	54
5–10 cm		
Acari		
Cryptostigmata	7,695	2,230
Prostigmata	6,353	7,862
Mesostigmata	1,013	766
Astigmata	237	17
Collembola		
Isotomidae	3,644	733
Entomobryidae	355	425
Onychiuridae	364	105
Sminthuridae	280	322
Poduridae	200	9

always slower in the cropping systems than in the bush-fallow (Table 9.2), reflecting restoration of the soil biota by bush-fallow. Interestingly, microarthropods contributed more to the decomposition of the *Senna* leaves over the period of 14 weeks of observation in the cropped plots than in the bush-fallow plots, although the microarthropods were lower in number in the cropped plots (data not shown). This difference could be more evident in the real field situation, because a litter bag with a 0.5 mm mesh-size as used in this study excluded microarthropods with a body size >0.5 mm, while a mesh-size of 0.08 mm will not prevent oviposition of microarthropod eggs in the litter bags, which will ultimately give rise to a population build-up inside the litter bags (Vreeken-Buijs and Brussaard, 1996).

Santos and Whitford (1981) incu-bated litter bags with shrub litter after vari-ous biocide treatments to exclude fungi, arthropods or arthropods and fungi in a semi-desert, where soil moisture is the most important limiting factor for decom-position, to determine the role of various biota in litter decomposition. Multiple regression involving seven abiotic factors showed that <50% of the variation in decomposition was accounted for by the abiotic factors in the litter bags with microarthropods versus 80–90% in those treated with biocides.

The higher proportion of micro-arthropod-mediated decomposition under cropping (own results) and the relatively higher influence of microarthropods on decomposition (Santos and Whitford, 1981) in a semi-desert is probably due to changes in interactions between soil fauna and soil microorganisms with changes in

Table 9.2. Effect of litter bag mesh-size on plant residue decomposition rate constants (k week⁻¹) under bush-fallow and cropping conditions and soil microarthropod-mediated decomposition (Tian *et al.*, unpublished data)

Treatments	Plant residue decomposition rate constants (k week^{-1}) litter bag mesh size		Soil microarthropod-mediated decomposition (%)
	0.08 mm	0.5 mm	
Cropped plots			
Continuous cropping	0.030	0.048	38
With *Leucaena*	0.041	0.052	21
With *Pueraria*	0.054	0.065	17
Bush-fallow plots			
3 month fallow	0.070	0.080	13
15 month fallow	0.075	0.083	10
27 month fallow	0.083	0.088	6

LSD$_{.05}$ for decomposition rate constants between mesh-size for the same treatment: 0.013.

soil moisture. It is well known that moisture stress negatively affects the microbial biomass and/or activity in soil. For example, one month of mildly drying a silt loam soil in a microplot experiment under outside rain-sheltered conditions from a water potential of -30 kPa to -120 kPa caused a significant decrease in O_2 consumption and N mineralization. After rewetting respiration increased 1.3–1.5 fold and N mineralization 3–5 fold (Bloem *et al.*, 1992). It is within this range of water potentials that the main group of bacterivorous nematodes, the rhabditids, become inactive (Verhoef and Brussaard, 1990). At further drying beyond -300 kPa common bacterivorous flagellates also stop growing (Zwart and Brussaard, 1991). On the assumption that certain, adapted, surface-dwelling (micro)arthropods and those which live in soil pores that have a relative humidity of near 100%, are less susceptible to drought, soil invertebrate-mediated decomposition will be relatively increased (i.e. less decreased) when the moisture conditions become unfavourable. Soil fauna may partly restore microbial activity by providing the microorganisms with a favourable microclimate and easily decomposable food, such as in the gut and excrements. Other factors than moisture, related to soil fauna-mediated decomposi-

tion include soil temperature and soil organic matter. Residue quality was the same in our own study and that of Santos and Whitford (1981).

Soil Fauna-mediated Decomposition as Related to Residue Quality

Quality of litter in terrestrial ecosystems may be defined in terms of C and nutrient availability to saprotrophs and the concentrations of modifying agents that inhibit or reduce organism and/or enzyme activities (Singh and Gupta, 1977; Swift *et al.*, 1979; Heal *et al.*, Chapter 1, this volume). The C sources in plant litters range from simple sugars, which are utilized by most organisms, to plant structural compounds such as cellulose and lignin, which are depolymerized by a more restricted range of fungi and bacteria. The nutrients theoretically include all elements required for biosynthesis by soil microorganisms and fauna. The modifiers consist of a wide range of secondary compounds, including terpenes, alkaloids, tannins and polyphenol compounds. However, the most important chemical factors influencing litter decomposition, nutrient release and soil organic matter dynamics are C and N, lignin and polyphenols (Swift *et al.*, 1979; Anderson and Flanagan, 1989; Tian *et al.*, 1992a, b). There is a well established relationship between litter decomposition and residue quality (Minderman, 1968; Palm and Sanchez, 1991; Tian *et al.*, 1992a,b).

Plants have higher C-to-N ratios and/or lignin contents under elevated CO_2 (e.g. Coûteaux *et al.*, 1991; Cotrufo *et al.*, 1994). Consequently, respiration rates of deciduous species were significantly decreased for litters grown under elevated CO_2, and reductions in mass loss rates were generally observed in litters derived from the elevated CO_2 treatment (Cotrufo *et al.*, 1994).

To investigate the effect of the decomposer community on the decomposition of litters with various qualities, Coûteaux *et al.* (1991) incubated sweet chestnut leaf litter with a high (75) and a low (40) C-to-N ratio, respectively, after inoculation with different assemblages of fauna for 24 weeks. The dry mass loss at the end of the experiment increased with increasing complexity of the food web for the high C-to-N litter, whereas there was little difference in dry mass loss after inoculation with soil fauna for low C-to-N litter. Cumulative

CO_2 per unit litter showed a similar trend as total dry mass loss (Table 9.3). The results indicate that the soil fauna contributed more to the decomposition of litter in the case of low quality substrate. The effect of soil fauna on litter decomposition was mainly observed at the later stages of incubation, and such effect was apparently earlier when the soil fauna was more complex (Fig. 9.1).

To establish a possible relationship between soil fauna-mediated decomposition and residue quality, Tian *et al.* (1995) used gnotobiotic microecosystems to measure the contribution of earthworms and millipedes to the decomposition of plant residues with contrasting chemical compositions (Table 9.4). Five plant residues: *Dactyladenia barteri*, *Gliricidia sepium* and *Leucaena leucocephala* prunings, maize stover and rice straw (cut to 10 cm long) were placed on the surface in large pots (38 cm in diameter and 27 cm in height) filled with Oxic Paleustalf surface (0–15 cm) soil (defaunated by sun drying and removing fauna by hand). The pots were installed to a depth of 25 cm in the soil. Eighteen mature earthworms (*Eudrilus eugeniae)* and/or three millipedes (Spirostreptidae) per pot were placed in the

Table 9.3. Soil fauna-mediated decomposition for sweet chestnut litters with two C-to-N ratios and various complexes of decomposer biota (from Coûteaux *et al.*, 1991).

Treatments	Cumulative CO_2 (mg per unit) (in brackets: % of initial content released)	Faunal contribution (%)
High C-to-N (75)		
P	196 (10)	–
NP	411 (20)	52
CNP	536 (26)	63
ICNP	643 (31)	69
Low C-to-N (40)		
P	491 (24)	–
NP	470 (23)	−5
CNP	450 (22)	−9
ICNP	483 (24)	−2

P, with inoculation of microflora + protozoa; NP, with inoculation of microflora + protozoa + nematodes; CNP, with inoculation of microflora + protozoa + nematodes + Collembola; and ICNP, with inoculation of microflora + protozoa + nematodes + Collembola + isopods.

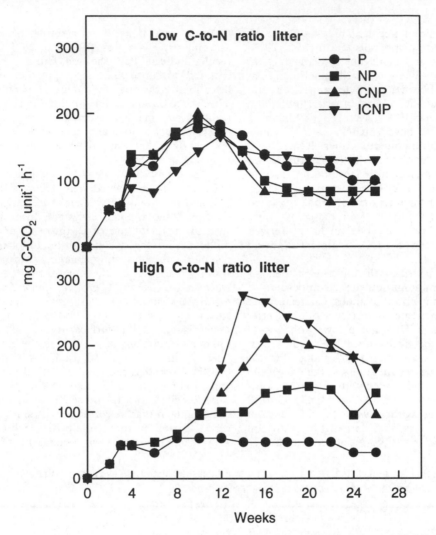

Fig. 9.1. Interaction of litter quality and soil fauna on litter decomposition during incubation. P, with inoculation of microflora + protozoa; NP, with inoculation of microflora + protozoa + nematodes; CNP, with inoculation of microflora + protozoa + nematodes + Collembola; and ICNP, with inoculation of microflora + protozoa + nematodes + Collembola + isopods (from Coûteaux *et al.*, 1991).

Table 9.4. Chemical composition* of prunings of woody species and crop residues (from Tian *et al.*, 1992b)

Plant residues	Lignin (%)*	Polyphenols (%)	C-to-N	SiO_2 (%)	Quality
Dactyladenia	47.6	4.09	28.0	2.71	Low
Gliricidia	11.6	1.62	13.1	0.59	High
Leucaena	13.4	5.02	12.8	0.53	High
Maize stover	6.8	0.56	42.6	2.22	Intermediate
Rice straw	5.2	0.55	42.3	11.35	Intermediate

* Lignin by acid detergent fibre method; polyphenols by Folin–Denis method; C by wet-combustion method; N by micro-Kjeldahl digestion method; and SiO_2 by spectrophotometry.

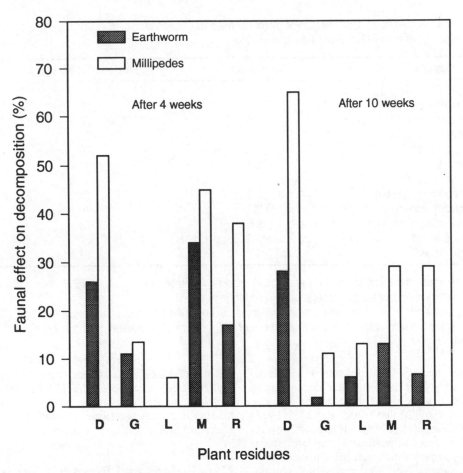

Fig. 9.2. Soil fauna-mediated decomposition of plant residues with various qualities. Faunal effect calculated as (A−B)/(100−B) × 100%, in which A = percentage of remaining plant residues without fauna and B = percentage of plant residues with fauna. D, *Dactyladenia barteri* prunings; G, *Gliricidia sepium* prunings; L, *Leucaena leucocephala* prunings; M, maize (*Zea mays*) stover and R, rice (*Oryza sativa*) straw (adapted from Tian *et al.*, 1995).

fauna treatments at the start of the experiment. At 4 and 10 weeks after addition, pots were removed. The remaining plant materials in the pot were determined to calculate decomposition rate and fauna-mediated decomposition (Tian *et al.*, 1995). There was a large variability in the effect of fauna on the breakdown of plant residues among different plant materials (Fig. 9.2). Both earthworms and millipedes contributed more to the breakdown of *Dactyladenia*, maize stover, and rice straw (lower quality) than to that of *Leucaena* and *Gliricidia* (higher quality).

The highest earthworm effect was observed on maize stover after 4 weeks of incubation and on *Dactyladenia* after 4 and 10 weeks of incubation. The highest millipede effect was observed on *Dactyladenia* after 4 and 10 weeks of exposure.

Soil fauna contributed relatively more to the decomposition of low quality residues, probably because they stimulated microbial activity. There was a small effect of fauna on the breakdown of the high quality residues, probably because these plant residues are easily decomposed by microorganisms without faunal stimulation

Table 9.5. 'Prime effect'* and 'Duration of effect'† of mulching on soil temperature and moisture during the cropping season of 1990 (from Tian *et al.*, 1993b).

Plant residues	Prime effect	Duration of effect (days)
Temperature (°C)		
Dactyladenia prunings	−2.6	90
Gliricidia prunings	−0.6	20
Leucaena prunings	−0.8	22
Maize stover	−1.7	60
Rice straw	−2.2	38
Moisture (% v/v)		
Dactyladenia prunings	5.5	63
Gliricidia prunings	2.1	21
Leucaena prunings	2.5	28
Maize stover	3.6	37
Rice straw	4.9	38

*Prime effect: the mean of soil temperature(°C) reduction or moisture (% v/v) surplus due to plant residue mulching over the first month following mulching.

†Duration of effect: the time period during which soil temperature (°C) reduction or soil moisture (% v/v) surplus due to mulching exceeds the mean of $LSD_{.05}$ between treatments over the experimental period.

of their activity. That the role of soil fauna is relatively greater in the decomposition of materials with high C-to-N ratios was also reported by Malone and Reichle (1973) and Seastedt (1984). High arthropod-mediated decomposition at the later stages of decomposition, when the substrate contains less easily decomposable components, also indicates the importance of soil fauna in the decomposition of low quality residues. Therefore, contrary to suggestions of some workers (e.g. Cotrufo *et al.*, 1994), when soil fauna are taken into consideration, soil organic matter accumulation in the soil due to the low quality input under elevated CO_2 may not be substantial.

Interactions between Residue Quality and the Environment

In the tropics, surface-application of plant residues in agricultural fields decreases soil temperature and increases soil moisture. This 'mulching effect' is controlled by residue quality (Tian *et al.*, 1993b). Compared to high quality residues, low

quality residues have strong and lasting mulching effects (Table 9.5). Hence, both mulching effects of plant residues on the microclimate and residue quality *per se* affect soil fauna-mediated decomposition.

Conclusions and Research Needs

Conversion of natural to agricultural systems in the tropics results in increased soil temperature fluctuation, decreased soil moisture content and reduction in diversity of plant residues. This leads to loss of soil fauna and changed quality of substrates for the decomposer biota, but the relative contribution of the soil fauna to plant residue decomposition increases. Under elevated CO_2, litter decomposition decreases due to the reduction in residue quality. However, the effect of soil fauna on the rate of decomposition of such litter is higher. Hence, manipulation of soil faunal activity can considerably modify carbon flow. It should be noted that the studies reported in this chapter included limited groups of soil fauna, and were mostly conducted under controlled conditions at time scales up to half a year. Longer-term studies with various groups of soil fauna under a greater variety of (field) conditions are recommended in view of the modifying role of the soil fauna on the decomposition of plant residues under constrained conditions.

References

Anderson, J.M. and Flanagan, P.W. (1989) Biological processes regulating organic matter dynamics in tropical soils. In: Coleman, D.C., Oades, J.M. and Uehara, G. (eds) *Dynamics of Soil Organic Matter in Tropical Ecosystems*. University of Hawaii Press, pp. 97–123.

Bloem, J., de Ruiter, P.C., Koopman, G.J., Lebbink, G. and Brussaard, L. (1992) Microbial numbers and activity in dried and rewetted arable soil under integrated and conventional management. *Soil Biology and Biochemistry* 24, 655–665.

Bouwman, L.A., Bloem, J., van den Boogert, P.H.J.F., Bremer, F., Hoenderboom, G.H.J. and de Ruiter, P.C. (1994) Short-term and long-term effects of bacterivorous nematodes and nematophagous fungi on carbon and nitrogen mineralization in microcosms. *Biology and Fertility of Soils* 17, 249–256.

Brussaard, L. and Juma, N.G. (1996) Organisms and humus in soils. In: Piccolo, A. (ed.) *Humic Substances in Terrestrial Ecosystems*. Elsevier, Amsterdam, pp. 329–359.

Brussaard, L., Noordhuis, R., Geurs, M. and Bouwman, L.A. (1995) Nitrogen mineralization in soil microcosms with or without bacterivorous nematodes and nematophagous mites. *Acta Zoologica Fennica* 196, 1–7.

Cotrufo, M.F., Ineson, P. and Rowland, A.P. (1994) Decomposition of tree leaf litters grown under elevated CO_2: Effect of litter quality. *Plant and Soil* 163, 121–130.

Coûteaux, M.M., Mousseau, M., Celerier, M.L. and Bottner, P. (1991) Increased atmospheric CO_2 and litter quality: decomposition of sweet chestnut leaf litter with animal food webs of different complexities. *Oikos* 61, 54–64.

Critchley, B.R., Cook, A.G., Critchley, U., Perfect, T.J., Russell-Smith, A. and Yeadon, R. (1979) Effects of bush clearing and soil cultivation on the invertebrate fauna of a forest soil in the humid tropics. *Pedobiologia* 19, 425–438.

Kevan, D.K.M. (1962) *Soil Animals*. H.F. & G. Witherby Ltd., London, 237 pp.

Kuikman, P.J. and van Veen, J.A. (1989) The impact of protozoa on the availability of bacterial nitrogen to plants. *Biology and Fertility of Soils* 8, 13–18.

Lal, R. and Cummings, D.J. (1979) Clearing a tropical forest, I. Effect on soil and microclimate. *Field Crop Research* 2, 91–107.

Lavelle, P. (1988) Earthworm activities and the soil system. *Biology and Fertility of Soils* 6, 237–251.

Lavelle, P., Blanchart, E., Martin, A., Spain, A.V., Toutain, F., Barois, I. and Schaefer, R. (1993) A hierarchical model for decomposition in terrestrial ecosystems: application to soils of the humid tropics. *Biotropica* 25, 130–150.

Lee, K.E. and Foster, R.C. (1991) Soil fauna and soil structure. *Australian Journal of Soil Research* 29, 745–775.

Lee, K.E. and Wood, T.G. (1971) *Termites and Soils*. Academic Press, London and New York, 251 pp.

Malone, C.R. and Reichle, D.E. (1973) Chemical manipulation of soil biota in fescue meadow. *Soil Biology and Biochemistry* 5, 629–639.

Melillo, J.M., Aber, J.D. and Muratore, J.F. (1982) Nitrogen and lignin control of hardwood leaf litter decomposition dynamics. *Ecology* 63, 621–626.

Minderman, G. (1968) Addition, decomposition and accumulation of organic matter in forests. *Journal of Ecology* 56, 355–362.

Mulongoy, K. and Bedoret, A. (1989) Properties of wormcasts and surface soils under various plant covers in the humid tropics. *Soil Biology and Biochemistry* 21, 197–203.

Palm, C.A. and Sanchez, P.A. (1990) Decomposition and nutrient release patterns of the leaves of three tropical legumes. *Biotropica* 22, 330–338.

Palm, C.A. and Sanchez, P.A. (1991) Nitrogen release from the leaves of some tropical legumes as affected by their lignin and polyphenolic contents. *Soil Biology and Biochemistry* 23, 83–88.

Seastedt, T.R. (1984) The role of microarthropods in decomposition and mineralization processes. *Annual Review of Entomology* 29, 25–46.

Santos, P. and Whitford, W.G. (1981) The effects of microarthropods on litter decomposition in a Chihuahuan desert ecosystem. *Ecology* 62, 654–663.

Singh, J.S. and Gupta, S.R. (1977) Plant decomposition and soil respiration in terrestrial ecosystems. *Botanical Reviews* 43, 449–528.

Syers, J.K., Sharpley A.N. and Keeney D.R. (1979) Cycling of nitrogen by surface-casting earthworms in a pasture ecosystem. *Soil Biology and Biochemistry* 11, 181–185.

Swift, M.J., Heal, O.W. and Anderson, J.M. (1979) *Decomposition in Terrestrial Ecosystems*, Studies in Ecology 5, Blackwell, Oxford, 372 pp.

Tian, G., Kang, B.T. and Brussaard, L. (1992a) Effect of chemical compositions on N, Ca and Mg release during incubation of leaves from selected agroforestry and fallow plant species. *Biogeochemistry* 16, 103–119.

Tian, G., Kang, B.T. and Brussaard, L. (1992b) Biological effects of plant residues with contrasting chemical compositions under humid tropical conditions: Decomposition and nutrient release. *Soil Biology and Biochemistry* 24, 1051–1060.

Tian, G., Brussaard, L. and Kang, B.T. (1993a) Biological effects of plant residues with contrasting chemical compositions under humid tropical conditions: effects on soil fauna. *Soil Biology and Biochemistry* 25, 731–737.

Tian, G., Kang, B.T. and Brussaard, L. (1993b) Mulching effect of plant residues with chemically contrasting compositions on maize growth and nutrients accumulation. *Plant and Soil* 153, 179–187.

Tian, G., Brussaard, L. and Kang, B.T. (1995) Breakdown of plant residues with contrasting chemical compositions: effect of earthworms and millipedes. *Soil Biology and Biochemistry* 27, 277–280.

Verhoef, H.A. and Brussaard, L. (1990) Decomposition and nitrogen mineralization in natural and agro-ecosystems: the contribution of soil animals. *Biogeochemistry* 11, 175–211.

Vreeken-Buijs, M.J. and Brussaard, L. (1996) Soil mesofauna dynamics, wheat residue decomposition and nitrogen mineralization in buried litter bags. *Biology and Fertility of Soils* (in press).

Wood, T.G. (1988) Termites and the soil environment. *Biology and Fertility of Soils* 6, 228–236.

Zwart, K.B. and Brussaard, L. (1991) Soil fauna and cereal crops. In: Firbank, L.G., Carter, N., Darbyshire, J.F. and Potts, G.R. (eds) *The Ecology of Temperate Cereal Fields*. Blackwell, Oxford, pp. 139–168.

10 Relationships between Litter Fauna and Chemical Changes of Litter during Decomposition under Different Moisture Conditions

C. Wachendorf[1], U. Irmler[1] and H.-P. Blume[2]

[1]Ecosystem Research Center, University of Kiel, Germany;
[2]Institute of Plant Nutrition and Soil Science, University of Kiel, Germany

Introduction

Litter decomposition plays a major role in the structure and dynamics of ecosystems, the transfer of elements and energy, and in control mechanisms and feedback processes. Microorganisms are directly responsible for most of the organic matter breakdown, but a diverse assemblage of soil and litter fauna greatly influences the decomposer flora as a result of their feeding activities. The influence of the fauna is due to the comminution of the litter (Anderson and Ineson, 1984) resulting in an increase in the surface area of the substrate (Seastedt, 1984). Furthermore, changes in the microorganism community occur after passage through the gut of the soil fauna (Ullrich et al., 1991). The influence of the fauna on the decomposition is usually estimated using litter bags with different mesh sizes. Investigations in northern Germany have shown that the contribution of fauna to the decomposition process ranges from 0% at a site with a raw humus up to 45% at a site with a mull humus form (Irmler, 1995).

The rate of litter decomposition depends on litter quality (McClaughterty et al., 1985; Meentemeyer and Berg, 1986; Blair, 1988), temperature and moisture regime (Moore, 1986; Donnelly et al., 1990; Berg et al., 1993) as well as on the nutrient status of the soil (Verhoeven and Toth, 1995). Many investigations prove the influence of soil fauna on the litter chemistry under laboratory conditions (Davidson, 1976; Walsch and Bolger, 1990; Schultz, 1991; Gunn and Cherrett, 1993). Little is known about the changes in litter quality in relation to the activity of soil organisms in the field. The results presented here show chemical changes in a mixture of deciduous litter during decomposition in connection with the successive activity of different litter dwelling organisms. Berg et al. (1993) and Irmler (1995) showed that with an increasing availability of water and nutrients the activity of fauna becomes increasingly important in the course of decomposition. In the work described here we estimated the contribution of fauna on the decomposition process at two sites in an alder forest, which mainly differ in their moisture regime and nutrient status.

Methodological Approach

Two field sites were chosen in an alder forest (Alnus glutinosa) at a lake margin in northern Germany, the 'Bornhöved Lakes Region', located about 30 km south of Kiel (59°97'N, 35°81'E) (Table 10.1). The climate in the research area is moderate oceanic with mean annual temperature of 8.1°C, and mean annual precipitation of 700 mm. One site is located along the shore of the lake (wet site), while the other is located at a distance of 35 m from the lake on the bottom of a slope (dry site) (Table 10.1). At the dry site the ground water level is 35 cm beneath land surface. The wet site is characterized by an annual lake water level of only 6 cm beneath land surface, and the site is periodically inundated by eutrophic lake water. Due to the influence of the lake water the wet site is characterized by a pH of 5.3. The low pH of the peat at the dry site is probably a consequence of the decomposition of the peat after the lowering of the lake water table in 1935 and 1936. In the understory, the fern, *Dryopteris dilatata,* is the dominant plant at the dry site, while the reed, *Phragmites australis,* is dominant at the wet site. The combined above-ground and below-ground litter deposit, 5600 and 7600 kg C ha^{-1} year^{-1} on the soil in the dry and wet sites respectively, with alder litter being dominant at both sites.

To compare the litter decomposition at both sites the falling litter was collected during the autumn of 1991 in the alder forest. The litter consisted on average of 61% alder, 13% oak, 8% beech, 8% hazel and some other leaf litter and twigs. Litter bags with 0.02 mm and 5 mm mesh size were each filled with 10 g dried (30°C) material. At the end of December 1991 the bags with the mixed litter were placed on the litter surface at each site. Bags were harvested every 1 or 2 months. In autumn 1992 the bags were covered with new litter.

The difference between 5 mm and 0.02 mm litter bags was used to estimate the contribution of the fauna to the litter breakdown. At each sampling macrofauna and mesofauna were extracted from four replicated 5 mm bags by hand sorting and heat extraction in a McFayden apparatus. Soil fauna from both sites was also determined by sampling 0.1 m^2 of soil and humus layer at depths up to 5 cm every month. The biomass of the macrofauna was determined directly by weight. The biomass of the mesofauna was estimated by group specific factors, which were once determined for each mesofauna group (Irmler, 1995). The respiration of the fauna was calculated by the formula of Ryszkowski (1975); $y = 0.357 \times x^{0.813}$ where y = rate of oxygen release (μl O$_2$ h^{-1}) and x = fresh weight of organism (mg). C loss by respiration of the fauna was estimated using a Q$_{10}$ of 2.5 and an equivalent of 1 ml O$_2$ to 0.577 mg C. This respiration activity was assumed for all faunal groups investigated. For estimations of microbial respiration in the litter bags during the course of decomposition, data from

Table 10.1. General characterization of two sites under alder at Bornhöved lake margin in northern Germany.

	Dry site	Wet site
Ground water level below land surface (cm)		
Mean value	35	6
Max and min	0−65	+3−30
Water content in 0−5 cm (% dry mass)		
Mean value 1992	54	131
Mean value 1993	84	301
Vegetation		
Overstory	*Alnus glutinosa*	*Alnus glutinosa*
Understory	*Dryopteris dilatata*	*Phragmites australis*
Litter input (kg C ha^{-1}year^{-1})	5600	7600
C-to-N ratio in upper peat layer	22	18
pH [CaCl$_2$] in upper peat layer	3.1	5.3
CEC eff. (mmol$_c$ kg^{-1}) upper layer	550	820
Soil unit	Dystri-fibric histosol	Eutri-fibric histosol

CEC eff., effective cation exchange capacity.

laboratory experiments (basal respiration at a temperature of 22°C and a water content of 750 mg g^{-1} dry litter) (Dilly et al., 1996) were calculated with a Q$_{10}$ of 3.6.

Between three and nine 5 mm litter bags were collected for chemical analysis, dried, bulked and milled. Chemical analysis was performed following Schlichting et al. (1995) and Beyer et al. (1993). Total carbon was determined after combustion using a Coulomat 702 (Ströhlein Instruments). Total nitrogen and phosphorus were determined after extraction with the micro-Kjeldahl procedure and colorimetric estimation in a flow injection analyser. Fat and wax were extracted with ethanol:benzene (1:1.9) at 80°C. Polysaccharides were sequentially extracted and hydrolysed. Sugar and starch were extracted with 0.025 M H$_2$SO$_4$, hemicellulose with 0.63 M HCl, and cellulose with 12 M HCl and hydrolysed with 0.38 M H$_2$SO$_4$. The extracted sugar was determined as reducing sugar. Lignin was estimated via determination of methoxylgroups. A mean content of methoxylgroups of 21% in the lignin of decidous litter was presumed. Protein was extracted in a mixture of hydrochloric acid and formic acid. Then α-amino acids were determined according to Stevenson and Cheng (1970) and multiplied by 6.25. CPMAS ^{13}C NMR spectra were obtained with a Bruker MSL 100 spectrometer at a frequency of 25.2 MHz. The subdivision of the spectra follows the commonly used scheme after Fründ et al. (1994).

Mass Loss in Litter Bags

Mass loss of different litters at the same site vary widely and are often explained by the chemical composition of the litter (Blair, 1988). Differences between the mass loss of the same litter under identical climatic conditions but at different sites are normally minor (McClaughtery et al., 1985) with the exception of the alder forest investigated in this study. After one year 40% dry matter of the deciduous litter was lost at the dry site, while 80% of the litter was lost at the wet site. The different loss

rates in the alder forest can be explained by different abiotic conditions such as water availability and nutrient status (Table 10.1). As the wet site is periodically inundated by eutrophic lake water, it is characterized by a higher pH and higher water availability, whereas at the dry site a lowering of the lake water table had resulted in the mineralization of the peat with a decrease in the pH. The different abiotic conditions may have led to a relative difference in the importance of processes causing mass loss, i.e. decomposition by microorganisms and fauna, dislocation of small particles through the mesh, and leaching of soluble organic matter, which are discussed below.

Influence of Fauna on Decomposition

The influence of fauna on the breakdown process can be analysed by using litter bags with different mesh sizes. In bags with mesh size of 0.02 mm both the mesofauna and macrofauna are excluded, whereas with mesh size of 5 mm all soil organisms can still be involved in the decomposition process. Small differences between the 5 and 0.02 mm mesh bags occurred at the dry site, whereas large differences were determined at the wet site (Fig. 10.1). At the end of the experiment, about 10% C-loss and 40% C-loss were attributed to the activity of the fauna at the dry and wet sites, respectively. The higher C-loss of litter at the wet site was thus mainly due to greater influence of the fauna on the breakdown process. Nevertheless, examination of the fauna in the litter bags proved that the faunal biomass in the litter bags was greater at the dry site, whereas the biomass of fauna sampled directly from the soil, mainly earthworms and diptera larvae, was greater at the wet site (Fig. 10.2). Fauna-induced C-loss at the wet site was primarily due to a higher percentage of soil organisms sampled directly from the soil, feeding on the litter in litter bags. The high consumption of litter by soil fauna at the wet site means that little substrate

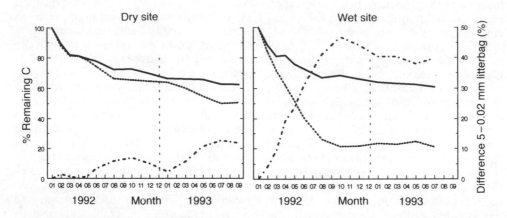

Fig.10.1. Amount of carbon remaining in litter bags (········ 5 mm and —— 0.02 mm) in two sites under alder, and the difference between 5 mm and 0.02 mm litter bags (-·--·-).

remained for the litter fauna. During dry periods, such as the summer of 1992, the activity of the litter fauna was probably reduced at both sites (Fig. 10.3), whereas the soil fauna of the wet site was not affected, because of the high water table at the wet site.

The loss of litter was divided into gaseous losses through respiration by fauna and microorganisms, by displacement of litter through the fauna, and by leaching (Table 10.2). The litter fauna contributed to only 3–7% of the decomposition related to the total C-loss after one year at the wet and dry sites respectively. The contribution of the soil fauna to C-loss was higher than the contribution of the micro-organisms at the wet site. At the dry site, the percentage of C-loss in 0.02 mm litter bags, i.e. losses due to microbial and microfauna respiration as well as leaching, were 88%. From the microbial biomass (Dilly *et al.*, 1996) the calculated respiration of microorganisms in the field was estimated to be 70% and 18% of the total C-loss at the dry site and wet site respectively. We conclude that leaching contributed to an 18% loss at the dry site and a 30% loss at the wet site. C-loss at the wet site was mainly due to the soil fauna, followed by C-loss due to leaching. C-loss at the dry site was predominantly due to microbial respiration.

Chemical Changes of Litter

The litter was characterized by a C-to-N ratio of 18 and a C-to-P ratio of 420 (Table 10.3). The mixed litter was similar in comparison to pure alder litter, characterized by a high content of protein, and fat and wax. After one year of decomposition the content of fat and wax, sugar and starch decreased at both sites. At the wet site the high mass- and C-loss was accompanied by an increase in % lignin and a decrease in protein content. The different C-losses at both sites indicated that the chemical composition of the remaining litter may have changed. The concentration of carbon in the 5 mm bags did not change significantly during decomposition (Fig. 10.3). Nitrogen concentration increased during the first 2 months of decomposition at both sites, but subsequently decreased at the wet site. The increase of N within the first month of incubation at both sites probably can be attributed to the relative enrichment of N by microorganisms, while the decrease of N at the wet site can be attributed to selective feeding of the fauna or leaching. A relative enrichment of P at the beginning of the experiment, as was observed with N, did not occur. Lignin concentration showed little change in the course of decomposition at the dry site, while it

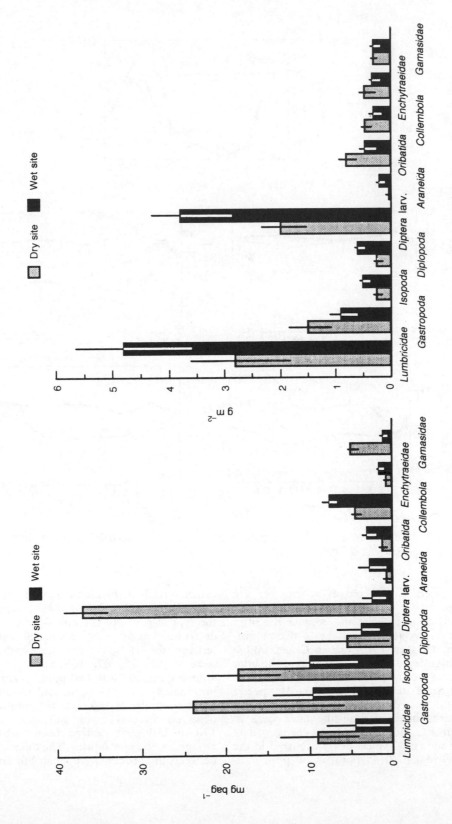

Fig. 10.2. Faunal biomass in 5 mm litter bags and sampled directly from the soil of two sites under alder (mean values of the years 1989–1992)

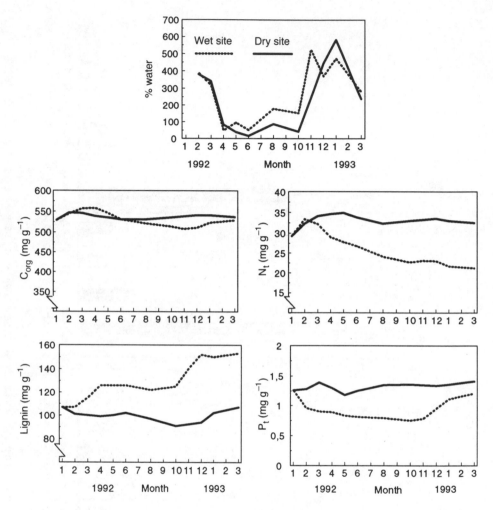

Fig.10.3. Water content, and content of organic carbon (C_{org}), total nitrogen (N_t), total phosphorus (P_t), and lignin of litter in 5 mm litter bags for two sites in an alder forest.

increased in two phases at the wet site (Fig. 10.3). Lignin is probably relatively enriched at the wet site because the litter was in an advanced stage of decomposition, as seen by a higher C-loss and a greater degree of fragmentation of the litter during the course of decomposition.

Solid-state CPMAS ^{13}C NMR spectra of the mixed litter after 4, 5, 8 and 12 months incubation in the litter bags at both sites showed only slight changes (data not shown). Typical peaks occurred in the alkyl region (0–45 ppm) at 30 ppm indi-

cating longchain paraffinic-carbon. In the O-alkyl chemical shift range (45–110 ppm) the typical peak for methoxyl-carbon was determined at 56 ppm, and peaks characterizing the carbon from carbohydrates were at 62, 72, 89, 105 ppm. In the aromatic region (110–160 ppm) peaks for protonated C at 119 ppm, and O-substituted aromatics at 130 and 150 ppm were observed. Carbon in the shift range from 170 to 180 ppm derived from carboxyl groups, esters, and amides. The differences between the spectra were small but could

Table 10.2. Absolute and relative C-loss in litter bags after one year in two sites under alder.

	Wet site		Dry site	
	%	% of total loss	%	% of total loss
Total C-loss[1]*	80.0	100	40.0	100
Loss by:				
Litter fauna[†]	2.0	2.5	2.7	6.8
Soil fauna[‡]	40.0	50.0	2.3	5.8
Microorganisms and leaching[§]	38.0	47.5	35.0	87.5
Microorganisms[¶]	14.4	18.0	28.0	70.0
Leaching[‖]	23.6	29.5	7.0	17.5

* Loss rate in litter bags with 5 mm mesh size.
[†] Estimated by the calculated respiration of the macro- and mesofauna in the litter bags.
[‡] Difference between 0.02 and 5 mm litter bags minus respiration of the litter fauna[†]
[§] Loss rate in 0.02 mm litter bag.
[¶] Estimated from measurements of microbial biomass (Dilly *et al.* 1996).
[‖] 4–5 Calculated by subtraction.

be observed when relative changes in the chemical shift ranges were shown as a function of mass loss (Fig. 10.4). At the wet site chemical changes were observed after 60% of the mass disappeared, with aromatic-C increasing and alkyl-C slightly decreasing. At the dry site, where the decomposition was slower, changes already occurred after only 20% of the litter disappeared. These changes were similar to those at the wet site, with aromatic-C increasing. A stimulatory effect on the microflora by the soil fauna was probably minor at the beginning of the decomposition at the wet site, and losses were probably mainly due to the displacement of litter by the soil fauna. We postulate that chemical changes, such as the relative enrichment of aromatic-C, were due to

microbial degradation, and can be detected with ^{13}C NMR-spectroscopy, whereas the relative enrichment of intact lignin by selective feeding was more accurately detected by chemical analysis. These interpretations of NMR-results are in accordance with the estimated contribution of microorganisms, fauna and leaching to the C-loss (Table 10.2).

Conclusions

Under specific conditions C-loss induced by soil fauna may exceed 50% of the total loss, whereas under more dry and nutrient poor conditions fauna contributed only 13% of the total C-loss of deciduous litter in an alder forest. Chemical changes of

Table. 10.3. Average chemical composition (litter compounds in mg g^{-1} dry wt.) of mixed litter before and after one year of decomposition in 5 mm litterbags at two sites in an alder forest.

Litter	C-to-N	C-to-P	Fat and wax	Sugar and starch	Hemicellulose	Cellulose	Lignin	Protein
Initial mixed (both sites)	18	420	93	19	140	129	107	117
Litter after one year								
Wet site	20	480	47	4	127	126	166	97
Dry site	16	414	48	6	142	92	93	112

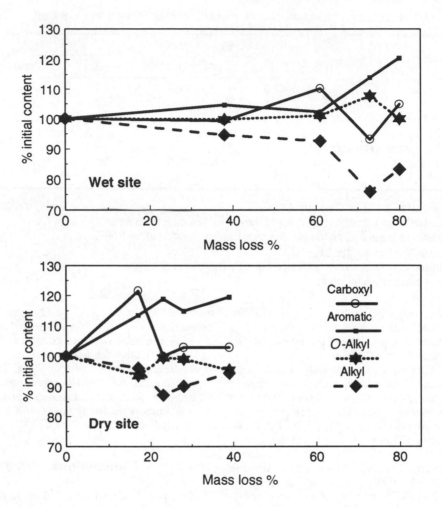

Fig.10.4. Relative percentage of the initial content of chemical shift areas of solid state CPMAS ^{13}C NMR spectra of litter in relation to the mass loss at two sites.

litter are due to a relative enrichment of resistant compounds, due to losses of soluble organic matter, and less resistant compounds, or due to transformation by microorganisms. Chemical changes were probably observed at the wet site because of resistant compounds like lignin, which were relatively enriched with increasing mass loss. With solid state CPMAS ^{13}C NMR-spectroscopy changes for the aromatic and alkyl-fraction were observed. Relative changes occurred at the beginning of decomposition at the dry site, where the proportion of total degradation attributed to microorganisms was higher. Whereas changes at the wet site were a result of the faster breakdown of the litter.

At the wet site there were higher decomposition rates and a greater input of litter. For estimation of the turnover of the litter, the displaced litter in deeper horizons as well as the decomposition rate of other plant material such as roots and the litter from reeds need to be considered.

Acknowledgements

We are grateful to Dr Heike Knicker and Prof. H.-D. Lüdemann for NMR analyses.

These studies were supported by the German Ministry of Science and Technology (BMBF).

References

Anderson, J.M. and Ineson, P. (1984) Interaction between microorganisms and soil invertebrates in nutrient flux pathways of forest ecosystems. In: Anderson J.M., Rayner A.D.M., Walton, D.W.H. (eds) *Invertebrate-Microbial Interactions.* Cambridge University Press, Cambridge, pp. 59–88.

Berg, B., Berg, M.P., Bottner, P., Box, E., Breymeyer, A., Calvo de Anta, R., Couteaux, M., Escudero, A., Gallardo, A., Kratz, W., Madeira, M., Mälkönen, E., McClaugherty, C., Meentemeyer, V., Munoz, F., Piussi, P., Remacle, J. and Vrizo de Santo, A. (1993) Litter mass loss rates in pine forests of Europe and Eastern United States: some relationships with climate and litter quality. *Biogeochemistry* 20, 127–159.

Beyer, L., Wachendorf, C. and Koebbemann, C. (1993) A simple wet chemical extraction procedure to characterize soil organic matter. I. Application and recovery rate. *Communications in Soil Science and Plant Analysis* 24, 1645–1663.

Blair, J.M. (1988) Nitrogen, sulfur and phosphorus dynamics in decomposing deciduous leaf litter in the Southern Appalachians. *Soil Biology and Biochemistry* 20, 5, 693–701.

Davidson, D.H., (1976) Assimilation efficiencies of slugs on different food materials. *Oecologia* 26, 267–273.

Dilly, O., Wachendorf, C., Irmler, U., Blume, H.P. and Munch, C. (1996) Changes of abiotic and biotic parameters in the course of decomposition of leaf litter in an black alder forest of a moranic landscape of Northern Germany. *EcosSys* (in press)

Donnelly, P., Entry, J.E., Crawford, D.L. and Cromack, K. (1990) Cellulose and lignin degradation in forest soils: Response to moisture, temperature, and acidity. *Microbial Ecology* 20, 289–295.

Fründ, R., Guggenberger, G., Haider, K., Kögel-Knabner, I., Knicker, H., Lüdemann, H.-D., Luster, J., Zech, W. and Spiteller, M. (1994) Recent advances in the spectroscopic characterization of soil humic substances and their ecological relevance. *Journal of Plant Nutrition and Soil Science* 157, 175–186.

Gunn, A. and Cherrett, J.M. (1993) The exploitation of food resources by soil meso- and macro invertebrates. *Pedobiologia* 37, 303–320.

Irmler, U. (1995) Die Stellung der Bodenfauna im Stoffhaushalt schleswig-holsteinischer Wälder. *Faunistische Ökologische Mitteilungen Supplement* 18, 184 pp.

Meentemeyer, V. and Berg, B. (1986) Regional variation in rate of mass loss of *Pinus sylvestris.* Needle litter in Swedish forests as influenced by climate and litter quality. *Scandinavian Journal of Forest Research* 1, 167–180.

McClaugherty, C.A., Pastor, J., Aber, J.D. and Melillo, J.M. (1985) Forest litter in relation to soil nitrogen dynamics and litter quality. *Ecology* 66, 266–275.

Moore, A.M. (1986) Temperature and moisture dependence of decomposition rates of hardwood and coniferous leaf litter. *Soil Biology and Biochemistry* 18, 427–435.

Ryszkowski, L. (1975) Energy and matter economy of ecosystems. In: Dobben, W.H.v., Lowe-Connel, R.H. (eds) *Unifying Concepts in Ecology.* Junk, The Hague, pp. 109–126.

Seastedt, T.R. (1984) The role of microarthropods in decomposition and mineralization processes. *Annual Review of Entomology* 29, 25–46.

Schlichting, E., Blume, H.-P., Stahr, K. (1995) *Bodenkundliches Praktikum.* Blackwell Wissenschafts-Verlag, Berlin, 295 pp.

Schultz, P.A. (1991) Grazing preferences ot two collembolan species, *Folsomina candida* and *Proisotoma minuta,* for ectomycorrhizal fungi. *Pedobiologia* 35, 313–325.

Stevenson, F.J. and Cheng, C.N. (1970) Amino acids in sediments: recovery by acid hydrolysis and quantitative estimation by colorimetric procedure. *Geochimica et Cosmochimica Acta* 31, 77 −88.

Ullrich, B., Storch, V. and Schairer, H.U. (1991) Bacteria on the food, in the intestine and on the faeces of the woodlouse *Oniscus asellus* (Crustacea, Isopoda). Species composition and nutritive value. *Pedobiologia 35, 41–52.*

Verhoeven, J.T.A. and Toth, E. (1995) Decomposition of *Carex* and *Sphagnum* litter in fens: effect of litter quality and inhibition by living tissue homogenates. *Soil Biology and Biochemistry* 27, 271–275.

Walsh, M. and Bolger, T. (1990) Effects of diet on the growth and reproduction of some Collembola in laboratory cultures. *Pedobiologia* 34, 161–172.

11 Metabiotic Interactions in Plant Litter Systems

J.S. Waid*

*School of Microbiology, La Trobe University,
Bundoora, Victoria 3083, Australia*

Introduction

Plant litter ecosystems provide habitats in which many organisms interact with one another in various ways. The outcomes of these interactions may determine the abundance and activities of the species that compose such ecosystems. Knowledge of how organisms in the litter and soil interact is fundamental to understanding how the ecosystems in which the organisms dwell have evolved and operate (Price, 1988).

In a thought-provoking article, Jones *et al.* (1994) pointed out that ecology textbooks summarize the most important interactions occurring in ecosystems as: interspecific and intraspecific competition for abiotic and biotic resources, predation, parasitism and mutualism. They rightly claimed that this list of key processes lacks an appreciation of how organisms create, modify and maintain habitats as well as the lack of terms to describe these processes and the organisms responsible for them. Jones *et al.* showed that these habitat-modifying activities are extremely important and widespread and the biological literature is replete with examples of habitat modification by organisms. Even so, population and community ecologists have still to fully define or systematically identify and study the role of organisms in the creation and maintenance of habitats.

My purpose is to identify, define and discuss metabiosis, which is a type of interaction between organisms that has received little systematic attention from biologists, and to suggest its possible role in the functioning of plant litter systems.

Definitions and Usage of the Term Metabiosis

The term *metabiosis* was proposed by the French chemist C. Garrè (1887). Metabiosis is 'a form of ecological dependence in which one organism must modify the environment before the second is able to live in it' (Brown, 1993). This term has been used by food and medical microbiologists but is rarely used, if at all, by other biologists and was (not surprisingly) overlooked by Jones *et al.* (1994).

The word metabiosis has usually been applied to describe one-way interrelationships between a pair of microorganisms or two dissimilar groups of microorganisms. Garrè (1887) found some pathogenic microorganisms did not grow in culture unless the chemical composition of the medium had been modified by the growth of another (donor) microorganism.

Meyer (1966) gave examples of metabiosis relevant to ecological situations. He mentioned microorganisms in soils depend to a large extent on the litter produced by plants; in turn the plants depend partly upon the mineralization of

---

*Present address: PO Box 760, Buderim, Queensland 4556, Australia.

the nutrients in the plant litter by non-symbiotic microorganisms and; NH_4-oxidizers depend on NH_4^+ produced by other organisms and the NO_2-oxidizers depend on the NO_2^- formed by the NH_4-oxidizers.

Metabiosis has been defined in *The New Shorter Oxford English Dictionary* (Brown, 1993) as 'a form of ecological dependence in which one organism must modify the environment before the second is able to live in it' and the adjective metabiotic as 'of, pertaining to, or of the nature of metabiosis'.

I propose to broaden the above definition of metabiosis (additions in italic) as follows: a form of ecological dependence in which one organism *or a functional group of organisms* must modify the environment before the second *organism or a functional group of organisms* is able to live *or thrive* in it. I will call the organisms responsible for metabiotic changes to the environment *metabionts*. Those metabionts that have a global influence through modifying the biosphere I call *panmetabionts*, e.g. oxygenic photosynthesizers, but a full discussion about panmetabionts is outside the scope of this chapter. Table 11.1 summarizes the characteristic features of metabiosis.

Characteristics of Metabiotic Interactions in Plant Litter Ecosystems

In this section the characteristics and examples of metabiotic interactions in plant litter ecosystems are described and discussed. Table 11.2 lists physical and chemical components of plant litter affected by metabionts.

Significance of relationship to the metabiont

Metabiosis does not fall neatly into the category of a commensal type of association where one species is neither harmed or benefited and the other benefits (Brown, 1993). A metabiont might be either unaffected, benefited or adversely affected by its interaction with dependent organisms.

The interaction of a metabiont with other organisms might maintain and prolong the existence of the microhabitat, ecosystem or environment in which the metabiotic population or community exists. For example, litter communities develop under forest trees (metabionts) and the continued existence of the litter ecosystems depends upon inputs (leaves, wood, roots) from the trees. In turn, the continued existence of the forest depends upon the decomposition of litter and the transformation and mineralization of nutrients by the decomposer community acting here as metabionts. Yet, a growth factor released from a plant root might have an instantaneous positive metabiotic effect on the growth of a rhizosphere organism without affecting the plant donor.

An example of long-term negative effects of a metabiotic interaction on a metabiont is provided by plant succession on N_2-deficient mineral soils, such as maritime sand dunes. There N_2-fixing plants, e.g. *Casuarina*, are early plant colonists and their litter and the underlying soil accumulate nutrients and organic matter. The litter-soil medium formed as a result of the metabiotic effects of the N_2-fixing plants will in time be colonized by other plants that compete with and replace the N_2-fixing metabionts. This is also an example of facilitation, an important factor in plant ecology where one plant species in a successional pathway improves conditions for the survival and growth of other plant species. Here the conceptual similarity to metabiosis suggests that facilitation is but a sub-set of metabiosis.

Positive and negative outcomes of metabiosis

Only the dependent organisms, by definition, necessarily benefit from the modification of the environment by metabionts. This contrasts with mutualism where there is 'An interaction between species that is beneficial to both' (Boucher *et al.*, 1982); 'A condition of (obligate or facultative) symbiosis in which two organ-

Table 11.1. Characteristics of metabiosis.

1. Significance of relationship to the metabiont
   - A metabiont can be unaffected, positively affected or negatively affected by its interaction with dependent organisms
   - The interaction may maintain and prolong the existence of the microhabitat, ecosystem or environment in which the metabiont lives
2. Positive and negative effects of metabiosis
   - An environment modified by a metabiont may favour some of the organisms or functional groups originally present but the remainder may be disadvantaged
3. Specificity of metabiotic relationships
   - Most metabiotic interactions are non-specific
4. Spatial relationships of metabionts to dependent organisms
   - Varies from close contact to remoteness
5. Timing, duration and scale of influence of metabiotic effects
   - Dependent organism can coexist with, survive or even develop after the death of the metabiont
   - The duration of a metabiotic effect can range from a momentary influence to one of 10^6 years or more
   - The dimensional scale of influence of metabionts can vary from μm to the biosphere (panmetabiotic)

Table 11.2. Features of the plant litter environment affected by metabionts.

Physical environment
  Provision of habitats, surfaces or refuges for growth
  Mixing organic detritus with underlying mineral soil
  Comminution of detritus
  Temperature regime
  Aeration of litter mass
  Drainage
Chemical environment
  Organic nutrients, quality and supply
  Gases, quality and supply
  Inorganic nutrients, quality and supply
  Growth factors etc.
  Bioactive substances, formation or destruction
  pH
  Water content and activity
  Oxidation-reduction potential

isms contribute mutually to the well-being of each other' (Brown, 1993).

Various groups and strains of microorganisms are the principal agents involved in the detoxification and mineralization of naturally-formed toxic inorganic (sulphide, nitrite, ammonia) or organic compounds (hydrocarbons, phenolics, terpenes) in soil and litter. Such microorganisms perform metabiotic activities that permit organisms susceptible to the toxins to survive or grow in the affected soils. The macroorganisms and microorganisms which benefit include plants and the biota of litter and soil. Likewise, biodegradable toxic man-made organic chemicals, e.g. 2,4-D, become detoxified by microbial metabionts enabling susceptible organisms to be re-established in treated soil.

Many macroorganisms modify the physical nature of soil habitats to favour the growth of other organisms and thus function as metabionts. For example, the root regions of plants support the spread and growth of rhizoplane and rhizosphere organisms. Earthworms form burrows and termites form voids in soil and these various structures provide well-ventilated habitats and channels for the movement of soil microfauna and aerobic microflora. The burrowing, mixing and casting by earthworms can have a considerable influence upon the mineral and organic composition of soils, nutrient cycling, soil hydrology and drainage. Many termite species relocate detritus in the litter-soil profile and some of these transfer organic debris to habitats or nests where the environments favour the growth of lignocellulose-degrading microorganisms (Anderson, 1988; Woomer et al., 1994). The processes and the organisms responsible for the *physical* modification, maintenance and creation of habitats were called ecosystem engineering and ecosystem engineers, respectively, by Jones et al. (1994). Lawton (1994) defined ecosystem engineers 'as organisms that directly or indirectly modulate the availability of resources (other than themselves) to other species, by causing physical state changes in biotic or abiotic materials. In so doing they modify, maintain and/or create habitats'. The activities of earthworms and termites, mentioned above, are examples of *allogenic engineering* (Jones et al., 1994), where the environment is altered by transforming living or non-living materials from one physical state to another, via mechanical or other means. Examples such as these of ecological engineering, where organisms cause physical state changes in biotic or abiotic materials that favour the growth of other organisms, can be considered as a sub-set of metabiosis. In Chapter 8 of this volume, Wardle and Lavelle discuss the effects of earthworms and termites as ecosystem engineers on the resource quality and digestibility of plant litter for other decomposer organisms, in particular microorganisms.

By modifying an environment a metabiont would not necessarily favour all of the organisms present. For example, an antibiotic-producing population (A) of a litter-dwelling microorganism would inhibit microbial strains susceptible (S) to the antibiotic it produces. However, microbial strains resistant (R) to the antibiotic could be at an advantage and thrive in the absence of competition from S strains. In this example, therefore, the antibiotic-producer acts as a metabiont to the R strains. In terms of interactions within the ecological community this could be considered to be an indirect effect (*sensu* Wootton, 1994) of A upon R.

An organism might function as a metabiont to a dependent organism under one set of conditions but under other conditions the same pair of organisms might interact in a different way. For example, a fungivorous arthropod might lightly graze a community of litter-decomposing fungal species and, in this case functioning as a metabiont, promote the growth or survival of one of the fungal species for one or more of the following reasons: the arthropod by preferential grazing of other species in the community releases the fungus from competitive or antibiotic stress; the arthropod faeces supply essential inorganic or organic nutrients for the growth of the fungus; the arthropod disperses the fungal propagules or hyphae to new resources or; hyphae of other members of the fungal community when grazed release essential nutrients (Moore, 1988; Parkinson, 1988; Couteaux and Bottner, 1994; Wardle and Lavelle, Chapter 8, this volume).

Under another set of conditions the balance between fungal growth and arthropod-grazing pressure might alter. The fungivorous arthropod, no longer functioning as a metabiont, could become a predatory antagonist and graze the fungal species heavily. The fungus would suffer negative effects, perhaps becoming decimated or locally exterminated if insufficient propagules or non-viable hyphae remained.

Sometimes litter comminution by fauna has to be preceded by a priming stage during which the detritus is colonized by microorganisms. Such microbial activity concentrates nutrients, increases substrate digestibility, removes toxins or degrades recalcitrant compounds. The microbial communities active during the priming stage can have different species compositions from the microbial communities of the post-comminution stage. The priming microbes can be consumed by arthropods and receive no benefit, apart from dispersal of their propagules, from their interaction with arthropods. What is clear is that during the priming phase the microbial communities function as metabionts and modify the litter environment so that dependent arthropod communities are enabled to occupy it.

During the comminution stage the arthropods and other fauna assume a metabiotic role by increasing the surface area of litter detritus. This permits dependent microbial communities to colonize and exploit the debris during the post-comminution stage. Predatory arthropods benefit by grazing on the degrader microbes and also consume nutrients released from the detritus. Some of the degrader microbes benefit by being dispersed by the arthropods and by consuming nutrients released when senescent hyphae and bacterial cells are grazed.

Although many arthropods and microorganisms belonging to functional groups involved in leaf litter decomposition might benefit (Moore, 1988) from arthropod–microbial interactions an unknown proportion of microbial species may not. Two examples demonstrate that the outcome of collembolan grazing can disadvantage one of a pair of fungi on decomposing litter while the second benefits from the metabiotic collembolan activity (c.f. keystone predation, *sensu* Menge, 1995). Parkinson *et al.* (1979) found that selective grazing by the collembolan (*Onychiurus subtenuis*) on sterile dark hyphae permitted better colonization of decomposing aspen litter by competing non-sporing hyphae of a basidiomycete. Newell (1984) found that Sitka spruce litter was decomposed by the basidiomycete *Mycena galopus* at one-third to one-half the rate of another basidiomycete *Marasmius androsaceus*. But selective grazing by the collembolan *Onychiurus latus* of *M. androsaceus* hyphae reduced its competitiveness towards *M. galopus*. The outcome of the altered inter-fungal competition was that the decomposition was carried out primarily by *M. galopus*. In contrast to the effects of Collembola, Hanlon and Anderson (1980) found that the grazing of decomposing oak leaves by woodlice (*Oniscus asellus*) resulted in considerable reductions in the standing crop of fungi. The change brought about by the macroarthropod was accompanied by proliferation of the bacterial community with a shift to bacterial dominance over the fungi. The 'microbial loop' in litter and soil (Clarholm, 1994) involves a similar set of interactions to the previous examples. The grazing of bacteria by naked amoebae (the metabionts) releases nutrients immobilized in the bacterial cells, which benefit ectomycorrhizal fungi and the bacterial communities that take up the regenerated nutrients.

Specificity of metabiotic relationships for dependent organisms

Most metabiotic relationships are probably non-specific, e.g. the dependence of microaerophilic and anaerobic microorganisms in plant litter habitats upon the reductions of O_2 concentrations brought about aerobic organisms, including plant roots yet some are specific, e.g. NO_2^--oxidizers have a specific relationship with NH_4^+-oxidizers. Examples of non-specific and specific nutrient transformations involving functional groups of metabionts are summarized in Table 11.3.

Spatial relationships of metabionts to dependent organisms

In some metabiotic relationships the metabionts coexist with the dependent organisms they benefit but in other relationships they can be separated from each other in space or time. A physically close interaction between metabiotic and dependent organisms is found when non-cellulolytic bacteria, non-cellulolytic 'sugar' fungi, protozoa and nematodes (potential beneficiaries) grow closely to cellulose- or lignin-degrading fungal hyphae in dead plant tissues. Here the beneficiaries utilize nutrients released from the degraded detritus. Price (1988) gives an example where toxins in the tissues of bald cypress are broken down by a fungus. This metabiotic activity enables symbiotic protozoa, involved in the release of nutrients from the detoxified plant tissues, to occupy and persist in the paunch of their host termite *Coptotermes*, which consumes the cypress wood.

The development of trees results in physical structures, which change the local environment by altering the microclimate, hydrology, soil stability, nutrient flows and provides physical structures upon which epiphytes grow and animals find refuge. When trees or parts of trees die they pro-

Table 11.3. Nutrient transformations involving functional groups of metabionts in the plant litter environment. Metabionts are shown in **bold** and dependent organisms in *italics*.

Non-specific interactions
- Organic substances released from **roots** support growth of *mycorrhizal hyphae* in soil and *rhizosphere microfloras*
- Lignocellulose decomposed by **fungi** releases organic C for secondary *microbial colonizers* of wood
- CO_2 formed by heterotrophic organisms in the **litter ecosystem** for *plant* photosynthesis
- O_2 formed by photosynthesis in the **plant canopy** as an electron acceptor for *aerobic organisms*
- Reduction of O_2 concentrations in litter and soil by **aerobes** favours *microaerophiles* and *anaerobes*
- **Microbial** mineralization of nutrients in litter assimilated by *plants*

Specific interactions
- NH_4^+ released by **bacterial deamination** oxidized by NH_4^+-*oxidizers*
- NO_3^- formed by **nitrifiers** used as terminal electron acceptor by *denitrifiers*
- CH_4 formed by **methanogens** oxidized by *methanotrophs*
- H_2S and other reduced inorganic-S compounds formed by **anaerobic microorganisms** as a substrate for *thiobacilli*
- SO_4^{2-} formed by **aerobic microorganisms** as an electron acceptor for *Desulfovibrio* and other SO_4^{2-}-*reducers*

vide habitats for the growth or shelter of many animal species. The vegetation creates spatially remote metabiotic effects on the physical environment in which decomposer organisms dwell beneath the plant canopy. The presence of dead standing trees prolongs such metabiotic effects of the vegetation on the decomposer communities but cease once the land has been cleared of live or dead vegetation. Trees are examples of autogenic engineers, defined by Jones *et al.* (1994) as organisms that change the environments of ecosystems via their own physical structures whether they are alive or dead.

Scale, timing and duration of metabiotic phenomena

The dimensional scale of metabiotic influence varies from micrometres, e.g. 'sugar' fungi dependent upon lignin-degrading fungi, to the scale of the biosphere. Lawton (1994) suggested that the 'very existence of some ecosystems depends on particular species – the major autogenic and allogenic engineers that create the system – and all ecosystems are probably modulated and modified to a significant extent by at least one species of engineer'.

Organisms dependent upon a metabiont might coexist with the metabiont, survive it or develop after its death. Dead forest trees, whether standing or fallen, provide habitats for a considerable range of organisms, e.g. wood decomposing fungi, many insect taxons (Price, 1988), epiphytes and refuges for many animal species.

Interactions between pairs of organisms (e.g. antagonism (amensalism), commensalism, competition, mutualism, parasitism, predation), at the most endure only for the duration of the life cycles of the participating organisms. In contrast, many metabiotic effects persist and have beneficial effects on dependent organisms even after the decease of the metabiont. Some ecological engineers bring about persistent environmental changes which last for as long as the constructs, artefacts and their effects persist in the absence of the original engineers (Jones *et al.*, 1994), e.g. higher plants whose body parts and dead tissues help form soil and create a medium whose effects on ecosystems persist for thousands of years.

The nature of metabiosis, where the 'second organism' depends upon the metabiont to modify the environment to make it habitable, can lead to the development of a continuing relationship. For example, in an N-deficient litter ecosystem an N_2-fixing community, by maintaining supplies of fixed N, has a prolonged metabiotic effect on the decomposer community.

Humans as Metabionts: Human Modification of the Plant Litter Environment

Humans attempt to improve soil fertility, crop growth and productivity in numerous ways. These practices modify the soil environment so that certain biological activities are promoted, Plants are sown or planted (agro-forestry, alley cropping); residues of plants or animals are added in various forms (mulch, compost, dung); plant-associated organisms (N_2-fixing symbionts, mycorrhizal fungi) are introduced; physical changes to the litter-soil are made (tillage, drainage, clear fallowing, irrigation); and chemicals are introduced for various reasons (to alter pH, supply nutrients, control weeds, pests or pathogens). These management practices are the unconscious application of metabiotic techniques to the litter–soil environment. Thereby the soil environment is modified so that certain biological activities are promoted (organic residue decomposition, nutrient mineralization, N_2-fixation, nitrification, formation of humus, soil aggregation) or to check others (denitrification, NH_3 loss, build-up of plant pathogens). Yet, some endeavours (excess tillage, slash and burn, burning-off, removing crop residues, irrigation with saline water) often lead to short-term increases in soil productivity accompanied by a long-term decline of soil quality.

An example of how soil management may adversely modify the litter–soil environment was provided by Kundu and Ladha (1995). They discussed a fall in wetland-rice productivity detected in intensely-cultivated soils. Biological N_2-fixation (BNF) has declined and farmers resort to fertilizer N to maintain productivity. In many continuously-wet soils shallow tillage (c. 15–20 cm) forms hard pans,

which rice roots cannot penetrate to use sub-soil N. The concentrations of O_2 in the sub-soil decline because there is less percolation of water to transport dissolved O_2 and no O_2 input from living rice roots. The rate of N mineralization in the sub-soil is reduced as there is insufficient O_2 to support BNF by microaerophilic N_2-fixation. Rice plant residues accumulate in the flooded plough layer and decompose under reducing conditions so that phytotoxic substances are formed. These include a considerable number of organic substances as well as Fe^{2+}, NO_2^- and S^{2-}, each of which adversely affect BNF organisms. This is a situation where the metabiotic influences of both the oxidizing rhizosphere region of the rice and the input of rice root litter to the sub-soil microflora of N_2-fixers and litter decomposers have been cut off. In turn, the metabiotic influences on the rice plants of microorganisms mineralizing organic-N and the N_2-fixers have been suppressed or diminished in the sub-soil and the flooded plough layer, respectively. Kundu and Ladha (1995) suggest that these problems may be overcome by deeper tillage, drying out the soils periodically and then remoistening to stimulate microbial activity, and the use of rice varieties that utilize N in the sub-soil.

Other human modifications of the plant-litter environment can affect the global environment, e.g. the destruction of litter, humus and forests is a major contributing factor to the build of CO_2 in the atmosphere. An example is provided by the methanotrophs (an ancient group of panmetabionts), which consume atmospheric CH_4, the second most important greenhouse gas. Methane has a mean residence time of 10 years in the atmosphere. As a greenhouse warming gas CH_4 is 20 times as potent as CO_2 (Schlesinger, 1991). Any assessment of the long-term value of particular environments or ecosystems, e.g. natural rain forest, wetlands, should include the capacity of their methanotrophic communities to destroy CH_4. Also the potential effects of proposed changes or uses leading to soil disturbance, e.g. clear-cutting, burning, cultivation of agricultural crops, drainage, on that capacity need to be understood. Tillage, types of plant and animal cropping and N fertilization can reduce, perhaps destroy, the methanotrophic communities of soil (Hütsch *et al.*, 1994). The survival of methanotrophic soil communities could depend on metabionts modifying or maintaining conditions in the soil environment to favour the growth and activity of methanotrophs, e.g. aerobic conditions, well-structured soil, continuous inputs of root litter. An understanding of such metabiotic interactions would assist the quest to conserve or to discover how to improve the CH_4-oxidizing activity of soil.

Conclusions

It is a central tenet of biological science that living organisms change and modify their environments. Many of such biological modifications to habitats, whether ephemeral or enduring, may alter ecosystems so that a selection of other species are enabled to live or to flourish in them while other species may become decimated or fail to grow. Menge (1995) analysed experimentally-based studies of direct and indirect effects in interaction webs in marine rocky habitats. Many of the interactions studied by Menge as direct or indirect effects can be considered to be metabiotic, e.g. direct effects – enhancement of recruitment, provision of habitat and shelter; indirect effects – keystone predation, trophic cascades, indirect mutualism, indirect commensalism and habitat facilitation. Indirect effects accounted for about 40% of the changes in community structure and of these keystone predation, where a predator indirectly increases the abundance of the competitors of its prey via consumption of the prey, was the most common (35% of the indirect effects). Further work may confirm the hypothesis that at some time during their life cycles, or thereafter, a large proportion of organisms act as metabionts and modify their habitats so other organisms are able to live or to thrive in them. This hypothesis concerning the universality of metabiosis as a characteristic of living organisms needs to be tested.

Metabiosis deserves belated recognition as a useful term to describe the role of organisms in modifying habitats to make them suitable for other organisms to inhabit or to flourish in. Metabiosis must rank with biotic interactions, such as competition, predation or mutualism, in plant litter ecosystems. Terms such as 'indirect mutualism – no contact' for priming of substrates or comminution and the term 'direct mutualism – contact between species' for dispersal and grazing (Boucher *et al.*, 1982; Moore, 1988) are ill-defined, clumsy and should be discarded. The term ecological engineering implies purpose-driven human-like activity. Do the organisms called 'ecological engineers' really have the purpose and intention as well as actually contrive to modify the environment for the benefit of other species?

The various types and scales of magnitude of metabiotic interactions need to be considered objectively, classified and named. The populations, communities and functional groups of metabionts involved in the modification of environments in plant litter ecosystems should be identified and investigated in this way. We need to know if by influencing the composition of decomposer communities, the quality of plant litter affects the types of metabiotic interactions that develop and, in turn, if these modify the pace and outcomes of the decomposition process. This approach may result in the development of new concepts about plant litter systems, which could then be investigated in a rigorous manner. The information gained might be germane to the question of maintaining the biodiversity and sustainability of plant litter ecosystems, particularly the effects of human disturbance or management upon these ecosystems. Much effort is needed to elucidate what happens when microbial, plant and animal populations are disturbed or lost from plant litter ecosystems. Perhaps by identifying key metabiotic interactions in plant litter ecosystems it might be possible to discover if and how different species, communities or functional groups of organisms can be manipulated to improve agricultural and silvicultural efficiency.

References

Anderson, J.M. (1988) Invertebrate-mediated transport processes in soil. *Agriculture, Ecosystems and Environment* 24, 5–19.

Boucher, D.H., James, S. and Keeler, K.H. (1982) The ecology of mutualism. *Annual Review of Ecology and Systematics* 13, 497–516.

Brown, L. (ed.) (1993) *The New Shorter Oxford English Dictionary*. Clarendon Press, Oxford, 3799 pp.

Clarholm, M. (1994) The microbial loop in soil. In: Ritz, K., Dighton, J. and Giller, K.E. (eds) *Beyond the Biomass: Compositional and Functional Analysis of Soil Microbial Communities*. John Wiley, Chichester, pp. 221–230.

Couteaux, M.-M. and Bottner, P. (1994) Biological interactions between fauna and the microbial community in soils. In: Ritz, K., Dighton, J. and Giller, K.E. (eds) *Beyond the Biomass: Compositional and Functional Analysis of Soil Microbial Communities*. John Wiley, Chichester, pp. 159–172.

Garrè, C. (1887) Ueber Antagonisten unter den Bacterien. *Correspondenz-Blatt für Schweizer Aerzte* 17, (13, 1 July 1887) 385–392.

Hanlon, R.D.G. and Anderson, J.M. (1980) Influence of macroarthropod feeding activities on microflora in decomposing oak leaves. *Soil Biology and Biochemistry* 12, 255–261.

Hütsch, B.W., Webster, C.P. and Powlson, D.S. (1994) Methane oxidation in soil as affected by land use, soil pH and fertilization. *Soil Biology and Biochemistry* 26, 1613–1622.

Jones, C.G., Lawton, J.H. and Shachack, M. (1994) Organisms as ecosystem engineers. *Oikos* 69, 373–386.

Kundu, D.K. and Ladha, J.K. (1995) Efficient management of soil and biologically fixed N_2 in intensively-cultivated rice fields. *Soil Biology and Biochemistry* 27, 431–439.

Lawton, J.H. (1994) What do species do in ecosystems? *Oikos* 71, 367–374.

Menge, B.A. (1995) Indirect effects in marine rocky interaction webs: patterns and importance. *Ecological Monographs* 65, 21–74.

Meyer, F.H. (1966) Mycorrhiza and other plant symbioses. In: Henry, S.M. (ed.) *Symbiosis*. Vol. 1, Academic Press, New York, pp. 171–255.

Moore, J.C. (1988) The influence of microarthropods on symbiotic and non-symbiotic mutualism in detrital-based below-ground food webs. *Agriculture, Ecosystems and Environment* 24, 147–159.

Newell, K. (1984) Interaction between two decomposer basidiomycetes and collembola under Sitka spruce: grazing and its potential effects on fungal distribution and litter decomposition. *Soil Biology and Biochemistry* 16, 235–239.

Parkinson, D. (1988) Linkages between resource availability, microorganisms and soil invertebrates. *Agriculture, Ecosystems and Environment* 24, 21–32.

Parkinson, D., Visser, S. and Whittaker, J.B. (1979) Effects of collembolan grazing on fungal colonization of leaf litter. *Soil Biology and Biochemistry* 11, 529–535.

Price, P.W. (1988) An overview of organismal interactions in ecosystems in evolutionary and ecological time. *Agriculture, Ecosystems and Environment* 24, 369–377.

Schlesinger, W.H. (1991) *Biogeochemistry: An Analysis of Global Change*. Academic Press, San Diego, 443 pp.

Woomer, P.L., Martin, A., Albrecht, D.V.S. and Scharpenseel, H.W. (1994) The importance and management of soil organic matter in the tropics. In: Woomer, P.L. and Swift, M.J. (eds) *The Biological Management of Tropical Soil Fertility*. John Wiley, Chichester, pp. 47–80.

Wootton, J.T. (1994) The nature and consequences of indirect effects in ecological communities. *Annual Reviews of Ecology and Systematics* 25, 443–466.

Part IV

Manipulation of Plant Litter Quality

12 Residue Quality and Decomposition: An Unsteady Relationship?

B. Vanlauwe[1,2], J. Diels[1], N. Sanginga[1] and R. Merckx[2]

[1]Soil Micobiology, IITA Nigeria, c/o Lambourn & Co, 26 Dingwall Road, Croydon CR9 3EE, UK; [2]Laboratory of Soil Fertility and Soil Biology, Faculty of Applied Agricultural Sciences, KU Leuven, K. Mercierlaan 92, 3001 Leuven/Heverlee, Belgium

Introduction

The limited availability or high cost of mineral fertilizers has generated improved tropical cropping systems that rely to a great extent on the proper management of plant residues as a source of nutrients. Proper residue management necessitates quantitative knowledge of the decomposition behaviour of those residues under the prevailing set of factors, influencing the decomposition process. In recent years, several attempts have been made to quantify residue quality and its relationship with residue decomposition, mostly in terms of N-mineralization (Palm and Sanchez, 1991; Oglesby and Fownes, 1992; Kachaka et al., 1993). Although general trends have been observed, for instance, a high residue lignin content or C-to-N ratio appears to reduce N-release, no unique relationship has been developed (Fig. 12.1). This is partly due to the different methodological approaches used to quantify the relationship between residue decomposition/N mineralization and its quality in terms of incubation methodology (temperature, leaching tube versus incubation jar technique) and various residue-related factors, such as addition rate, particle size, dryness, or method of application (mixed or surface applied).

Information on field estimates of the relationship between residue decomposition or N release and quality is less available and more essential for implementation of residue management practices. In contrast with laboratory incubation techniques, decomposition under field conditions is affected by climate (Vanlauwe et al., 1995) and faunal activity (Tian et al., 1992a) and addresses several aspects of the interaction between residue quality and decomposition more realistically (e.g. interaction between polyphenols and residue derived N may be reduced under field conditions due to polyphenol leaching (Handayanto et al., 1994; Chapter 14, this volume), or mineralized residue N may be lost beyond the reach of the decomposer community due to leaching).

The objectives of this chapter are: (i) to indicate some methodological problems associated with residue characterization, especially in terms of residue particle size; (ii) to evaluate relationships between residue quality and various aspects of the decomposition process under controlled laboratory conditions; (iii) to assess the impact of climate on the relationship between residue quality and decomposition; and finally (iv) to propose some degree of standardization for the development of models relating residue quality and decomposition.

© 1997 CAB INTERNATIONAL *Driven by Nature: Plant Litter Quality and Decomposition.*
(eds G. Cadisch and K.E. Giller)

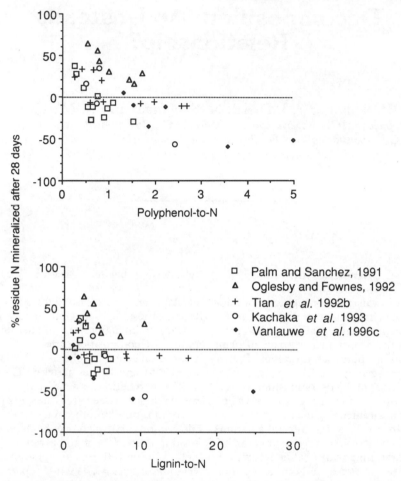

Fig. 12.1. Relationships between the proportion of residue N released after 28 days in laboratory incubations and the residue polyphenol-to-N and lignin-to-N ratios (adapted from Palm and Sanchez, 1991, Oglesby and Fownes, 1992, Tian *et al.*, 1992b, Kachaka *et al.*, 1993 and Vanlauwe *et al.*, 1996c).

Some Methodological Problems Associated with Residue Quality Determination

At present, residue quality is generally measured as a set of biochemical plant constituents, such as lignin, total C, total N, polyphenol (or protein binding capacity, Handayanto *et al.*, 1994), neutral detergent fibre or water-soluble components. All these methods, except for the total C and N determination, are based on a hot or cold chemical extraction scheme.

As such, a reduced residue particle size that results in an increased contact between the extractant and residue leads to a significant increase in the quantity of components extracted (Table 12.1). Moreover, the impact of varying residue particle size on various quality parameters was shown to be dependent on both the extraction procedure and plant species. The most significant relationships between the 'traditionally' measured residue characteristics, such as lignin-to-N or (lignin+polyphenol)-to-N ratio, and the C

Table 12.1. Determination of polyphenol (King and Heath, 1967) and lignin content (NDF-lignin, van Soest, 1963), as affected by particle size. Values in the same row followed by the same letter are not significantly different (5% level).

	2–4 mm	0.5–2 mm	<0.5 mm	Ball-milled
Lignin content (%)				
Leucaena leucocephala	21.6[a]	13.2[b]	10.1[c]	9.0[d]
Senna siamea	14.2[a]	13.2[b]	12.0[c]	10.1[d]
Gliricidia sepium	26.6[a]	21.7[b]	14.5[c]	8.9[d]
Flemingia macrophylla	34.3[a]	29.4[b]	21.4[c]	15.7[d]
Dactyladenia barteri	17.7[a]	15.5[b]	10.8[c]	9.1[d]
Polyphenol content (%)				
Leucaena leucocephala	5.26[d]	5.51[c]	5.77[b]	5.84[a]
Senna siamea	2.80[c]	3.38[b]	3.57[b]	4.20[a]
Gliricidia sepium	1.90[d]	2.04[c]	2.73[b]	2.83[a]
Flemingia macrophylla	2.17[d]	2.49[c]	3.21[b]	3.60[a]
Dactyladenia barteri	4.49[c]	4.90[b]	5.36[a]	5.44[a]

or N mineralization of leaf litter of five agroforestry species was found for ball-milled residues (B. Vanlauwe, N. Sanginga and R. Merckx, unpublished).

Based on the foregoing, differences in quality determination due to varying extraction efficiencies between coarsely ground leaf and root residues could be expected, as the planar leaf particles will have a different surface:volume ratio than the cylindrical root particles.

The above observations strongly indicate the need for a standardized scheme for residue characterization. Even so, residues must be characterized as a function of their age (Table 12.2), plant part composition (leaflets, complete leaves, prunings, complete tree canopy), growth conditions (nutritional conditions, ambient CO_2 concentration (Cotrufo *et al.*, 1994)) and residue treatment, e.g. Mafongoya *et al.* (Chapter 13) found a significant impact of the residue drying regime on its polyphenol content).

Table 12.2. Quality determination of *Leucaena leucocephala* and *Senna siamea* residues (ground <0.5 mm), collected during four pruning activities (P1 to P4) in an alley cropping system. Polyphenols were determined with the King and Heath method (1967), while the cellulose and lignin content were determined following van Soest and Wine (1967). The water-soluble fraction was determined with a cold extraction method. Values in the same column, followed by the same letter are not significantly different (Duncan mean separation test).

	Age (Weeks)	N	Polyphenols	Water-soluble	Cellulose	Lignin
				(% of dry matter)		
Leucaena						
P1	29	4.21[a]	4.09[a]	37.4[ab]	11.40[ab]	10.53[a]
P2	6	5.33[c]	1.85[d]	40.7[c]	11.98[a]	7.46[cd]
P3	14	4.32[a]	5.02[e]	40.6[c]	10.18[a]	5.88[d]
P4	8	4.76[f]	1.52[d]	40.4[c]	9.80[a]	5.85[d]
Senna						
P1	29	2.68[b]	2.90[b]	35.8[b]	14.52[c]	8.30[bc]
P2	6	3.40[de]	3.31[bc]	39.7[ac]	12.34[a]	6.99[cd]
P3	14	3.22[e]	3.38[bc]	39.2[ac]	12.20[a]	7.88[bc]
P4	8	3.63[d]	3.47[c]	36.4[b]	12.65[a]	9.71[ab]

Relating Residue Quality with Decomposition under Controlled Laboratory Conditions

Laboratory incubations normally control the temperature and soil moisture content at an optimal level for microbial activity. Because of the exclusion of varying climatological conditions, soil faunal activity and nutrient leaching, this technique simplifies the study of various aspects of the decomposition process, but at the same time generates data which are not easily transferable to field conditions. However, in order to shed some light on the complex relationship between residue quality and decomposition, it is useful to first consider the decomposition process under controlled conditions.

Any effect of residue quality on its decomposition should be clear from differences in residue C or N release, soil microbial biomass dynamics after residue addition or the amount of undecomposed material remaining in the soil organic matter pool. In laboratory experiments with leaf litters of few agroforestry species, reported by Kachaka (1993) and Kachaka *et al.* (1993), a linear relationship could be observed between the increase in microbial biomass C and the residue (polyphenol+lignin)-to-N ratio for values lower than 8 (Fig. 12.2a). Residues with a larger ratio did not cause a substantial change in microbial biomass C. The proportion of residue C or N released after 28 days followed similar trends for three laboratory incubations with above and below ground residues of different agroforestry species (Fig. 12.2b). In a greenhouse pot experiment with ^{15}N labelled leaf and root residues from three agroforestry species, the proportion of leaf litter N remaining in the soil litter pool (organic material >0.25 mm, separated by floatation in water (Anderson and Ingram, 1993)) was related with the (lignin+polyphenol)-to-N ratio after 3 months' incubation. Residual root litter N, on the other hand, was not affected by changes in (lignin+polyphenol)-to-N ratio (Fig. 12.2c).

Relationships between residue quality and decomposition appear to be linear only for a limited range of residue qualities. This is not surprising, as low quality materials – for instance with a (lignin+polyphenol)-to-N ratio larger than 8 or 10 in Figs 12.2a and 12.2b – will not cause substantial changes in the short term microbial C or CO_2 production dynamics. Values for the increase in microbial C, increase or in the proportion of residue C released appear to approach 0 as residue quality parameters which are negatively correlated with decomposition become large. N mineralization, expressed as a proportional release of the amount of added residue N – which gives negative values in the case of net N immobilization – might become less negative for very low quality material. In the case of pure lignin N immobilization may be negligible after 28 days, because of the low decomposability of lignin. Similarly, the proportion of undecomposed residue C, will approach 100% for very low quality materials.

In aerobic laboratory incubations, the end-products of the decomposition process, such as mineral C (CO_2) or N, are most often measured. CO_2 and mineral N release from aerobically decomposing fresh plant materials is not linear in time. Usually, the CO_2-C production or mineral N release rate decreases with time until a rather constant rate, slightly higher or not different from the unamended control soil, is reached. It is important to take note of the incubation time after which the proportion of released residue C or N is related with specific residue characteristics. First, the (lignin+polyphenol)-to-N ratio for which the C release or microbial biomass C increase is low (Fig. 12.2) might increase with incubation periods larger than 28 days. Second, in a laboratory experiment with roots and leaves of three hedgerow species, the slope of the regression between the proportion of residue C released and the C-to-N ratio was constant and the intercept increased linearly after 14 days of incubation (Vanlauwe *et al.*, 1996c).

Soil factors affecting the activity or composition of the microbial decomposer community will have a significant impact

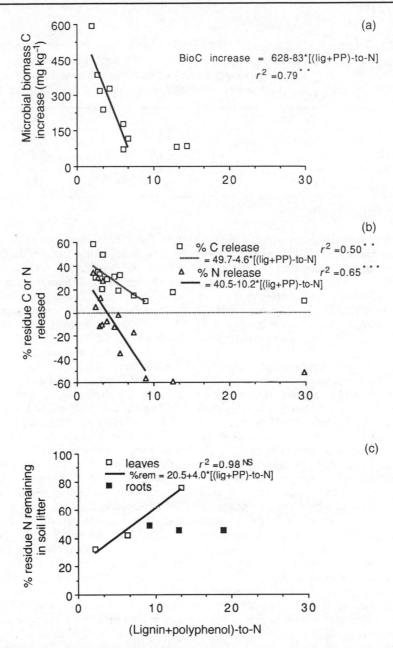

Fig. 12.2. Relationship between (lignin+polyphenol)-to-N ratio and (a) the increase in microbial biomass after 28 days of laboratory incubation at 25°C (adapted from Kachaka (1993) and Kachaka *et al.* (1993)); (b) the proportion of residue C and N released after 28 days in three laboratory incubations at 25°C (two sets adapted from Kachaka *et al.*, 1993 and Vanlauwe *et al.*, 1996c); and (c) the proportion of leaf and root residue N remaining in the soil litter fraction (>0.25 mm), as separated with a floatation technique on water (adapted from Vanlauwe *et al.*, 1996b), in a greenhouse pot experiment (16–32°C) after 3 months. In Fig. 12.2b, the (lignin+polyphenol)-to-N ratio for the residues used by Kachaka *et al.* (1993) was measured on residues <2 mm and was recalculated to represent the values for ball-milled residues, according to the factors of Table 12.1.

Table 12.3. Coefficients of determination and their significance for the linear regressions between the decomposition (k_{DM}) and N release rate (k_N), calculated versus the number of days where rainfall exceeds pan evaporation, and selected residue characteristics for the first three prunings of an alley cropping experiment. Lignin and polyphenols were determined as stated in Table 12.2.

	k_{DM}	k_N
C-to-N	0.74*	0.61
lignin-to-N	0.68	0.42
polyphenol-to-N	0.54	0.76*
(lignin+polyphenol)-to-N	0.77*	0.68*

* $P < 0.05$.

on relationships between residue quality and decomposition. Mulongoy and Gasser (1993) showed substantial differences between N release from *Leucaena leucocephala* and *Dactyladenia barteri* residues in soils with contrasting pH. Jensen (1994) proposed that decomposition products derived from smaller residue particles may be stabilized by clay particles.

All above relationships between residue quality and any aspect of the decomposition process are trends rather than clear-cut linear relationships. This is not surprising, as the attempt to characterize a biochemically and structurally complex substrate such as plant material in terms of two or three independently measured parameters seems rather ambitious. Furthermore, the question remains whether chemical agents are able to mimic the activity of complex enzyme systems, produced by the microbial decomposer community. Finally, one could debate whether the development of equations relating residue quality and decomposition for a whole range of above as well as belowground organic inputs under laboratory conditions is essential, as few of these relationships are likely to be transferable to field conditions.

Relating Residue Quality with Decomposition under Field Conditions

Transferring the residues from a laboratory incubation jar to litter bags or decomposi-

tion tubes in the field causes a drastic shift in the functioning of the decomposition subsystem (Swift *et al.*, 1979). In addition to the impact of varying climatological conditions, soil fauna will also interact with the microbial component of the decomposer community. The latter aspect, however, is discussed in Chapter 9 (Tian *et al.*, this volume) and beyond the scope of this chapter.

The litter bag technique has been used most often for studying residue decomposition in the field, because of its simple, easily replicable, non-destructive character and its ability to exclude certain classes of soil fauna. Two problems associated with this technique, however, may cause biases in relating residue N release from litter bags with its availability for the crops. First of all, confining the residues surely alters the microclimate in the bags and the soil–residue contact. Second, N released from the litter bags is not necessarily available, but may have been reimmobilized by soil organisms or entered the soil litter pool through through-fall. Alternative methods, maintaining a realistic soil–residue contact and including a soil litter determination or a soil organic matter fractionation scheme, eventually relying on ^{15}N-labelled materials, may give better estimates of N availability than the litter bag technique (Fig. 12.3). For belowground residues, buried bag techniques may be more suitable (Anderson and Ingram, 1993). However, for relative comparisons between the decomposition of different residues, litter bags are a useful tool.

Climatic factors modifying the decom-

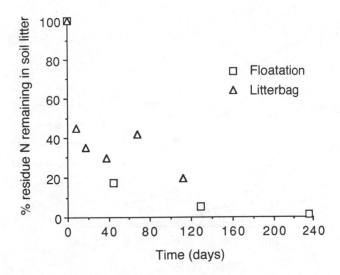

Fig. 12.3. *Leucaena* N simultaneously remaining in litter bags (1.4 mm mesh-size) and in the soil litter fraction (>0.25 mm), as measured with a floatation method on water (Vanlauwe *et al.*, 1996a).

position process include temperature, rainfall and evaporation, which in its turn is a function of air relative humidity, solar radiation and wind speed. Under tropical conditions, temperature will be of a minor importance, as it remains rather constant, except during the dry season. Rainfall influences decomposition through leaching and the maintenance of an optimal litter moisture content (Swift *et al.*, 1979). Evaporation, usually measured with the pan evaporation technique, is a measure of moisture loss from the soil surface.

In a litter bag field experiment, the impact of artificially applied precipitation on the decomposition of leaf litter from three agroforestry species was studied (Vanlauwe *et al.*, 1995). Residue dry matter loss was better correlated with the number of rainfall events than with the total amount of rainfall. Significant relationships (*P* <0.05) between the sensitivity of residue decomposition to a varying number of rainfall events and selected residue characteristics (polyphenol-to-N, C-to-N, and water-soluble fraction) were demonstrated.

The decomposition and N release of four *Leucaena* and *Senna siamea* prunings (Table 12.2) in alley cropping systems was studied during the 1994 growing season (maize in the first and cowpea in the second season). The third pruning decomposed partly during the dry season, while dry matter loss from the residues of the fourth pruning, which was mainly decomposing during the dry season, was lower than 10%. Fitting a first order equation through the decomposition and N release of the third pruning vs the number of 'rainy days' (days where rainfall exceeds pan evaporation) gave better fits than versus time, although both regressions were highly significant (B. Vanlauwe, N. Sanginga and R. Merckx, unpublished). Both the decomposition and N release rates, calculated versus the number of 'rainy days', were significantly correlated (*P* <0.05) with various residue quality parameters (Table 12.3).

While in laboratory incubations, authors usually express the extent of decomposition in terms of a proportion of C or N mineralized after a number of days, decomposition or N release rates are more frequently used in litter bag studies (Melillo *et al.*, 1982, Tian *et al.*, 1992a). Both variables are not linearly related, which may lead to mathematically different relationships between residue quality and

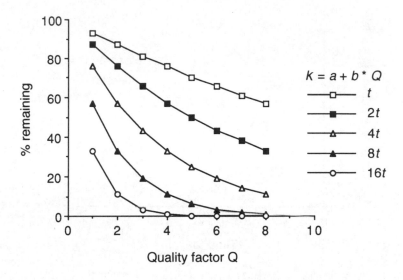

Fig. 12.4. Theoretical relationship between the proportion of remaining residue and its quality *Q* for different times *t* to 16*t*, supposing that the decomposition rate *k* is linearly related to *Q*.

decomposition expressed as a proportional release or a release rate (Fig. 12.4).

Conclusions, Suggestions for Standardization and Future Research Needs

As residue particle size significantly affected its quality determination, depending on species and extraction procedure, standardization of residue particle size for quality determination is needed. We found the best correlation between quality and decomposition for ball-milled residues, but this needs to be validated for a larger range of residues and for decomposition under field conditions.

Laboratory incubations aim at understanding the decomposition process, rather than developing practically useful equations for N release as a function of quality. Residue quality had an impact on various aspects of the decomposition process under controlled laboratory conditions (C and N mineralization, microbial biomass, residue N remaining in the soil litter) for a limited range of residue qualities. Although mechanistic decomposition

simulation models may diminish the need for standardization of laboratory incubations in terms of amount of residue applied and incubation temperature (mostly, the soils are incubated at optimal soil moisture contents) standardization will exclude speculations, especially when dealing with residues which have not been previously tested. Residue addition rates should be equal and within the range of observed biomass productions in the field, surely avoiding exuberant addition rates. Although the Q_{10} for most soil biological processes is close to 2 (Scholes *et al.*, 1994), a standardized incubation temperature of 25°C for tropical conditions would exclude speculation on observed differences for residues decomposing at different temperatures.

Standardization of incubation procedures could also clarify the impact of soil characteristics, such as microbiological activity, pH or clay content, as discussed above, on the relationship between residue quality and decomposition.

Climate, and especially the occurrence of dry spells has a marked impact on decomposition in the field under subhumid tropical conditions and thus on the relationship between residue decomposition and quality. Calculating the decomposition or N release rates versus the number of 'rainy days' matched the experimental data better than versus time, when dry spells occurred during part of the decomposition period. The decomposition or N release rates can afterwards be related to residue quality.

For field estimates, techniques combining residue decomposition with a soil organic matter fractionation scheme, eventually using labelled residues, should improve quantification of the real residue N availability as a function of time.

Above mentioned field decomposition experiments were carried out with surface applied residues. However, mixing the residues with the top soil will drastically affect their decomposition and thus the relationship between residue decomposition and quality. Moreover, root residues are always intimately mixed with the soil matrix. Field estimates of the relationship between residue quality and decomposition for buried residues are needed.

For laboratory incubations, decomposition is usually expressed as a proportional release of mineralized C or N, while for field techniques, the proportion of remaining residue is measured. A proportional release of mineral C or N will increase with time. Moreover, a proportional release is not linearly related with a decomposition rate. It is important to note this when comparing residue quality-decomposition models from different authors.

Acknowledgements

The authors are grateful to ABOS, the Belgian Administration for Development Cooperation, for sponsoring this work as part of the cooperative project between KU Leuven and IITA on 'Process-based studies of soil organic matter dynamics in relationship to the sustainability of agricultural systems in the tropics'.

References

Anderson, J.M. and Ingram, J.S.I. (1993) *Tropical Soil Biology and Fertility: A Handbook of Methods.* CAB International, Wallingford, 221 pp.

Cotrufo, M.F., Ineson, P. and Rowland, A.P. (1994) Decomposition of tree leaf litters grown under elevated CO_2: Effect of litter quality. *Plant and Soil* 163, 121–130.

Handayanto, E., Cadisch, G. and Giller, K.E. (1994) Nitrogen release from prunings of legume hedgerow trees in relation to quality of the prunings and incubation method. *Plant and Soil* 160, 237–248.

Jensen, E.S. (1994) Mineralization-immobilization of nitrogen in soil amended with low C:N ratio plant residues with different particle sizes. *Soil Biology and Biochemistry* 26, 519–521.

Kachaka, S. (1993) Decomposition and N-mineralization of prunings of various quality and age. PhD thesis, Katholieke Universiteit Leuven, Leuven, Belgium.

Kachaka, S., Vanlauwe, B. and Merckx, R. (1993) Decomposition and nitrogen mineralization of prunings of different quality. In: Mulongoy, K. and Merckx, R. (eds) *Soil Organic Matter Dynamics and Sustainability of Tropical Agriculture.* John Wiley and Sons, Chichester, UK, pp. 199–208.

King, J.G.C. and Heath, G.W. (1967) The chemical analysis of small samples of leaf material and the relationship between the disappearance and composition of leaves. *Pedobiologia* 7, 192–197.

Melillo, J.M., Aber, J.D. and Muratore, J.F. (1982) Nitrogen and lignin control of hardwood leaf litter decomposition dynamics. *Ecology* 63, 621–626.

Mulongoy, K. and Gasser, M.O. (1993) Nitrogen supplying capacity of leaves of *Dactyladenia barteri* (Hook ex Owl) and *Leucaena leucocephala* (Lam.) de Wit in two soils with different acidity from southern Nigeria. *Biology and Fertility of Soils* 16, 57–62.

Oglesby, K.A. and Fownes, J.H. (1992) Effects of chemical composition on nitrogen from green manures of seven tropical leguminous trees. *Plant and Soil* 143, 127–132.

Palm, C.A. and Sanchez, P.A. (1991) Nitrogen release from the leaves of some tropical legumes as affected by their lignin and polyphenolic contents. *Soil Biology and Biochemistry* 23, 83–88.

Scholes, R.J., Dalal, R. and Singer, S. (1994) Soil physics and fertility: The effects of water, temperature and texture. In: Woomer, P.L. and Swift, M.J. (eds) *The Biological Management of Tropical Soil Fertility*. John Wiley and Sons, Chichester, UK, pp. 117–136.

Swift, M.J., Heal, O.W. and Anderson, J.M. (1979) *Decomposition in Terrestrial Ecosystems*. Studies in Ecology Vol. 5, University of California Press, Berkeley, California, 372 pp.

Tian, G., Kang, B.T. and Brussaard, L. (1992a) Biological effects of plant residues with contrasting chemical compositions under humid tropical conditions – decomposition and nutrient release. *Soil Biology and Biochemistry* 24, 1051–1060.

Tian, G., Kang, B.T. and Brussaard, L. (1992b) Effects of chemical composition on N, Ca, and Mg release during incubation of leaves from selected agroforestry and fallow plant species. *Biogeochemistry* 16, 103–119.

Vanlauwe, B., Vanlangenhove, G. and Merckx, R. (1995) Impact of rainfall regime on the decomposition of leaf litter with contrasting quality under subhumid tropical conditions. *Biology and Fertility of Soils* 20, 8–16.

Vanlauwe, B., Swift, M.J. and Merckx, R. (1996a) Soil-litter dynamics and N use in a leucaena (*Leucaena leucocephala* Lam. (de Wit)) alley cropping system in Southwestern Nigeria. *Soil Biology and Biochemistry* 28, 739–749.

Vanlauwe, B., Van den Bosch, S., Van Gestel, M. and Merckx, R. (1996b) Soil litter dynamics and N cycling in alley cropping systems. *Proceedings of the 8th Nitrogen Workshop*, Gent, Belgium, September 5–8, 1994 (in press).

Vanlauwe, B., Nwoke, O.C., Sanginga, N. and Merckx, R. (1996c) Impact of residue quality on the C and N mineralization of leaf and root residues of three agroforestry species. *Plant and Soil* (in press).

Van Soest, P.J. (1963) Use of detergents in the analysis of fibrous feeds. II. A rapid method for the determination of fiber and lignin. *Journal of the Association of Official Agricultural Chemists* 46, 829–835.

Van Soest, P.J. and Wine, R.H. (1967) Use of detergents in the analysis of fibrous feeds. IV. Determination of plant cell-wall constituents. *Journal of the Association of Official Agricultural Chemists* 50, 50–55.

13 Effect of Multipurpose Trees, Age of Cutting and Drying Method on Pruning Quality

P.L. Mafongoya[1], B.H. Dzowela[1] and P.K. Nair[2]

[1] SADC-ICRAF Agroforestry Project, c/o Department of Research & Specialist Services, PO Box CY594, Causeway, Harare, Zimbabwe; [2] University of Florida, Department of Forestry, 118 Newins-Ziegler Hall, Gainesville, 32611, USA

Introduction

In many developing countries, smallscale farmers have limited access to inorganic fertilizers. In such situations multipurpose tree (MPT) prunings applied to the soil are used to meet the nitrogen (N) requirements of annual food crops. In order to manage N availability to annual crops from MPT prunings we need to understand and predict N mineralization and immobilization patterns in relation to chemical composition of the prunings. Several studies have shown that both soluble polyphenol and lignin contents are inversely related to N release from MPT prunings (Palm and Sanchez, 1991; Oglesby and Fownes, 1992). The polyphenol-to-N ratio was shown to be a good predictor of N release (Palm and Sanchez, 1991; Oglesby and Fownes 1992) although, other studies have found the (lignin + polyphenol)-to-nitrogen ratio to be a better index of N release over the long-term (Fox et al., 1990; Handayanto et al., 1994; Constantinides and Fownes, 1994). There has been also disagreement on which index to use as this may depend on the type of polyphenols found in the prunings and the range of species studied (Palm, 1995).

Plant tannins are a diverse group of naturally occurring polyphenolic metabolites, defined by their relatively high molecular weight and ability to complex proteins and carbohydrates (see Harborne, Chapter 4, this volume). These naturally occurring polyphenols, however, differ widely in their solubility in water, and their capacity to bind proteins. Hence, use of total polyphenol content as a predictor of nitrogen release from MPT prunings may be of limited use since a mixture of polyphenolic compounds with different biological activity are present in plant tissues. Oven drying of prunings before incorporation into the soil is a common procedure (Gilmour et al., 1985) and sun-drying has been used in some cases (Herman et al., 1977). Elevated drying temperatures have been shown to reduce the recovery of soluble polyphenols (Hagerman, 1988; Scalbert, 1992) and increase the concentration of lignin (van Soest, 1982). Oven-drying at 50°C has been shown to lower extractability of condensed tannins in a range of tropical legumes (Cano et al., 1994). This reduced extractability of condensed tannins (proanthocyanidins) has been associated with increases in protein-bound and fibre-bound condensed tannins, resulting in little change in total condensed tannin content. This implies that certain condensed tannins in plants are extractable (i.e. reactive tannins) and other fractions are bound (i.e. unreactive tannins) and the relative distribution of these may be affected by the drying method used, the MPT species and the age of species at pruning. For example, the concentration of tannins was found to be higher in new growth of browse than in mature leaves (Lee and Lowry, 1980). There have been few studies that have looked at the interactions between age of prunings, drying method and MPT species on percentage soluble polyphenols and lignin (Dzowela *et al.*, 1995). The rationale of this study was to examine possible interactions between MPT species, drying method and age of cutting on the chemical composition of MPT prunings.

Methodological approach

The effects of age of cutting and drying method on the chemical composition of seven MPT species were examined. The species studied were *Acacia angustissima*, *Calliandra calothyrsus*, *Flemingia macrophylla*, *Cajanus cajan*, *Gliricidia sepium*, *Leucaena leucocephala* and *Sesbania sesban*. The first prunings were cut after 12 months of growth and regrowth prunings were cut after 8 weeks of growth from the first-cut. Prunings, consisting of leaves + pinnae only, were cut from various MPTs grown at the International Centre for Research in Agroforestry (ICRAF) Domboshawa research site in Zimbabwe. These samples were collected from a replicated trial where several MPTs were being screened for local adaptability. Prunings were either sun-dried for 2–3 days on a concrete floor at 25°C or oven-dried at 55°C for 48 h.

Results of the chemical analysis are expressed on a percentage dry matter (DM) basis. The plant materials were analysed for N, by micro-Kjeldahl, neutral detergent fibre N (NDF-N) by micro-Kjeldahl on neutral detergent fibre (NDF) fraction, lignin by the acid-detergent fibre method (Goering and van Soest, 1970), and soluble polyphenols by the gravimetric method of precipitating with trivalent ytterbium acetate (Reed *et al.*, 1985). Insoluble proanthocyanidins in NDF were determined by heating NDF at 95°C for 1 h in *n*-butanol (5 ml) containing concentrated HCl (5% v/v) (Reed *et al.*, 1982).

Effect of MPT Species on Pruning Quality

The distribution of tannins in the prunings varied between the MPT species (Tables 13.1 and 13.2). Tree species like *Gliricidia* and *Cajanus* had low soluble polyphenols whereas *A. angustissima* and *C. calothyrsus* had high soluble polyphenols. A high proportion of the tannins in *G. sepium* and *F. macrophylla* were bound in the cell wall as insoluble proanthocyanidins (Tables 13.1 and 13.2) whereas condensed tannins in *A. angustissima*, *C. calothyrsus* and *Sesbania* were highly extractable and appeared in the soluble fraction.

These results demonstrate that a certain fraction of condensed tannins in plants is extractable (reactive tannins) and the other fraction is bound to the cell wall as insoluble proanthocyanidins which are not reactive. The distribution of these fractions in prunings was influenced by MPT species and the drying method (Table 13.2). The distribution of these tannins has implications on the nitrogen mineralization patterns of the MPT species as shown in Fig. 13.1. Species like *Gliricidia* and *Cajanus* with low soluble polyphenols had high rates of N release compared with species like *Acacia*, *Calliandra* and *Flemingia*.

Effects of Drying Methods

There is some evidence that the method of drying of legume plant material (i.e. freeze drying, oven or sun-drying) has an effect on lignin and tannin contents (Terrill *et al.*, 1994). Oven-drying prunings at 55°C reduced the concentrations of soluble polyphenols while sun-drying at 25°C yielded higher concentrations of polyphenols (Table 13.2) in agreement with the results of Hagermann (1988), Constantinides and Fownes (1994) and Dzowela *et al.* (1995). Oven-drying has been shown to cause polymeralization of soluble tannins with other cell macromolecules (Terrill *et al.*, 1990; Scalbert,

Table 13.1. Effect of tree species and age of prunings on % soluble polyphenols, % lignin, % nitrogen, % neutral detergent fibre N (NDF-N) and insoluble proanthocyanidins in oven-dried prunings.

MPT species	% Soluble polyphenols		% Lignin		% Nitrogen		% NDF-N		Insoluble proanthocyanidins (absorbance units)	
	First cut	Regrowth	First cut	Regrowth	First cut	Regrowth	First cut	Regrowth	First cut	Regrowth
Acacia angustissima	8.0	7.0	17.1	17.7	3.1	4.1	1.6	1.8	17.6	39.7
Gliricidia sepium	1.5	3.0	15.1	11.2	2.4	3.6	1.1	1.8	162.7	102.4
Flemingia macrophylla	5.5	8.3	22.5	21.5	2.0	2.9	1.4	2.0	116.3	128.8
Sesbania sesban	9.2	8.1	6.5	6.6	3.1	3.9	1.1	1.2	16.0	16.9
Calliandra calothyrsus	6.9	7.2	20.9	19.0	2.9	3.4	1.8	1.8	69.1	41.9
Cajanus cajan	1.2	3.9	18.9	15.0	3.3	4.4	2.1	1.8	78.0	58.1
Leucaena leucocephala	2.6	4.3	13.6	14.1	3.1	4.0	1.7	1.7	53.6	35.4
SED	0.4		0.8		0.4		0.1		7.7	

Table 13.2. Effect of MPT species and drying method on % soluble polyphenols, % lignin, % nitrogen, % neutral detergent fibre N (NDF-N) and insoluble proanthocyanidins in prunings.

MPT species	% Soluble polyphenols		% Lignin		% Nitrogen		% NDF-N		Insoluble proanthocyanidins (absorbance units)	
	Sun-dried	Oven-dried	Sun-dried	Oven-dried	Sun-dried	Oven-dried	Sun-dried	Oven-dried	Sun-dried	Oven-dried
Acacia angustissima	12.2	8.0	14.0	17.1	3.1	2.6	1.9	1.6	40.1	16.9
Gliricidia sepium	2.3	1.5	11.1	15.1	2.4	1.8	0.9	1.1	170.8	94.2
Flemingia macrophylla	10.5	5.5	19.3	22.5	2.0	1.9	1.1	1.4	119.8	125.3
Sesbania sesban	11.2	9.2	6.7	6.5	3.1	3.2	0.5	1.1	14.8	18.1
Calliandra calothyrsus	15.4	6.9	11.4	21.0	2.9	2.4	1.2	1.8	39.4	71.6
Cajanus cajan	4.2	1.4	14.0	18.9	3.3	3.4	1.2	2.1	79.5	56.7
Leucaena leucocephala	12.2	2.7	12.1	14.0	3.1	2.5	1.1	1.6	37.0	52.0
SED	0.4		0.8		0.2		0.1		7.7	

1992) resulting in lower extractability of tannins and increases protein-bound and fibre-bound condensed tannins. This may partly explain the fact that some MPT species had higher NDF-N in oven-dried material compared with sun-dried material (Table 13.2). Oven-dried materials yielded higher concentrations of lignin with tree species compared with sun-drying (Table 13.2) as previously reported (van Soest, 1982; Papachristou and Nastis, 1994; Dzowela *et al.*, 1995). This can be attributed to the production of artificial lignin via non-enzymatic browning reactions that involve carbohydrate degradation products, proteins and amino acids to form insoluble polymers which will increase lignin content (Goering and van Soest, 1970).

Effects of Age of Prunings

Regrowth prunings had higher concentrations of soluble polyphenols (Table 13.1) except for *Acacia* and *Sesbania* and had a higher concentration of insoluble proanthocyanidins for *Flemingia* and *Acacia* whereas for species like *Calliandra*, *Cajanus*, *Gliricidia* and *Leucaena* first cut materials had a higher concentration of insoluble proanthocyanidins compared with regrowth prunings (Table 13.1). This result can be explained by the fact that different MPT species may contain different types of tannins. Future research should focus on characterization of tannins within provenances of the same species. Higher concentrations of condensed tannins in regrowth materials agrees with the results of Lee and Lowry (1980). Such a result is attributed to the fact that young prunings with a higher concentration of soluble cell contents (e.g. proteins and sugars) need to have a higher tannin content for protection against herbivory (Zucker, 1983; see Harborne, Chapter 4, this volume).

Effect of Pruning Quality on Nitrogen Release

Many studies have shown that N mineralization from MPT prunings can be pre-dicted using (lignin + polyphenol)-to-N ratio (Fox *et al.*, 1990; Handayanto *et al.*, 1994; Constantinides and Fownes, 1994; Lehmann *et al.*, 1995). Incubation studies with sun-dried prunings of the seven MPT species in this study showed that (lignin + soluble polyphenol)-to-N ratio was the chemical quality index which was highly correlated with cumulative N release (Fig. 13.1) (Mafongoya, 1995). MPT species like *Gliricidia* and *Cajanus* with low soluble polyphenols, low lignin and high total N hence a low (lignin + polyphenol)-to-N ratio which was less than 10 released N very rapidly compared with species like *Calliandra*, *Acacia* and *Flemingia* which had high soluble polyphenols and lignin and hence a high (lignin + polyphenol)-to-N ratio and thus released N slowly. Total soluble polyphenols and insoluble proanthocyanidins on their own were not significantly correlated to N release. This stresses the fact that the type of tannins is important in their capacity to bind protein and reduce the degradability of that bound protein N by microbes. Insoluble proanthocyanidins are not reactive hence, they have no effect on N release although they may affect the degradation of the cell wall by microbes. This effect of insoluble tannins on carbon mineralization needs further research.

Soluble polyphenols may be reactive and have a high capacity to bind protein (see Handayanto *et al.*, Chapter 14, this volume). As shown by results of the present study, species like *Flemingia*, *Calliandra* and *Acacia* which had high soluble polyphenol contents released N slowly compared with *Gliricidia*, *Cajanus* and *Sesbania*. *S. sesban* had high soluble polyphenols although it released N very rapidly. These results are in agreement with those of Woodward and Reed (1995), who have also shown that despite the high soluble polyphenol content of *Sesbania*, its dry matter and nitrogen were highly digested by ruminants. These observations may suggest that these polyphenols were not very reactive in terms of protein binding capacity. Similar results were found by Lehmann *et al.* (1995) working with *Gliricidia* and *Calliandra* using prunings of

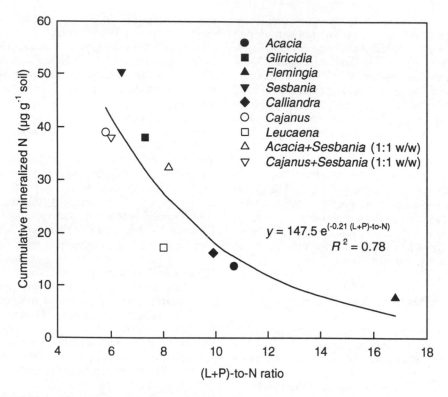

Fig. 13.1. The relationship between cumulative mineralized nitrogen and the (lignin + soluble polyphenol)-to-N ratio of leaves of multipurpose trees.

leaves, twigs and roots of these species. Therefore, it appears that the rate of N release from MPT prunings is a function of the amount of reactive soluble polyphenols rather than the total tannin content of prunings. If oven-drying affects the distribution of soluble polyphenols and insoluble tannins, then the use of oven-dried prunings in decomposition and N mineralization studies may lead to erroneous conclusions about N mineralization and the use of (lignin + polyphenol)-to-N ratio as a predictor of N release. This is especially important to farmers who use fresh or sun-dried prunings as a source of nitrogen to their crops.

Future Research Needs

The following research themes are recommended to elucidate factors that affect quality of MPT prunings and their rate of N release upon application to soils:

- effects of soil pH on stability of plant tannin-protein complexes as these complexes are pH dependent;
- effects of artificially reducing reactive polyphenols by using polyethylene glycol (PEG) on rate and extent of pruning decomposition;
- the influence of fibre-bound or insoluble condensed tannins on rate of prunings decomposition and N release;
- effect of reactive condensed tannins on the rate and extent of N release from MPT prunings.

Conclusion

The chemical composition of prunings differed depending on the drying method, age of prunings and the MPT species examined. Total soluble polyphenols and insoluble tannin contents were not significantly correlated to the amount of N released by prunings of various MPT species. It was the lignin plus polyphenol-to-N ratio that was consistently related to the amount of N released. Partitioning of total tannins into soluble and insoluble tannins appears to be more useful when relating the polyphenol content to N mineralization. The drying method did not alter the total polyphenolic content of most MPT species. However, it significantly affected the distribution of the polyphenols into soluble (reactive) polyphenols and insoluble (non-reactive) polyphenols. This effect was dependent on the MPT species and possibly the type of polyphenols found in that MPT species. The use of (lignin + polyphenol)-to-N ratio to predict N mineralization rate should be used with caution when comparing studies with prunings of different ages which have been dried differently.

Acknowledgements

The work reported in this chapter was supported by a scholarship to the senior author from the International Centre for Research in Agroforestry (ICRAF) through the funding of the Canadian International Development Agency (CIDA) for which the authors are grateful. The authors would also like to thank Dr J. Reed of University of Wisconsin, Madison, USA for the use of his laboratory for chemical analyses.

References

Cano, R., Carulla, J. and Lascano, C. (1994) Métodos de conservaciùn de muestreos de forraje de leguminosas tropicales y su efecto en el nivel y la actividad biologica de los taninos. *Pasturas Tropicales* 16, 2–7.

Constantinides, M. and Fownes, J.H. (1994) Tissue to solvent ratio and other factors affecting determination of soluble polyphenolics in tropical leaves. *Communications in Soil Science and Plant Analysis* 25, 3221–3227.

Dzowela, B.H., Hove, L. and Mafongoya, P.L. (1995) Effect of drying method on chemical composition and *in vitro* digestibility of multipurpose tree and shrub fodders. *Tropical Grasslands* 29, 263–269.

Fox, R.H., Myers, R.J.K. and Vallis, I. (1990) The nitrogen mineralization rate of legume residues in soil as influenced by their polyphenol, lignin and nitrogen content. *Plant and Soil* 129, 251–259.

Goering, H.K. and van Soest, P.J. (1970) *Forage Analyses.* ARS, USDA Agriculture Handbook No. 379, 20 pp.

Gilmour, J.T., Clark, M.D. and Sigua, G.C. (1985) Estimating net nitrogen mineralization from carbon dioxide evolution. *Soil Science Society of America Journal* 49, 1398-1402.

Handayanto, E., Cadisch, G. and Giller, K.E. (1994) Nitrogen release from prunings of legume hedgerow trees in relation to quality of prunings and incubation method. *Plant and Soil* 160, 237–248.

Hagerman, A.E. (1988) Extraction of tannins from fresh and preserved leaves. *Journal of Chemical Ecology* 14, 453–461.

Herman, W.A., McGill, W.B. and Dorman, J.F. (1977) Effects of initial chemical composition on decomposition of roots of three grass species. *Canadian Journal of Soil Science* 51, 205–215.

Lee, D.W. and Lowry, J.B. (1980) Young leaf anthocyanadin and solar ultraviolet. *Biotropica* 12, 75–76.

Lehmann, J., Schroth, G. and Zech, W. (1995) Decomposition and nutrient release from leaves, twigs and roots of three alley-cropped tree legumes in Central Togo. *Agroforestry Systems* 29, 21–36.

Mafongoya, P. (1995) Multipurpose tree prunings as source nitrogen to maize (*Zea mays L.*)

under semi-arid conditions in Zimbabwe. PhD thesis, University of Florida, USA.

Oglesby, K.A. and Fownes, J.H. (1992) Effects of chemical composition on nitrogen mineralization from green manures of seven tropical leguminous trees. *Plant and Soil* 143, 127–132.

Palm, C.A. and Sanchez, P.A. (1991) Nitrogen release from the leaves of some tropical legumes as affected by their lignin and polyphenolic content. *Soil Biology and Biochemistry* 23, 83–88.

Palm, C.A. (1995) Contribution of agroforestry trees to nutrient requirements of intercropped plants. *Agroforestry Systems* 30, 105–124.

Papachristou, T.G. and Nastis, A.S. (1994) Change in chemical composition and *in vitro* digestibility of oesophageal fistula and hand picked forages samples due to drying method and stage of maturity. *Animal Feed Science and Technology* 46, 87–95.

Reed, J.D., MacDowell, R.E., van Soest, P.J. and Horvath, P.J. (1982) Condensed tannin a factor limiting the use of cassava forage. *Journal of the Science of Food and Agriculture* 33, 213–220.

Reed, J.D., Horvath, P.J., Allen, M.S. and van Soest, P.J. (1985) Gravimetric determination of soluble phenolics including tannins from leaves by precipitation with trivalent ytterbium. *Journal of the Science of Food and Agriculture* 36, 255–261.

Scalbert, S. (1992) Quantitative methods for the estimation of tannins in plant tissues. In: Hemingway, R.W. and Laks, P.E. (eds) *Plant Polyphenols, Synthesis Properties and Significance*. Plenum Press, New York, pp. 259–286.

Terrill, T.H., Windham, W.R., Evans, J.E. and Hoveland, C.S. (1990) Condensed tannins in *Sericea lespedeza*: Effect of preservation method on tannin concentration. *Crop Science* 30, 219–224.

Terrill, T.H., Windham, W.R., Evans, J.E. and Hoveland, C.S. (1994). Effect of drying method and condensed tannin on detergent fibre analyses of *Sericea lespedeza*. *Journal of the Science of Food and Agriculture* 66, 337–343.

Van Soest, P.J. (1982) *Nutritional Ecology of the Ruminant*. O&B Books, Corvallis, Oregon, 374 pp.

Woodward, A. and Reed, J.D. (1995) Intake and digestibility for sheep and goats consuming supplementary *Acacia brevispia* and *Sesbania sesban*. *Animal Feed Science and Technology* 56, 207–216

Zucker, W.V. (1983) Tannins: does structure determine function? An ecological perspective. *The American Naturalist* 121, 335–365.

14 Regulating N Mineralization from Plant Residues by Manipulation of Quality

E. Handayanto[1], G. Cadisch[2] and K.E. Giller[2]

[1] *Fakultas Pertanian, Universitas Brawijaya, Jalan Veteran, Malang 65145, Indonesia;*
[2] *Department of Biological Sciences, Wye College, University of London, Wye, Ashford, Kent TN25 5AH UK*

Introduction

Rates of decomposition and nutrient release from legume tree prunings determine the short-term benefits of the prunings to crop N uptake. The rate of N mineralization of a specific type of plant residues, given favourable environmental conditions, has been shown to be largely determined by a variety of chemical and physical plant quality parameters (see Heal *et al.*, Chapter 1; Chesson, Chapter 3; Vanlauwe *et al.*, Chapter 12; this volume). The importance of plant polyphenols has been emphasized in regulating decomposition of both forage legume litter (Vallis and Jones, 1973) and tree legume prunings (Fox *et al.*, 1990; Palm and Sanchez, 1991). The role of polyphenols has been further refined by the recognition that the degree of retardation of decomposition depends on the form of polyphenols present; hydrolysable polyphenols have less effect than condensed polyphenols because they are less reactive with N (Harborne, Chapter 4, this volume; Handayanto *et al.*, 1995a,b). Plant residues poor in N, but with large concentrations of lignin and active polyphenols decompose and release N slowly, so that little of the plant N applied is available for the succeeding crop although it remains in the soil (Cornforth and Davis, 1968). Residues rich in N, but with small lignin and polyphenol concentrations decompose rapidly and supply a large amount of nitrogen during the early periods of crop growth, but may not contribute much to the maintenance of soil organic matter. Under field conditions the contribution of pruning N to crop nutrition is often small and crop N recoveries from prunings are often less than 20% (Giller and Cadisch, 1995). The poor efficiency of nitrogen use from high quality prunings, from trees such as *Gliricidia sepium*, is often attributed to a lack of synchrony between crop N demand and nitrogen release coupled with losses of N due to leaching (Myers *et al.*, Chapter 17, this volume). In this chapter we explore possible management interventions for altering the N release patterns by manipulation of the quality of plant residues in order to achieve a better synchrony of N supply and crop demand.

Approaches to Manipulate Plant Quality

Modifying residue structure

The quality of plant materials can be altered physically by fractionation of materials into different sizes (Vanlauwe *et al.*, Chapter 12, this volume). Grinding (<2 mm) of residues significantly enhanced decomposition (measured as CO_2 evolution) of low quality (high C-to-N ratio) straw in the early stages of decomposition but had little effect on high quality (low C-to-N ratio) lentil green manures (Bremer *et al.*, 1991). The increased decomposition of low quality residues may be attributed to the increased accessibility of moderately available C (e.g. cellulose, hemicellulose) to microbial attack by providing a greater available surface area and by reducing the protection of degradable C compounds by recalcitrant constituents such as cutin or lignin (see Chesson, Chapter 3, this volume). Jensen (1994) confirmed that with narrow C-to-N ratio residues, decomposition of different sized materials varied only slightly. Despite the small effect of residue size on CO_2 evolution from high quality residues, the particle size had a profound effect on the duration of the initial immobilization phase during decomposition. Immobilization of soil N was stronger and lasted for a longer time with finely-ground rather than large particles while the final amount of net N release may have been similar (an important prerequisite if the N use efficiency by crops is to be maximized by manipulating residue quality). The larger N immobilization observed with small particles may be due to a more intimate mixing with soil, resulting in greater stabilization of microorganisms and the products of degradation by clay particles. While there is some potential for manipulating crop residue decomposition by reducing particle size in order to conserve N, fine grinding of plant residues is rarely likely to be a feasible practice in most agricultural systems.

Alterations in chemical quality of residues

Quality of tree prunings can most easily be manipulated by pruning trees at different ages. The stage of growth affects the degree of lignification as well as other constituents, such as the polyphenol content of the prunings (see Vanlauwe *et al.*, Chapter 12, this volume). Alleviation of environmental stress conditions, such as drought or soil acidity, is also likely to lead to changes in the chemical composition of plants and their residues (see Heal *et al.*, this volume). Pest and disease attack can induce plant responses resulting in changes in plant tissue chemistry, such as the increased lignification of cell walls found in response to fungal attack. Drying conditions for residues, once they are removed from the plant, can alter the amount and balance of types of polyphenols. High temperature drying can reduce the amount of 'active' polyphenols able to bind proteins (Mafongoya *et al.*, Chapter 13, this volume). New techniques for genetic manipulation of plant species open further possibilities to alter the chemistry of lignins and polyphenols in plant tissues (Bavage *et al.*, Chapter 16, this volume), although in many plant families there is still a vast variability of both species and ecotypes which offers a wide range of plant residue qualities to be explored in future.

Manipulating plant residue quality by varying N supply

Determination of key factors regulating litter decomposition in studies involving several different plant species is complicated because several quality factors, including physical properties such as leaf thickness, can differ at the same time. Therefore, we manipulated the quality of prunings of individual species by growing saplings of tropical legume trees in a glasshouse with different N concentrations supplied as $K^{15}NO_3$ in the nutrient solution (Handayanto *et al.*, 1995b). Two species were used: *Calliandra calothyrsus* which had been reported to have high N and polyphenol concentrations, and *Gliricidia sepium* which is known to have low polyphenol but high N concentrations (Handayanto *et al.*, 1994; Little *et al.*, 1988). After 7 months growth, the trees were pruned and the prunings analysed after drying at 60°C and grinding to <2 mm. In both *Calliandra* and *Gliricidia* prunings, the tissue N concentration increased with increasing N concentration in the nutrient solution whereas total extractable polyphenol concentration (measured according to Anderson and Ingram, 1989), protein-binding capacity (Handayanto *et al.*, 1994) and the C-to-N ratio all decreased (Fig. 14.1). Acid-detergent lignin concentration

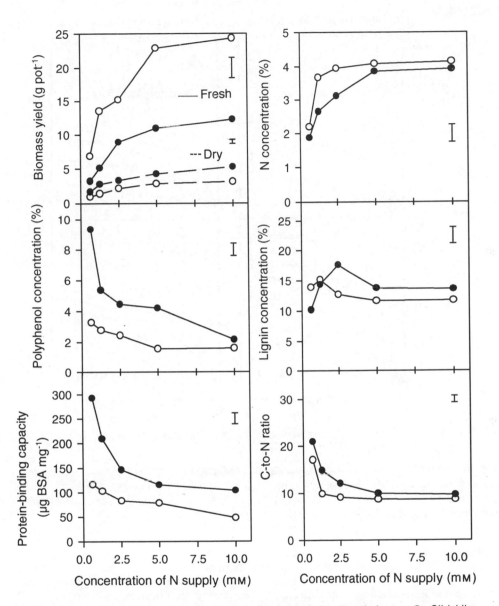

Fig. 14.1. Chemical composition of legume tree prunings (●, *Calliandra calothyrsus*; ○, *Gliricidia sepium*) grown with different N supply. Bars represent SED (from Handayanto *et al.*, 1995b).

(Goering and Van Soest, 1970) was not significantly affected by the N treatment.

The N mineralization behaviour of the legume tree prunings was then tested upon incorporation in soil in a laboratory incubation experiment. As some polyphenols are water-soluble compounds that are capable of binding protein (Haslam,

1989), incubation experiments were performed under controlled leaching and non-leaching conditions. The %N mineralized from *Calliandra* prunings ranged between 19% (from the prunings of plants which had been grown with low N supply) to 44% (from prunings of plants which had been grown with high N supply) after 14

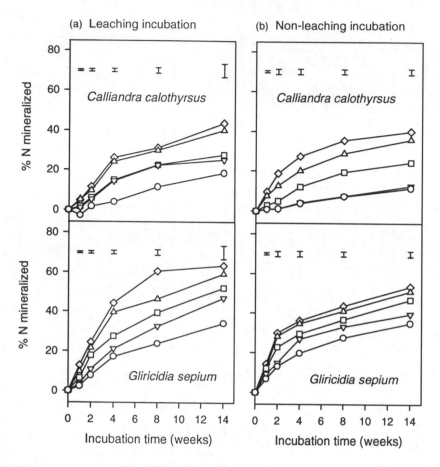

Fig. 14.2. Percent of the initial added N of *Calliandra calothyrsus* and *Gliricidia sepium* mineralized when mixed with soil under leaching (a) and non-leaching (b) conditions. Prunings derived from plants grown under different N regimes (○ 0.625, ▽ 1.25, □ 2.5, △ 5.0,◇ 10.0 mм N). Bars represent SED (from Handayanto *et al.*, 1995b).

weeks of incubation under leaching conditions (Fig. 14.2a). Greater amounts were mineralized under leaching conditions than under non-leaching conditions (Fig. 14.2b). The reason for this is unclear but the reduced overall net mineralization of N may be due to greater N immobilization stimulated by soluble C compounds which were removed during the leaching incubation. The N mineralization rate constants (k) were significantly related to all quality parameters, except lignin (Table 14.1). The regression between these k rate constants and the protein-binding capacity had consistently the largest coefficient of determination when the prunings were incubated under non-leaching conditions, suggesting that soluble polyphenols present actually interfere with net mineral-N release from the residues in soil. Under leaching conditions, however, the (lignin + polyphenol)-to-N ratio had an even greater coefficient of determination than the protein-binding capacity (Table 14.1). This could be due to leaching of soluble polyphenols during the early stage of decomposition resulting in greater amounts of N mineralization than when incubations were conducted under non-leaching conditions. Protein complexes

Table 14.1. Coefficients of determination for linear regressions between N release rate constants (k) and composition of *Calliandra* and *Gliricida* (N supply experiment; Handayanto *et al.*, 1995b) and *Gliricida* and *Pelthophorum* (mixture experiment) prunings under leaching or non-leaching conditions.

Composition of legume tree prunings	Coefficients of determination (R^2) N mineralization rate constants, k (week^{-1})			
	Leaching incubation		Non-leaching incubation	
	N supply	Mixture	N supply	Mixture
N (%)	0.67**	0.74*	0.69**	0.71*
Polyphenol (%)	0.66**	0.77*	0.78***	0.70*
Protein-binding capacity (μgBSA mg^{-1})	0.72**	0.87*	0.87***	0.81*
Lignin (%)	0.09	0.78*	0.04	0.75*
C-to-N ratio	0.62**	0.77*	0.65**	0.76*
Polyphenol-to-N ratio	0.54*	0.79*	0.66*	0.76*
Ligning-to-N ratio	0.71**	0.81*	0.64**	0.81*
(Lignin+polyphenol)-to-N ratio	0.77***	0.81*	0.81***	0.81*

*, **, *** Significance of the respective correlation coefficients at $P < 0.05$, <0.01, <0.001.

formed by phenolic compounds in solution can be either soluble or insoluble (Mole and Waterman, 1985). It is thus also possible that the soluble complexes formed had been dissociated by excess of water under leaching conditions, thus liberating proteins which were previously bound by the polyphenols. Mitaru *et al.* (1984) reported that a reduction in tannin concentration of high-tannin sorghum grains after soaking with water was due to solubilization of the tannins. This was followed by increased polymerization which resulted in a large increase in the molecular weight of tannins, and reduced their reaction with proteins.

To verify the relevance of the incubation mineralization tests in evaluating N availability to a growing crop, recovery of N from incorporated ^{15}N-labelled residues by maize as a catch crop was assessed in pots. In accordance to the mineralization results the recovery of pruning N by maize was greater with *Gliricidia* prunings than with *Calliandra* (data not presented). Prunings of *Calliandra* and *Gliricidia* which had been grown with better N supply gave greater %N recovery in maize than prunings which had been grown under N stress even though the same amount of N had been added in each case. There was a par-

ticularly poor recovery of N (4.2%) from the *Calliandra* prunings which had been grown with limited N supply, which was closely related to its high protein-binding capacity and the small proportion of N released during decomposition (Fig. 14.2b). Thus, while we were able to alter the chemical quality of prunings and reduce the speed of the initial N release, the total amount of N mineralized was also strongly reduced.

Mixing of Residues of Different Quality

Mixing of legume tree prunings from different species

A second approach to manipulate plant quality is mixing prunings of high quality with prunings of low quality. For our experiments *Peltophorum dasyrrachis* was chosen as a low quality pruning (low N, high lignin and polyphenol contents) and mixed with high quality *Gliricidia sepium*. Prunings (small stems and leaves) of the two species, grown in the glasshouse or collected from the field in Lampung, Indonesia were mixed in different proportions (Fig. 14.3). The N concentration of

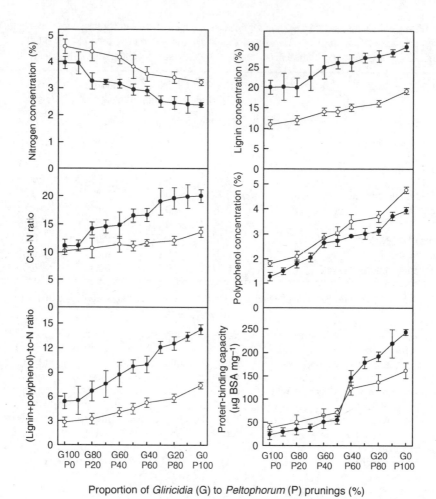

Proportion of *Gliricidia* (G) to *Peltophorum* (P) prunings (%)

Fig. 14.3. Changes in chemical composition of prunings mixtures upon mixing different proportions of *Gliricidia sepium* and *Pelrophorum dasytrachis* pruning materials. ○ prunings from the glasshouse, ● prunings from the field. Bars represent SEM.

the pruning mixtures decreased with increasing proportion of *Peltophorum* prunings to *Gliricidia* prunings, whereas lignin concentration, polyphenol concentration and C-to-N ratio increased. The protein-binding capacities of the pruning mixtures up to 50:50 w/w were almost constant and relatively close to the protein-binding capacity of pure *Gliricidia*. When more than 50% of *Peltophorum* prunings was mixed with *Gliricidia* prunings, the protein-binding capacity of the pruning mix-

tures increased considerably (Fig. 14.3). This could be due to the saturation of protein with soluble reactive polyphenols of the *Peltophorum* prunings when the mixture consisted of more than 50% of the prunings. The differences in quality between the prunings collected from the field and those from the glasshouse were due to the difference in age of the prunings and the different growth conditions. These complex, non-linear interactions in quality found by mixing residues of different

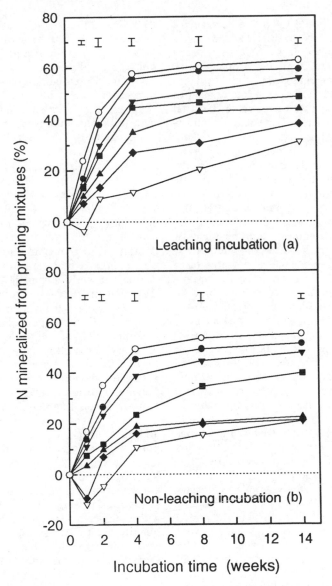

Fig. 14.4. Percentages of the initial added N of *Gliricidia sepium* : *Peltophorum dasyrrachis* mixtures (○ 100:0, ● 80:20, ▼ 60:40, ■ 50:50, ▲ 40:60, ◆ 20:80, ▽ 0:100) mineralized when mixed with soil under leaching (a) and non-leaching (b) conditions. Bars represent SED.

species indicate that attempts to manipulate residue quality using polyphenol-rich prunings may yield unexpected results. As polyphenol-protein interactions may be specific for different polyphenols as well as for different proteins (Asquith and Butler, 1986), it may be speculated that mixtures of different pruning materials will also

result in different patterns of quality interactions.

After 14 weeks of incubation, the proportions of N mineralized from the pruning mixtures which were recovered as mineral-N in soil ranged from 31% to 63% (leaching incubation) and from 21% to 55% (non-leaching incubation) of the initial added N (Fig. 14.4). The decrease of the resulting N mineralization with increasing proportion of *Peltophorum* prunings in the mixtures indicated that N mineralization of prunings can be manipulated by mixing different quality materials but again at the expense of reducing the total amount of N released, at least in the short-term. By mixing with slow N release legume prunings, such as *Peltophorum*, losses of N mineralized from the *Gliricidia* prunings may be minimized and the *Peltophorum* prunings can contribute to the maintenance of soil organic matter and hence increase the residual benefits of the prunings in subsequent cropping seasons. This will depend, however, on whether the N remaining is in easy decomposable or in recalcitrant forms. As observed in the previous experiment, the protein-binding capacity and the (lignin + polyphenol)-to-N ratio were among the best quality indicators for N mineralization (Table 14.1). However, in the mixing experiment the k value was best correlated with the protein-binding capacity under leaching conditions. In these pruning mixtures a degree of leaching may have been important in ensuring the movement of soluble polyphenols from the *Peltophorum* prunings so that they were able to bind proteins from the *Gliricidia* prunings and thus influence N mineralization.

The recovery of ^{15}N labelled *Gliricidia* N by maize in a subsequent pot experiment was reduced, from 40% of the added N when *Gliricidia* prunings alone were applied down to 8% when 80% of the applied material consisted of *Peltophorum*. This indicates that there was a significant reduction in the availability of N from the *Gliricidia* prunings caused by the presence of increasing amounts of *Peltophorum* prunings. The non-linear changes in protein-binding capacity (Fig. 14.3) in the pruning mixtures was closely related to N release in incubation experiments (Fig. 14.4) and therefore the reduced N recovery in mixtures with *Peltophorum* were probably caused by the binding of protein-N by active polyphenols. In an earlier study, Mulongoy *et al.* (1993) showed that by mixing prunings of *Leucaena leucocephala*

(3.3%N, 14% lignin and 1.5% polyphenols) with prunings of *Senna siamea* (2.4% N, 26% lignin and 2.2% polyphenols) the N release rate constant of *Leucaena* decreased whereas that of *Senna* increased. Thus while the N release from prunings can be manipulated by mixing with different quality materials, interactions may be unexpected.

Mixing of different plant parts

Instead of mixing prunings of different species Collins *et al.* (1990) mixed different plant parts of wheat residues (stems, leaf sheath, chaff and leaf blades). They demonstrated that cumulative CO_2 evolution of mixed materials was 25% greater during the early stages of decomposition than that predicted by summing cumulative CO_2 evolved from the individual components. They hypothesized that the decomposition of the mixture was stimulated by fungal hyphal extensions from residue components with high available substrate concentration to adjacent components with low substrate concentration. CO_2 evolution interactions between single or mixed leaves and stems of rye, wheat and oat during the early stages of decomposition were also observed by Quemada and Cabrera (1995). Their decomposition test further highlighted that the increased C decomposition of mixtures of leaves and stems (50:50 weight basis) was associated with a higher microbial activity that consequently led to an increase in N immobilization, that is the cumulative N mineralized of the mixture was smaller than that predicted from the incubation of isolated leaves and stems (Fig. 14.5). The strong initial retardation of N release from the wheat mixture was caused by the high soluble C content (13.4% C, cold water soluble) of the wheat stems compared with the wheat leaves (5.5%). In order to manipulate synchrony, such interactions in N mineralization are particularly interesting if they occur early in the decomposition process as in the case of the mixed wheat residues, as this is the time when plant demand is low but leaching losses may often be high.

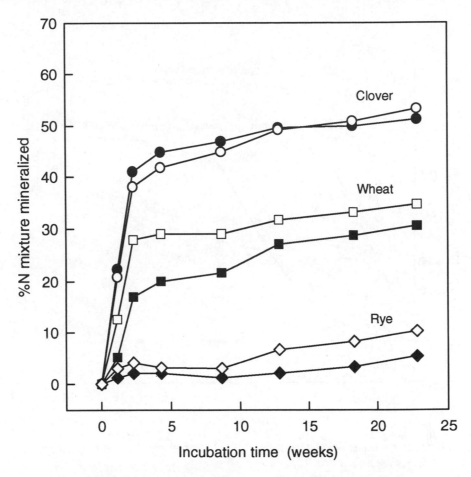

Fig. 14.5. Cumulative net mineralization of the initial added N in mixed leaves and stems of clover, wheat or rye residues applied to soil (closed symbols, actual mineralization). Predicted mineralization (open symbols) was calculated as the sum of the mineralization results of isolated leaf and stem incubations (adapted from Quemada and Cabrera, 1995).

Conclusion

The results of these studies demonstrate that chemical and physical attributes of plant residue quality can be manipulated in order to alter the rate of N release and N uptake by crops. Whereas grinding and N supply to growings plants led to more predictable changes in mineralization patterns mixing of residues of differing quality resulted in unexpected complex interactions in quality and decomposition behaviour of the mixtures. The impact mixing has on N release is increased where one of the residues has a high soluble C content or a high content of soluble, active polyphenols.

In the case of residues with a high content of soluble, 'active' polyphenols, the protein-binding capacity and the combined (lignin + polyphenol)-to-N ratio were consistently among the best quality indicators of N release. The circumstances (degree of leaching) under which 'active' polyphenols most strongly modified N release depended on whether single species prunings or pruning mixtures of different species were applied. In the field leaching and non-leaching conditions usually alternate and consequently change the impor-

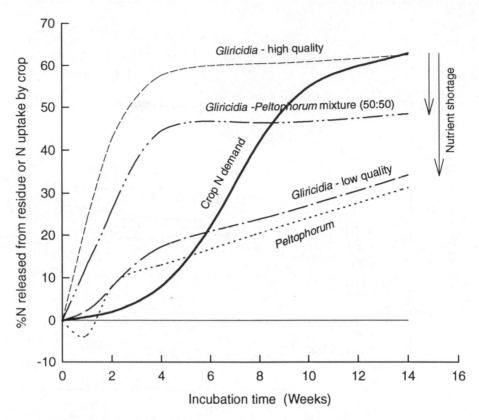

Fig. 14.6. Effect of manipulated plant residue quality on the hypothesized degree of synchrony between residue N release and crop N uptake.

tance of different chemical characteristics. When plant residues have high polyphenol concentrations, a direct measurement of the capacity of polyphenols in the prunings to bind protein is worth including as a quality factor.

Our primary aim, using manipulation of plant quality and hence decomposition, was to improve synchrony between N release from high quality prunings and crop N demand (Fig. 14.6) in order to avoid nutrient losses under high rainfall conditions. Most manipulation approaches while reducing the initial high N release also reduced the total amount of available N in the longer term and thereby potentially leading to a crop N shortage (Fig. 14.6). Thus synchrony of N release and plant N demand would only partly be achieved. It remains to be tested if the approaches explored may be more successful for increasing N recovery efficiency under field conditions.

Acknowledgements

We thank the European Community and the Government of the Republic of Indonesia for financial support.

References

Anderson, J.M. and Ingram, J.S.I. (1989) *Tropical Soil Biology and Fertility: A Handbook of Methods.* CAB International, Wallingford, 171 pp.

Asquith, T.N. and Butler, L.G. (1986) Interactions of condensed tannins with selected proteins. *Phytochemistry* 25, 1591–1593.

Bremer, E., van Houtum, W. and van Kessel, C. (1991) Carbon dioxide evolution from wheat and lentil residues as affected by grinding, added nitrogen and the absence of soil. *Biology and Fertility of Soils* 11, 221–227.

Collins, H.P., Elliot, L.F., Rickman, R.W., Bezdicek, D.F. and Papendick, R.I. (1990) Decomposition and interactions among wheat residue components. *Soil Science Society of America Journal* 54, 780–785.

Cornforth, I.S. and Davies, J.B. (1968) Nitrogen transformations in tropical soils. I. The mineralization of nitrogen-rich organic materials added to soil. *Tropical Agriculture (Trinidad)* 45, 211–221.

Fox, R.H., Myers, R.J. and Vallis, I. (1990) The nitrogen mineralization rate of legume residues in soils as influenced by their polyphenol, lignin and nitrogen contents. *Plant and Soil* 129, 251–259.

Giller, K.E. and Cadisch, G. (1995) Future benefits from biological nitrogen fixation in agriculture: An ecological approach. *Plant and Soil* 174, 255–277.

Goering, H.K. and van Soest, P.J. (1970) *Forage fibre analyses (Apparatus, reagents, procedures, and some applications). Agriculture Handbook*, USDA, No. 379, 1–19.

Handayanto, E., Cadisch, G. and Giller, K.E. (1994) Nitrogen release from prunings of legume hedgerow trees in relation to quality of the prunings and incubation method. *Plant and Soil* 160, 237–248.

Handayanto, E., Cadisch, G. and Giller, K.E. (1995a) Decomposition and nitrogen mineralization of selected hedgerow tree prunings. In: Cook, H.F. and Lee, H.C. (eds) *Soil Management in Sustainable Agriculture.* Wye College Press, Wye, Ashford, Kent, UK, pp. 113–122.

Handayanto, E., Cadisch, G. and Giller, K.E. (1995b) Manipulation of quality and mineralization of tropical legume tree prunings by varying nitrogen supply. *Plant and Soil* 176, 149–160.

Haslam, E. (1989) *Plant Polyphenols: Vegetable Tannins Revisited.* Cambridge University Press, Cambridge, 230 pp.

Jensen, E.S. (1994) Mineralization-immobilization of nitrogen in soil amended with low C:N ratio plant residues with different particle sizes. *Soil Biology and Biochemistry* 26, 519–521.

Little, D.A., Kompiang, S. and Petheram, R.J. (1988) Mineral composition of Indonesia ruminant forages. *Tropical Agriculture (Trinidad)* 66, 33–37.

Mitaru, B.N., Reichert, R.D. and Blair, R. (1984) The binding of dietary protein by sorghum tannins in the digestive tract of pigs. *Journal of Nutrition* 114, 1787–1796.

Mole, S. and Waterman, P.G. (1985) Stimulatory effects of tannins and cholic acid on tryptic hydrolysis of proteins: ecological implications. *Journal of Chemical Ecology* 11, 1323–1332.

Mulongoy, K., Ibewiro, E.B., Oseni, O., Kilumba, N., Opara-Nadi, A.O. and Osonubi, O. (1993) Effect of management practices on alley-cropped maize utilization of nitrogen derived from prunings on a degraded Alfisol in south-western Nigeria. In: Mulongoy, K. and Merckx, R. (eds) *Soil Organic Matter Dynamics and Sustainability of Tropical Agriculture.* John Wiley & Sons, Chichester, UK, pp. 223–230.

Palm, C.A. and Sanchez, P.A. (1991) Nitrogen release from the leaves of some tropical legumes as affected by their lignin and polyphenolic contents. *Soil Biology and Biochemistry* 23, 83–88.

Quemada, M. and Cabrera, M.L. (1995) Carbon and nitrogen mineralized from leaves and stems of four cover crops. *Soil Science Society of America Journal* 59, 471–477.

Vallis I., and Jones, R.J. (1973) Net mineralization of nitrogen in leaves and leaf litter of *Desmodium intortum* and *Phaseolus atropurpureus* mixed with soil. *Soil Biology and Biochemistry* 5, 391–398.

15 Climate Change: The Potential to Affect Ecosystem Functions through Changes in Amount and Quality of Litter

W.J. Arp[1], P.J. Kuikman[2] and A. Gorissen[2]

[1] *Department of Terrestrial Ecology and Nature Conservation, Wageningen Agricultural University, Bornsesteeg 69, 6708 PD Wageningen, The Netherlands;* [2] *DLO-Research Institute for Agrobiology and Soil Fertility (AB-DLO), PO Box 14, 6700 AA Wageningen, The Netherlands*

Introduction

A higher concentration of CO_2 in the atmosphere and an associated temperature rise are now considered highly likely to take place in the next 50–100 years (IPCC, 1995). Less well documented are potential changes in the terrestrial ecosystems. It is these ecosystems that most likely sequester C that is released by human activities but cannot be accounted for in global C budgets (Tans *et al.*, 1990).

Storage of C and nutrients in the various compartments of terrestrial ecosystems and the fluxes between these compartments not only depend on abiotic conditions but also on biotic processes like fixation of CO_2 in the photosynthetic process, conversion to plant mass, transfer to litter (or harvested products), decomposition by heterotrophic biota and conversion into soil organic matter.

The success of plant and decomposer communities depends on the balance between gains and losses of nutrients and the possibility of meeting their respective nutrient requirements both in the short (1–5 years) and long term (>10 years). Turnover rates of different C and nutrient stores range from days (microbes) to centuries (soil organic matter). Since the changes in the climate will be gradual, we might expect ecosystems to change gradually as well, and possibly adapt to the new

conditions. Usually, only short term ecosystem responses to climate change and elevated CO_2 concentration are measured through experimentation. Long-term predictions are obtained from calculations with simulation models.

Uncertainties relate to whether a gradually rising atmospheric CO_2 concentration will alter ecosystem functioning in the long term by:

1. changing the C and N content, and productivity of terrestrial ecosystems and its components, resulting in gains or losses of C and N from soils and vegetation;
2. adaptations of plant and decomposer communities resulting in a different ecosystem succession.

It has been stated that CO_2 fertilization lowers the concentration of N in the plant organs and increases the C-to-N ratio of the plant material (Bottner and Couteaux, 1991). This altered quality then slows down the decomposition rate. This theory actually consists of *three* separate hypotheses comprising not only changes in plant litter quality but also changes in the quantities of C and N in plant litter and the effects of both quantity and quality on the rate of decomposition and mineralization. The decomposability of residues and litter is, at least in part, a function of their chemical composition. Predictors or indicators of decomposability include C-to-N ratio and lignin-to-N ratio for the initial (<1 year) decomposition phase.

We will examine how elevated CO_2 can directly affect plant functioning, how litter produced in a high CO_2 environment influences soil functioning, and finally how interactions between plant and soil could determine the functioning of the ecosystem in the long term. Unknowns will be identified and suggestions made for further monitoring and experimentation. Since a quantitatively important flow of C and nutrients to soils is through root littering and other losses from live roots (in agricultural systems most of the shoots are removed) and the most pronounced effects of a higher CO_2 concentration are probably belowground (Rogers *et al.*, 1994), we will explicitly discuss roots.

Plant Functioning

At an elevated CO_2 concentration, plant biomass is often increased in the short term (Kimball, 1983). The capacity to maintain a higher productivity at elevated CO_2 in the long term depends on how the plants interact with other ecosystem factors, such as the availability of nutrients and water. The balance between uptake, efficiency of use, and loss of nutrients or water determines the success of a plant species and its competitive strength. A high CO_2 concentration can directly affect these three processes, and consequently influence this balance.

Acquisition of nutrients

The possibility that elevated CO_2 may stimulate nitrogen and water uptake is often overlooked in long term predictions (Rastetter *et al.*, 1991). Increased root growth and an increased capacity for nutrient uptake per unit root may increase the uptake per plant (Field *et al.*, 1992). Plants may also allocate more carbon to the fine roots, which could stimulate mycorrhizal infection (O'Neill *et al.*, 1987; Diaz *et al.*, 1993). Ericoid mycorrhiza, mutualistic associations in many nutrient limited ecosystems, are able to actively participate

in the mobilization of nitrogen from organic matter, and supply their hosts with nitrogen (Leake and Read, 1989; 1990). Allocation of carbohydrates to *Rhizobium* or other nitrogen fixing symbionts may provide another effective way to permanently increase the nutrient availability (Norby, 1987). These different processes may all enhance the nitrogen uptake capacity and some also the nitrogen availability at elevated CO_2.

Nitrogen use efficiency and plant tissue quality

One of the most frequently reported effects of growing plants in an elevated CO_2 concentration is the increase in the C-to-N ratio, or rather a decrease in the nitrogen concentration of the plant (Cure *et al.*, 1988; Johnson and Lincoln, 1990; Bottner and Couteaux, 1991; Owensby *et al.*, 1993). However, when the C-to-N ratios of different plant parts are compared, it is often found that only the C-to-N ratio of the leaves is significantly increased by elevated CO_2 (Cure *et al.*, 1988; Den Hertog and Stulen, 1990).

There are several mechanisms by which elevated CO_2 can cause an increase in the C-to-N ratio. When plant growth is limited by carbon rather than by nitrogen, enhanced growth at elevated CO_2 is possible by diluting the nitrogen in all plant parts. Under nitrogen limiting conditions at elevated CO_2, nitrogen can be used more efficiently in photosynthesis, allowing some nitrogen to be allocated to other processes, or to increased growth (Pearcy and Bjorkman, 1983; Hilbert *et al.*, 1991). This will decrease the N content of leaves, but not necessarily decrease the nitrogen concentration of non-photosynthetic plant parts. CO_2 may increase the C-to-N ratio of these tissues if the production of nitrogen-poor compounds such as lignin, cellulose, and secondary metabolites is stimulated by an increased availability of carbohydrates. The C-to-N ratio will also increase when non-structural carbohydrates accumulate under elevated CO_2. Due to the increased carbon fixation, more carbohydrates will be allocated to storage

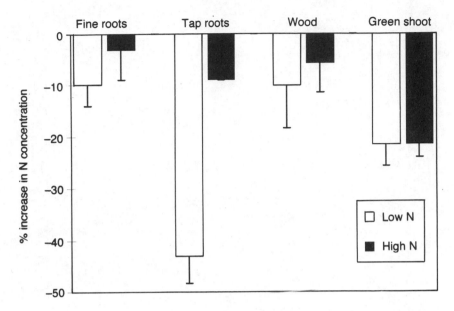

Fig. 15.1. The effect of elevated CO_2 on the nitrogen concentration of four different plant parts (% reduction in N concentration relative to ambient atmospheric CO_2). The 'tap root' category consists of data for both tap roots and coarse roots, while data for green stems, leaves and green shoots are combined under 'green shoot'. For each category the data are separated for plants grown under low and high nitrogen supply. The mean values and the standard errors are shown. The complete set of data and references are given in Table 15.1.

organs such as tap roots or rhizomes, or when the sink for carbohydrates is limited, starch may also accumulate in the leaves (DeLucia *et al.*, 1985).

When these different effects are combined, one would expect an increase in the C-to-N ratio of leaves and other photosynthetically active parts, caused by an increased nitrogen use efficiency (NUE) and a possible accumulation of starch. The C-to-N ratio of storage organs will increase because of increased accumulation of carbohydrates, while increased lignification may primarily affect the woody parts of the plant. The nitrogen concentration of all plant parts will also depend on the nitrogen availability. At a high nitrogen supply growth may be balanced by increased nitrogen uptake, at a lower supply of N, growth may increase without a corresponding uptake of nitrogen, while at severely limiting nitrogen supply neither growth nor N uptake may be increased.

A survey of the available data confirms that nitrogen content depends on plant part and on nitrogen supply (Fig. 15.1, Table 15.1). The reduction in nitrogen concentration is larger in leaves and green stems than in woody stems. The difference in N concentration is smallest in the fine roots, but very large (10–40%) in tap roots and coarse roots. For this reason, the separate effects on the different plant parts must be distinguished when referring to the effect of elevated CO_2 on the C-to-N ratio of plants. A consequence of this finding is that a change in the biomass distribution pattern at elevated CO_2 will also influence the effect on the C-to-N ratio of the whole plant. Another difference which becomes apparent from Table 15.1 is that for roots and stems the relative reduction in N content is larger for plants grown in low N.

Nutrient loss and plant litter quality

Often the connection is made between an increased C-to-N ratio in the plant and its effect on decomposition, while ignoring

Table 15.1. A compilation of data from the literature on the nitrogen concentration of plants grown under elevated CO_2. Data are shown for different plant parts, and for plants grown under high and low nitrogen. The elevated CO_2 concentration is given in ppm. The percent increase in N content is defined as $((\text{N content of plants grown at high } CO_2 / \text{N content of ambient } CO_2 \text{ grown plants}) - 1) \times 100$.

			% increase in N concentration at high CO_2															
			Fine roots		Coarse root		Tap root		Woody stem		Green stem		Leaf		Green shoot		Leaf litter	
Author	Species	CO_2	Low N	High N	Low N	High N	Low N	High N	Low N	High N	Low N	High N	Low N	High N	Low N	High N	Low N	High N
Couteaux *et al.* 91	Chestnut	700	6.0		−33.0		−45.0		−25.3								−47.9	
Curtis *et al.* 89	Scirpus olneyi	686														−15.4		
Den Hertog and Stulen 90	Plantago major	700		1.8												−22.1		
	Urtica dioica	700		3.4												−9.2		
Diaz *et al.* 93	Rumex obtusifolius	700				−8.9												
Gries 93	sour orange	650												−13.0				
Hocking and Meyer 85	Xanthium occidentale	1500	−14.7	28.1											−23.4	−40.8		
		1500									−23.7	−18.2						
Hocking and Meyer 91	wheat	650											−26.5	−41.4	−21.8	−36.7		
Johnson and Lincoln 91	sagebrush	650											−15.3	−25.8				
Larigauderie *et al.* 88	Bromus mollis	650	−17.2				−51.1						−27.1	−19.8				
Norby *et al.* 86	Quercus alba	690	−8.9	−3.6					−26.5				−19.2	−13.5			−4.9	−1.4
Norby and O'Neill 91	Liriodendron	493							−6.9	−5.3			−13.5	−17.2				
	tulipifera	787	−14.4	−27.7					−10.3	−17.5			−28.1	−33.3				
														−34.5				
Oberbauer *et al.* 86	Carex bigelowii	675											38.9	1.9				
	Betula nana	675											−18.7	−5.6				
	Ledum palustre	675											−14.3	−4.8				
Reddy *et al.* 89	Soybean	800		−9.1								0		−22.9				
														−24.4				
														−28.9				
Williams *et al.* 86	Mixed trees	500		0.5				0		9.5								0
		700		−18.6						−9.7								
Wong 79	Cotton	640													−52.3	−43.4		
Wong *et al.* 92	Eucalyptus camaldulensis	660											−31.2	−26.2				
	Eucalyptus cypellocarpa	660											−29.8	−25.0				
	Eucalyptus pulverulenta	660											−18.1	−17.5				
	Eucalyptus pauciflora	660											−22.0	−21.0				
Woodin *et al.* 92	Calluna vulgaris	570							18.2				−40.4					
Mean			**−9.8**	**−3.1**	**−33.0**	**−8.9**	**−48.1**	**0**	**−10.2**	**−5.7**	**−23.7**	**−9.1**	**−19.0**	**−20.7**	**−32.5**	**−27.9**	**−26.4**	**−0.7**
Number of observations			5	8	1	1	2	1	5	4	1	2	14	18	3	6	2	2
Standard error			4.2	5.9		3.0	3.0		8.1	5.7		9.1	4.9	2.6	9.9	5.8	21.5	0.7

the possibility that the C-to-N ratio of litter may differ from that of living plants. Plants will try to conserve the scarce nutrients, and withdraw nitrogen from the senescing leaves (Table 15.2). Both the initial C-to-N ratio and the efficiency with which nitrogen is withdrawn from the dying plant part determine the C-to-N ratio in the litter. If this efficiency is affected by elevated CO_2 the increase in tissue C-to-N ratio of plants grown at elevated CO_2 may not be retained in the litter. Nambiar (1987) reported that, in contrast to leaves, nitrogen is not withdrawn from fine roots before senescence.

Why would the efficiency of reallocating nitrogen during senescence be different at elevated CO_2? According to O'Neill (1994) an enhanced efficiency of nutrient use under CO_2 enrichment may increase reallocation and result in a further decline of litter quality. However, it is more likely that the increase in NUE enables the plant to allocate less nitrogen to the photosynthetic apparatus, and more to new permanent structures. While nitrogen in photosynthetic enzymes can easily be re-used by the plant, it may not be possible to recycle nitrogen allocated to permanent plant structures. Assuming the extreme case that all of the non-structural N is reused and none of the structural N, there will be no difference between litter grown in ambient and elevated CO_2. While reallocation of scarce nutrients is essential in a nutrient limited environment, the plant may not benefit from reallocation in situations where nitrogen is not a limiting factor, and the response could then be unpredictable (Woodrow, 1994). Plants adapted to nutrient poor conditions invest relatively more in structural material and less in the photosynthetic apparatus compared with fast growing plants (Berendse and Elberse, 1990). Elevated CO_2 may therefore cause only a relatively small shift in the C-to-N ratio of slow growing plants because the pool from which they can reallocate nitrogen is smaller.

Only few publications in the literature deal with the effects of elevated CO_2 on the C-to-N ratio of both the green leaves and the litter. However, the available data show a clear trend, with nitrogen withdrawal being consistently less efficient at elevated CO_2 than in ambient CO_2, with the exception of a C4 species (Fig. 15.2, Table 15.2). This lower reallocation efficiency may even cause the difference in leaf N concentration between treatments to disappear during senescence (Larigauderie et al., 1988; Curtis et al., 1989). The statement that a low N status induced by elevated CO_2 either has no effect or enhances the N reallocation efficiency (Field et al., 1992) is therefore not correct.

Unfortunately, nothing is known about the effect of elevated CO_2 on the efficiency of withdrawal of nutrients from senescing roots. Owensby et al. (1993) reported higher C-to-N ratios in roots from tall grass prairie species grown under elevated CO_2. Curtis et al. (1990) reported a 22% higher C-to-N ratio for roots of Scirpus olneyi grown under an elevated CO_2 concentration. Pregitzer et al. (1995) studied fine root dynamics in Populus trees in relation to atmospheric CO_2 and soil nitrogen and concluded that the availability of nitrogen overruled the effects of CO_2 concentration on root quality. Changes in the atmospheric CO_2 did however, increase the rates of fine root production and mortality. Elevated CO_2 may affect the loss of nutrients through enhanced root turnover or root exudation, which would also change the amount and quality of carbon input into the soil. Very little is known about the effects of CO_2 enrichment on these processes.

Data on quality of dead roots are very scarce, since most studies used killed roots or combinations of live and dead roots. We measured the amount and the C-to-N ratio in roots of a wheat crop grown at ambient and ambient + 350 ppm CO_2 in open top chambers after harvest at maturity of the crop (Table 15.3). Under elevated CO_2, 43% more root mass with a 6% lower nitrogen concentration was left in soil after harvest. However, the roots from the elevated CO_2 concentration contained 33% more nitrogen. Here, elevated CO_2 resulted in more N in litter but with a reduced quality for decomposition.

A decrease in the nitrogen concentration in litter is still very likely for plant

Table 15.2. A compiliation of data from the literature showing the effect of elevated CO_2 on the nitrogen reallocation efficiency.

			Plant quality														Reallocation efficiency	
			Live material						Dead material								Resorption of N (%)	
			Low CO_2		High CO_2		% increase		Low CO_2		High CO_2		% increase					
Author	Species	comment	%N	C:N	%N	C:N	%N	C:N	%N	C:N	%N	C:N	%N	C:N			Low CO_2	High CO_2
Curtis et al. 89	Scirpus olneyi	Pure	1.13	36.7	0.95	44.2	−15.4	20.6		83.1		86.0		3.6			74.1	68.7
		Mixed	1.23	33.2	0.87	47.3	−28.8	42.6		47.5		51.7		8.9			49.3	26.4
	Spartina patens	C4	0.72	58.1	0.62	69.1	−14.5	19.0										
Larigauderie et al. 88	Bromus mollis	Low N	2.21		1.61		−27.1		2.05		1.95		−4.9				7.2	−21.1
		High N	3.84		3.08		−19.8		2.78		2.74		−1.4				27.6	11.0
Norby et al. 86	Quercus alba		1.20		0.97		−19.2		0.94		0.92		−2.1				21.7	5.2
Reddy et al. 89	Soybean	Leaves	3.50		2.70		−22.9		1.80		1.80		0.0				48.6	33.3
		Stems	1.60		1.60		0.0		0.60		0.80		33.3				62.5	50.0

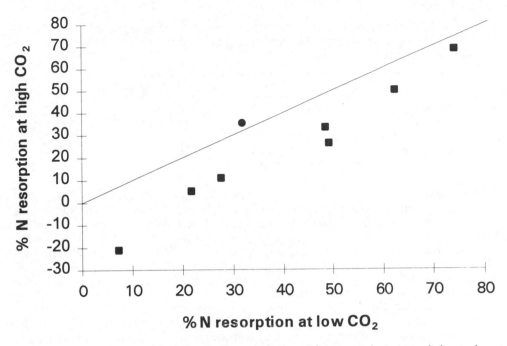

Fig. 15.2. A comparison of the efficiency of nitrogen reallocation between plants grown in low and high CO_2. The squares (■) represent C3 species and the circle (●) a C4 species, respectively. The drawn line shows the situation where there is no difference in resorption between high and low CO_2 treatments. The complete set of data and references are given in Table 15.2.

parts which accumulate non-structural carbohydrates if, due to the abundance of carbohydrates, these carbon compounds are not or only partly reallocated during senescence. In this case the lower nitrogen concentration is the result of a dilution effect.

Although in natural ecosystems the chemical composition of litter from the different plant species often does not change under elevated CO_2 concentrations (Curtis et al., 1989), a shift in species composition (Arp et al., 1993) may well alter the quality of the mixed species litter and change decomposition rates (Kemp et al., 1994; O'Neill, 1994) and ultimately feed back on competition between species.

Soil Functioning

Three aspects which relate soil and decomposer functioning at an elevated atmospheric CO_2 concentration are: (i)

the amount of nutrients in litter and the availability to microbes during microbial production and decomposition, (ii) the quality of the soil organic matter produced or used as substrate by microbes and (iii) the composition of the decomposer community.

Cotrufo and Ineson (1995) grew birch and Sitka spruce under CO_2 concentrations of either 350 or 600 ppm and under two nutrient regimes and studied the decomposition of the fine roots in laboratory microcosms. The C-to-N ratio of roots increased from 20.9 to 24.6 (18%) for birch and from 38.5 to 54.2 (40%) for spruce, respectively, as a response to elevated CO_2 only in unfertilized soil. However, the incubation of these roots yielded no conclusive evidence that decomposition was affected by CO_2 concentration. In a parallel experiment, birch was grown under ambient CO_2 and a range of fertilization rates. Decomposition of these roots was negatively correlated to

Table 15.3. The effect of elevated CO_2 concentration in open top chambers on root mass (g.m^{-2}), N-concentration (%N), C-to-N ratio and N-content (g m^{-2}) of wheat roots after harvest at maturity.

	350 ppm CO_2	700 ppm CO_2
Dry mass root (g m^{-2})	35.1	50.3
% N	1.24	1.16
C-to-N ratio	27	29
Amount of N in roots (g m^{-2})	0.435	0.583

their C-to-N ratio. The combination of these results lead Cotrufo and Ineson (1995) to suggest that the increased C-to-N ratio of plant tissue by elevated CO_2 can result in a reduction of the decomposition rate, with a resulting increase in forest C stores in soil.

A similar response was found by Gorissen *et al.* (1995b) for *Lolium perenne* grown under ambient or elevated CO_2 during decomposition of killed roots under laboratory conditions. They found a 30% decrease of the decomposition rate in roots from elevated CO_2 and suggested that this effect was due to the higher C-to-N ratio (32 versus 18, respectively). Additional tissue quality measures such as lignin-to-N ratio were not determined.

An increased production of carbohydrates may promote synthesis of carbon-based secondary metabolites such as lignin, cellulose and phenols (Field *et al.*, 1992). These products give rise to low quality, slowly decomposing litter (van Vuuren, 1992). So far, there is little evidence for a direct effect of elevated CO_2 on the production of secondary metabolites at elevated CO_2. However, indirect effects are possible when nitrogen becomes more limiting, resulting in increased production of secondary metabolites (Lambers, 1993). The production of allelopathic compounds (which inhibit mycorrhizal symbiosis and root growth of other species) can affect the interaction between species in nutrient poor environments (Nilsson *et al.*, 1993). It is unknown what the effects of elevated CO_2 will be on the production of these compounds.

More likely than the increase in secondary metabolites is the accumulation of

non-structural carbohydrates in the plant (Körner and Arnone, 1992). When the carbohydrate sinks of the growing plant are limited, this will end up in the litter. Although accumulation of starch will increase the C-to-N ratio, its effect on decomposition will be different from that of structural carbon. While secondary metabolites will suppress the decomposition process, non-structural carbohydrates will be quickly used by decomposers, and may not affect the overall decay process (Loehle, 1995). Microbes will prefer these energy rich and easily decomposable compounds over recalcitrant organic matter, but to meet their requirement of nitrogen they are forced to decompose the relatively nitrogen-rich soil organic matter (van de Geijn and van Veen, 1993). Simultaneously, the availability of this energy source may facilitate the decomposition of lignin. The net effect may be a stimulation of the decomposition rate.

Effect of elevated CO_2 on N availability

Greater below ground C inputs through root exudation and turnover at elevated CO_2 have been shown to increase the soil microbial biomass (Diaz *et al.*, 1993; Zak *et al.*, 1993). It is not clear how an increase in the microbial biomass in the soil will affect the availability of nutrients for the plants, but stimulation of microbial activity and recycling of microbial N at elevated CO_2 may increase N availability (Lekkerkerk *et al.*, 1990; Zak *et al.*, 1993).

While the decomposition rate will increase, more nitrogen is required by an increased soil microbial biomass. In one study under N-limited conditions this resulted in an increased N availability (Zak et al., 1993), while in a nutrient rich soil an increased soil microbial biomass reduced the N availability (Diaz et al., 1993). How a rapid growth of microorganisms will affect the availability of nitrogen in the soil may be dependent on rate of release of nitrogen from the microbial biomass: faster availability through grazing by other soil organisms (De Ruiter et al., 1993) or slower availability as a result of death or return to dormancy of microbes after the substrate has been depleted (Wu et al., 1993).

The decomposition rate appears to be dependent on the complexity of the soil fauna and the duration of the experiment. Couteaux et al. (1991) have shown that plant litter produced under elevated CO_2 decomposed slower than material produced in ambient CO_2 during the first months. While this could be attributed to a change in litter quality, it could also be that the decomposer community needed time to adapt to the new situation. In the long term, a complex community of soil organisms decomposed litter with a high C-to-N ratio grown at elevated CO_2 faster than litter grown in ambient CO_2, while in a system with only a few species of decomposers the opposite response was observed (Couteaux et al., 1991).

Ecosystem Functioning

To determine how ecosystems will function in a high CO_2 atmosphere, not only the effects of high CO_2 on the individual segments of the nitrogen cycle must be understood, but the cycle must also be completed, with the nitrogen made available by the soil processes influencing the uptake of nitrogen by the plants. This allows potential feedback and feedforward mechanisms.

The effect of elevated CO_2 on the nitrogen cycle

Will nutrient limited ecosystems respond to an increase in CO_2 concentration? Some of the available data show that elevated CO_2 will enhance growth even when nitrogen is limited (Norby et al., 1986), although transient responses are also reported under nutrient-poor conditions (Oechel et al., 1994; Gorissen et al., 1995a). Ultimately, the response of nutrient limited ecosystems to elevated CO_2 will depend on the availability of nitrogen in the soil, the nitrogen use efficiency of the plant and the uptake and loss of nutrients by the plant. We can distinguish two scenarios for ecosystem responses.

In the first scenario, an increase in the C-to-N ratio and a higher concentration of secondary compounds will slow down the decomposition rate, and in due time will reduce the nutrient availability. Even if the total loss of nitrogen from plants does not increase, the growth will eventually be limited by the reduced availability of nutrients, now tied up in soil organic matter.

However, as shown in Tables 15.1 and 15.2, the C-to-N ratio in some plant parts may not increase, while the reallocation efficiency of nitrogen from leaves is reduced at elevated CO_2. So, in a second scenario, the C-to-N ratio of leaf litter will therefore not increase as much as the C-to-N ratio of green leaves. Hence, the combination of a 'relatively high' nitrogen concentration and an increased growth rate at elevated CO_2 will result in more nitrogen being lost through the production of litter (Curtis et al., 1989). Plants will have to take up an equivalent amount of nutrients to attain a similar biomass during the following growing season (Berendse et al., 1987). Because nitrogen in litter will only become available to the plant through the relatively slow decomposition process, this nitrogen lost will not be immediately accessible again. This scenario, where the loss of nitrogen from the plant increases and the rate at which nitrogen becomes available to the plant remains unchanged, results in an increased nutrient limitation during the following years and so limit the CO_2 response. This will last until a new equilibrium will be reached when losses equal mineralization of nitrogen again. Only in agricultural, fertilized ecosystems (Table 15.2), an increased fertilization rate may sustain the CO_2 response and increase soil organic matter production (van de

Geijn and van Veen, 1993).

In both scenarios, the expected initial growth response cannot be not sustained because the required nutrients are tied up in the decomposition process. The long-term responses are dominated by the slow changes in turnover of soil C and N pools (Rastetter *et al.*, 1991).

The consequences for long-term ecosystem functioning

The long-term effects of an elevated CO_2 concentration on ecosystem functioning depend on the sustained availability and possible limitation of the ecosystem by nutrients, carbon and water. For a lasting effect of CO_2 on growth, either an external input of nutrients is required (i.e. atmospheric deposition or biological nitrogen fixation), or the release of nutrients from soil organic matter through decomposition must be accelerated, for instance as a result of climate change (i.e. temperature rise or water availability). Such an increase in nitrogen availability will directly stimulate production in nutrient limited ecosystems (Berendse and Elberse, 1990), while the potential for increased production will be even higher in combination with elevated CO_2 (Field *et al.*, 1992). This means that nitrogen must not be tied up in the decomposition process for a longer period, and the accumulation of soil organic matter (which reduces the amount of nitrogen available) should not exceed the rate of input of nitrogen into the system. As a consequence, an increased accumulation of carbon by nitrogen limited ecosystems is inherently impossible without an increase in the input of nitrogen into the system (van de Geijn and van Veen, 1993), unless the C-to-N ratio of soil organic matter would increase.

Field data show the interactions between CO_2, nitrogen and water. In a tall-grass prairie nitrogen limitation negated the increase in productivity at elevated CO_2 in wet years, while in a dry year the biomass was increased through an enhanced water use efficiency (Owensby *et al.*, 1994). In a nutrient-poor and wet tundra ecosystem increasing the CO_2 concen-

tration had very little effect on biomass production (Tissue and Oechel, 1987).

In systems where carbon is limiting growth rather than nitrogen (high nutrient or light limited systems) a sustained increase in biomass is likely and an increased accumulation of organic matter is possible during a relatively long period (Arp *et al.*, 1993). An improved water use efficiency at elevated CO_2 can increase biomass in water limited systems, but elevated CO_2 could also increase soil water content with consequences for the decomposition process. In a prairie system the microbial activity was greater under elevated CO_2 because of better soil water conditions (Rice *et al.*, 1994).

Experiments

We have seen that several hypotheses of ecosystem functioning with respect to climate change cannot be evaluated because no data are available. It is therefore necessary to evaluate whether the appropriate experiments are performed, i.e. do we correctly include future conditions with respect to (gradual) changes in quality of litter, decomposer communities and soil organic matter in our experiments and whether the necessary measurements are made.

Experiments which aim to study the effects of elevated CO_2 on decomposition and mineralization often suffer from one or more of the following shortcomings:

1. Green leaves and live roots are used for decomposition experiments. Where possible naturally senesced litter should be used for reasons explained above.
2. Only litter from a single plant part (mostly leaves) is used. However, the composition of the plants total litter will also be affected by a change in the allocation of biomass over different plant organs at elevated CO_2. Therefore the 'whole plant' response depends on the results for decomposition of the different plant parts, and on the ratio at which litter from these parts are produced.
3. Differences in the quantity of litter produced are not taken into account, which

would change the total carbon and nitrogen supply to the soil, affecting the balance between 'new' and 'old' litter.

4. Decomposer communities are used which are adapted to the old litter quality and which may require time to adapt to the new litter.

Final remarks

The increase in the C-to-N ratio and the level of secondary metabolites in plant litter as a result of growth at elevated CO_2 is probably not as large as often expected based on plant tissue data. Moreover, a large fraction of the increase in the C-to-N ratio could be caused by starch accumulation which may have a different effect on soil processes than changes in recalcitrant carbon compounds. Litter produced under elevated CO_2 may therefore have only a slightly lower ratio of nitrogen to secondary compounds, which is considered a measure of decomposability. However, the total amount of nitrogen in litter will increase with a higher biomass production and the content of energy rich, easily decomposable compounds is enhanced. The ability to predict long-term climate change effects on ecosystem functioning is limited by our knowledge of the actual amounts of nitrogen lost through litter and roots, and the consequences of changes in litter quality on decomposition. A possible stimulation of plant symbionts must also be taken into account. This leads to the conclusion that the general notion that elevated CO_2 will retard the decomposition process needs reconsideration.

The consequences of global change for nutrient limited ecosystems depend on the balance between the increase in nutrient loss from plants, the availability of nutrients in soil and the capacity for uptake by plants, and the reduction of the nutrient requirement of the plants at elevated CO_2. In the long term, the availability of nitrogen will determine the response of nitrogen limited ecosystems to elevated CO_2. Only an increase in available nitrogen caused by a higher nitrogen deposition, stimulation of nitrogen fixation or an enhanced decomposition rate caused by a higher temperature would allow a permanent increase in biomass. Ultimately, an enhanced sequestration of soil organic matter, which would counteract the increasing CO_2 concentration in the atmosphere, requires a simultaneous net input of nitrogen into the system or a shift in the C-to-N ratio of soil organic matter.

References

Arp, W.J., Drake, B.G., Pockman, W.T., Curtis, P.S., and Whigham, D.F. (1993) Interactions between C_3 and C_4 salt marsh plant species during four years of exposure to elevated atmospheric CO_2. *Vegetatio* 104/105, 133–143.

Berendse, F., Berg, B. and Bosatta, E. (1987) The effect of lignin and nitrogen on the decomposition of litter in nutrient-poor ecosystems: a theoretical approach. *Canadian Journal of Botany* 65, 1116–1120.

Berendse, F. and Elberse, W.T. (1990) Competition and nutrient availability in heathland and grassland ecosystems. In: Grace, J.B. and Tilman, D. (eds) *Perspectives on Plant Competition*. Academic Press, San Diego, California, pp. 93–116.

Bottner, P. and Couteaux, M.M. (1991) Effect of plant activity on decomposition: soil-plant interactions in response to increasing atmospheric CO_2 concentration. In: van Breemen, N. (ed.) *Decomposition and Accumulation of Organic Matter in Terrestrial Ecosystems: Research Priorities and Approaches*. Doorwerth, The Netherlands, pp. 39–45.

Cotrufo, M.F. and Ineson, P. (1995) Effects of enhanced atmospheric CO_2 and nutrient supply on the quality and subsequent decomposition of fine roots of *Betula pendula* Roth. and *Picea sitchensis* (Bong.) Carr. *Plant and Soil* 170, 267–277.

Couteaux, M.M., Mousseau, M., Célérier, M.-L. and Bottner, P. (1991) Increased atmospheric CO_2 and litter quality: decomposition of sweet chestnut leaf litter with animal food webs of different complexities. *Oikos* 61, 54–64.

Cure, J.D., Israel, D.W. and. Rufty, T.W (1988) Nitrogen stress effects on growth and seed yield of nonnodulated soybean exposed to elevated carbon dioxide. *Crop Science* 28, 671–677.

Curtis, P.S., Drake, B.G., Leadley, P.W., Arp, W.J. and Whigham, D.F. (1989) Growth and senescence in plant communities exposed to elevated CO_2 concentrations on an estuarine marsh. *Oecologia* 78, 20–26.

Curtis, P.S., Balduman, L.M., Drake, B.G. and Whigham, D.F. (1990) Elevated atmospheric CO_2 effects on belowground processes in C_3 and C_4 estuarine marsh communities. *Ecology* 71, 2001–2006.

De Ruiter, P.C., Moore, J.C., Zwart, K.B., Bouwman, L.A., Hassink, J., Bloem, J. de Vos, J.A., Marinissen, J.C.Y., Didden, W.A.M., Lebbink, G. and Brussaard, L. (1993) Simulation of nitrogen mineralization in the below-ground food webs of two winter wheat fields. *Journal of Applied Ecology* 30, 95–106.

DeLucia, E.H., Sasek, T.W. and Strain, B.R. (1985) Photosynthetic inhibition after long-term exposure to elevated levels of atmospheric carbon dioxide. *Photosynthesis Research* 7, 175–184.

Den Hertog, J. and Stulen, I. (1990) The effects of an elevated atmospheric CO_2 concentration on dry matter and nitrogen allocation. In: Goudriaan, J., van Keulen, H. and van Laar, H.H. (eds) *The Greenhouse Effect and Primary Productivity in European Agro-ecosystems*. Pudoc, Wageningen, the Netherlands, pp. 27–30.

Diaz, S., Grime, J.P., Harris, J. and McPherson, E. (1993) Evidence of a feedback limiting plant response to elevated carbon dioxide. *Nature* 364, 616–617.

Field, C.B., Chapin, F.S., Matson, P.A. and Mooney, H.A. (1992) Responses of terrestrial ecosystems to the changing atmosphere: a resource-based approach. *Annual Review of Ecological Systems* 23, 201–235.

Gorissen, A., Kuikman, P.J. and van de Beek, H. (1995a) Carbon allocation and water use in juvenile Douglas fir under elevated CO_2. *New Phytologist* 129, 275–282.

Gorissen, A., van Ginkel, J.H., Keurentjes, J.J.B. and van Veen, J.A. (1995b) Grass root decomposition is retarded when grass has been grown under elevated CO_2. *Soil Biology and Biochemistry* 27, 117–120.

Gries, C. (1993) Nutrient uptake during the course of a year by sour orange trees growing in ambient and elevated atmospheric carbon dioxide concentrations. *Journal of Plant Nutrition* 16, 129–147.

Hilbert, D.W., Larigauderie, A. and Reynolds, J.F. (1991) The influence of carbon dioxide and daily photon-flux density on optimal leaf nitrogen concentration and root:shoot ratio. *Annals of Botany* 68, 365–376.

Hocking, P.J. and Meyer, C.P. (1985) Responses of Noogoora Burr (*Xanthium occidentale* Bertol.) to nitrogen supply and carbon dioxide enrichment. *Annals of Botany* 55, 835–844.

Hocking, P.J. and Meyer, C.P. (1991) Effects of CO_2 enrichment and nitrogen stress on growth, and partitioning of dry matter and nitrogen in wheat and maize. *Australian Journal of Plant Physiology* 18, 339–356.

IPCC (1996) *Climate Change 1995 – The Science of Climate Change*. Contribution of Working Group I to the Second Assessment Report of the Intergovernmental Panel on Climate Change. Houghton, J.T., Meira Filho, L.G., Callander, B.A., Harris, N., Kattenberg, A. and Maskell, K. (eds). Cambridge University Press, Cambridge, 564 pp.

Johnson, R.H. and Lincoln, D.E. (1990) Sagebrush and grasshopper responses to atmospheric carbon dioxide concentration. *Oecologia* 84, 103–110.

Johnson, R.H. and Lincoln, D.E. (1991) Sagebrush carbon allocation patterns and grasshopper nutrition: the influence of CO_2 enrichment and soil mineral limitation. *Oecologia* 87, 127–134.

Kemp, P.R., Waldecker, D.G., Owensby, C.E., Reynolds, J.F. and Virginia, R.O. (1994) Effects of elevated CO_2 and nitrogen fertilization on decomposition of tallgrass prairie leaf litter. *Plant and Soil* 165, 115–127.

Kimball, B.A. (1983) Carbon dioxide and agricultural yield: an assemblage and analysis of 430 prior observations. *Agronomy Journal* 75, 779–788.

Körner, C. and Arnone, J.A. III. (1992) Responses to elevated carbon dioxide in artificial tropical ecosystems. *Science* 257, 1672–1675.

Lambers, H. (1993) Rising CO_2, secundary plant metabolism, plant-herbivore interactions and litter decomposition. *Vegetatio* 104/105, 263–271.

Larigauderie, A., Hilbert, D.W. and Oechel, W.C. (1988) Effect of CO_2 enrichment and nitrogen availability on resource acquisition and resource allocation in a grass, *Bromus mollis*. *Oecologia* 77, 544–549.

Leake, J.R. and Read, D.J. (1989) The biology of mycorrhiza in the Ericaceae. XIII. Some characteristics of the extracellular proteinase activity of the ericoid endophyte *Hymenoscyphus ericae*. *New Phytologist* 112, 69–76.

Leake, J.R. and Read, D.J. (1990) Proteinase activity in mycorrhizal fungi. I. The effect of extracellular pH on the production and activity of proteinase by ericoid endophytes from soils of contrasted pH. *New Phytologist* 115, 243–250.

Lekkerkerk, L.J.A., van Veen, J.A. and van de Geijn, S.C. (1990) Influence of climatic change on soil quality; consequences of increased atmospheric CO_2-concentration on carbon input and turnover in agro-ecosystems. In: Goudriaan, J., van Keulen, H. and van Laar, H.H. (eds), *The Greenhouse Effect and Primary Productivity in European Agro-ecosystems*. Pudoc, Wageningen, The Netherlands, pp. 46–47.

Loehle, C. (1995) Anomalous responses of plants to CO_2 enrichment. *Oikos* 73, 181–187.

Nambiar, E.K.S. (1987) Do nutrients retranslocate from fine roots? *Canadian Journal of Forest Research* 17, 181–187.

Nilsson, M.-C., Högberg, P., Zackrisson, O. and Fengyou, W. (1993) Allelopathic effects by *Empetrum hermaphroditum* on development and nitrogen uptake by roots and mycorrhiza of *Pinus sylvestris*. *Canadian Journal of Botany* 71, 620–628.

Norby, R.J., O'Neill, E.G. and Luxmoore, R.J. (1986) Effects of atmospheric CO_2 enrichment on the growth and mineral nutrition of *Quercus alba* seedlings in nutrient-poor soil. *Plant Physiology* 82, 83–89.

Norby, R.J. (1987) Nodulation and nitrogenase activity in nitrogen-fixing woody plants stimulated by CO_2 enrichment of the atmosphere. *Physiologia Plantarum* 71, 77–82.

Norby, R.J. and O'Neill, E.G. (1991) Leaf area compensation and nutrient interactions in CO_2-enriched seedlings of yellow-poplar (*Liriodendron tulipifera* L.). *New Phytologist* 117, 515–528.

O'Neill, E.G. (1994) Responses of soil biota to elevated atmospheric carbon dioxide. *Plant and Soil* 165, 55–65.

O'Neill, E.G., Luxmoore, R.J. and Norby, R.J. (1987) Elevated atmospheric CO_2 effects on seedling growth, nutrient uptake, and rhizosphere bacterial populations of *Liriodendron tulipifera* L. *Plant and Soil* 104, 3–11.

Oberbauer, S.F., Sionit, N., Hastings, S.J. and Oechel, W.C. (1986) Effects of CO_2 enrichment and nutrition on growth, photosynthesis, and nutrient concentration of Alaskan tundra plant species. *Canadian Journal of Botany* 64, 2993–2998.

Oechel, W.C., Cowles, S., Grulke, N.E., Hastings, S.J., Lawrence, B., Prudhomme, T., Riechers, G., Strain, B., Tissue, D. and Vourlitis, G. (1994) Transient nature of CO_2 fertilization in arctic tundra. *Nature* 371, 500–503.

Owensby, C.E., Coyne, P.I. and Auen, L.M. (1993) Nitrogen and phosphorus dynamics of a tallgrass prairie ecosystem exposed to elevated carbon dioxide. *Plant, Cell and Environment* 16, 843–850.

Owensby, C.E., Auen, L.M. and Coyne, P.I. (1994) Biomass production in a nitrogen-fertilized, tallgrass prairie ecosystem exposed to ambient and elevated levels of CO_2. *Plant and Soil* 165, 105–113.

Pearcy, R.W. and Björkman, O. (1983) Physiological effects. In: Lemon, E.R. (ed.) *CO_2 and Plants: The Response of Plants to Rising Levels of Atmospheric Carbon Dioxide*. Westview Press, Inc. Boulder, Colorado, pp. 65–105.

Pregitzer, K.S., Zak, D.R., Curtis, P.S., Kubiske, M.E., Teeri, J.A. and Vogel, C.S. (1995)

Atmospheric CO_2, soil nitrogen and turnover of fine roots. *New Phytologist* 129, 579–585.

Rastetter, E.B., Ryan, M.G., Shaver, G.R., Melillo, J.M., Nadelhoffer, K.J., Hobbie, J.E. and Aber, J.D. (1991) A general biogeochemical model describing the responses of the C and N cycles in terrestrial ecosystems to changes in CO_2, climate and N deposition. *Tree Physiology* 9, 101–126.

Reddy, V.R., Acock, B. and Acock, M.C. (1989) Seasonal carbon and nitrogen accumulation in relation to net carbon dioxide exchange in a carbon dioxide-enriched soybean canopy. *Agronomy Journal* 81, 78–83.

Rice, C.W., Garcia, F.O., Hampton, C.O. and Owensby, C.E. (1994) Soil microbial response in tallgrass prairie to elevated CO_2. *Plant and Soil* 165, 67–74.

Rogers, H.H., Runion, G.B. and Krupa, S.V. (1994) Plant responses to atmospheric CO_2 enrichment with emphasis on roots and the rhizosphere. *Environmental Pollution* 83, 155–189.

Tans, P.P., Fung, I.Y. and Takahashi, T. (1990) Observational constraints on the global atmosphere CO_2 budget. *Science* 247, 1431–1438.

Tissue, D.T. and Oechel, W.C. (1987) Response of *Eriophorum vaginatum* to elevated CO_2 and temperature in the Alaskan tussock tundra. *Ecology* 68, 401–410.

Van de Geijn, S.C. and van Veen, J.A. (1993) Implications of increased carbon dioxide levels for carbon input and turnover in soils. *Vegetatio* 104/105, 283–292.

Van Vuuren, M.M.I. (1992) Effects of plant species on nutrient cycling in heathlands. PhD Thesis, Utrecht University, The Netherlands, 159 pp.

Williams, W.E., Garbutt, K., Bazzaz, F.A. and Vitousek, P.M. (1986) The response of plants to elevated CO_2. IV. Two deciduous-forest tree communities. *Oecologia* 69, 454–459.

Wong, S.C. (1979) Elevated atmospheric partial pressure of CO_2 and plant growth. I. Interactions of nitrogen nutrition and photosynthetic capacity in C_3 and C_4 plants. *Oecologia* 44, 68–74.

Wong, S.C., Kriedemann, P.E. and Farquhar, G.D. (1992) $CO_2 \times$ nitrogen interaction on seedling growth of four species of eucalypt. *Australian Journal of Botany* 40, 457–472.

Woodin, S., Graham, B., Killick, A. and Skiba, U. (1992) Nutrient limitation of the long term response of heather [*Calluna vulgaris* (L.) Hull] to CO_2 enrichment. *New Phytologist* 122, 635–642.

Woodrow, I.E. (1994) Optimal acclimation of the C_3 photosynthetic system under enhanced CO_2. *Photosynthesis Research* 39, 401–412.

Wu, J., Brookes, P.C. and Jenkinson, D.S. (1993) Formation and destruction of microbial biomass during the decomposition of glucose and ryegrass in soil. *Soil Biology and Biochemistry* 25, 1435–1441.

Zak, D.R., Pregitzer, K.S., Curtis, P.S., Teeri, J.A., Fogel, R. and Randlett, D.L. (1993) Elevated atmospheric CO_2 and feedback between carbon and nitrogen cycles. *Plant and Soil* 151, 105–117.

16 Progress and Potential for Genetic Manipulation of Plant Quality

A. Bavage, I.G. Davies, M.P. Robbins and P. Morris

Institute of Grassland and Environmental Research,
Plas Gogerddan, Aberystwyth, Dyfed, UK

Introduction

It has been reported that the lignin and polyphenolic contents of plant material may have a crucial role in litter degradation and recycling processes. In litters derived from some plant species the lignin-to-N content correlates with the rate of nitrogen mineralization (Stump and Binkley, 1993), in others (lignin+polyphenol)-to-N is the critical factor (Oglesby and Fownes, 1992; Constantinides and Fownes, 1994; Handyanto *et al.*, Chapter 14, this volume; Vanlauwe *et al.*, Chapter 12, this volume).

The lignin content of plants also affects their digestibility, is important in the paper making process and in determining timber characteristics. Lignin biosynthesis has therefore been targeted for genetic manipulation. Similarly polyphenols have a role in the dietary properties of plants (Harborne, Chapter 4, this volume; Robbins *et al.*, 1996). In particular polymeric proanthocyanins (condensed tannins (CTs)) in forages are involved in bloat amelioration, palatability and nutritive value. We are using genetic manipulation to alter the quantity, structure and tissue distribution of CTs in forage legumes.

Advances in the understanding of the biosynthesis of CTs (for which some enzymes are common to anthocyanin production) and of lignins in plants provide an opportunity to manipulate plant quality. Many genes encoding enzymes involved in lignin and condensed tannin biosynthesis have been cloned and characterized. The technologies of gene tagging, antisense gene expression and sense gene expression can now be added to traditional plant breeding methods to generate plant material with novel qualities.

Here we will examine the advances made, in particular in the manipulation of lignins and tannins. The implications for litter degradation and utilization will be discussed.

Manipulation of Plant Characteristics

Traditionally modified/improved plant varieties have been obtained by selection from breeding programmes. The success of such strategies is largely dependent on the available genepool. With increasing pressure on the habitats of wild plant populations and the dominance of particular crop plants and varieties the opportunities to recover novel phenotypes are diminishing. Mutagenesis of existing species and cultivars, by chemical means or by disrupting genes with transposable elements (gene tagging), is limited to loss/modification of function of pre-existing genes. The advances made in molecular biology provide an alternative means of introducing new traits into many species. Protocols have been developed to introduce genetic material, from any source,

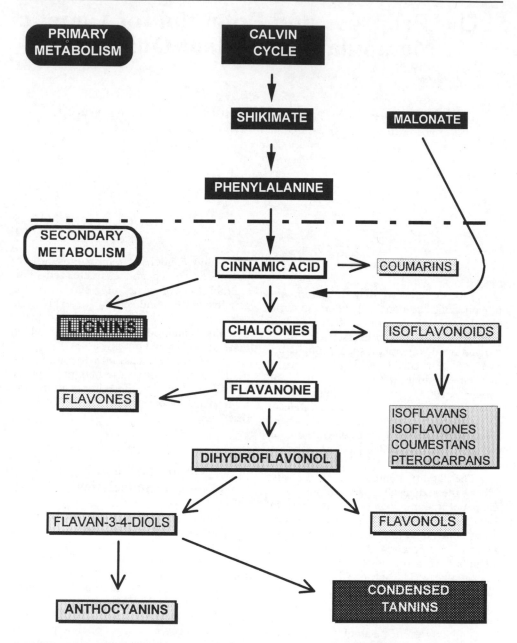

Fig. 16.1. Schematic representation of major components of phenylpropanoid metabolism.

into most of the major crop plants, as well as many other species.

Phenylpropanoid metabolism in plants is responsible for the production of a broad range of complex molecules (Fig. 16.1, Harborne, Chapter 4, this volume) however, only specific end products will appear in any given cell. There is also temporal regulation of accumulation. For example lignification is greatest in mature xylem vessels and isoflavonoid phytoalexins are often only produced in response to a pathogen. To manipulate these processes either the core phenylpropanoid metabolism or specific branch pathways may be targeted.

The manipulation of lignins and condensed tannins in plants has been attempted using 'sense' and 'antisense' techniques. Sense gene expression involves the introduction of the gene of interest, driven by a promoter, in its correct orientation such that a functional protein can be produced. In antisense gene expression the gene is reversed relative to the promoter. In this case an antisense mRNA is produced which interacts with the mRNA from the native gene causing a loss of the gene product. In cases where genes already present in plants are introduced as transgenes in sense orientation a proportion of the resulting transgenic plants display characteristics of antisense gene expression. The mechanism(s) responsible for this phenomenon, termed co-suppression, is poorly understood.

Genetic Manipulation of Lignin Biosynthesis

The process of lignification of plant tissues has been studied in great detail although some areas of doubt remain, notably with regard to the polymerization and deposition processes (reviewed by Boudet et al., 1995). Fig. 16.2 shows one interpretation of the biosynthesis of the lignin precursors p-coumaroyl CoA, feruloyl CoA and sinapoyl CoA. The majority of the genes involved in these steps in lignin formation have been cloned although the enzymes (if they exist) which catalyse the final polymerization remain elusive. There remains some doubt as to whether caffeic acid O-methyltransferase (COMT) is exclusively responsible for the methylation reactions. The enzyme caffeoyl Co enzyme A O-methyltransferase (CCoACOMT), which has been cloned from parsley (Schmitt et al., 1991), could provide an alternative route to the lignin precursors (Fig. 16.3). The later stages of lignin biosynthesis (Fig. 16.4) are unique and this part of the pathway has therefore been targeted for genetic manipulation. In contrast genetic manipu-

lation of early stages in phenylpropanoid metabolism using phenylalanine ammonia-lyase (Elkind et al., 1990) has serious consequences for the plant with a broad range of pleiotropic effects, including reduction of lignification, alterations in morphology, reduced pollen viability and modified flower pigmentation, being reported.

Reports on the antisense expression of COMT are contradictory. In one case, when an alfalfa COMT gene was introduced into tobacco, some of the transgenic tobacco plants showed a reduction in lignin content but no change in the lignin sub-unit composition (Ni et al., 1994). However, in tobacco transformed with sense or antisense tobacco COMT, strong suppression of COMT resulted in an unaltered lignin content but modified lignin sub-unit composition (Legrand et al.,1994; Atanassova et al., 1995). This suggests that manipulation of lignification at this step can be successful but that the possible presence of an alternative pathway through CCoACOMT may be an added complication.

Manipulation of the later stages of lignin synthesis (Fig. 16.4) has concentrated on cinnamyl alcohol dehydrogenase (CAD). Although present in normal quantities, lignins were more easily extracted from tobacco plants with very low CAD activity, generated by antisense CAD expression, apparently due to an increase in cinnamyl aldehyde residues (Halpin et al., 1994). Transgenic tobacco plants with reduced CAD activity had red/brown coloured vascular tissue similar to that seen in the 'brown mid-rib mutants' of sorghum and maize, which also have reduced CAD activity (Bucholtz et al., 1980). Similar results, including the red/brown coloration, were obtained with white poplar transformed with an antisense poplar CAD cDNA (Tollier et al., 1994).

To date the mechanism of polymerization of p-coumaryl, coniferyl, and sinapyl alcohols (Fig. 16.4) (or their corresponding aldehydes in antisense plants) is not understood. No genes involved in the regulation of lignin biosynthesis genes have been isolated to date. Hopefully, further

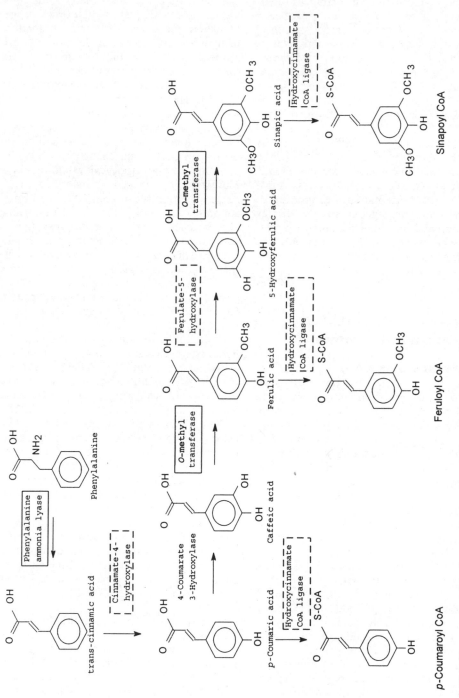

Fig. 16.2. The common phenylpropanoid pathway leading to precursors of lignins. Genes boxed ——— have been cloned and used to produce transgenic plants, genes boxed – – – have been cloned, unboxed genes have not been reported to be cloned (Redrawn from Boudet *et al.*, 1995).

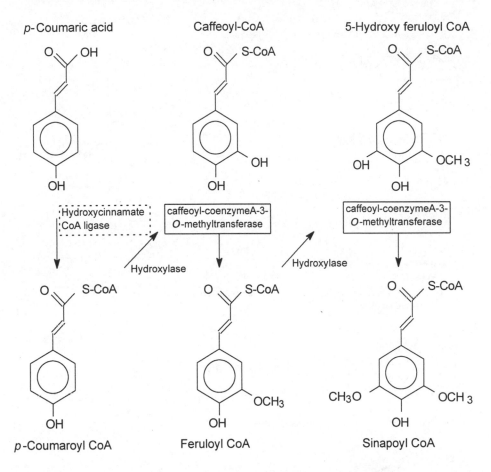

Fig. 16.3. The alternative methylation pathway for lignin precursor synthesis. Genes boxed —— have been cloned and used to produce transgenic plants, genes boxed – – – have been cloned, unboxed genes have not been reported to be cloned (redrawn from Boudet *et al.*, 1995).

examination of mutants will begin to overcome this limitation in the future.

Genetic Manipulation of Condensed Tannins

Like lignins, condensed tannins (CTs) are produced by a branch of phenylpropanoid metabolism (Figs 16.1 and 16.5). They are an important factor in forage quality, due to their effects on bloat, nutritive value and palatability. Many of the enzymes which catalyse the production of the tannin subunits are common with anthocyanin biosynthesis. Several have been cloned from a variety of plant species (Fig. 16.5)

facilitating the manipulation of CTs in the legume *Lotus corniculatus* (bird's foot trefoil). We have been working on the manipulation of chalcone synthase (CHS) and dihydroflavonol reductase (DFR).

CHS is an enzyme of the core phenylpropanoid pathway. It is present in most species as a relatively large gene family. Introduction, by *Agrobacterium rhizogenes* mediated transformation, of a stress responsive CHS gene from bean in antisense into *L. corniculatus* produced unexpected effects. Transgenic plants had increased levels of CTs, but other phenylpropanoid pathway products were significantly reduced (Colliver *et al.*, 1994; Colliver, 1995).

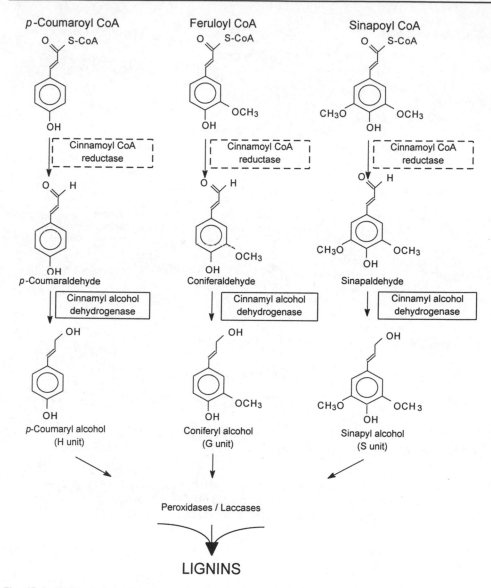

Fig. 16.4. The lignin branch pathway. Genes boxed —— have been cloned and used to produce transgenic plants, genes boxed – – – have been cloned and cloning of unboxed genes has not been reported (from Boudet *et al.*, 1995).

An antisense construct comprising the 5′ half of the *Antirrhinum majus* dihydroflavonol reductase gene has also been introduced into three isogenic *L. corniculatus* lines, by the same method. This gave rise to transgenic hairy root cultures from which plants were regenerated. Root cultures showed an antisense phenotype with reduced CTs content and altered proportions of procyanidin and prodelphinidin hydrolysable from the CTs (Carron *et al.*, 1994). Plants regenerated from root cultures also showed reduced CTs content plus other pleiotropic effects including dwarfing and delayed flowering (Robbins *et al.*, 1994). Surprisingly, the phenotypes of regenerated plants did not always correspond to the phenotype of the root culture

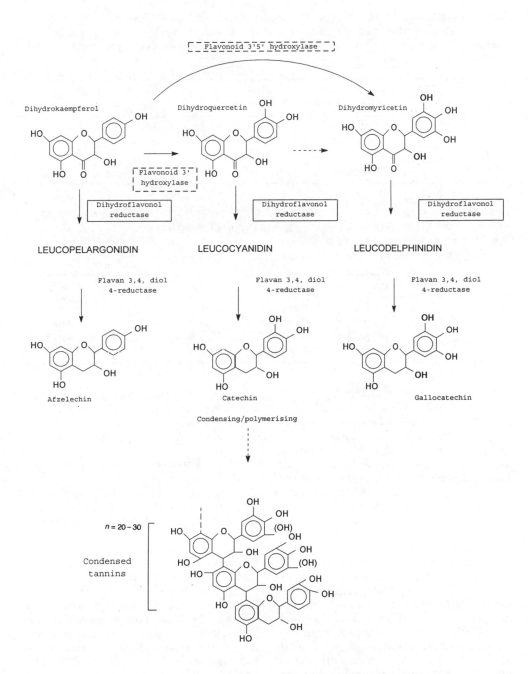

Fig. 16.5. Condensed tannin biosynthesis. Genes boxed —— have been cloned and used to produce transgenic plants, genes boxed – – – have been cloned, cloning of unboxed genes has not been reported. The polymerization of condensed tannins from precursors is not understood.

Table 16.1. Comparison of the occurrence of antisense phenotypes between 'hairy root' cultures of *Lotus corniculatus*, transformed with an antisense dihydroflavonol-4-reductase gene, and plants regenerated from the same cultures (+ antisense phenotype observed, – wild type phenotype observed).

Line	Type	Root culture phenotype	Regenerated plant phenotype
C26	Control	Wild-type	Wild-type
RFD7	Antisense	–	+
RFD8	Antisense	+	+
RFD19	Antisense	+	–

Table 16.2. Phenotypes of 'hairy root' cultures of *Lotus corniculatus* transformed with *Antirrhinum majus* dihydroflavonol-4-reductase gene. Condensed tannin (CT) was measured by butanol/HCl hydrolysis after 21 days growth. The percentages of procyanidin (PC), prodelphinidin (PD) and propelargonidin (PP) were calculated by HPLC analysis.

Line	Phenotype	CT mg g^{-1} Fwt	%PC	%PD	%PP
G1	Control	0.28±0.07	81±0.3	13±0.7	4±0.3
ADFR02	Suppressed	0.02±0.01	78±2.0	18±2.1	4±0.3
ADFR07	Suppressed	0.07±0.02	83±1.2	12±1.3	4±0.3
ADFR10	Increased	0.60±0.04	73±2.4	13±1.8	14±8.0

they were derived from (Table 16.1).

The *A. majus* DFR gene is believed to have a different substrate specificity from the *L. corniculatus* gene. The *A. majus* DFR has a high affinity for dihydrokaempferol and dihydroquercetin, producing pelargonidin and cyanidin pigments respectively. The *L. corniculatus* gene is presumed to have greater affinity for dihydroquercetin and dihydromyricetin because CTs are a polymer of procyanidin and prodelphinidin derivatives. When the *A. majus* gene was introduced, in a sense construct, into *L. corniculatus* the transgenic root cultures had a range of CT levels (Table 16.2). Interestingly, in a culture with increased CTs, polymer sub-unit composition was also changed. High performance liquid chromatography analysis of the proanthocyanin constituents of these lines detected significant amounts of propelargonidin in this transgenic line, in line with the substrate specificity of the introduced gene (Table 16.2). Plants regenerated from these cultures are being analysed. 'Hairy root' cultures with reduced tannin contents, apparently due to co-suppression, were similar to their antisense derived counterparts.

Future Prospects

The options for manipulating plant quality and hence litter characteristics have been increased by the techniques of molecular biology. In general, targeting later stages in biosynthesis pathways appears to produce more consistent results with less severe pleiotropic effects. This may be due to multigene families, encoding closely related but differentially expressed genes, making the common phenylpropanoid pathway more complicated than is apparent from consideration of the biochemical processes alone. Attempts to manipulate COMT in lignin biosynthesis have demonstrated the difficulties encountered when manipulating pathways which may contain a number of routes to the same end product. Genetic manipulation of the dedicated stages of the pathways for lignin and condensed tannin biosynthesis have both been successful, although the outcomes were not completely predictable. These approaches would be enhanced by a greater understanding of the processes of polymerization, which remain obscure in both pathways. Similarly, although genes involved in the regulation of the common phenylpropanoid pathway have been identified and some understanding of the their regulation gained (van der Meer *et al.*, 1993; Dixon *et al.*, 1995), the regulation of the lignin and CTs branch pathways remains largely obscure.

Genetic manipulations can produce profound changes in the accumulation of secondary metabolites. It is worth remembering however, that environmental factors also have an important role to play in the accumulation of these compounds. Soil moisture and temperature can affect CTs accumulation (Anuraga *et al.*, 1993; Carter *et al.*, 1995). The growth stage of the plant can also affect CTs (Koupai-Abyazani *et al.*, 1993) and deposition of cell wall components (Bidlack and Buxton, 1992). Such factors are responsible for the variation in litter characteristics seen in prunings taken at different times from the same plants (Handyanto *et al.*, 1994; Mafongoya *et al.*, Chapter 13; Tian *et al.*, Chapter 9; and Vanlauwe *et al.*, Chapter 12, this volume).

Genetic manipulation has produced a range of transgenic plant material with modified phenylpropanoid characteristics. What repercussions such alterations will have in the field are yet to be assessed. Transgenic tobacco plants with suppressed levels of pre-formed phenylpropanoids are more susceptible to the fungal pathogen *Cercospora nicotianae* (Maher *et al.*, 1994). In *Triticum aestivum* varieties with reduced Mn uptake capability, root lignin accumulation per unit phenolics is reduced and susceptibility to take-all fungus (*Ophiobolus graminis*) increased (Rengel *et al.*, 1994). The manipulation of litter quality must therefore take into account factors influencing the success of the modified plants in their environment.

The spatial and temporal regulation of plant quality characteristics provides a way around some of these potential problems.

If genes involved in tissue, development-stage or temporal specificity of gene expression are cloned, it may be possible to use them to manipulate litter characteristics. For example, a gene involved in senescence of maize leaves has been isolated (Smart *et al.*, 1994, 1995). The promoter from this gene could be used to promote the expression of transgenes in senescing tissue. If extant phenylpropanoids can be metabolized or modified by the plant, as appears to be the case for CTs (Koupai-Abyazani *et al.*, 1993), then litter quality could be altered without affecting the plant earlier in its growth.

The most powerful capability of molecular biology is the ability to introduce genes from unrelated species or kingdoms into plants, the best known example of this being the introduction of the *Bacillus thuringiensis* toxin into maize. In a similar way cloned genes for fungal ligninases or other enzymes can be introduced into plants. If the expression of these genes can be controlled spatially and temporally it may be possible to modify the characteristics of plant material in an entirely novel way. For example the expression of a ligninase in senescing leaves might modify the lignin content, leading to more rapid degradation in the soil.

The elucidation of the genetics of phenylpropanoid metabolism has potential for the manipulation of plant, and hence litter, quality. Successful experiments have been conducted producing material with novel characteristics. How this material will behave as a component of litter remains to be investigated. Similarly, environmental factors may enhance or diminish these changes. There is very real potential to modify the characteristics of plant litter through a combination of plant breeding and genetic manipulation. It is important to bear in mind, however, that such changes may profoundly affect the performance of the plants in the field/environment.

Acknowledgements

The authors wish to thank Dr Eunice Carter and other members of the Cell

Manipulation group for useful discussions. We acknowledge financial support from Biotechnology and Biological Sciences Research Council (BBSRC) during execution of the studies on condensed tannins. Adrian Bavage is funded by a grant from the BBSRC Plant Molecular Biology Initiative (PG203/536).

References

Anuraga, M., Duarsa, P., Hill, M.J. and Lovett, J.V. (1993) Soil moisture and temperature affect on condensed tannin concentrations and growth in *Lotus corniculatus* and *Lotus pedunculatus*. *Australian Journal of Agricultural Research* 44, 1667–1681.

Atanassova, R., Favet, N., Martz, F., Chabbert, B., Tollier, M.-T., Monties, B., Fritig, B. and Legrand, M. (1995) Altered lignin composition in transgenic tobacco expressing O-methyltransferase sequences in sense and antisense orientation. *The Plant Journal* 8, 465–477.

Bidlack, J.E. and Buxton, D.R. (1992) Content and deposition rates of cellulose, hemicellulose and lignin during regrowth of forage grasses and legumes. *Canadian Journal of Plant Science* 72, 809–818.

Boudet, A.M., Lapierre, C. and Grima-Pettenati, J. (1995) Tansley review No. 80. Biochemistry and molecular biology of lignification. *New Phytologist* 129, 203–236.

Bucholtz, D.L., Cantrell, R.P., Axtell, J.D. and Lechtenberg, V.L. (1980) Lignin biochemistry of normal and brown midrib mutant of *Sorghum*. *Journal of Agricultural and Food Chemistry* 28, 1239–1241.

Carron, T.R., Robbins, M.P. and Morris, P. (1994) Genetic modification of condensed tannin biosynthesis in *Lotus corniculatus*. 1. Heterologous antisense dihydroflavonol reductase down-regulates tannin accumulation in 'hairy root' cultures. *Theoretical and Applied Genetics* 87, 1006–1015.

Carter, E., Morris, P. and Theodorou, M. (1995) Impact of climate change on anti-nutritive compounds in forage legumes. In: *Adaptation in Plant Breeding, XIV EUCARPIA Congress*, University of Jyväskylä, Jyväskylä, Finland. Jyväskylä University Printing House, pp. 23.

Colliver, S.P., Robbins, M.P. and Morris, P. (1994) An antisense strategy for the genetic manipulation of condensed tannin and isoflavonoid phytoalexin accumulation in transgenic *Lotus corniculatus* L. *Acta Horticulturae* 381, 148–151.

Colliver, S.P. (1995) The genetic manipulation of flavonoids and isoflavonoids in transgenic *Lotus corniculatus* L. PhD Thesis. The University of Wales, University College Aberystwyth, Aberystwyth, Wales, UK.

Constantinides, M. and Fownes, J.H. (1994) Nitrogen mineralization from leaves and litter of tropical plants – relationship to nitrogen, lignin and soluble polyphenol concentrations. *Soil Biology and Biochemistry* 26, 49–55.

Dixon, R.A., Harrison, M.J. and Paiva, N.L. (1995) The isoflavonoid phytoalexin pathway: from enzymes to genes to transcription factors. *Physiologia Plantarum* 93, 385–392.

Elkind, Y., Edwards, R., Mavandad, M., Hedrick, S.A., Ribak, O., Dixon, R.A. and Lamb, C.J. (1990) Abnormal plant development and down-regulation of phenylpropanoid biosynthesis in transgenic tobacco containing a heterologous phenylalanine ammonia-lyase gene. *Proceedings of the National Academy of Sciences USA*, 87, 9057–9061.

Halpin, C., Knight, M.E., Foxon, G.A., Campbell, M.M., Boudet, A.M., Boon, J.J., Chabbert, B., Tollier, M.T. and Schuch, W. (1994) Manipulation of lignin quality by down regulation of cinnamyl alcohol dehydrogenase. *The Plant Journal* 6, 339–350.

Handayanto, E., Cadisch, G. and Giller, K.E. (1994) Nitrogen release from prunings of legume hedgerow trees in relation to quality of the prunings and incubation method. *Plant and Soil* 160, 237–248.

Koupai-Abyazani, M.R., McCallum, J., Muir, A.D., Bohm, B.A., Towers, G.H.N. and Gruber, M.Y. (1993) Developmental changes in the composition of proanthocyanidins from leaves of sainfoin (*Onobrychis viciifolia* Scop.) as determined by HPLC analysis.

Journal of Agricultural and Food Chemistry 41, 1066–1070.

Legrand, M., Atanassova, R., Favet, N., Martz, F., Chabbert, B., Tollier, M.T., Monties, B. and Fritig, B. (1994) Inhibition of O-methyltransferase (OMT) activity in transgenic tobacco plants – modified lignin monomeric composition. *International Plant Molecular Biology Meeting* Amsterdam. International Society of Plant Molecular Biology. Abstract No. 592.

Maher, E.A., Bate, N.J., Ni, W., Elkind, Y., Dixon, R.A. and Lamb, C.J. (1994) Increased disease susceptibility of transgenic tobacco plants with suppressed levels of preformed phenylpropanoid products. *Proceedings of the National Academy of Sciences (USA)*, 91, 7802–7806.

Meer, I.M. van der, Stuitje, A.R. and Mol, J.N.M. (1993) Regulation of general phenylpropanoid and flavonoid gene expression. *Control of Plant Gene Expression*. CRC Press Florida, USA, pp. 125–155.

Ni, W.T., Paiva, N.L. and Dixon, R.A. (1994) Reduced lignin in transgenic plants containing caffeic acid O-methyltransferase antisense gene. *Transgenic Research* 3, 120–126.

Oglesby, K.A. and Fownes, J.H. (1992) Effects of chemical composition on nitrogen mineralization from green manures of seven tropical leguminous trees. *Plant and Soil* 143, 127–132.

Rengel, Z., Graham, R.D. and Pedler, J.F. (1994) Time course of biosynthesis of phenolics and lignin in roots of wheat genotypes differing in manganese efficiency and resistance to take-all fungus. *Annals of Botany* 75, 471–477.

Robbins, M.P., Carron, T.R., Colliver, S.P. and Morris, P. (1994) A study on the genetic manipulation of flavonoids and condensed tannins in the *Lotus corniculatus* using antisense technology. In: *Proceedings of the First International Lotus Symposium*, University of Missouri-Columbia, St. Louis, University Extension Press, Missouri, USA, pp. 118–121.

Robbins, M.P., Bavage, A.D. and Morris, P. (1996) Options for the genetic manipulation of astringent and antinutritional metabolites in fruit and vegetables. In: Thomas Barberan, F.A. and Robins, R.J. (eds) *Phytochemistry of Fruit and Vegetables. Proceedings of the Phytochemical Society of Europe.* 41 (in press).

Schmitt, D., Pakusch, A.E. and Matern, U. (1991) Molecular cloning, induction and taxonomic distribution of caffeoyl-CoA 3-O-methyltransferase, an enzyme involved in disease resistance. *Journal of Biological Chemistry* 266, 17416–17423.

Smart, C.M., Hoskens, S.E., Thomas, H., Greaves, J.A., Blair, B.G. and Schuch, W. (1994) Gene expression and maize leaf senescence; characterisation of senescence-related cDNAs. *Journal of Experimental Botany* 45(May supplement), 10.

Smart, C.M., Hoskens, S.E., Thomas, H., Greaves, J.A., Blair, B.G. and Schuch, W. (1995) The timing of maize leaf senescence and characterisation of senescence related cDNAs. *Physiologia Plantarum* 93, 673–682.

Stump, L.M. and Binkley, D. (1993) Relationships between litter quality and nitrogen availability in rocky mountain forests. *Canadian Journal of Forest Research* 23, 492–502.

Tollier, M.T., Chabbert, B., Lapierre, C., Monties, B., Francesch, C., Rolando, C., Jouanin, L., Pilate, G., Cornu, D., Boucher, M. and Inze, D. (1994) Lignin composition in transgenic poplar shoots with modified cinnamyl alcohol dehydrogenase activity with reference to dehydropolymer models of lignin. In: *Polyphenols 94* INRA Editions, Paris. pp. 339–340.

Part V

Synchrony and Soil Organic Matter

———————————

17 Synchrony of Nutrient Release and Plant Demand: Plant Litter Quality, Soil Environment and Farmer Management Options

R.J.K. Myers[1], M. van Noordwijk[2] and Patma Vityakon[3]

[1] ICRISAT, Patancheru 502 324, Andhra Pradesh, India; [2] ICRAF, PC Box 161, Bogor 16001, Indonesia; [3] Department of Soil Science, Khon Kaen University, Khon Kaen, Thailand

Introduction

Synchrony refers to the matching, through time, of nutrient availability and crop demand. When supply and demand do not match, there may be periods of plant nutrient deficiency as well as periods of temporary surplus, even if total supply equals total demand. Lack of synchrony is of concern in two situations: when the supply comes too late for the demand, and when the supply comes earlier than demand in a situation where available nutrients in excess of current plant demand are at risk of loss from the system or of being converted into unavailable forms.

In its broad sense, synchrony has been promoted by varying the formulation, placement and timing of fertilizer inputs, and by other management options such as tillage, time of sowing or type of crop. The Tropical Soil Biology and Fertility program (TSBF; Woomer and Swift, 1994) has focused on the synchrony theme to find options for improving nutrient use efficiency by better management of plant litter and other organic inputs.

Myers *et al.* (1994) reviewed the relevant literature and found evidence for absence of synchrony in many crop production systems, but failed to find many unequivocal examples of improved synchrony through better management of organic inputs. None the less, the synchrony principle has attracted widespread interest among scientists as evidenced by the increasing use of the term in publications. The reasons for such interest have included the idea of finding management practices that might be attractive to poor, smallholder farmers in developing countries, and of avoiding some of the undesirable environmental effects of existing practices in both developed and developing countries.

Buresh (1995) and Van Noordwijk and Garrity (1995) reviewed literature on nutrient use efficiency of tropical cropping systems, with an emphasis on agroforestry. Four main aspects are:

1. Uptake efficiency: Plant nutrient uptake from stored, as well as recently added, organic and/or inorganic resources. Non-available nutrient sources can stay in the soil by chemical occlusion and the equivalent soil biological and physical phenomena, or can be lost to other environmental compartments, by leaching to deeper layers, beyond the reach of crop roots, or as losses to the atmosphere as gas, dust, or particulate ash.

2. Physiological efficiency: Internal redistribution of nutrients during plant growth and yield formation,

3. Processes that lead to spatial heterogeneity of nutrient supply, thereby reducing overall efficiency (Cassman and Plant, 1992; van Noordwijk and Wadman, 1992), for example, horizontal nutrient transfer by trees, crops or farmers' practices, creating depletion and enrichment zones, or soil loss and displacement by erosion/deposition cycles.

4. Economic efficiency: Removal of harvest products, their exchange for external inputs and the recycling of harvest residues in the system.

Litter, or other organic input, quality and the time pattern of mineralization are important especially for aspects 1 and 3, but contributions of the synchrony concept to the overall nutrient use efficiency at farm scale and to the farmers' objectives depend on the integration of all aspects under local conditions.

Despite widespread recognition of the synchrony concept, there are few published data based on rigorous experimental evaluation. At a TSBF meeting in Watamu, Kenya in 1992, a working group developed a series of testable synchrony hypotheses (TSBF unpublished meeting report, 1992). In this chapter we report and discuss some of these testable hypotheses that relate principally to litter quality and the environment in which decomposition takes place. We then try to discuss some areas of the synchrony principle that have not so far been adequately developed. We also discuss some ideas about whether farmers might share the scientists' enthusiasm for synchrony.

Twelve Synchrony Hypotheses

Here we examine a testable set of hypotheses based on the synchrony concept, and outline a research framework that can be used to test many of the hypotheses. A key feature of the framework is flexibility which permits either detailed process studies or applied research. The general objective of synchrony-related research is to test the potential for maximizing nutrient capture in the soil–plant system by optimizing the timing, quantity, quality and location of inorganic and organic nutrient inputs. The synchrony hypotheses

include some relatively untested concepts, whereas others are well-known and are included for completeness. The relevance of the hypotheses will vary with soil, climate, plant and farm conditions.

S1. The maximum crop yield achievable by the use of inorganic (mineral fertilizer) inputs can be approached or exceeded by optimizing the time of application, placement and quality of organic nutrient sources.

S2. In environments where significant leaching or denitrification occurs, plant uptake of mineral N applied at planting can be increased by simultaneous application of a low N organic material which temporarily immobilizes N early in the crop growth cycle and remineralizes N later on.

S3. Stabilization of organic matter in the soil is enhanced by the addition of mineral nitrogen simultaneously with the addition of organic materials of high C-to-N ratio.

S4. Residues high in lignin will result in a low net mineralization and plant uptake in the first cropping season, but will produce a greater residual effect in subsequent seasons.

S5. Residues high in tannins exhibit delayed nutrient release, but will after a lag period release nutrients rapidly.

S6. Immobilization of P by microbes, or blocking of P sorption and fixation sites, following addition of organic material can prevent fixation of P, thereby improving medium-term availability of P. This phenomenon would be best exhibited in P-fixing soils that are poor in organic matter and high in Fe.

S7. Nutrient uptake efficiency increases with the longevity of the plant. Implicit here also is the notion that relays of short-lived plants may act in the same way as long-lived plants.

S8. The nutrient uptake efficiency of the system will be increased by plants that have more rapidly growing, deeper and more extensive root systems.

S9. Incorporation of organic inputs, as opposed to surface application, accelerates the release of nutrients, thereby providing another option for modifying nutrient use efficiency.

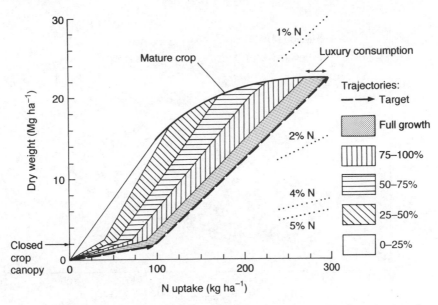

Fig. 17.1. Relationship between cumulative N uptake and cumulative dry matter production of annual crops; the different shadings indicate whether or not growth depressions will ocur if the trajectory of an actual crop development course enters a certain zone. The bold convex line shows the typical relationship at harvest time when crops are compared which grew at different N supply (based on De Willigen and Van Noordwijk, 1989).

S10. Improvement of nutrient uptake efficiency due to the use of organic inputs is more likely when crop growth and soil processes are less constrained by water deficits.

S11. Quality and quantity of organic inputs can influence faunal composition and activity, and thus affect the synchrony of nutrient supply and crop demand.

S12. The need for exact synchrony and crop demand can be reduced by storage of nutrients within the crop in excess of the crop's immediate requirement for growth.

With reference to the topic of this volume, note that hypotheses S1, S2, S4, S5 and S11 refer specifically to litter quality.

Plant Demand and the 'Internal Buffer'

Plant nutrient uptake can be viewed in two ways. Often it is considered to be directly determined by the nutrient concentrations in the root zone. Where such concentrations are just sufficient to maintain un-restricted plant growth, synchrony is ensured. Alternatively, plants are seen as regulators which control their uptake rates despite widely fluctuating external conditions. The truth is likely to be closer to the second than the first, but in practice, the regulation will be less than perfect. A lack of synchrony between N mineralization and N demand which leads to a build-up of mineral N in the rooting zone is not a problem so long as crop uptake occurs before the N is leached from the root zone.

In examining the role of the plant in synchrony, one needs to consider the exponential growth phase up to the closed crop canopy, during which high N concentrations in the leaves are maintained (if external conditions allow), and the linear crop growth phase, in which internal redistribution maintains high N concentrations in the photosynthetically active leaves, but the average N concentration in above-ground biomass can decrease. The relationship between crop dry weight and total N uptake at harvest time typically follows a convex (quadratic) approach to a plateau value (Fig. 17.1), but the trajec-

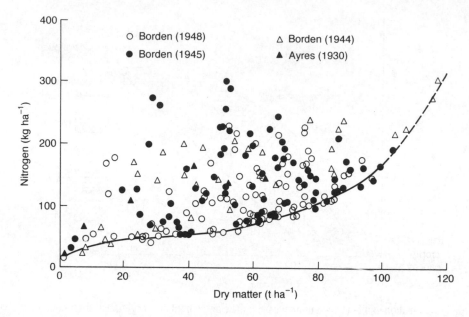

Fig. 17.2. Relationship between N uptake and dry matter yield of sugarcane showing the minimum required to achieve certain yields (R.J.K. Myers and D.R. Ridge, unpublished manuscript).

tories though time at different N supply are concave. Lines through the origin in Fig. 17.1 represent constant average N concentrations in the biomass: maximum yield is achieved by starting off at about 5% N in the tissue and switching over to about 1% N in all dry weight accumulated after the crop canopy has closed (at a biomass of about 2 Mg ha⁻¹); the 'target' uptake includes a certain degree of 'luxury consumption', which allows for temporary shortfalls in uptake without direct effects on dry matter accumulation. The figure is schematized for modelling purposes; versions with real data can be found in De Willigen and van Noordwijk (1987). According to their model, there is a target for the regulation of uptake, and growth is reduced if the target is not met. However, some reduction of uptake below the target can occur without reduction of growth, due to the internal buffer (otherwise referred to as luxury uptake). The existence of this internal buffer reduces the need for exact synchrony since it means that there can be some storage of nutrients within the plant and reduction of the risk of nutrient loss by leaching and gaseous

loss. The potential of this internal buffer is substantial, with some plants able to contain double the nutrient concentration necessary for maximum growth (Fig. 17.2). In practice, the concentrations are usually only 10–20% higher than needed.

In semi-arid systems in particular, excessive early nutrient uptake may be a disadvantage if there is excessive use of water and there is insufficient water available to complete floral development and grain filling. When grain sorghum is grown where water stored in the profile before sowing is important, use of wide rows helps achieve balanced water use (Myers *et al.*, 1986) which implies also a synchrony for nutrients. However, when water is less limiting, or in semi-arid areas where rainfall distribution is favourable, a considerable proportion of the mobile nutrients can be translocated from vegetative parts into grain. As a result, a crop could perform quite well in completing its growth cycle from use of subsoil water low in nutrients, when dry topsoil conditions make most of the nutrients inaccessible. In these cases, the plant has a mechanism for ensuring its well-being when nutrients are highly available early in the growth cycle but in short supply as the crop matures. In the absence of this mechanism, available nutrients would remain in the soil for longer and there would be the risk of loss or reduced availability.

Plant species differ in the rate and degree to which later nutrient uptake can compensate for deficiencies early in growth. Such compensation is likely to be limited in crops where the vegetative parts are harvested.

Leaching Rates, Root Development and the External Buffer

Roots contribute to synchrony through variation between species and cultivars in their capacity to explore the soil volume. Root depth is most important for mobile nutrients, whereas intensity of exploration and mycorrhizae are particularly seen as important with the immobile nutrients. Active root exploration and mycorrhizal development are important to the rapid acquisition of nutrients during the period when the plant's needs are greatest. Root depth is important when mobile nutrients are at risk of loss by leaching (de Willigen and van Noordwijk, 1989), or when nutrients are naturally present in subsoils in significant concentrations, or in mixed systems where deep-rooted plants can assist the mixture to achieve synchrony. Wetselaar and Norman (1960) observed that a lack of synchrony resulted in leaching of nitrate which reduced N uptake and yield in a shallow-rooted crop (grain sorghum). However, in a deep-rooted crop (pearl millet) the root system recovered nitrate that had leached below one metre, and N uptake and dry matter production were higher than with the sorghum. A similar difference was observed between shallow-rooted maize and deep-rooted upland rice on an acid soil in Lampung, Indonesia (H. van Noordwijk, unpublished).

Leaching losses of mobile nutrients are not necessarily closely related to infil-tration rates. Bypass flow of water through macropores or through old root channels (Van Noordwijk et al., 1991) can leave behind mobile nutrients protected within the soil matrix, and thus reduce the need for synchrony. Surface heterogeneity by ridge tillage is also effective in increasing bypass flow and increasing N-use efficiency of organic N applied in the ridge.

Root system architecture is undoubtedly important. In acid soils that are high in soil solution aluminium, the shallowness of the root system of crops such as corn is a cause of lack of synchrony. With such crops, synchrony must be close to perfect to avoid losses of N below the root system. One of the claimed advantages of agroforestry systems is the supposed mix of deep and less-deep root systems providing a more efficient capture of mobile nutrients. In contrast, monocropping in a range of environments has been shown to be 'leaky' with respect to mobile nutrients as reported by Campbell et al. (1975) in a temperate environment, Catchpoole (1992) in the subtropics, and Wetselaar (1962) in the tropics.

Aspects of Synchrony Relevant to Nitrogen

In the review by Myers et al. (1994), much of the discussion of synchrony was about nitrogen. Here we will restrict ourselves to some areas that were not covered in detail in that chapter, and some areas where new ideas have developed.

Plants may have characteristics which avoid the need for synchrony, or provide buffering against asynchrony. Many plants have the capacity to accumulate more N than needed for growth as seen above for sugarcane. Perennial species are good examples, particularly those with seasonal leaf drop preceded by a substantial withdrawal of N from leaves (Table 17.1). In this example, the eucalypt tree achieves the maximum standing crop of N early in the life cycle, and thereafter conserves the N within the plant or recycles it through litter.

Table 17.1. Comparison of N concentration in young, fully expanded leaves (outer canopy) and old leaves approaching senescence (inner canopy) and the N concentration in litter collected below the same trees, in a *Eucalyptus grandis* plantation in southern Queensland (R.J.K. Myers, D.M Cameron and S.J. Rance, unpublished manuscript).

Age of trees (months)	Canopy part	Fertilized %N	Unfertilized %N
6	Outer	2.54	1.92
	Inner	1.93	1.36
	Litter	1.16	
7.5	Outer	2.55	1.92
	Inner	1.75	1.50
	Litter	0.66	
21	Outer	1.94	1.96
	Inner	1.44	1.73
	Litter	0.58	

To date, most interest in the synchrony idea has focused on manipulating the immobilization–remineralization cycle. Myers *et al.* (1994) concluded that quality factors influenced the rate of decomposition of litter, and the quantity of nutrients such as N that were immobilized. It was less clear that the nutrients were then remineralized sufficiently rapidly to supply plant needs. It was also considered that the published information found did not adequately test some of the synchrony hypotheses. According to hypothesis S2, the low quality litter immobilizes the nutrient, but later the nutrient is mineralized and eventually the available nutrient accumulation equals that of the soil without litter. However, there would be no advantage nor disadvantage to the synchrony of supply and demand unless some of the nutrient would otherwise have been lost or become converted to an unavailable form. There is anecdotal evidence for such situations in the real world, but we lack solid scientific evidence. Particularly lacking is evidence for rapid enough remineralization of nutrients. Most observations are that remineralization of nutrients occurs relatively slowly and that only a fraction of the immobilized nutrient is remineralized during one crop cycle. There seems to be a difference between the relatively slow rate of remineralization of immobilized N and

the relatively rapid release of N from high quality materials such as green manures. The absence of solid supporting evidence for substantial enough remineralization is partly due to the absence so far of experiments specifically designed to test the hypothesis. In particular, many past studies have not run for long enough to truly evaluate the rate of remineralization.

Another mechanism contributing to synchrony is that whereby chemical constituents of litter slow or delay the release of nutrients. Constituents with such properties have included the lignins and polyphenols. In some cases the rate of mineralization from a material was dependent on the lignin-to-N ratio (Becker *et al.*, 1994; Becker and Ladha, Chapter 18, this volume), or the polyphenol-to-N ratio (Palm and Sanchez, 1991; Oglesby and Fownes, 1992), or the (lignin + polyphenol)-to-N ratio (Fox *et al.*, 1990; Handayanto *et al.*, 1994). We believe that a large part of the differences between these various results stems from the different ranges of materials used by the different researchers (see Palm and Rowland, Chapter 28; Handayanto *et al.*, Chapter 14, this volume). Thus when one group used materials with a wide range of lignin and N but a narrow range of polyphenol concentration, lignin-to-N was the best predictor. When another group used

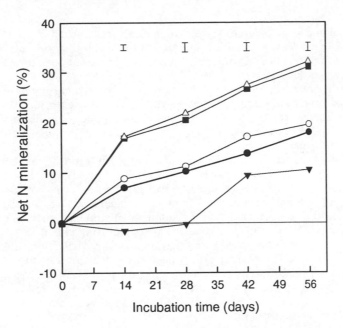

Fig. 17.3. Percentage net N mineralization of N from legume tree prunings (redrawn from Handyanto *et al.*, 1994). ●, *Calliandra calothyrus*; ▼, *Peltophorum dasyrachis*; ■, *Leucaena leucocephala*; △, *Gliricidia Sepium*; ○, *Gliricidia and Peltophorum*.

materials with a range of polyphenol and N concentration, but a narrower range of lignin, then polyphenol-to-N was the best predictor. There is still a need for a comprehensive study in which a wide range of all three components is included. There is also a need for some longer term studies. The way that short-term studies are difficult to interpret is shown well in several studies, of which that of Handayanto *et al.* (1994) provides an example (Fig. 17.3). One interpretation of this figure is that between 14–56 days, all materials behaved similarly – the slopes are reasonably close together. Then if we believe the idea of lignin and polyphenols inhibiting mineralization, there are delays of a few days to about 4 weeks in the onset of mineralization. However, it would be very interesting to know what happens after 56 days – do the curves continue parallel to each other, or do they converge? This is important because in this example, a loss event occurring at either 14 or 28 days would provide no advantage due to synchrony for the materials with the delayed start to mineralization. Only longer-term studies

can answer this question. Handayanto *et al.* (1994) emphasized the role of leaching of water-soluble polyphenolics out of litters, but their data show that a certain, perhaps cell-wall bound, fraction of the polyphenolics resists leaching and remains active.

Research into the importance of polyphenols has suffered from analytical uncertainties. Questions have been raised regarding the most suitable solvent for extraction, the standard used, and the tissue-to-solvent ratio. Constantinides and Fownes (1994) have recommended a tissue-to-solvent ratio of 1.0–2.0 mg ml^{-1}, the use of aqueous methanol rather than water as solvent, and the use of the Folin–Ciocalteu reagent rather than the Folin–Denis reagent (see Palm and Rowland, Chapter 28, this volume). They also observed that air drying gave higher concentrations than oven drying. The group of materials referred to as polyphenols is diverse, and may well be differentially soluble in the solvents used. The usual standards represent only one type of polyphenol, and one can envisage errors if materials are high in a different type. Further, there is considerable variation in the protein-binding capacity of different polyphenols. Handayanto *et al.* (1994; Chapter 14, this volume) explored the idea that an assay of protein-binding capacity might have predictive value for the N mineralization of different materials.

Another niche where synchrony may be achievable may be acid soils. We have already mentioned the asynchrony associated with the shallow root systems of some crops growing in strongly acidic soils. Strongly acid soils are frequently deficient in nutrients such as N, P and S whose availability depends at least partly on organic cycling. For such nutrients, synchrony may be promoted by organic inputs whereby decomposition products could complex Al and thereby increase soil microbial activity, which in turn may increase the supply of these nutrients. Decomposition products high in low molecular weight aliphatic and aromatic acids are active in complexing Al. However, these compounds are likely to have only transient existence in soil solutions, and fulvic and humic acids are more probably detoxifiers of Al in acid soils (Edwards *et al.*, 1994).

Placement of fertilizers has long been recognized as a means of promoting synchrony. However, placement of residues may also have a role. This could apply to placement of residues relative to that of the fertilizers provided that it was established that there was an improvement in nutrient use efficiency. In the case of P synchrony, it might be necessary to make maximum use of the residues to block adsorption, in which case it may not work unless there is localized placement. In the case of residues containing polyphenols, it may be desirable to place the residues in such a way that leaching of the polyphenols is minimized, for example along ridges. This is another area for research.

Aspects of Synchrony Related to Phosphorus

Synchrony with respect to P refers to the promotion of processes whereby P is main-tained in the soil solution for longer. In agricultural production systems which become P deficient sooner or later, despite large stocks of total soil P, P is out of synchrony. The asynchrony is expressed as deficiency, and is even more evident when fertilizer is applied. Fertilizers in which P is relatively soluble produce highly asynchronous situations where the P is highly available early in crop growth, but due to essentially irreversible soil chemical changes, it is less available later in growth. This asynchrony can be reduced by P application strategies such as split application, or applying the P in bands (locally swamping the P fixation sites), and the need for synchrony can be reduced if the plant is capable of taking up P more rapidly than its need for current growth (luxury uptake). There are several other possibilities for improving synchrony of P.

The rate of the P fixation processes can be reduced by some organic materials which react with potential phosphate sites in the soil. Hue *et al.* (1994) observed that P sorption by an Ultisol was reduced by the addition of a yard-waste compost (Fig. 17.4). Rebafka *et al.* (1994) found that fertilizer P utilization was increased from 6% to 14% by adding crop residues, and to 25% after repeated crop residue applications. This was attributed to increased root length density, increased mobility of P and influx of P per unit root length, and perhaps to reduction in the rate of reaction of P with fixation sites. Wang *et al.* (1995) found that adding humic acids to soil with P fertilizer significantly increased the amount of water-soluble P, strongly retarded the formation of occluded P and increased P uptake and yield of wheat by 25%. Singh *et al.* (1991), reviewing research on green manures, concluded that green manure increased the availability of P in waterlogged soils by a range of mechanisms including that of reduction of P sorption capacity.

A second soil biological mechanism that might improve synchrony with respect to P is that of immobilization–remineralization. In this case microorganisms attacking an organic input with a suitably high C-to-P ratio would require an exter-

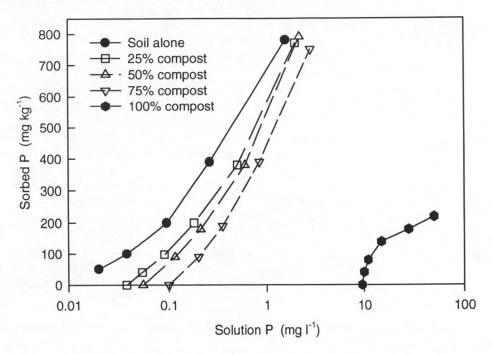

Fig. 17.4. Phosphate sorption isotherms of an Ultisol, a yardwaste compost, and soil/compost mixtures (Hue *et al.*, 1994).

nal source of P, and would therefore remove P from the soil solution, and reduce the opportunity for soil solution P to react with adsorption sites. Hue *et al.* (1994), though not specifically addressing this question, found that immobilized P in a compost had the capacity to mineralize P and maintain the soil solution P of an Ultisol at higher levels than in the unamended soil.

A third mechanism would be that whereby organic anions may reduce soil solution Al by complexation, and as a consequence Al toxicity is decreased and N mineralization is increased, and P demand is increased due to increased plant vigour. Whereas such ideas have logic, it is not so easy to devise experiments that test them exclusively, nor is it easy to interpret the results of past work as related to this sequence of events to the exclusion of other explanations.

The existence of several P-synchrony mechanisms is easy to establish, but it is not easy to ascribe relative importance. In practice, it is more important to identify the management options. Apart from the 'motherhood' options of maintaining soil organic matter and retaining residues, there are some specific questions:

- Can placement of organic inputs together with P fertilizer, as for example in a band or a 'pocket', enhance P-synchrony?
- Can immobilization and remineralization of P improve synchrony, and if so is it necessary to immobilize before growing the crop, as for example would occur with composting?
- What are the quality factors of organic materials that contribute to the complexing of potential P adsorption sites or which optimize the immobilization–remineralization processes?

Can Synchrony be Farmer-friendly?

Technologies that follow the synchrony concept need to be farmer acceptable or

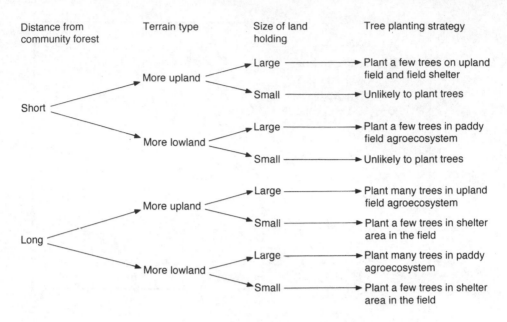

Fig. 17.5. Decision tree for northeast Thailand farmers' strategy to integrate trees into their farms (after Vityakon *et al.*, 1994).

they will remain merely textbook phenomena. Many factors influence farmers in their decisions (Fujisaka, 1992) and researchers need to understand them in order to see which technologies can be adapted and adopted by farmers.

It is a basic TSBF hypothesis that mature natural systems have a tight nutrient cycling as uptake demand normally exceeds the available supply and that they thus are in synchrony.

Many traditional farming systems, particularly those practising polyculture with a considerable tree component, may mimic natural systems and show synchrony of uptake and supply. This explains the research interest in trying to understand complex traditional systems such as the Indonesian home garden system (Michon and Mary, 1994) as well as the slash and burn system of farming (Kleinman *et al.*, 1995). However, increasing pressure on land, plus other factors, has led many farmers to intensify their farming and frequently this has resulted in adoption of higher input monocropping, lacking synchrony, as alternative to the low

input traditional systems. Increasing nutrient supply may or may not increase the 'economic efficiency' of the farm, but it often reduces the 'uptake efficiency'. Achieving a high uptake efficiency without incurring plant nutrient stress is not impossible, but it requires additional efforts. Even in the most technically advanced horticultural production systems, a wide range of nutrient use efficiencies can be found (van Noordwijk, 1990).

The likelihood of soil improvement technologies being adopted and adapted by farmers depends on a number of factors. These factors pertain to physical, biological, economic, social and cultural aspects of the farmer's livelihood. Soils, water sources, terrain, climate and infrastructure are among the physical factors. Biological factors include the plants (annual crops, tree crops and cropping systems) and animals (livestock). Among the economic factors are amount of land holding, labour, marketing, on-farm and off-farm income, availability of credit, access to information, and education and experience of farmers. Cultural factors include the farmers' beliefs and traditions. A combination of these factors influence the farmers' decision whether to adopt and adapt a technology. In northeast Thailand, rice cultivation is the rural people's way of life, and constitutes their food security. Rice cultivation therefore strongly influences decisions regarding other farming operations and off-farm activities. Any new technology should not therefore interfere, for labour or for time, with rice growing if it is to be adopted by the farmers. It is helpful to understand the process of farmer decision making in deciding whether to adopt an introduced technology. The decision tree given in Fig. 17.5 illustrates the process in selecting the best strategy for integrating trees into farms. Distance from the community forest, which reflects accessibility, is the key indicator. The major factors influencing farmers' decision making then are biological (forest availability), physical (terrain type) and socioeconomic (size of land holding).

Some existing technologies of soil improvement, whether or not they are traditional, can be developed further. Technologies that are built from existing ones stand a better chance of adoption by farmers than those that are totally new and unknown. In northeast Thailand, soil improvement by use of compost made in the livestock pen has been practised for decades, whereas new compost-making techniques promoted by government extension agents have not been adopted. The farmers' technique is simpler, less tedious, and less expensive (Wongsamun et al., 1989). Also in northeast Thailand, the effects of trees on soil fertility and growth of associated rice have been studied (Sae-Lee et al., 1992). Traditionally trees and agricultural crops, such as rice and cassava, grow in association. Socioeconomic studies (Vityakon et al., 1994) have identified farmers' criteria for adopting trees into the farming system. Such studies bring about better understanding of factors controlling the existence of a farming practice, and can improve the chances of adoption of new technologies.

The transfer of technologies from an area where they work well to another area depends on whether the environment is suitable for the technology, as well as the constraints of farmers. This can be seen in some areas of northeast Thailand where farmers practise mixed cultivation of rainfed paddy rice and mungbean. The seeds of both are broadcast onto untilled soil at the beginning of the rains. A layer of rice straw is then added to maintain soil moisture and prevent weed growth. The paddies are flooded when the rice is about 10 cm tall. This kills the mungbean which then acts as a green manure for the rice. This practice will work only where there is adequate rice straw for the mulch, and where rains come in sufficient quantity and at the right time to kill the mungbean (Vichiensanth et al., 1992).

Provided that the technology is economically sound, it should be adopted by at least some farmers. Those who adopt it will do so to different degrees; some will adopt it wholeheartedly and others not at all. Success will be greater if the farmers are categorized according to some appropriate criteria, and the technology can then be modified to suit the different categories.

Adoption of soil improvement technologies frequently depends on farmers' economic status. Well-to-do farmers are often more willing to try new things, and are less concerned about risk than poorer farmers. Often this is related to size of land holding and this also indicates potential availability of resources. Extension often follows this approach, initially aiming at adoption by well-to-do farmers followed by diffusion to the poorer farmers. Synchrony may not be exciting to farmers by itself, but it may be acceptable if promoted together with some other beneficial technology. Currently, integrated packages of agroforestry, cropping systems, soil erosion control, and soil fertility maintenance are being promoted. The synchrony concept could readily be built into such packages. However, the risk exists that in adapting the package to their needs, farmers may remove the components giving synchrony.

Potentially there are many reasons why synchrony may not be adopted. Let us examine some scenarios:

Table 17.2. Farmer management options involving the use of organic inputs.

Management option	Hypothesis	Drawback/constraint
Quality		
Choice of species, especially if a cover crop/green manure	S1, S2, S4, S5	Direct economic value usually dominates choice of species; potential conflict where there is more than one use.
Management of plants grown as green manure (low N gives less and slowly decomposing litter)	S1, S2, S4	
Mixing organic sources in combination with mulch/manure transfer)		Labour requirements
Location		
Placement, e.g. surface-applied versus incorporated, flat versus ridged, band placement	S1, S9	Erosion risks of tillage; volatilization losses from surface mulch
In situ mulch production – intercropping or sequential systems		Competition in simultaneous systems
Mulch/manure transfer		High labour requirements
Timing		
Adjusting crop species to existing mineralization pattern	S7, S8	Reduced choice of crop species
Timing of organic and inorganic inputs	S1	Labour requirements; crop growth cycle; knowledge and lack of predictability

- What if a technology required a farmer to collect cereal straw from his field, take it somewhere where it could be chopped finely, then mixed with a fertilizer, then carry it back to the field, and that this had to be done at a time when there were other important farm operations to be done, and he had no other labour available?
- What if a technology required the preparation of a compost, but that it was critical for the farmer to maintain the composting process at a particular temperature and within a narrow pH range?
- What if a particular quality of litter was found to promote synchrony due to its content of polyphenols, but the farmer also wanted to use the litter for feeding to farm animals, and it proved to have

very poor digestibility?
- What if the measure that promoted synchrony required 3 years of no-response before the benefits became apparent, and the farmers were cash-poor farmers very dependent in the next crop for survival?

These scenarios are not presented in order to be negative, but rather to make it clear that researchers' enthusiasm for synchrony may not necessarily be shared by farmers, and that unless researchers recognize the need to consider farmers' constraints, the synchrony technology will go the way of many other so-called improved technologies.

Table 17.2 lists some farmer management options that involve the use of organic inputs. This shows that there are almost always drawbacks or constraints to the use of these inputs and that the individual options do not specifically address individual hypotheses. That is, a successful experiment or on-farm demonstration of such management options may not necessarily result in adoption, even though the results may well be consistent with one or more of the hypotheses.

Conclusions

Research relevant to the synchrony concept far outnumbers research specifically testing synchrony hypotheses. Much of the research being done therefore neither proves or disproves that the synchrony hypotheses are valid. There is a need for strategic research to examine the various synchrony hypotheses. The role of the plant in contributing to synchrony is reaffirmed, particularly in respect of the

plant's internal buffer and the root system reducing the need for exact synchrony. With nitrogen there is a need for further research into the potential to achieve synchrony through litter quality factors such as lignin and polyphenols. There is also scope for improving synchrony with respect to phosphorus through the use of organic inputs which reduce the rate of P fixation. Finally there is a need to identify practices that incorporate appropriate components of synchrony that would likely be acceptable to farmers. Such practices need to be taken to farmers through either or both adaptive and on-farm research.

Acknowledgements

We acknowledge the contribution of our colleagues, including Bob Scholes and Cheryl Palm, in the development of the set of testable hypotheses.

References

Becker, M., Ladha, J.K., Simpson, I.C. and Ottow, J.C.G. (1994) Parameters affecting residue nitrogen mineralization in flooded soils. *Soil Science Society of America Journal* 58, 1666–1671.

Buresh, R.J. (1995) Nutrient cycling and nutrient supply in agroforestry systems. In: Dudal, R. and Roy, R.N. (eds) *Integrated Plant Nutrient Systems*. FAO, Rome, pp. 155–164.

Campbell, C.A., Nicholaichik, W. and Warder, F.G. (1975) Effects of wheat-summer-fallow rotation on subsoil nitrate. *Canadian Journal of Soil Science* 55, 279–286.

Cassman, K.G. and Plant, R.E. (1992) A model to predict crop response to applied fertilizer nutrients in heterogeneous fields. *Fertilizer Research* 31, 151–163.

Catchpoole, V.R. (1992) Nitrogen dynamics of oats, sorghum, black gram, green panic and lucerne on a clay soil in south-eastern Queensland. *Australian Journal of Experimental Agriculture* 32, 1113–1120.

Constantinides, M. and Fownes, J.H. (1994) Tissue-to-solvent ratio and other factors affecting determination of soluble polyphenols in tropical leaves. *Communications in Soil Science and Plant Analysis* 25, 3221–3227.

De Willigen, P. and van Noordwijk, M. (1987) Roots, plant production and nutrient use efficiency. PhD Thesis, University of Wageningen, 282 pp.

De Willigen, P. and van Noordwijk, M. (1989) Rooting depth, synchronization, synlocalization and N-use efficiency under humid tropical conditions. In: van der Heide, J. (ed.) *Nutrient Management for Food Crop Production in Tropical Farming Systems*. Institute for Soil Fertility, Haren, pp. 145–156.

Edwards, D.G., Bell, L.C. and Grundon, N.J. (1994) ACIAR Project 8904 – Research at the University of Queensland 1992–1993. In: *Network Developments in the Management of Acid Soils (IBSRAM/ASIALAND)*. *Network Document No. 11*. IBSRAM, Bangkok, pp. 9–29.

Fox, R.H., Myers, R.J.K. and Vallis, I. (1990) The nitrogen mineralization rate of legume residues in soil as influenced by their polyphenol, lignin, and nitrogen contents. *Plant and Soil* 129, 251–259.

Fujisaka, S. (1992) Thirteen reasons why farmers do not adopt innovations intended to improve the sustainability of upland agriculture. In: *Evaluation for Sustainable Land Management in the Developing World.* IBSRAM Proceedings No. 12. IBSRAM, Bangkok, pp. 509–522.

Handayanto, E., Cadisch, G. and Giller, K.E. (1994) Nitrogen release from prunings of legume hedgerow trees in relation to quality of the prunings and incubation method. *Plant and Soil* 160, 237–248.

Hue, N.V., Ikawa, H. and Silva, J.A. (1994) Increasing plant-available phosphorus in an Ultisol with a yard-waste compost. *Communications in Soil Science and Plant Analysis* 25, 3291–3303.

Kleinman, P.J.A., Pimentel, D. and Bryant, R.B. (1995) The ecological sustainability of slash-and-burn agriculture. *Agriculture, Ecosystems and Environment* 52, 235–249.

Michon, G. and Mary, F. (1994) Conservation of traditional village gardens and new economic strategies of rural households in the area of Bogor, Indonesia. *Agroforestry Systems* 25, 31–58.

Myers, R.J.K., Foale, M.A., Thomas, G.A., French, A.V. and Hall, B. (1986) How row spacing affects water use and root growth of grain sorghum. *Proceedings of the Australian Sorghum Conference*, 5.82–5.92.

Myers, R.J.K., Palm, C.A., Cuevas, E., Gunatilleke, I.U.N. and Brossard, M. (1994) The synchronisation of mineralisation and plant nutrient demand. In: Woomer, P.L. and Swift, M.J. (eds) *The Biological Management of Tropical Soil Fertility.* John Wiley and Sons, Chichester, UK, pp. 81–116.

Oglesby, K.A. and Fownes, J.H. (1992) Effects of chemical composition on nitrogen mineralization from green manures of seven tropical leguminous trees. *Plant and Soil* 142, 127–132.

Palm, C.A. and Sanchez, P.A. (1991) Nitrogen release from the leaves of some tropical legumes as affected by their lignin and polyphenolic contents. *Soil Biology and Biochemistry* 23, 83–88.

Rebafka, F.-P., Hebel, A., Bationo, A., Stahr, K. and Marschner, H. (1994) Short- and long-term effects of crop residues and of phosphorus fertilization on pearl millet yield on an acid sandy soil in Niger, West Africa. *Field Crops Research* 36, 113–124.

Sae-Lee, S., Vityakon, P. and Prachaiyo, B. (1992) Effects of trees on paddy bund on soil fertility and rice growth in Northeast Thailand. *Agroforestry Systems* 18, 213–223.

Singh, Y., Khind, C.S. and Singh, B. (1991) Effective management of leguminous green manures in wetland rice. *Advances in Agronomy* 45, 135–222.

Van Noordwijk, M. (1990) Synchronization of supply and demand is necessary to increase efficiency of nutrient use in soilless horticulture. In: van Beusichem, M.L. (ed.) *Plant Nutrition – Physiology and Applications.* Kluwer Academic, Dordrecht, pp. 525–531.

Van Noordwijk, M. and Garrity, D.P. (1995) Nutrient use efficiency in agroforestry systems. In: *Potassium in Asia: Balanced Fertilization to Increase and Sustain Agricultural Production Proc. 24th IPI Colloquium*, Chiang Mai, International Potash Institute, Basel 1995, pp. 245–279.

Van Noordwijk, M. and Wadman, W. (1992). Effects of spatial variability of nitrogen supply on environmentally acceptable nitrogen fertilizer application rates to arable crops. *Netherlands Journal of Agricultural Science* 40, 51–72.

Van Noordwijk, M., Widianto, Heinen, M. and Hairiah, K. (1991) Old tree root channels in acid soils in the humid tropics: important for crop root penetration, water infiltration and nitrogen management. *Plant and Soil* 134, 37–44.

Vichiensanth, P., Polthanee, A. and Suphanchaiyamat, N. (1992) *Farmers' Constraints in the Application of Modern Technologies for Improving Soil Fertility in the Northeast. A Research Report.* Research and Development Institute, Khon Kaen University, Khon Kaen University, Thailand, 152 pp. (in Thai with English abstract).

Vityakon, P., Polthanee, A. and Grisanaputi, W. (1994) *Factors Influencing Number of Trees and Farmers' Conditions for Tree Integration in the Farming System. A Case Study of District of Kranuan, Khon Kaen, Northeast Thailand.* 66 pp. (in Thai with English abstract).

Wetselaar, R. (1962) Nitrate distribution in tropical soils. 3. Downward movement and accumulation of nitrate in the subsoil. *Plant and Soil* 16, 19–31.

Wetselaar, R. and Norman, M.J.T. (1960) Recovery of available soil nitrogen by annual fodder crops at Katherine, Northern Territory. *Australian Journal of Agricultural Research* 11, 693–704.

Wang, X.J., Wang, Z.Q. and Li, S.G. (1995) The effect of humic acids on the availability of phosphorus fertilizers in alkaline soils. *Soil Use and Management* 11, 99–102.

Wongsamun, C., Sorn-srivichai, P., Ayuwat, D. and Pakdee, P. (1989) *Indigenous Agricultural Technology: Making Compost in Livestock Pen*. A research report. Farming systems research project. Faculty of Agriculture, Khon Kaen University, Khon Kaen, Thailand. 40 pp. (in Thai with English abstract).

Woomer, P.L. and Swift, M.J. (eds) (1994) *The Biological Management of Tropical Soil Fertility*. John Wiley and Sons, Chichester, UK, 243 pp.

18 Synchronizing Residue N Mineralization with Rice N Demand in Flooded Conditions

M. Becker[1] and J.K. Ladha[2]

[1] West Africa Rice Development Association (WARDA), BP 2551, Bouaké 01, Côte d'Ivoire; [2] International Rice Research Institute (IRRI), PO Box 933, Manila, Philippines

Introduction

To meet future demand, world rice production must increase by more than 60% in the next 25 years (IRRI, 1989). This is only possible if soil and water resources and inputs are used more efficiently. Nitrogen is the single most important input limiting rice production worldwide. Rice can use inorganic N derived from both mineral and organic N sources, but use efficiency of mineral N fertilizers is generally low due to large losses as N gases (Buresh and De Datta, 1991). If mineral fertilizers were always applied at efficient dose-response rates, N losses from lowland rice fields can be minimized. Multiple split applications of mineral N reduce N losses and increase N use efficiency and grain yield of rice (Cassman et al., 1995), but increase operating costs.

Leguminous green manure crops or N rich plant residues are promising alternatives to mineral fertilizers in rice and other crops (Becker et al., 1990). In addition, farm residues such as weeds, rice straw and stubble, and left-over dry season crops, may be available in significant amounts at the beginning of the rice cropping season (George et al., 1992). Unlike mineral fertilizers, such organic residues must decompose before their N becomes available to the rice plant in mineral form. Thus, residues acting as slow-release fertilizers, might match nutrient supply with crop demand. Achieving synchronization of N mineralization from applied organic fertilizers and N demand for a given target yield level may reduce N losses and increase N use efficiency (McGill and Myers, 1987).

Flooded or lowland rice culture is unique in the sense that the soil remains submerged for the major part of the growing season. As a result, the soil redox potential declines and reaches a steady state which is governed by the temperature, the availability of electron acceptors (NO_3^-, Mn^{2+}, Fe^{3+}), and the amount of decomposable organic matter (Ponnamperuma, 1978). Carbon and nitrogen pools seem to be sustained in flooded rice soils (Kundu and Ladha, 1995), indicating that residue decomposition dynamics and mecanisms in flooded soils may be different from those in upland soils. Although considerable progress has been made in understanding the dynamics of N in flooded soils, little is known regarding the processes governing the decomposition and mineralization of diverse residues. A quantitative understanding of residue mineralization is needed to achieve synchronization of N supply from applied residues and crop N demand.

This chapter addresses three questions relevant for the N supply–demand synchrony concept in flooded rice:

- What parameters affect residue N mineralization in flooded soils?

- How to achieve supply demand synchrony for a given target yield? and
- What are short-term and potential long-term benefits of supply–demand synchrony?

Parameters Affecting Residue N Mineralization in Flooded Soils

The following factors primarily determine the dynamics of residue breakdown: temperature, cultural practices, soil properties, and chemical composition of organic material (Becker et al., 1994a). Temperature is usually not limiting in tropical lowland rice. Cultural practices include duration of submergence, method of incorporation, and application of mineral N (Broadbent, 1979). The influence of soil characteristics on N mineralization kinetics in flooded systems is still largely unknown. Particularly the activity of soil fauna is usually not considered in incubation experiments, but may affect N mineralization. Becker et al. (1994a) showed that in the absence of fauna from flooded rice soils, residue N mineralization during the first 6 weeks was 30% lower than in plots containing soil fauna. The influence of residue chemical composition on decomposition in tropical upland soils is well studied (Fox et al., 1990; Handayanto et al., Chapter 14; Vanlauwe et al., Chapter 12, this volume). Factors influencing N mineralization include amounts of water, carbon (C), nitrogen (N), lignin (L), polyphenol (P_p), and their ratios such as C-to-N ratio, L-to-N ratio, P_p-to-N ratio, and L+P_p-to-N ratio. Recent studies revealed that the P_p-to-N and the L+P_p-to-N ratios are probably the most reliable parameters to predict residue N mineralization in upland soils (Fox et al., 1990; Palm and Sanchez, 1991).

Although several experiments have used a variety of residues in flooded rice systems, only few have examined N mineralization patterns (Nagarajah et al., 1989, Diekmann et al., 1992). Until recently, there has been no attempt to identify the relative importance of the dominant factors governing organic matter

decomposition in flooded soils in so far as they impact on N mineralization and rice production, nor as to how mixtures of legumes and rice straw may decompose.

Becker et al. (1994a) conducted a phytotron study to determine plant chemical parameters that affect decomposition and N mineralization of residues in different flooded rice soils. Seven legume species, three legume–rice straw mixtures, rice straw alone and an Azolla sp. were incorporated in different flooded rice soils at 100 mg N kg^{-1} dry soil and incubated for 6 weeks. Large differences in mineralization patterns were observed from different soils. Net residue N mineralization (NH_4^+-N in amended minus that in unamended control plots) in a clay Mollisol was about twice that of the sandy Entisol. Nitrogen mineralization was not significantly correlated with N content of the residues, C-to-N ratio, moisture content, or polyphenol contents (Table 18.1). Net N mineralization was correlated to lignin-to-N ratio. In contrast to recent evidence from tropical upland soils (Palm and Sanchez, 1991), polyphenols showed no apparent interaction with N mineralization in flooded soils and combining soluble polyphenols with non-soluble lignin in the L-to-N ratio (L+P_p-to-N ratio) did not improve correlation coefficients. The dilution of these water-soluble compounds in soil solution and floodwater of rice soils apparently limits their reaction with N compounds (Clement et al., 1995).

For residues low in polyphenols, the L-to-N ratio seems to suitably predict their N mineralization rate in flooded soils, but this relation varies from soil to soil. Manipulating residue L-to-N ratio may allow control of the rate of net-N mineralization in a given soil. This hypothesis was tested in the field (Becker et al, 1994b). Organic substrates were adjusted to L-to-N ratios of 2, 4, 6, and 12 by combining material of different origin, age, and composition, incorporated in a flooded lowland field (clay soil) at 60 kg N ha^{-1} for legumes and 30 kg N ha^{-1} for rice straw, and incubated for 70 days in the absence of plant cover. Initial net NH_4^+-N mineralization rates were higher from residues with rela-

Table 18.1. N release and correlation of residue composition (12 organic materials) with net N mineralization in two flooded rice soils (phytotron incubation study, adapted from Becker *et al.*, 1994a).

Plant material/ parameter	Range	Correlation with N release	
		Clay	Sand
Moisture content	0–92%	0.26	0.32
Nitrogen (N)	1.8–3.8%	0.22	0.36
Polyphenol (P_p)[†]	0.2–1.9%	0.55	0.31
Lignin (L)[‡]	5.5–18.5%	0.70*	0.38
C-to-N ratio	10.2–22.7	0.39	0.57
L-to-N ratio	1.4–5.0	0.83**	0.71*
P_p-to-N ratio	0.2–1.0	0.42	0.39
$L+P_p$-to-N ratio	1.7–5.6	0.76*	0.59
Range of N release after 6 weeks (mg NH_4^+-N kg^{-1} soil)		53–86	22–55

* $P < 0.05$.

** $P < 0.01$.

[†] Water-soluble tannins (Folin–Denis method; Burns, 1963).

[‡] Acid-detergent fibre (van Soest, 1968).

tively low L-to-N ratio (*Sesbania rostrata*, L-to-N ratio 2) than from material with higher L-to-N ratio *(S. emerus*, L-to-N ratio 6, and *S. rostrata*–rice straw mixture, L-to-N ratio 6). Rice straw alone (L-to-N ratio 12) resulted in an initial net N immobilization (Fig. 18.1). Manipulating the L-to-N ratio of an organic material by mixing high-quality material (e.g. legumes) with crop residues (e.g. straw) that are available in rice fields is a simple and effective way to control residue N mineralization in flooded soils. The role of soil parameters on the amounts of N mineralized from residues requires further investigation.

Potential Benefits from Supply Demand Synchrony

The previous section indicated that residue quality can be manipulated (e.g., by mixing residues of different nature) in order to synchronize N supply with rice N demand. In this section we pose the question: What are the benefits from supply–demand synchrony and are there costs involved in order to achieve these benefits? It was speculated that synchronizing N supply

from incorporated plant residues with N demand of a crop may increase N use efficiency, reduce N losses and may thus have beneficial effects on yield and soil productivity (Myers *et al.*, Chapter 17, this volume). Ghai *et al.* (1988) showed that reducing the time interval between green manure incorporation and rice transplanting from 15 to 5 days increased rice yields and N use efficiency. The authors attributed these beneficial effects to a tighter coupling of residue N mineralization and rice N demand. We tested this hypothesis further in a three-season field experiment. Leguminous green manures and rice straw–*S. rostrata* mixtures with various L-to-N ratios as described above (L-to-N ratios 2, 4, 6, and 12) and urea (L-to-N ratio 0) were compared using soil N mineralization in unplanted plots, rice N uptake, grain yield, and total ^{15}N balance as criteria. As expected, basally applied urea (60 kg N ha^{-1}) resulted in high initial soil NH_4^+. Where residues were applied (60–90 kg N ha^{-1}), exchangeable NH_4^+-N varied as a function of L-to-N ratio as discussed earlier.

Rice N uptake patterns reflected the amounts of net NH_4^+-N mineralization in

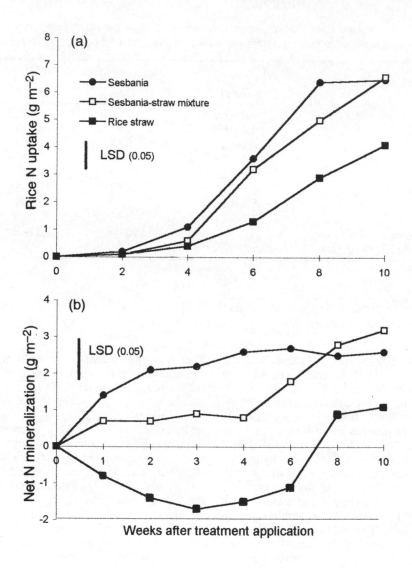

Fig. 18.1. Rice N uptake (a) and net N mineralization (b) in unplanted plots (soil exchangeable NH_4^+-N in amended minus that in unamended control plots) from organic materials with different Lignin-to-N ratios (2, 6, and 12 for *Sesbania rostrata*, *S. rostrata*-rice straw mixture, and rice straw, respectively) incorporated in flooded clay soil in the field.

the unplanted soil. Fig. 18.1 illustrates this trend using the examples of three materials with contrasting L-to-N ratios. Other residues used in the study showed interme-diate N mineralization and uptake dynam-ics as determined by their L-to-N ratios (data not shown). At 4 weeks after trans-planting (maximum tillering stage), rapidly mineralizing N sources (urea and *S. rostra-ta*) showed the highest cumulative N uptake. A comparatively slow N mineral-ization from materials with L-to-N ratio of 4 and 6 resulted in relatively low initial rice N uptake, but daily N uptake was equiva-lent to that of 'fast-release' N sources at panicle initiation stage (6 weeks). Only a

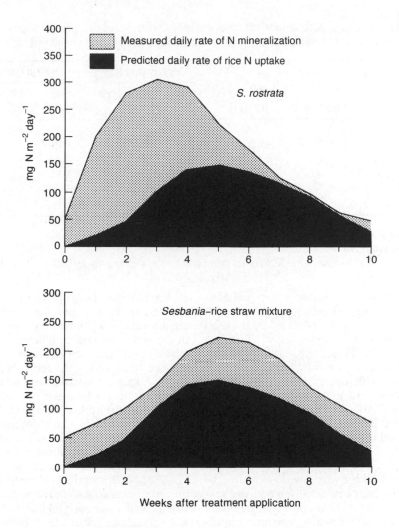

Fig. 18.2. Daily rate of soil exchangeable NH_4^+-N mineralization (measured in unplanted plots in the field) and daily rate of rice N uptake (simulated for a target yield of 6 Mg ha^{-1}, using ORYZA1 model) with application of two materials with different Lignin-to-N ratios (2 and 6 for *Sesbania rostrata, S. rostrata*-rice straw mixture, respectively).

poor match of soil NH_4^+-N with daily rice N uptake (simulated daily uptake for a target yield of 6 Mg ha^{-1}, using ORYZA1 model) was achieved in urea and *S. rostrata* treatments (L-to-N ratio 2), while matching was close with *S. emerus* (L-to-N ratio 4) and in the *S. rostrata*–rice straw mixture with a L-to-N ratio of 6. Figure 18.2 illustrates these differences in supply–demand synchrony using the examples of *S. rostrata* and the *S. rostrata*-rice straw mixture dur-

ing the 1992 dry season. Nitrogen-15 balance indicates that the mismatch between supply and demand may have caused the measured N losses of 35% from urea and of about 10% from *S. rostrata* (Table 18.2). Synchronized N supply and rice N uptake resulted in negligible N loss and increased the portion of applied N remaining in the soil. However, in contrast to the study by Ghai *et al.* (1988), a close match of N supply and N uptake did not satisfy N

Table 18.2. Effect of fertilizer source and residue quality as determined by Lignin-to-N ratio on ^{15}N recovery, grain yield and residual effect on an unfertilized crop of rice (field experiment, clay soil).

N source	L-to-N ratio	N rate (kg ha^{-1})	^{15}N recovery (%) Plant	Soil	Loss	Yield (Mg ha^{-1}) 1992 DS	1992 WS	1993* DS
			(DS 1992)					
Control	–	0	–	–	–	4.1^d	3.3^c	3.8^b
Urea	0	60	27^{ab}	38^c	35^a	6.0^a	4.0^b	3.6^b
Sesbania rostrata	2	60	31^a	59^b	10^b	6.4^{ab}	4.3^{ab}	3.8^b
Sesbania emerus	4	60	32^a	62^{ab}	7^b	6.6^a	4.2^{ab}	3.7^b
Sesbania rostrata + rice straw	6	90	24^b	75^a	1^b	5.5^c	4.4^a	4.2^a

* Residual effect in unfertilized crop.
DS, dry season; WS, wet season.
Data followed by the same letter within a column do not differ significantly by DMRT ($P \leqslant 0.05$).

demand in the first season's crop and consequently rice yields were reduced. However, since N not taken up by the crop was retained in the soil, positive effects on grain yield in the second season (wet season 1992) and a residual effect of 10% compared with the urea treatment in the third crop (dry season 1993) were observed (Table 18.2). At a given quantity of applied residue N, a close match of supply and uptake may not satisfy the crop demand for a given yield. Some excess soil N seems to be required at early growth stages to build a sink size that is able to effectively use N released later in the season. This excess N will, however, be prone to losses. In other words, synchrony minimizes N losses and improves long-term soil fertility, but this may happen at the expense of the short-term yield.

A Conceptual Model

These relationships are summarized in the conceptual model presented in Fig. 18.3. At a given quantity of applied N, fast-release N sources (as determined by the L-to-N ratio) provoke a mismatch of N supply and demand. The resulting N losses may reduce plant N uptake and thus reduce the yield. A slow but steady mineralization of N throughout the rice growing season, on the other hand, will be

closely matched by crop N uptake but may not satisfy the demand for a given target yield level. In this situation, N is retained in the soil pool, and insufficient N may be available at early crop growth stages to achieve the target yield level. Where synchrony of supply and demand is optimal, a balance is struck between N losses and N retained in the soil, thus optimizing plant N uptake and yield. This optimum, however, varies from soil to soil as discussed earlier and it changes over time. When slow-mineralizing N sources are continuously applied, residual N will gradually become available. This increasing N supply will improve yield over time and will gradually shift the L-to-N ratio required for supply-demand synchrony of added residues towards lower quality material. The complexity and dynamic nature of these interactions calls for the use of simulation models. The development of such models, however, is constrained by our current lack of quantitative understanding as to how soil parameters affect residue decomposition dynamics.

We conclude that the L-to-N ratio of residues with low polyphenol content is a suitable indicator for predicting N mineralization rates in flooded soils. Mineralization may be manipulated by mixing high quality (low L-to-N ratio) material with lower quality rice field residues in view of achieving N

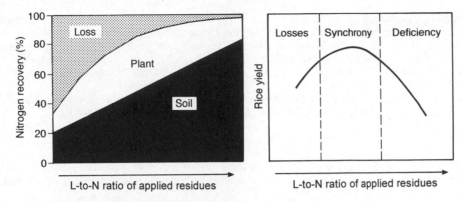

Fig. 18.3. Conceptual relationships between residue quality, residue N recovery, and rice yield.

supply–demand synchrony. Synchrony does not necessarily increase short-term yields, but it greatly reduces N losses, it improves long-term soil fertility. In situations where a short-term yield reduction is the price for long-term benefits, the application of N supply–demand synchrony may not be attractive to farmers. Research has to develop decision-making tools to optimize synchrony for a range of soil and residue scenarios and propose, where required, options to overcome the apparent short-term yield depression.

References

Becker, M., Ladha, J.K. and Ottow, J.C.G. (1990) Growth and nitrogen fixation of two stem-nodulating legumes and their effect as green manure on lowland rice. *Soil Biology and Biochemistry* 22, 1109–1119.

Becker, M., Ladha, J.K., Simpson, I.C. and Ottow, J.C.G. (1994a) Parameters affecting residue nitrogen release in flooded soil. *Soil Science Society of America Journal* 58, 1666–1671.

Becker, M., Ladha, J.K. and Ottow, J.C.G. (1994b) Nitrogen losses and lowland rice yield as affected by residue N release. *Soil Science Society of America Journal* 58, 1660–1665.

Broadbent, F.E. (1979) Mineralization of organic nitrogen in paddy fields. In: *Nitrogen in Rice*, IRRI, Manila, Philippines, pp.105–117.

Buresh, R.J. and De Datta, S.K. (1991) Nitrogen dynamics and management of rice-legume cropping systems. *Advances in Agronomy* 45, 1–59.

Burns, R.E. (1963) Method of tannic acid analysis for forage crop evaluation. *Georgia Agricultural Experimental Technical Bulletin* 32, 1–14.

Cassman, K.G., Kropff, M.J. and Yan Zhende (1995) A conceptual framework for nitrogen management of irrigated rice in high yield environments. In: *Proceedings of the International Rice Research Conference 1992*. IRRI, Manila, Philippines (in press).

Diekmann, K.H., DeDatta, S.K. and Ottow, J.C.G. (1992) Nitrogen-15 balance in lowland rice as affected by green manure and urea amendment. *Plant and Soil* 148, 91–99.

Clement, A., Ladha, J.K. and Chalifour, F.P. (1995) Crop residue effects on nitrogen mineralization, microbial biomass, and rice yield in submerged soils. *Soil Science Society of America Journal* 59, 1595–1603.

Fox, R.H., Myers, R.J.K. and Vallis, I. (1990) The nitrogen mineralization rate of legume residues in soil as influenced by their polyphenol, lignin, and nitrogen contents. *Plant and Soil* 129, 251-259.

George, T., Ladha, J.K., Buresh, R.J. and Garrity, D.P. (1992) Managing native and legume-fixed nitrogen in lowland rice-based cropping systems. *Plant and Soil* 141, 69–91.

Ghai, S.K., Rao, D.L.N. and Batra, L. (1988) Nitrogen contribution to wetland rice by green manuring with *Sesbania* spp. in alkaline soil. *Biology and Fertility of Soils* 6, 22–25.

IRRI-International Rice Research Institute (1989) *IRRI Towards 2000 and Beyond.* IRRI, Manila, Philippines, 66 pp.

Kundu, D.K. and Ladha, J.K. (1995) Efficient management of soil and biologically fixed N_2 in intensively-cultivated rice fields. *Soil Biology and Biochemistry* 27, 431–439.

McGill, W.B. and Myers, R.J.K. (1987) Controls on dynamics of soil and fertilizer nitrogen. Soil fertility and organic matter as critical components of production systems. *Soil Science Society of America Special Publication* No. 19, pp. 73–99.

Nagarajah, S., Neue, H.U. and Alberto, M.C.R. (1989) Effect of *Sesbania, Azolla* and rice straw incorporation on the kinetics of NH_4, K, Fe, Mn, and P in some flooded rice soils. *Plant and Soil* 116, 37–48.

Palm, C.A. and Sanchez, P.A. (1991) Nitrogen release from the leaves of some tropical legumes as affected by their lignin and polyphenolic contents. *Soil Biology and Biochemistry* 23, 83–88.

Ponnamperuma, F.N. (1978) Electrochemical changes in submerged soils and the growth of rice. In: *Soils and Rice.* The International Rice Research Institute (IRRI), PO Box 933, Manila, Philippines pp. 421–441.

Van Soest, P.J. (1968) Use of detergents in the analysis of fibrous feeds: a rapid method for the determination of fibers and lignin. In: *Association of Official Analytical Chemists,* 14th edn. AOAC Inc., Arlington, Virginia, USA, pp. 829–835.

19 Management of Leguminous Leaf Residues to Improve Nutrient Use Efficiency in the Sub-humid Tropics

R.B. Jones[1], S.S. Snapp[2] and H.S.K. Phombeya[3]

[1] Department of Natural Resource Sciences, Washington State University, Pullman, WA 99164-6226, USA; [2] The Rockefeller Foundation, PO Box 30721, Lilongwe 3, Malawi; [3] Agroforestry Commodity Team, Chitedze Agricultural Research Station, PO Box 158, Lilongwe, Malawi

Introduction

Two approaches can be recognized in managing soil fertility (Sanchez, 1995). First, soil constraints can be alleviated so that plant requirements are satisfied through the application of purchased inputs. The 'Green Revolution' is based on this approach which is responsible for the bulk of food produced in the world (Pinstrup-Andersen, 1993). The second approach relies more on biological processes to optimize nutrient cycling, minimize external inputs and maximize the efficiency of their use. The second approach is of great importance for the old and already highly leached soils of humid and sub-humid zones in Africa which all have inherently poor capacity to supply nutrients (Kumwenda et al., 1995).

In southern and eastern Africa, where the dominant smallholder cropping systems are based on maize (Zea mays), farming systems were traditionally based around extended fallows and the harvesting of nutrients stored in woody plants (Ruthenburg, 1980). Today, in most areas of Malawi, Zimbabwe and Kenya, fallowing has almost disappeared as a result of growing population pressure. Continuous cropping of maize with little or no use of purchased inputs is now the norm (see Kumwenda et al., 1995). The net result is

low crop productivity from land that is being further degraded by erosion and the continuous removal of crops.

The use of purchased inputs has had a mixed success for smallholder farmers because of the poor profitability of inorganic fertilizer use (Blackie and Jones, 1993; Kumwenda et al., 1995). Legume-based agriculture has been the focus of well-meaning efforts to transform smallholder agriculture since the turn of the century but again with limited success (Metelerkamp, 1988). P deficiency is known to reduce legume growth and so P must be added to boost productivity (Giller and Cadisch, 1995). The labour is often not available to manage the legume crop and there are difficulties in obtaining seed (Giller et al., 1994). In densely populated areas the land is needed for the production of food crops so that legumes are frequently intercropped. In the unimodal rainfall areas the growing season is limited by moisture and/or temperature. Current organic material inputs are insufficient to maintain soil organic matter (SOM) contents of agricultural soils. Attempts to increase organic matter production by intercropping in space and time with appropriate crops have produced mixed results (Ong, 1994). It is difficult to obtain large quantities of high quality organic matter and competition can reduce crop yields. Another constraint is the difficulty in measuring the effect of crop management on SOM dynamics. Physical fractionation, which is used to obtain the fraction of soil that is 'light' (floats in water) and 'large' (greater than 50 μm), has shown promise. The light–large fraction is a labile fraction of soil, made up primarily of recently decomposed residues (Cambardella and Elliot, 1994). This fraction can be expected to be predictive of the effects of leaf residues on the active organic matter, and thus on plant nutrient availability, yields and soil resource quality, over the short and medium term (see Barrios *et al.*, 1996; Saka *et al.*, 1995). The lack of organic inputs and the poor understanding of effects of adding organic residues on soil are some of the major challenges facing efforts to improve crop productivity and fertilizer use efficiency in the tropics.

There is increasing evidence that the most promising route to improving crop yields in smallholder cropping systems is by increasing inorganic fertilizer efficiency through the addition of small amounts of high quality organic matter (Ladd and Amato, 1985; Larbi *et al.*, 1993; Snapp, 1995). Jones *et al.* (1996) and Wendt *et al.* (1996) working with established alleys of *Leucaena leucocephala* showed that although the application of leaf prunings alone increased maize yields, leaf management strategy × P and fertilizer-N × P interactions were highly significant. The yield increase due to inorganic N and P supplements was high, averaging 41 kg maize for each kilogram of applied nutrient. Farmers can obtain 17–25 kg maize

for each kg N applied with good fertilizer management in the absence of leaf prunings. Soil analyses revealed that plots receiving leaf prunings had significantly greater organic C, total N, pH, exchangeable Ca, Mg, K and S, and a reduced C-to-N ratio compared with plots where leaves had been removed. In contrast to the results obtained from *Leucaena* residues, MacColl (1990) showed that the application of maize residues over several years had no appreciable benefit in terms of grain yield. In Côte d'Ivoire, experiments conducted for over a decade also indicated that there are few or no benefits from low quality residues (Traore and Harris, 1995). Farmers are reluctant to incorporate non-leguminous crop residues because they tend to immobilize N.

In this chapter, the results are reported from a series of trials that were initiated to investigate the effect of leaf residues from leguminous multi-purpose trees (MPTs) that are well adapted to the upland ecology. The overall goal was to determine residue management strategies appropriate for smallholder farmers to increase maize yields in a biomass transfer system.

Methodological Approach

The residue and fertilizer management experiment was initiated in 1993 on a sandy soil at Naminjiwa Residential Training Centre (15° 45′S, 35° 40′E, altitude 876 m). The soil is representative of the Phalombe Plain and is classified as a Eutric Fluvisol (FAO, 1988; Table 19.1). The objectives were to investigate the

Table 19.1. Soil (0–15 cm) analyses from sites used in residue and fertilizer management experiment (Naminjiwa) and residue management experiment (Dedza).

Site	pH*	P[†] (ppm)	K	Mg	Ca	Organic C[‡] (g kg⁻¹)	Texture
			(cmol₊ kg⁻¹)				
Dedza	5.7	16	0.4	2.9	1.1	13	Sandy loam
Naminjiwa	5.5	39	0.14	0.32	1.4	6.6	Loamy sand

* 2:1 (w/v) in water.
[†] P and exchangeable cations in Mehlich III extract.
[‡] Organic C by Walkley Black.

effect of residue quality (*Gliricidia sepium* compared with *Leucaena leucocephala*), residue application method (banded in ridges, point placement in the ridge and surface placement on the ridge), residue application rate (1.5 and 3 t ha^{-1} dry weight) and rate of fertilizer-N (0 and 40 kg N ha^{-1}). The trial was laid out as a 3 × 2^3 factorial design in randomized blocks, with a single control plot in each block which did not receive any leaf residues or fertilizer-N. It was not possible to accommodate an additional fertilizer-N control and maintain a balanced experimental design.

The residue management experiment was also initiated in 1993 on a sandy loam soil on a farmer's field in the Dedza Hills (14° 22′S, 34° 19′E, altitude 1600 m). The soil is representative of the Dedza Hills and is classified as a Chromic Luvisol (FAO, 1988). The objectives were to investigate the effect of residue quality (*Tephrosia vogelii* compared with *Leucaena leucocephala*) and residue application method (point placement in the ridge and surface placement on the ridge). Residues of both species were applied at the rate of 3 t ha^{-1} dry weight. The trial was laid out as a 2^2 factorial design in four randomized full blocks with a control plot in each block.

Both sites experience a tropical continental climate with three seasons. The cool season lasts from May to August and is dry with relatively low temperatures. Temperatures rise steadily from September to November, but conditions remain dry. During December the inter-tropical convergence zone becomes established and rainfall occurs from December to March with moderately hot temperatures (Fig. 19.1). Temperatures vary largely with altitude and so the Naminjiwa site at 876 m experiences higher temperatures than the Dedza site at 1600 m. The very dry conditions from May through to November effectively halt all decomposition of organic matter although fauna, mainly termites, can be very active during this period.

Leaf residues for both experiments were obtained by pruning the first flush of leaves from established trees in late October just before the start of the 1993/94 rainy season. The prunings were air dried in the sun and then separated, by shaking, from the woody stems which were discarded. Prunings were obtained from trees growing at locations other than those used in the field trials. Leaf quality (Table 19.2) may be affected by the season, the time of harvesting, the drying method and by the environment in which the trees were grown (Mafongoya *et al.*, Chapter 13, this volume).

In both experiments, hybrid maize was planted together with 40 kg P$_2$O$_5$ ha^{-1} banded in the form of triple super phosphate. Inorganic N was applied as a top dressing in the form of urea when the maize reached a height of 45 cm. The urea was surface applied along the top of the ridge and then covered with soil.

Plots of both experiments were maintained but re-ridged in the following season. Maize was again monocropped on all plots and the residual effects on maize grain yield, from the treatments applied in the previous season, determined.

In the residue management experiment soil organic matter fractionation was carried out on samples collected after one rainy season and one dry season just before the start of the second rainy season. Ten sub-samples of soil were collected at random from each plot from the 0–15 cm layer of the ridge and then composited. Whole soil was dried, sieved to pass a 2 mm mesh, and analysed for organic C and total N by standard wet oxidation methods (modified Walkey-Black and Kjeldhal procedure, see Anderson and Ingram, 1993). Fractionation of whole soil was also conducted to collect the light–large fraction and determine organic C and total N of this fraction. Fractionation methodology was modified from Okalebo *et al.* (1993), by the addition of a density fractionation step to largely remove the sand fraction (see Cambardella and Elliot, 1994). The whole soil was dispersed by shaking for five minutes in 0.5% sodium hexametaphosphate. Dispersed samples were wet sieved to collect the fraction between 2 mm and 53 µm, designated as the large fraction. The heavy fraction

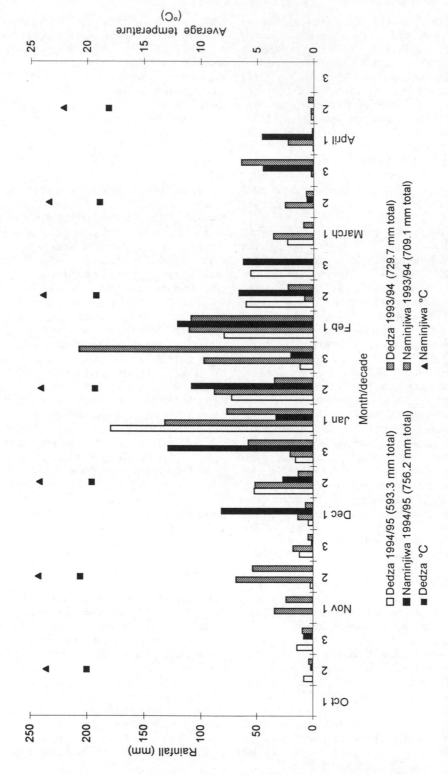

Fig. 19.1. Rainfall (10-day periods) for the 1993/94 and 1994/95 seasons with average monthly temperatures (°C) for sites used in residue and fertilizer management experiment (Naminjiwa) and residue management experiment (Dedza).

Table 19.2. Percentage lignin, polyphenols, C and N in leaf residues of *Leucaena leucocephala*, *Tephrosia vogelii* and *Gliricidia sepium*, and total N (kg ha^{-1}) applied in leaf residues in two treatments (1.5 and 3 t ha^{-1}) in the 1993/94 season for both experiments.

Plant residues	Lignin*	Polyphenols[†]	C	N	N applied in leaf residues (kg ha^{-1})	
		(%)			1.5	3
Leucaena	7.0	3.3	44.4	3.10	47	93
Tephrosia	8.0	1.77	45.2	3.42	N/A	103
Gliricidia	7.7	1.34	42.4	3.16	47	95

* Acid-detergent lignin.
[†] Folin–Denis method.

was discarded after suspension in water (20 s mix and 20 s settling period).

Residue and Fertilizer Management Experiment

Interactions between residue quality and fertilizer-N recovery

The application of residues without fertilizer-N significantly increased maize yields in both the season of application and in the subsequent growing season (Fig. 19.2). The residual effect, in terms of maize grain yield, was less than that in the original season of application. In the first season, *Gliricidia* residue was superior to that of *Leucaena*. The application of fertilizer-N had a greater effect when combined with *Leucaena* than with *Gliricidia* although the yield from *Leucaena* + N was significantly less than that from *Gliricidia* + N. Under field conditions the N release constant was found to be higher for *Gliricidia* than for *Leucaena* leaf residues in the humid tropics (Budelman, 1988; Tian *et al.*, 1992). The N release constant of *Gliricidia* residues was found to be larger than the decomposition rate constant and that during decomposition of *Leucaena* prunings, N release was slower and some N was even immobilized in *Leucaena* as compared to *Gliricidia*. This observation was attributed to the higher polyphenol concentrations in *Leucaena* (5.02%) than in *Gliricidia* (1.62%) (Tian *et al.*, 1992) which confirms

our observations. The N contents of the prunings from the two species used in the residue and fertilizer management experiment were almost identical (Table 19.2) suggesting that decomposition of *Gliricidia* residue was in better synchrony with the demands of the maize crop than that of *Leucaena* residue. Soil nitrate at a site in central Malawi (1100 m) amended with *Gliricidia* and *Leucaena* residues also suggest that N is more rapidly available from *Gliricidia* residues compared with *Leucaena* (Snapp and Materechera, 1994).

Interactions between season and residue application method

In the season of application, surface application of residue gave a significantly higher yield than either banding in the ridge or point placement in the soil (Fig. 19.3). In the subsequent season, there was no significant difference between any of the residue application methods. These data suggest that synchrony of nutrient release with maize demand was better achieved with surface application where residues were not in such intimate contact with the soil. Xu *et al.* (1993b) working with *Leucaena* residues, compared surface placement with incorporation into the top 5 cm of soil and found that placement had little effect on the availability of N to maize plants over a 2-month period. However, the apparent loss of *Leucaena* ^{15}N was increased from 27%, when the *Leucaena* prunings were mulched, to 41%, when

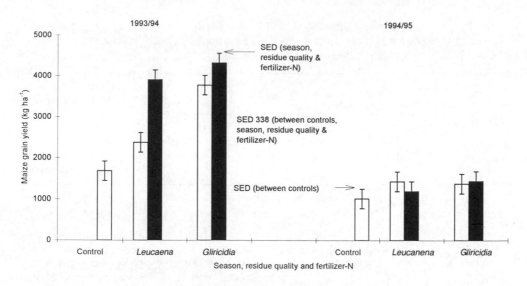

Fig. 19.2. Effect of season, residue quality and fertilizer-N (($\square$) 0 or ($\blacksquare$) 40 kg N ha^{-1}) on maize grain yield (kg ha^{-1}) in the residue and fertilizer management experiment.

they were incorporated and this increased loss was ascribed to denitrification. It is probable that the same effect was observed in the residue and fertilizer management experiment.

Interactions between residue application rate and fertilizer-N

In the absence of fertilizer-N, the high rate of residue application (3 t ha^{-1}) gave a significantly higher maize grain yield than the low rate of 1.5 t ha^{-1} in the first season (Fig. 19.4). The combination of fertilizer-N and residues increased maize yield above that measured from residues alone at the 1.5 t ha^{-1} residue application rate but had no significant effect at the higher residue application rate of 3 t ha^{-1}. There was a suggestion that in the second season, maize yields were higher where residues had been applied but only at the high rate.

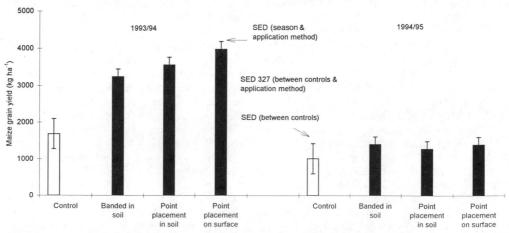

Fig. 19.3. Effect of season and residue application method on maize grain yield (kg ha^{-1}) in the residue and fertilizer management experiment.

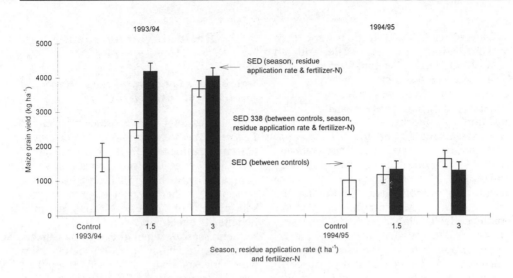

Fig. 19.4. Effect of season, residue application rate (t ha⁻¹) and fertilizer-N (($\square$) 0 or ($\blacksquare$) 40 kg N ha⁻¹) on maize grain yield (kg ha⁻¹) in the residue and fertilizer management experiment.

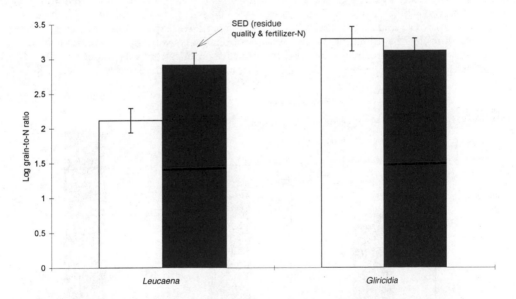

Fig. 19.5. Effect of residue quality and fertilizer-N (($\square$) 0 or ($\blacksquare$) 40 kg N ha⁻¹) on maize grain-to-total applied N ratio (data transformed to natural log scale) in the residue and fertilizer management experiment.

Residue quality and fertilizer-N on grain-to-N ratio

One way of examining the effect of applied N is to examine the grain-to-N ratio. These were calculated on an individual plot basis using the control yields from each block for the 1993/94 season, by dividing the yield increment from the control by the total N added from residues and fertilizer-N. The ratios were then transformed to the log scale and an ANOVA done on the transformed values. The grain-to-N ratio resulting from *Gliricidia* prunings was not significantly changed when fertilizer-N was added. The addition of fertilizer-N with *Leucaena* prunings increased the maize grain-to-N ratio to the same level as that for *Gliricidia* (Fig. 19.5). Both Jenkinson *et al.* (1985) and Xu *et al.* (1993a), found that large amounts of added plant N entering the mineralization and immobilization cycle enhance the availability of additional soil N. The residue quality × fertilizer-N interaction supports the hypothesis that N use efficiency can be increased by the combination of inorganic and organic fertilizers. This is of interest to farmers using leaf residues where the N release content may not be in synchrony with the growing crop. The observed difference between the *Leucaena* and *Gliricidia* leaf residues is probably due to the different N release constants that have been measured for these residues.

Residue Management Experiment

Interactions between season and residue application method

The application of leaf residues from either *Leucaena leucocephala* or *Tephrosia vogelii* had no significant effect on maize grain yield in the season of application when compared with the control plots (data not presented). There was a suggestion that soil placement of residues had a small beneficial effect on yield. In the subsequent season, when the residual effects from the previous season's application were investigated, the surface applied residue treatment significantly outyielded the soil applied treatment which was superior to the control (Fig. 19.6). These findings confirm the trends observed in the 1993/94 season in the residue and fertilizer management experiment although the residual effects were not so clearly evident. Surface application was superior at both sites suggesting that losses by denitrifica-

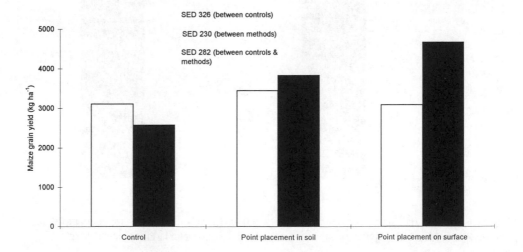

Fig. 19.6. Effect of season (($\square$) 1993/94 and ($\blacksquare$) 1994/95) and residue application method on maize grain yield (kg ha^{-1}) (average of *Leucaena* and *Tephrosia*) in the residue management experiment.

Table 19.3. Carbon and nitrogen characteristics of the light–large (LL) fraction g kg^{-1} and whole soil of treatments with and without residues (residue management experiment).

Treatment	Dry wt of LL fraction	Concentration in LL fraction		Concentration in whole soil	
		C	N	C	N
No residues applied	13.5	88.2	4.37	12.5	0.90
Residues applied	19.4	113.3	5.94	14.1	1.10
SED	2.8	7.3	0.32	1.2	0.07
CV%	28	12	10	16	13

Values are expressed in g kg^{-1}.

tion and possibly leaching losses were exacerbated by incorporation (cf. Xu et al. 1993a).

There were no significant interactions between residue quality and application method or between the residues of the two species tested. These results suggest that residues of Tephrosia vogelii and Leucaena leucocephala have similar decomposition characteristics. The application of residues had little effect in the first season in Dedza but a substantial effect in Naminjiwa. The major differences between the two sites are the soil type and temperature. The influence of environmental conditions on decomposition is well known (see Scholes et al., 1994) and can play a strong regulatory role in residue decomposition and management in tropical agroecosystems (Mazzarino et al., 1993). The lower temperatures at the Dedza Site would have resulted in a lower residue respiration rate which, for a residue like Leucaena, where catabolism is relatively more important than leaching of water-soluble materials, would have delayed the release of N.

The results from organic matter fractionation carried out on soil samples collected from the plots in November 1994 confirm that the application of residues increased both N and C in the light–large fraction (Table 19.3). The increase in N was sufficient to increase the N content in the whole soil but the C fraction of whole soil was not affected. The fractionation procedure was not sensitive enough to

show up differences between application methods which were reflected in the maize yields. The data suggest that the light–large fraction can provide an indicator of effects of nutrient management and cropping practices on the 'active' fraction of the soil organic matter, in agreement with earlier findings (Janzen et al., 1992). A light–large fraction from a silica suspension fractionation procedure has also been found to be indicative of differences in residue inputs and soil nutrient dynamics (Barrios et al., 1996); however, the suspension in water methodology, as used in this study, produces a 10-fold larger light–large fraction than the method used by Barrios et al. (1996). It is easier to conceptualize how a larger fraction could have a significant role influencing C and N dynamics. Fractionation procedures are a promising tool for providing indicators of changes in soil fertility although much methodological development remains to be done.

Short Term Timing

The soils used in this study are typical of those used by smallholders for maize production. They tend to have inherently poor nutrient contents and are often severely degraded from erosion and continuous cropping with little or no input of nutrients (see Kumwenda et al., 1995). Under such conditions, the timing of nutrient release from leaf residues is critically important. Early maize growth is compromised when fertilization is delayed much beyond 3 weeks after planting (see Mengel and Barber, 1974; Jones, 1993). The choice of fertilizer and the application method are recognized by farmers as factors affecting nutrient availability. For example, point placement of urea was found to be inferior to surface application while calcium ammonium nitrate is preferred by farmers over urea because it is faster acting (Jones, 1993).

That such differences are important with highly soluble nutrient sources, would tend to mitigate against less available organic sources whose nutrient release patterns are affected by several factors that are largely beyond the control of farmers. This study has shown how quality and management can be manipulated to improve synchrony and increase yield. However, in the residue management experiment, the release of nutrients was not in synchrony with the demands of the maize in the season of application even though there was some evidence that point placement in soil resulted in slightly faster decomposition than surface placement. The relatively small residual effect in the residue and fertilizer management experiment that was conducted on a loamy-sand soil (see Table 19.1) suggests that if the benefits are not obtained in the season of application, there is a danger that nutrients will be lost before the next crop can be planted.

Conclusions

Two field experiments examined residue management of differing quality residues and N efficiency in a maize cropping system. Residues of *Gliricidia*, *Tephrosia* and *Leucaena* were applied to maize using a range of application methods and rates. *Gliricidia* was superior to *Leucaena* in one experiment while *Leucaena* and *Tephrosia* were similar in the second experiment. There was evidence that N from *Gliricidia* was rapidly available to the maize crop. At both sites there was a residual effect in the second year, enhancing maize yields and increasing soil C and N in the labile SOM fraction. Residue decomposition rate and residue effects on yields were delayed at the higher altitude site, with cooler temperatures and heavier soils, compared with the lower altitude site. Residue application on the surface compared with incorporated had a positive effect on yield. Overall, the data suggested N efficiency gains from combining organic and inorganic nutrient sources, and that residue management must take into account rapid availability of N from *Gliricidia* residues and slower availability of N from *Tephrosia* and *Leucaena*. The difference between *Gliricidia* and *Leucaena* may suggest that increased N use efficiency can be obtained by combining low-medium quality residue with fertilizer-N whereas high quality residue does not benefit from additional N. Further research on residue interactions with N fertilizer will clarify these findings and allow recommendations to be developed for farmer management of combined organic and inorganic nutrient sources.

References

Anderson, J.M. and Ingram, J.A.S. (1993) *Tropical Soil Biology and Fertility: A Handbook of Methods*, 2nd edn. CAB International, Wallingford, 221 pp.

Barrios, E., Buresh, R.J. and Sprent, J.I.(1996) Organic matter in soil particle size and density fractions from maize and legume cropping systems. *Soil Biology and Biochemistry* 28, 185–193.

Blackie, M.J. and Jones, R.B. (1993) Agronomy and increased maize productivity in eastern and southern Africa. *Biological Agriculture and Horticulture* 9, 147–160.

Budelman, A. (1988) The decomposition of the leaf mulches of *Leucaena leucocephala*,

Gliricidia sepium and *Flemingia macrophylla* under humid tropical conditions. *Agroforestry Systems* 7, 33–45.

Cambardella, C.A. and Elliot, E.T. (1994) Carbon and nitrogen dynamics of soil organic matter fractions from cultivated grassland soils. *Soil Science Society of America Journal* 58, 123–130.

FAO (1988) *FAO/UNESCO Soil Map of the World*, Revised Legend. World Soil Resources Report No. 60. FAO, Rome.

Giller, K.E. and Cadisch, G. (1995) Future benefits from biological nitrogen fixation: An ecological approach to agriculture. *Plant and Soil* 174, 255–277.

Giller, K.E., McDonagh, J.F. and Cadisch, G. (1994) Can nitrogen fixation sustain agriculture in the tropics? In: Syers, J.K. and Rimmer, D.L. (eds) *Soil Science and Sustainable Land Management in the Tropics*. CAB International, Wallingford, pp. 173–191.

Janzen, H.H., Campbell, C.A., Brandt, S.A., Lafond, G.P. and Townley-Smith, L. (1992) Light-fraction organic matter in soils from long term crop rotations. *Soil Science Society of America Journal* 56, 1799–1806.

Jenkinson, D.S., Fox, R.H. and Rayner, J.H. (1985) Interaction between fertilizer nitrogen and soil nitrogen – the so-called 'priming' effect. *Journal of Soil Science* 36, 425–444.

Jones, R.B. (1993) Improving the efficiency of inorganic fertilizers in Malawi. In: Munthali, D.C., Kumwenda, J.D.T. and Kisyombe, F. (eds) *Proceedings of a Conference on Agricultural Research for Development*. University of Malawi, Zomba, pp. 165–170.

Jones, R.B., Wendt, J.W., Bunderson, W.T. and Itimu, O.A. (1996) Maize yield and soil changes as affected by N and P application in alley cropping. *Agroforestry Systems* (in press).

Kumwenda, J.D.T., Waddington, S.R., Snapp, S.S., Jones, R.B. and Blackie, M.J. (1995) Soil fertility management research for the smallholder maize-based cropping systems of southern Africa: a review. *Soil Fertility Network for Maize-Based Cropping Systems in Selected Countries of Southern Africa*, Network Research Working Paper Number 1, CIMMYT, Harare. 34 pp.

Ladd, J.N. and Amato, M. (1985) Nitrogen cycling in legume–cereal rotations. In: Kang, B.T. and Van der Heide, J. (eds) *Nitrogen Management in Farming Systems in Humid and Sub-Humid Tropics*. Institute for Soil Fertility Research, Haren, pp. 105–127.

Larbi, A., Jabbar, M.A., Atta-Krah, A.N. and Cobbina, J. (1993) Effect of taking a fodder crop on maize grain yield and soil chemical properties in *Leucaena* and *Gliricidia* alley farming systems in western Nigeria. *Experimental Agriculture* 29, 317–321.

MacColl, D. (1990) The effects of lime, minimum tillage and return of stover to the soil on yields of maize over seven years. *Bunda Journal of Agricultural Research* 2, 85–98.

Mazzarino, M.J., Szott, L., Jimenez, J.M. and Szott, L.T. (1993) Dynamics of soil total C and N, microbial biomass, and water soluble C in tropical agroecosystems. *Soil Biology and Biochemistry* 25, 205–214.

Mengel, D.B. and Barber, S.A. (1974) Rate of nutrient uptake per unit of corn root under field conditions. *Agronomy Journal* 66, 399–402.

Metelerkamp, H.R.R. (1988) Review of crop research relevant to the semiarid areas of Zimbabwe. In: *Proceedings of a Workshop on Cropping in the Semiarid Areas of Zimbabwe*. Agritex/DR&SS/GTZ, Harare, pp. 190–315.

Okalebo, J.R, Gathua, K.W. and Woomer, P.L. (1993) *Laboratory Methods of Soil and Plant Analysis: A Working Manual*. KARI and TSBF, Nairobi, 88 pp.

Ong, C. (1994) Alley cropping – ecological pie in the sky? *Agroforestry Today* 6, 8–10.

Pinstrup-Andersen, P. (1993) *World Food Trends and How They May be Modified*. International Food Policy Research Institute, Washington D.C.

Ruthenburg, H. (1980) *Farming Systems in the Tropics*, 3rd edn. Clarendon Press, Oxford, 424 pp.

Saka, A.R., Phombeya, H.S.K., Jones, R.B., Minae, S. and Snapp, S.S. (1995) *Agroforestry Commodity Team Annual Report 1993/94*. Department of Agricultural Research, Lilongwe.

Sanchez, P.A. (1995) Tropical soil fertility research: Towards the second paradigm. In: *Transactions 15th World Congress of Soil Science*. ISSS 1, 65–88.

Scholes, R.J., Dalal, R. and Singer, S. (1994) Soil physics and fertility: the effects of water, temperature and texture. In: Woomer, P.L. and Swift, M.J. (eds) *The Biological Management of Tropical Soil Fertility.* Wiley-Sayce, Chichester, pp. 117–136.

Snapp, S.S. (1995) Improving fertilizer efficiency with small additions of high quality organic inputs. In: Waddington, S.R. (ed.) *Report on the First Meeting of the Network Working Group.* Soil Fertility Network for Maize-Based Cropping Systems in Selected Countries of Southern Africa, CIMMYT, Harare, pp. 60–65.

Snapp, S.S. and Materechera, S.A. (1994) Soil organic matter quality: Evaluating the relationship to N dynamics and maize yields. *Agronomy Abstracts,* American Society of Agronomy, Madison. pp. 275.

Tian, G., Kang, B.T. and Brussaard, L. (1992) Biological effects of plant residues with contrasting chemical compositions under humid tropical conditions – decomposition and nutrient release. *Soil Biology and Biochemistry* 24, 1051–1060.

Traore, S. and Harris, P.J. (1995) Long-term fertilizer and crop residue effects in Côte d'Ivoire. In: Lal, R. and Stewart, B.A. (eds) *Soil Management Experimental Basis for Sustainability and Environmental Quality.* CRC Press, Boca Raton, pp. 141–180.

Wendt, J.W., Jones, R.B., Bunderson, W.T. and Itimu, O.A. (1996) Residual effects of *Leucaena* leaf management and P application on alley cropped maize and soil properties. *Agroforestry Systems* (in press).

Xu, Z.H., Saffigna, P.G., Myers, R.J.K. and Chapman, A.L. (1993a) Nitrogen cycling in *Leucaena (Leucaena leucocephala)* alley cropping in semi-arid tropics. I. Mineralization of nitrogen from *Leucaena* residues. *Plant and Soil* 148, 63–72.

Xu, Z.H., Saffigna, P.G., Myers, R.J.K. and Chapman, A.L. (1993b) Nitrogen cycling in *Leucaena (Leucaena leucocephala)* alley cropping in semi-arid tropics. II. Response of maize growth to addition of nitrogen fertilizer and plant residues. *Plant and Soil* 148, 73–82.

Part VI

Building Soil Organic Matter

20 Characterization of Soil Organic Matter by Solid-state ^{13}C NMR Spectroscopy

J.O. Skjemstad[1,2], P. Clarke[3], A. Golchin[3], and J.M. Oades[2,3]

[1] Division of Soils, CSIRO, Glen Osmond, SA 5064;
[2] CRC for Soil and Land Management, Glen Osmond, SA 5064;
[3] Department of Soil Science, Waite Agricultural Research Institute,
The University of Adelaide, Glen Osmond, SA 5064, Australia

Introduction

^{13}C NMR spectroscopy has been widely applied to soils and soil fractions in studies of soil organic matter chemistry and dynamics for at least 15 years. Early studies utilized solution techniques applied to humic materials but in these studies the largest fraction of organic matter, the insoluble humin fraction, was by necessity ignored. With the advent of solid state techniques, the chemistry of the insoluble materials, whole soils and soil fractions derived by physical rather than chemical fractionation schemes could be addressed. Humic fractionation schemes based on solubility had suggested that soil organic matter was highly aromatic in nature (Hatcher et al., 1981) and although soils containing highly aromatic humus were recognized (Calderoni and Schnitzer, 1984; Zech et al., 1989), most soils and soil fractions were soon shown by solid-state ^{13}C NMR spectroscopy to be relatively low in aromaticity (Wilson et al., 1981a, b; Preston and Ripmeester, 1982).

Since these early studies, a more comprehensive view of soil organic matter is emerging with respect to its chemistry, its associations with other soil materials, its distribution in the soil matrix at the micron level and the role of litter quality. Much of this information has been derived with the aid of solid-state ^{13}C NMR spectroscopy which is non-destructive and has proven itself to be an informative and convenient method for soil organic matter analysis. In this chapter, we briefly discuss many of the applications of this technique to the study of organic matter in soils and soil fractions and decomposition studies. Some of the limitations of this technique and methods used to address these limitations are also discussed.

The Solid-State ^{13}C NMR Technique

Characterization using cross polarization with magic angle spinning

Cross polarization with magic angle spinning and high-power decoupling are the essence of modern solid state ^{13}C NMR spectroscopy in soil science. High-power proton decoupling is used to overcome dipolar and scalar interactions while magic angle spinning (MAS) at 54.7° largely overcomes the chemical shift anisotropy. Cross polarization (CP) between the abundant ^{1}H and dilute ^{13}C spins is used to overcome the unacceptably long ^{13}C spin lattice (T_1) relaxation times and to enhance signal intensity. The theory, application and technology behind these techniques are widely reported in the literature (see for example Alemany et al., 1983a,b;

1997 CAB INTERNATIONAL Driven by Nature: Plant Litter Quality and Decomposition.
(eds G. Cadisch and K.E. Giller)

253

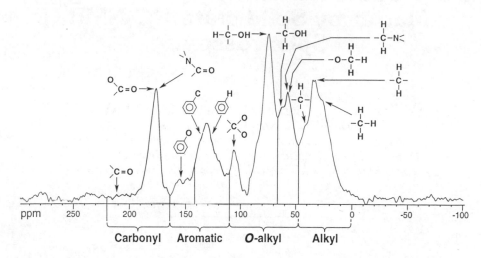

Fig. 20.1. Four general chemical regions and typical chemical shift assignments for soil humus.

Wilson, 1987) and will not be repeated here except where they may impact on quantitation.

Chemical shift information

One of the main reasons for performing solid-state ^{13}C NMR spectroscopy on soil and plant materials is to obtain information on the chemistry of the organic materials present. The position in a spectrum at which a nucleus resonates is called the chemical shift of that nucleus and provides information about the chemical environment within which that nucleus exists. As shown in Fig. 20.1, four major regions in the spectrum are generally recognized representing alkyl- (0–45 ppm), O-alkyl- (45–110 ppm), aromatic- (110–165 ppm) and carbonyl-C (165-190 ppm). These chemical structures can be differentiated as a result of the electron clouds that surround nuclei. NMR measurements are made in a strong magnetic field and neighbouring groups which strongly withdraw electrons from around a nucleus increase the effective magnetic field experienced by that nucleus. This reduced electron shielding causes the nucleus to resonate at a higher chemical shift. Conversely, electropositive groups cause a change towards a lower chemical shift.

The same principles apply within each of the regions. Within the alkyl region, the 1° carbons resonate closer to 0 ppm while the 2° and 3° carbons generally resonate at ever increasing chemical shifts up to 50 ppm or higher. In soil and plant derived materials, the dominant peak in this region is generally around 30 ppm and represents polymethylene (-CH$_2$-) carbon. A peak at 33 ppm is also often observed and represents polymethylene carbon in highly rigid structures (Stark and Garbow, 1992).

The O-alkyl region is generally dominated by signals from cellulose or other polysaccharide structures. The major band from 60 to 95 ppm is due to C$_2$-C$_5$ of cellulose with the peak at 65 ppm attributed to the C$_6$ 1° alcohol group and the sharp peak near 105 ppm to the C$_1$ anomeric carbon. Another peak near 55 ppm is the combination of two peaks, a sharp peak due to methoxyl carbon and a much broader peak attributed to the α-carbon to nitrogen in structures such as proteins and peptides (Skjemstad *et al.*, 1983).

The aromatic region reflects a combination of aryl and O-aryl structures and possibly olefins. The main peak centred near 130 ppm is due to unsubstituted and alkyl substituted carbons in aromatic rings with the bulk of the peak on the lower chemical shift side due to unsubstituted carbons. In lignin structures such as the guaiacyl unit, the phenolic and methoxyl substituted C_3 and C_4 carbons resonate between 145 and 150 ppm, the C_1 alkyl substituted carbon resonates near 135 ppm and the remaining unsubstituted carbons near 120 ppm (Maciel et al., 1981). The chemical shifts of aromatic carbons depend greatly on the degree, location and type of substitution in the ring. In the syringyl unit for example, the C_4 position, although substituted with an oxygen group, resonates at 135 ppm (Hatcher, 1987) while in tannin structures, the non-protonated C_{4a} carbon resonates near 106 ppm (Wilson and Hatcher, 1988).

The carbonyl region, although often appearing as a single relatively sharp peak, is made up of a combination of carboxyl, ester and amide groups. These groups are not readily distinguished from one another but in soil and plant materials, amide groups are the largest contributors to this peak. A small broad band near 200 ppm can sometimes be seen and can be attributed to aldehydes, ketones or quinones.

Quantitative considerations

An assessment of the relative contribution of each type of carbon can be made from the area under individual peaks or regions. Fig. 20.1 shows the regions representing the four major functional groups found in soil organic matter. It should be noted that in solid-state spectra, peak heights do not accurately reflect the intensities of each type of carbon because peaks can vary in width and shape. However, it is clear from spin counting techniques and other comparisons with standard materials that the NMR technique does not 'see' all of the nuclei present in a sample. This is mostly due to interferences such as the presence of magnetic materials or free radicals which drastically reduce T_1 to the extent that the nuclei are generally not observed. In the cross polarization experiment, an additional problem is encountered. A carbon nucleus can only cross polarize with protons that are in close proximity, probably within 4 to 5 bond lengths (Alameny et al., 1983b). Carbon

nuclei that are further removed from protons cannot undergo cross polarization and therefore will not be observed in the CP experiment (Snape et al., 1989). Furthermore, the rate at which magnetization is transferred from the proton to the carbon nucleus is a function of distance so that the carbons not directly bonded to hydrogens, such as substituted aromatic carbons and carbonyls, will require longer contact times than carbons such as methyl groups, which have abundant protons. In order to obtain quantitative data, it is essential that the contact time be optimized for each type of carbon nucleus present. In practice this is not possible and a compromise is struck between build up (T_{CH}) and loss of magnetization ($T_{1\rho}H$) for all nuclei. If these differ greatly between functional groups, corrections can be made in the integrals for each group provided a contact time experiment has been performed for each sample (Malcolm, 1992).

It is also important that all nuclei involved in the measurement be allowed to relax back to their equilibrium states between measurements and recycle times should be at least seven times longer than the longest $T_1(H)$ in the sample (Wilson, 1987). For soil and plant material, recycle times between 0.3 and 1.0 s are usually adequate but longer recycle times may sometimes be necessary.

For samples that contain significant amounts of carbon isolated from protons, such as in charcoal, CP/MAS cannot give quantitative data. In this case, it is necessary to carry out a Bloch decay experiment which does not utilize CP but magnetizes the carbon nuclei directly. Relaxation is now through the carbons so that theoretically all of the carbon nuclei are observed. This technique still suffers from problems associated with free radicals and magnetic materials and requires very long recycle times because of the typically very long carbon T_1s. Recycle times of between 30 and 90 s are usually required resulting in the collection of very few transients within a reasonable time frame or very long accumulation times to acquire acceptable signal to noise. Figure 20.2 shows a comparison between the CP/MAS and Bloch

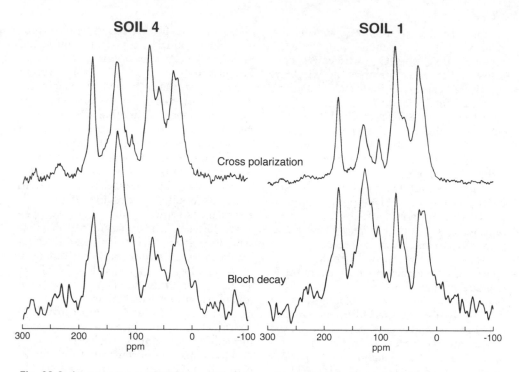

Fig. 20.2. A comparison of CP/MAS and Bloch decay spectra of the <53 μm fractions from two soils containing appreciable amounts of finely divided charcoal.

decay spectra from two soil fractions (<53 μm) which contain very finely divided natural charcoal. It is clear from these spectra that the CP technique underestimates the aromatic carbon in these samples by a factor between 30% and 50%. This does not occur for soils that do not contain charcoal and thus underestimation errors in CP spectra are minimal.

Comparisons with other Techniques

There are very few studies reported in the literature that compare the data from NMR experiments with those obtained by other methods. Some that have been reported include wet chemical degradation studies, infrared and pyrolysis mass spectra. Many studies show that these techniques are often complementary, providing similar data but more often providing detailed information on different aspects of

the chemistry of the sample. Some examples of this include the use of infrared spectroscopy to differentiate between carboxylic acids and amides which appear as a single carbonyl band in solid-state ^{13}C NMR spectra.

^{13}C NMR provides an overview of the chemical structure of samples. Many wet chemical techniques therefore do not provide data that agree directly with ^{13}C NMR since all of these techniques are specific to individual chemical groups while ^{13}C NMR determines all of the structures within a general functional group. Several studies have shown that ^{13}C NMR will quantitatively determine the various chemical structures within model compounds of varying complexity (Alemany *et al.*, 1983a,b; Hawkes *et al.*, 1993) while Bates and Hatcher (1992) showed that solid-state ^{13}C NMR spectroscopy could be used to accurately determine the content of carbohydrate carbon in wood samples. Love *et al.* (1994) demonstrated

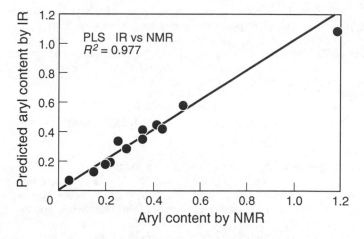

Fig. 20.3. The predicted aryl content of the <53 μm fractions from 12 soils by Fourier transform IR and Partial Least Squares (PLS) against the aryl content determined on the same samples by CP/MAS ¹³C NMR but after HF treatment.

with solid-state ¹³C NMR spectroscopy that classical methods overestimate the lignin content of flax fibres. Similar difficulties confront pyrolysis mass spectroscopy and many studies have found this technique and ¹³C NMR to be highly complementary.

Another approach which shows considerable promise is the use of ¹³C NMR spectroscopy in conjunction with multivariate data analysis methods such as partial least squares (PLS). This enables NMR data to be correlated with other techniques using a set of chemically defined standards and then using this correlation to predict specific chemical properties in other 'unknown' samples from ¹³C NMR spectra. Nordén and Berg (1990) successfully used this approach to predict the concentration of lignin in decomposing litter from ¹³C NMR spectra against classical methods. Fig. 20.3 shows the aryl carbon content of some <53 μm samples predicted from Fourier transform infrared (FTIR) spectra using PLS against aryl contents determined by CP/MAS ¹³C NMR spectroscopy. The good correlation between the two techniques suggests that FTIR can be used to predict the aryl content of samples with reasonable accuracy provided sufficient ¹³C NMR analyses are available for calibration.

Limitations in Solid-State ¹³C NMR Studies of Soil Organic Matter

Sensitivity

¹³C NMR spectroscopy is a relatively insensitive technique because of the low natural abundance of ¹³C nuclei and the wide range of chemical groups in which carbon is found. Without the advantages of CP, the application of ¹³C NMR to the study of soil organic matter would be limited. The major limitations are the low levels of organic carbon in many mineral soils and the low recovery of many fractionation techniques. In practice, useful spectra cannot be obtained on samples with a carbon content <0.5% and if interfering species such as free radicals and Fe (III) are present in significant concentrations, even higher levels of carbon would be required. However, with the advent of wide-bore magnets, larger sample spinners are now available and it should in future be possible to obtain useful spectra on samples with carbon contents approaching 0.1%.

Presence of magnetic materials

Magnetic materials severely reduce the proton rotating frame relaxation times

$(T_{1\rho}H)$ thereby reducing the signal to noise of a spectrum and in severe cases preventing a spectrum from being obtained at all. In particular, iron (III) which is a common element in soils, reduces the capacity of NMR to be a useful technique in studying soils and soil derived fractions. The effect can be localized (Pfeffer *et al.*, 1984; Preston *et al.*, 1984), reducing the sensitivity of selected functional groups or more generally (Vassalo *et al.*, 1987), reducing the sensitivity of all the nuclei in the sample.

To overcome this problem, various chemical treatments have been used to reduce the Fe(III) to the more soluble Fe(II) form after which the soluble Fe can be removed. The two most common treatments have been dithionite over a range of pH (Vassalo *et al.*, 1987; Arshad *et al.*, 1988; Skjemstad *et al.*, 1992; Preston *et al.*, 1994) or hydrofluoric acid (HF) solution at varying concentrations (Calderoni and Schnitzer, 1984; Skjemstad and Dalal, 1987; Preston *et al.*, 1989; Skjemstad *et al.*, 1994). If ferromagnetic materials are present, even a strong magnet can remove sufficient magnetic material to allow useful spectra to be obtained.

Of these various techniques, repeated treatment with HF at a concentration of 2% has been shown to be by far the most effective and reliable (Skjemstad *et al.*, 1994). These authors found that HF could improve the relative visibility of carbon in a soil sample by as much as 25 times that of the best dithionite treatment. HF also substantially increases sensitivity by removing a considerable proportion of the inorganic matrix thereby increasing the carbon concentration in the sample. The problems of contamination of the sample from citrate, required as an essential part of the dithionite procedure, were also absent.

HF treatment, although very effective in improving the signal to noise of most ^{13}C NMR spectra, must be used with some caution. Some soil organic materials are also soluble in HF and if this soluble fraction is significant, the spectrum obtained after HF treatment may not be representative of the organic matter originally present in the sample. In all of the surface soil samples that we have treated, we have found no significant effect on the final spectra due to loss of carbon. It should be noted however that in many subsoils the loss of carbon can be large and HF treatment is not recommended if quantitative data are required.

Decomposition Studies

Many laboratory studies on the microbial decomposition of plant residues and manures have been made using solid state ^{13}C NMR spectroscopy to observe changes in organic chemistry during decomposition. All of these studies consistently demonstrate that the first materials to be affected are the polysaccharides which decompose rapidly while the alkyl, aromatic and carbonyl materials decompose more slowly. The net effect is a relative increase in these latter components as decomposition proceeds, although overall chemical changes within each component appear to be slight (Wilson *et al.*, 1983a; Inbar *et al.*, 1989; Holmgren *et al.*, 1990; Vinceslas-Akpa and Loquet, 1994; Knicker and Lüdemann, 1995). Many studies attribute high alkyl contents in highly decomposed materials to selective preservation (Theng *et al.*, 1992). Kögel-Knabner *et al.* (1992) suggested however that selective preservation is not the dominant process leading to the high alkyl content in forest soil organic matter but rather an increase in cross linking of the long-chained alkyls occurring during humification. Similar trends are also evident between soils and in soil profiles with the most decomposed materials containing significantly higher levels of alkyl and aryl materials (Kögel-Knabner *et al.*, 1988; Fox *et al.*, 1994).

The chemical nature of the input can also significantly influence the chemistry of soil organic matter (Krosshavn *et al.*, 1992; Golchin *et al.*, 1994b). For example, Golchin *et al.* (1995a) showed that the high alkyl carbon content of brigalow vegetation, as indicated by the litter light fraction (<1.6 Mg m^{-3}, brown), was carried through to other soil organic matter fractions which also showed a high alkyl con-

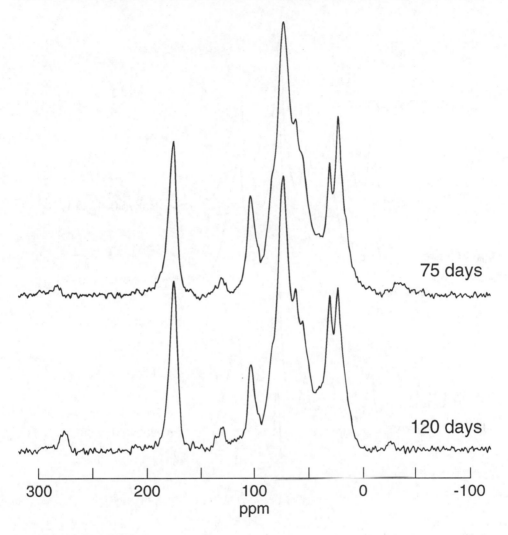

Fig. 20.4. CP/MAS ^{13}C NMR spectra of ^{13}C-glucose after incubation in an Al saturated montmorillonite matrix for 75 and 120 days.

tent. Golchin *et al.* (1995a) were also able to demonstrate that, despite extended periods of cultivation, if the inputs returned to a cultivated soil had significantly higher carbohydrate contents than those returned to the equivalent uncultivated soil, cultivation can lead to a relative increase in carbohydrate levels, even though soil organic matter declines overall.

Another application of solid state ^{13}C NMR spectroscopy in decomposition studies has been the incorporation of simple ^{13}C-labelled substrates, the decom-

position of which can then be followed using solid state ^{13}C NMR spectroscopy. Because the ^{13}C nucleus has a natural abundance of only 1.1%, additions of only small amounts of ^{13}C-labelled materials can significantly increase the number of ^{13}C nuclei in the sample to the extent that the resultant NMR signal is almost entirely derived from the added substrate and its decomposition products. This approach was first shown to be feasible by Preston and Ripmeester (1983) using ^{13}C-acetate and subsequently by Baldock *et al.* (1989,

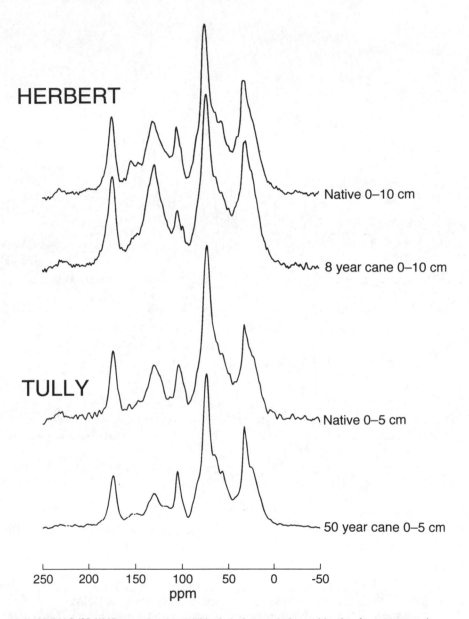

Fig. 20.5. CP/MAS ^{13}C NMR spectra of two Ultisols before and after cultivation for sugarcane in northern Australia. The Herbert soil was originally under eucalypt forest and was cultivated for 8 years. The Tully soil was originally under native grass and was cultivated for in excess of 50 years.

1990a,b) who followed the decomposition of ^{13}C-glucose in soil. In these studies, the resulting ^{13}C concentration must be high enough to give a signal above the native soil carbon signal but not so high that homonuclear J-coupling of adjacent ^{13}C nuclei significantly broaden signals. Baldock *et al.* (1990c) were able to demonstrate that a considerable amount of polymethylene carbon was derived from the microbial biomass with bacteria producing relatively more alkyl carbon than fungi. Figure 20.4 shows a high quality spectrum obtained from incubating ^{13}C -glucose in an Al-montmorillonite clay matrix for 75 and 120 days. Methyl, polymethylene, O-alkyl, anomeric, aryl and carbonyl groups are clearly visible and the contents of polymethylene and carbonyl groups increased with increasing time of incubation. These data clearly demonstrate the potential for this type of study in soil organic matter dynamics.

Effect of Cultivation on Decomposition

Cultivation invariably accelerates decomposition and generally results in a rapid decline in soil organic matter content. Studies on whole soil samples using solid-state ^{13}C NMR spectroscopy generally conclude that cultivation results in minor changes in the distribution of functional groups in soil organic matter (Skjemstad *et al.*, 1986). As demonstrated previously in laboratory incubations, increased decomposition results in a greater loss in O-alkyl carbon and this effect is also seen after long-term cultivation. Preston *et al.* (1987), for example, report a doubling in the relative alkyl content of a peat soil after 15 years of cultivation. The amount of input, however, can also affect the carbohydrate content and where organic matter levels do not fall appreciably during cultivation, the nature of the soil organic matter can remain relatively unchanged despite high turnover rates.

Figure 20.5 shows the solid state ^{13}C NMR spectra of two podzolic (Ultisol) soils (Herbert and Tully) under sugarcane for 8 years and in excess of 50 years respectively in tropical north Queensland. In the Herbert soil, which was originally under eucalypt forest, the organic carbon content has been reduced from 6.3% to 4.2% by 8 years of cultivation; however, only minor changes in the contents of aryl and O-aryl carbon were noted. The Tully soil was originally under grass and here

very little change has occurred except for a slight increase in the content of carbohydrate carbon as a result of 50 years of cane production. It should also be noted that the light fractions (<1.6 Mg m^{-3}) in these two soils were not different before and after cultivation.

For heavy clay soils from a temperate region of Queensland (Fig. 20.6), large differences in the changes in chemical composition of soil organic matter induced by long-term cultivation can be seen. The grey clay (Chromustert) loses carbon rapidly on cultivation and after 45 years contains only about one-third of its original carbon while the black earth (Pellustert) retains about two-thirds, even after 50 years of exploitive cultivation. Both soils are similar in mineralogy and clay content and were sampled within a few kilometres of one another. The solid-state ^{13}C NMR spectra of these soils under different periods of cultivation show that the chemistry of the grey clay has barely changed; however, for the black earth a substantial increase in aromatic carbon at the expense of the other functional groups is observed. In fact, when calculated on the basis of the amount of carbon present at any stage during cultivation, the aromatic carbon remains static while the other functional groups decline. In the black earth therefore, aromatic carbon appears to be selectively preserved while in the grey clay all functional groups decline at about the same rate. The aromatic carbon in the black earth, as will be demonstrated later in this chapter, is contained in particles of charcoal with little, if any, contribution from lignin or its decomposition products.

Fractionation Studies

Solid state ^{13}C NMR spectroscopic studies of whole soil samples can be useful in determining and following general changes in soil organic matter with time and between samples. The heterogeneous nature of soil organic matter in its chemistry, degree of decomposition and spatial distribution within the soil matrix, how-

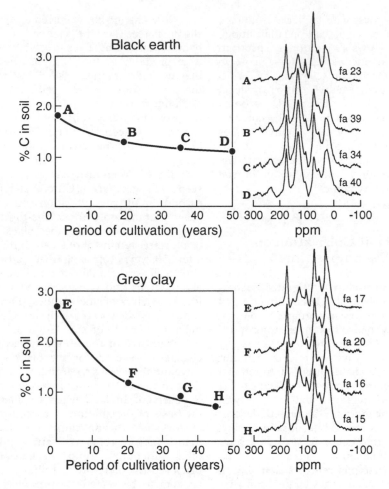

Fig. 20.6. The change in carbon concentration with time and the CP/MAS ^{13}C NMR spectra of two heavy clay soils at different times under exploitive cultivation. The percentage of the carbons that are aromatic (fa) is also given.

ever, limit the usefulness of whole soil studies. When considering the dynamic processes of soil organic matter, fractionation procedures that isolate biologically meaningful fractions are essential and solid-state ^{13}C NMR spectroscopy applied to these fractions can provide considerable insight into their role in these processes.

For many years, chemical extraction procedures were extensively used to study soil organic matter. These procedures were popular because until recently most analytical techniques, including NMR, required the sample to be in solution. With the advent of solid state ^{13}C NMR spectroscopy and other modern techniques, solid samples could be effectively utilized and new fractionation techniques based on soil physical properties could be more effectively used to study the composition of soil organic matter. Chemical fractionations that use acids and strong bases can still be useful where the mobility of organic matter is of interest and studies have been successfully carried out on soil processes such as podzolization using this approach (Skjemstad, 1984) as well as studies concerned with newly formed products of decomposition (Kögel-Knabner and Ziegler, 1988).

Separations on the basis of particle size

Physical separations based on density and particle size are also widely used to reduce the chemical complexity of the sample and are believed to separate 'pools' of substrate of varying microbial availability in mineral soils. Oades *et al.* (1988) for example, demonstrated that in a red-brown earth after a considerable time under pasture or exploitive cultivation, major chemical changes only occurred in the larger particles rich in carbohydrates and that the chemistry of the highly alkyl organic matter intimately associated with the clay fraction remained largely unchanged by cultivation. They concluded that in a particular soil, changing input does not change the chemical composition of the soil organic matter which is controlled by the microbial biomass and the interactions of microbial decomposition products with the mineral matrix. Baldock *et al.* (1992), using solid-state ^{13}C NMR spectroscopy, investigated the sand, silt and clay fractions from a number of soil types from a range of environments and proposed a three-stage model of oxidative decomposition with organic matter moving from the largest particle sizes (>20 μm) to the intermediate (2–20 μm) and finally to the fraction containing the most resistant organic matter (>2 μm). During decomposition, carbohydrates and proteins were lost in transition from the first to the intermediate stage while lignin was lost from the intermediate stage containing partially decomposed residues to produce the highly alkyl and recalcitrant clay-associated organic matter. These conclusions are strongly supported by Nordén *et al.* (1992) who demonstrated the same changes in chemistry with decreasing particle size in peats.

Such clearly defined changes in chemical composition with changing particle size are not always evident, however, as indicated by the NMR spectra of the two soils in Fig. 20.7. Both soils were saturated with Na$^+$ and dispersed with mild ultrasonic treatment. The prairie soil shows the presence of lignin in all fractions >2 μm with the alkyl carbon content being similar throughout. The black earth on the other hand shows an increase in aromatic and a decrease in polymethylene carbon with decreasing particle size. It is clear therefore, that the chemistry of the organic matter associated with different particle classes are not always predetermined by size alone but may be influenced by soil type, land management and the chemistry of the input organic matter.

Separations on the basis of density

Separation of soil into different density classes has also been used to isolate soil organic matter with varying degrees of association with the mineral matrix. Skjemstad *et al.* (1986), using solid state ^{13}C NMR spectroscopy, demonstrated that the lightest fraction (<1.6 Mg m^{-3}) separated from a grey clay was chemically indistinguishable from litter while increasing density led to fractions with decreasing carbohydrate signals and increasing alkyl signals. Recently, Golchin *et al.* (1994a) demonstrated that two light fractions (<1.6 Mg m^{-3}) could be separated from soils if a normal light fractionation was followed by ultrasonic treatment and a second <1.6 Mg m^{-3} fractionation. This latter fraction was considered to be organic matter occluded between microaggregates but not strongly bound to mineral matter and through solid-state ^{13}C NMR spectroscopy was shown to be considerably more aromatic and alkyl in nature compared with the original litter-like light fraction. Extending this approach into more dense soil fractions, Golchin *et al.* (1994b) were able to demonstrate the chemical changes that organic materials undergo as they enter the soil, are enveloped by clay to form microaggregates and finally are incorporated into the microbial biomass and metabolites which become intimately associated with clay. From these observations and carbon turnover data, Golchin *et al.* (1994b) were able to propose a model linking organic matter turnover, chemistry and aggregate stability and clearly linked aggregate stability with interactions

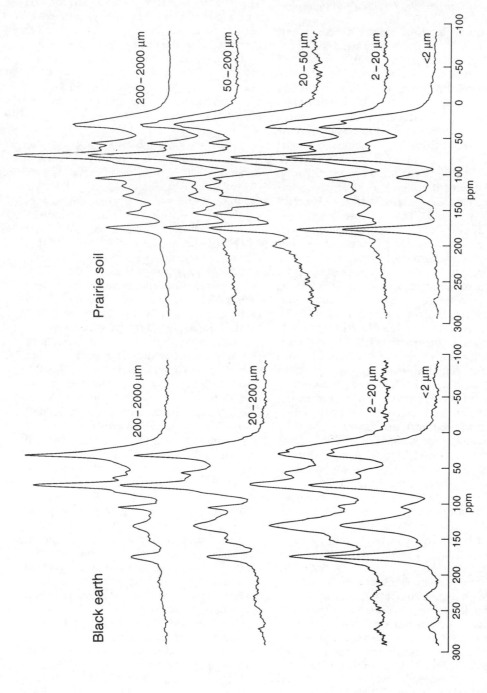

Fig. 20.7. CP/MAS ^{13}C NMR spectra of different particle size fractions from two soils from eastern Australia.

between the microbial biomass and young, readily decomposable soil organic matter. Golchin *et al.* (1995b) were also able to demonstrate that in an Oxisol, changing vegetation from rain forest to pasture resulted in little change in the chemistry of the >2.0 Mg m^{-3} fraction, which contains the carbon most intimately associated with clay, but did result in substantial chemical changes in the lighter fractions. The largest changes involved a loss of lignin structures although the occluded fractions under pasture still contained aromatic carbon. Golchin *et al.* (1995a) also demonstrated on a number of soils that cultivation had little effect on the chemistry of the 'heavy' >2.0 Mg m^{-3} fraction with significant

chemical changes only occuring in the 'light' fractions. Functional data from density separations also agree well with data published for particle size separations in that alkyl carbon tends to be concentrated in the clay and the clay-associated heavy fractions while the carbohydrates mostly occur in the sand and light fractions.

Spatial Heterogeneity of Soil Organic Matter

Proton spin relaxation editing

In heterogeneous organic materials, spatial separation of spin populations and limited

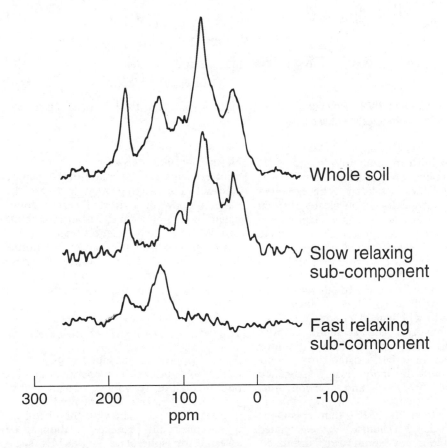

Fig. 20.8. The fast and slowly relaxing organic fractions in an Andisol separated by Proton Spin Relaxation Editing (PSRE) compared with the normal whole-soil CP/MAS ^{13}C NMR spectrum.

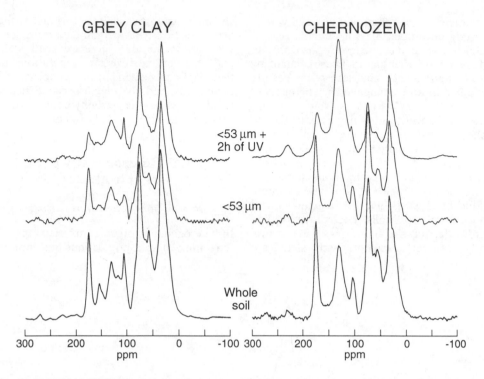

Fig. 20.9. CP/MAS ^{13}C NMR spectra of the whole soil, and the <53 µm fraction before and after 2 h of high energy UV photo-oxidation of two soils.

spin diffusion can result in a range of ^{1}H and ^{13}C relaxation parameters. Differences in $T_1(H)$ values in solids can be exploited to select subspectra of physically distinct domains on a scale of tens of nanometers within soil samples. Preston and Newman (1992) used proton spin relaxation editing (PSRE) to separate organic domains with short and long $T_1(H)$ values in three HF treated humin samples. In all cases, the materials with long $T_1(H)$ values were dominated by polymethylene carbon while the carbohydrate and aromatic carbons gave short $T_1(H)$ values. They concluded that these materials existed in separate domains with Fe more closely associated with the aromatic and carbohydrate materials. Newman (1992) was able to demonstrate using PSRE that crystalline and disordered cellulose and acetylated hemicellulose and lignin were intimately mixed in samples of wood. Using a similar approach to that described by Preston and

Newman (1992), separations of two distinct domains can be made in a Japanese Andosol (Fig. 20.8) but in this case the material showing a short $T_1(H)$ is dominated by aromatic and carbonyl carbons and is most likely due to finely-divided charcoal. The aromatic-rich type A humic acids isolated from some Japanese Andosols has been attributed to a long history of vegetation burning (Tate *et al.*, 1990).

High energy ultraviolet photo-oxidation

Another approach developed by Skjemstad *et al.* (1993) utilizes high energy ultraviolet photo-oxidation to remove organic matter external to strongly cemented micro-aggregates thereby allowing the chemistry of the physically protected carbon to be assessed by solid-state ^{13}C NMR spectroscopy. The oxidation relies only on UV light and excess oxygen, supplied by bubbling air into a dispersed sample, so that only organic matter exposed directly to the UV light can undergo oxidation. Solid-state ^{13}C NMR spectroscopy, however, revealed that the organic material remaining after UV photo-oxidation not only contained protected organic matter but also finely divided charcoal. Fig. 20.9 shows the chemistry of the organic matter in two soils, a grey clay and a Chernozem (Mollisol), their <53 μm soil fractions and the unoxidized organic materials remaining in the <53 μm fractions after 2 h of photo-oxidation. Photo-oxidative treatment of the grey clay shows that the most physically protected organic matter consists of polysaccharide and alkyl groups with proteinaceous materials being largely absent. In the Chernozem, the highly aromatic material remaining is largely charcoal which, as demonstrated by Skjemstad et al. (1996), is also resistant to photo-oxidation. Further studies have demonstrated that charcoal is present in almost all Australian soils and represents a significant proportion of the organic carbon in the >53 μm fractions of many soil types. This material therefore is responsible for much of the aromatic signal in many soil fractions where it can easily be confused with partially decomposed lignin and humic substances.

Interrupted Decoupling

Interrupted decoupling (ID) or dipolar dephasing (DD) can be used to determine the proportion of protonated and non-protonated carbons in organic materials (Oppella and Frey, 1979; Alemany et al., 1983c). The pulse sequence used is the same as for the conventional CP/MAS experiment except that between the CP and the free induction decay (FID) acquisition, the high power decoupler is turned off for a short period of time and then turned back on. This is followed by a 180° refocusing pulse applied to the ^{13}C nuclei after which the FID is acquired. During the interuption time, the net magnetization of the carbons directly coupled to protons will undergo rapid dephasing

due to ^1H-^{13}C dipolar coupling, whereas carbons remote from protons will dephase more slowly due to the weaker dipolar interactions. Carbons that undergo rapid rotational motion, such as methoxy groups attached to aromatic rings, will also experience a weaker dipole interaction with the protons due to the averaging of the ^1H-^{13}C dipole. Thus by choosing an appropriate delay so that the protonated carbons have lost nearly all of their magnetization and the non-protonated carbons still retain significant magnetization, then it is possible to obtain a spectrum containing only non-protonated and rotationally fast protonated carbons. This approach has been successfully used in structural studies of lignins (Hatcher, 1987; 1988), tannins (Wilson and Hatcher, 1988; Newman and Porter, 1992) and humic materials (see for example Wilson et al., 1983b; Skjemstad et al., 1992) and provides considerable insight into the structure of these materials and the chemical changes that occur during decomposition.

Measurement of Organic Domains

By monitoring the mixing of proton spins (MOPS) with CP/NMR, the rates of proton spin diffusion can provide information about the size and geometry of domains in heterogeneous organic solids (Newman 1991). Differences in proton spin relaxation time constants of domains can be exploited to measure the average sizes of like domains. For the $T_1(H)$ and $T_{1\rho}(H)$ relaxation, the constants are typically up to 1 s and in the range of milliseconds respectively and differences in these constants over these timescales may yield information on dimensions of greater than 30 nm and 2–30 nm respectively. Newman (1992) used this technique on samples of wood to establish that microfibrils of cellulose about 14 nm thick were encrusted with a mixture of hemicellulose and lignin. This approach may prove valuable in litter quality and decomposition studies since the proximity, size and packing of various substrates in distinct domains would be expected to influence decomposition rates.

Conclusions

It is clear from the data reviewed in this paper that soil organic matter is a mixture of residual plant material in various stages of decomposition and the microbial biomass and its products. The latter are concentrated in the >2.0 Mg m^{-3} fraction and appear to be closely associated with the mineral matrix. Since the nature of microbial products does not directly reflect the nature of the substrate from which they are derived, litter quality (chemistry) can only have an impact on the nature of soil organic matter where the 'light' <2.0 Mg m^{-3} fraction is significant. In soils where decomposition rates are low, accumulating litter will dominate the overall organic chemical nature of the soil organic matter. During cultivation, decomposition rates are increased and the 'light' fraction in the soil is rapidly decomposed resulting in a relative decrease in carbohydrate content. Because of the rapid turnover of the input organic matter during cultivation, the nature of the input has little impact on the nature of the soil organic matter. However, significant changes in organic matter chemistry can occur following cultivation where charcoal is present. Charcoal is relatively inert to microbial activity and as total soil organic matter declines due to cultivation, the charcoal fraction increases as a proportion of the total carbon and results in a relative increase in the aromatic nature of the organic matter.

The pulse sequences described in this chapter have considerable potential for elucidating soil organic matter structure and determining the associations of soil organic matter *in situ*. These techniques warrant further investigation and their application to soils and soil fractions should considerably advance our understanding of the chemical and spatial heterogeneity of soil organic matter and the role that litter quality plays in humus chemistry.

References

Alemany, L.B., Grant, D.M., Pugmire, R.J., Alger, T.D. and Zilm, K.W. (1983a) Cross polarization and magic angle sample spinning NMR spectra of model organic compounds. 1. Highly protonated molecules. *Journal of the American Chemical Society* 105, 2133–2141.

Alemany, L.B., Grant, D.M., Pugmire, R.J., Alger, T.D. and Zilm, K.W. (1983b) Cross polarization and magic angle sample spinning NMR spectra of model organic compounds. 2. Molecules of low or remote protonation. *Journal of the American Chemical Society* 105, 2142–2147.

Alemany, L.B., Grant, D.M., Alger, T.D. and Pugmire, R.J. (1983c) Cross polarization and magic angle sample spinning NMR spectra of model compounds. 3. Effect of the ^{13}C-^{1}H dipolar interaction on cross-polarization and carbon-proton dephasing. *Journal of the American Chemical Society* 105, 6697–6704.

Arshad, M.A., Ripmeester, J.A. and Schnitzer, M. (1988) Attempts to improve solid state ^{13}C NMR spectra of whole mineral soils. *Canadian Journal of Soil Science* 68, 593–602.

Baldock, J.A., Oades, J.M., Vassallo, A.M. and Wilson, M.A. (1989) Incorporation of uniformly labelled ^{13}C-glucose carbon into the organic fraction of a soil. Carbon balance and CP/MAS ^{13}C NMR measurements. *Australian Journal of Soil Research* 27, 725–746.

Baldock, J.A., Oades, J.M., Vassallo, A.M. and Wilson, M.A. (1990a) Solid state CP/MAS ^{13}C NMR analysis of particle size and density fractions of a soil incubated with uniformaly labelled ^{13}C-glucose. *Australian Journal of Soil Research* 28, 193–212.

Baldock, J.A., Oades, J.M., Vassallo, A.M. and Wilson, M.A. (1990b) Solid state CP/MAS ^{13}C NMR analysis of bacterial and fungal cultures isolated from a soil incubated with glucose. *Australian Journal of Soil Research* 28, 213–225.

Baldock, J.A., Oades, J.M., Vassallo, A.M. and Wilson, M.A. (1990c) Significance of microbial activity in soils as demonstrated by solid-state ^{13}C NMR. *Environmental Science and Technology* 24, 527–530.

Baldock, J.A., Oades, J.M., Waters, A.G., Peng, X., Vassallo, A.M. and Wilson, M.A. (1992) Aspects of the chemical structure of soil organic matter materials as revealed by solid-state ^{13}C NMR spectroscopy. *Biogeochemistry* 16, 1–42.

Bates, A.L. and Hatcher, P.G. (1992) Quantitative solid-state ^{13}C nuclear magnetic resonance spectrometric analyses of wood xylem: effect of increasing carbohydrate content. *Organic Geochemistry* 18, 407–416.

Calderoni, G. and Schnitzer, M. (1984) Effects of age on the chemical structure of paleosol humic acids and fulvic acids. *Geochimica et Cosmochimica Acta* 48, 2045–2051.

Fox, C.A., Preston, C.M. and Fyfe, C.A. (1994) Micromorphological and ^{13}C NMR characterization of a Humic, Lignic and Histic Folisol from British Columbia. *Canadian Journal of Soil Science* 74, 1–15.

Golchin, A., Oades, J.M., Skjemstad, J.O. and Clarke, P. (1994a) Study of free and occluded particulate organic matter in soils by ^{13}C CP/MAS NMR spectroscopy and scanning electron microscopy. *Australian Journal of Soil Research* 32, 285–309.

Golchin, A., Oades, J.M., Skjemstad, J.O. and Clarke, P. (1994b) Soil structure and carbon cycling. *Australian Journal of Soil Research* 32, 1043–1068.

Golchin, A., Oades, J.M., Skjemstad, J.O. and Clarke, P. (1995a) The effects of cultivation on the composition of organic matter and structural stability of soils. *Australian Journal of Soil Research* 33, 975–993.

Golchin, A., Oades, J.M., Skjemstad, J.O. and Clarke, P. (1995b) Structural and dynamic properties of soil organic matter as reflected by ^{13}C natural abundance, pyrolysis mass spectrometry and solid-state ^{13}C NMR spectroscopy in density fractions of an Oxisol under forest and pasture. *Australian Journal of Soil Research* 33, 59–76.

Hatcher, P.G. (1987) Chemical structural studies of natural lignin by dipolar dephasing solid-state ^{13}C nuclear magnetic resonance. *Organic Geochemistry* 11, 31–39.

Hatcher, P.G. (1988) Dipolar-dephasing ^{13}C NMR studies of decomposed wood and coalified xylem tissue: Evidence for chemical structural changes associated with defunctionalization of lignin structural units during coalification. *Energy and Fuels* 2, 48–58.

Hatcher, P.G., Schnitzer, M., Dennis, L.W. and Maciel, G.E. (1981) Aromaticity of humic substances in soils. *Soil Science Society of America Journal* 45, 1089–1094.

Hawkes, G.E., Smith, C.Z., Utley, J.H.P., Vargas, R.R. and Viertler, H. (1993) A comparison of solution and solid state ^{13}C NMR spectra of lignins and lignin model compounds. *Holzforschung* 47, 302–312.

Holmgren, A., Wikander, G. and Borden, B. (1990) Spectroscopic investigations of peat-water interactions: an ESR, FT-IR, and NMR study. *Soil Science* 149, 279–291.

Inbar, Y., Chen, Y. and Hadar, Y. (1989) Solid-state carbon-13 nuclear magnetic resonance and infrared spectroscopy of composted organic matter. *Soil Science Society of America Journal* 53, 1695–1701.

Knicker, H. and Lüdemann, H.-D. (1995) N-15 and C-13 CPMAS and solution NMR studies of N-15 enriched plant material during 600 days of microbial degradation. *Organic Geochemistry* 23, 329–341.

Kögel-Knabner, I., de Leeuw, J.W. and Hatcher, P.G. (1992) Nature and distribution of alkyl carbon in forest soil profiles: implications for the origin and humification of aliphatic biomacromolecules. *The Science of the Total Environment* 117/118, 175–185.

Kögel-Knabner, I., Zech, W. and Hatcher P.G. (1988) Chemical composition of the organic matter in forest soils: The humus layer. *Zeitschrift für Pflanzenernährung und Bodenkunde* 151, 331–340.

Kögel-Knabner, I. and Ziegler, F. (1988) Carbon distribution in different compartments of forest soils. *Geoderma* 56, 515–525.

Krosshavn, M., Southon, T.E. and Steinnes, E. (1992) The influence of vegetational origin and degree of humification of organic soils on their chemical composition, determined by solid-state ^{13}C NMR. *Journal of Soil Science* 43, 485–493.

Love, G.D., Snape, C.E., Jarvis, M.C. and Morrison, I.M. (1994) Determination of phenolic structures in flax fibre by solid-state 13C NMR. *Phytochemistry* 35, 489–491.

Maciel, G.E., O'Donnell, D.J., Ackerman, J.J.H., Hawkins, B.H. and Brutuska, V.J. (1981) A ^{13}C NMR study of four lignins in the solid and solution states. *Macromolekulare Chemie* 182, 2297–2304.

Malcolm, R.L. (1992) ^{13}C-NMR spectra and contact time experiment for Skjervatjern fulvic and humic acids. *Environment International* 18, 609–620.

Newman, R.H. (1991) Proton spin diffusion monitored by ^{13}C NMR. *Chemical Physics Letters* 180, 301–304.

Newman, R.H. (1992) Nuclear magnetic resonance study of spacial relationships between chemical components in wood cell walls. *Holzforschung* 46, 205–210.

Newman, R.H. and Porter, L.J. (1992) Solid state ^{13}C NMR studies on condensed tannins. In: Hemingway, R.W. and Laks, P.E. (eds) *Plant Polyphenols: Biogenesis, Chemical Properties, and Significance*. Plenum, New York. pp. 339–348.

Nordén, B. and Berg, B. (1990) A non-destructive method (solid state ^{13}C NMR) for determining organic chemical components of decomposing litter. *Soil Biology and Biochemistry* 22, 271–275.

Nordén, B., Bohlin, E., Nilsson, M., Albano, A. and Rockner, C. (1992) Characterization of particle size fractions of peat. An integrated biological, chemical, and spectroscopic approach. *Soil Science* 153, 382–396.

Oades, J.M., Waters, A.G., Vassallo, A.G., Wilson, M.A. and Jones, G.P. (1988) Influence of management on the composition of organic matter in a Red-brown earth as shown by ^{13}C nuclear magnetic resonance. *Australian Journal of Soil Research* 26, 289–299.

Oppella, S.J. and Frey, M.R. (1979) Selection of non-protonated carbon resonances in solid-state nuclear magnetic resonance. *Journal of the American Chemical Society* 101, 5854–5856.

Pfeffer, P.E., Gerasimowicz, W.V. and Piotrowski, E.G. (1984) Effect of paramagnetic iron on quantitation in carbon-13 cross polarization magic angle spinning nuclear magnetic resonance spectroscopy of heterogeneous environmental matrices. *Analytical Chemistry* 56, 734–741.

Preston, C.M. and Newman, R.H. (1992) Demonstration of spatial heterogeneity in the organic matter of de-ashed humin samples by solid-state ^{13}C CPMAS NMR. *Canadian Journal of Soil Science* 72, 13–19.

Preston, C.M. and Ripmeester, J.A. (1982) Application of solution and solid-state ^{13}C NMR to four organic soils, their humic acids, fulvic acids, humins and hydrolysis residues. *Canadian Journal of Spectroscopy* 27, 99–105.

Preston, C.M. and Ripmeester, J.A. (1983) ^{13}C-labelling for NMR studies of soils: CPMAS NMR observation of [13]C -acetate transformation in a mineral soil. *Canadian Journal of Soil Science* 63, 495–500.

Preston, C.M., Dudley, R.L., Fyfe, C.A. and Mathur, S.P. (1984) Effects of variations in contact times and copper contents in a ^{13}C CPMAS NMR study of samples of four organic soils. *Geoderma* 33, 245–253.

Preston, C.M., Shipitalo, S.-E., Dudley, R.L., Fyfe, C.A., Mather, S.P. and Levesque, M. (1987) Comparison of ^{13}C CPMAS NMR and chemical techniques for measuring the degree of decomposition in virgin and cultivated peat profiles. *Canadian Journal of Soil Science* 67, 187–198.

Preston, C.M., Schnitzer, M. and Ripmeester, J.A. (1989) A spectroscopic and chemical investigation on the de-ashing of a humin. *Soil Science Society of America Journal* 53, 1442–1447.

Preston, C.M., Newman, R.H. and Rother, P. (1994) Using ^{13}C CPMAS NMR to assess effects of cultivation on the organic matter of particle size fractions in a grassland soil. *Soil Science* 157, 26–34.

Skjemstad, J.O. (1984) The nature and distribution of organic carbon and nitrogen in podzols developed on siliceous sands in south east Queensland. M. Appl. Sci. Thesis, Queensland Institute of Technology, Brisbane, Australia.

Skjemstad, J.O. and Dalal, R.C. (1987) Spectroscopic and chemical differences in organic matter of two Vertisols subjected to long periods of cultivation. *Australian Journal of Soil Research* 25, 323–335.

Skjemstad, J.O., Frost, R.L. and Barron, P.F. (1983) Structural units in humic acids from south-eastern Queensland soils as determined by ^{13}C NMR spectroscopy. *Australian Journal of Soil Research* 21, 539–547.

Skjemstad, J.O., Dalal, R.C. and Barron, P.F. (1986) Spectroscopic investigations of cultivation effects on organic matter of Vertisols. *Soil Science Society of America Journal* 50, 354–359.

Skjemstad, J.O., Waters, A.G., Hanna, J.V. and Oades, J.M. (1992) Genesis of podzols on coastal dunes in southern Queensland. IV. Nature of the organic fraction as seen by ^{13}C nuclear magnetic resonance spectroscopy. *Australian Journal of Soil Research* 30, 667–681.

Skjemstad, J.O., Janik, L.J., Head, M.J. and McClure, S.G. (1993) High energy ultraviolet photo-oxidation: a novel technique for studying physically protected organic matter in clay- and silt-sized aggregates. *Journal of Soil Science* 44, 485–499.

Skjemstad, J.O., Clarke, P., Taylor, J.A., Oades, J.M. and Newman, R.H. (1994) The removal of magnetic materials from surface soils. A solid state ^{13}C CP/MAS NMR study. *Australian Journal of Soil Research* 32, 1215–1229.

Skjemstad, J.O., Clarke, P., Oades, J.M., Taylor, J.A. and McClure, S.G. (1996) The chemistry and nature of protected carbon in soil. *Australian Journal of Soil Research* 34, 251–271.

Snape, C.E., Axelson, D.E., Botto, R.E., Delpuech, J.J., Tekely, P., Gerstein, B.C., Pruski, M., Maciel, G.E. and Wilson, M.A. (1989) Quantitative reliability of aromaticity and related measurements on coals by ^{13}C NMR. A debate. *Fuel* 68, 547–560.

Stark, R.E. and Garbow, J.R. (1992) Nuclear magnetic resonance relaxation studies of plant polyester dynamics. 2. Suberized potato cell wall. *Macromolecules* 25, 149–154.

Tate, K.R., Yamamoto, K., Churchman, G.J., Meinhold, R. and Newman, R.H. (1990) Relationships between the type and carbon chemistry of humic acids from some New Zealand and Japanese soils. *Soil Science and Plant Analysis* 36, 611–621.

Theng, B.K.G., Tate, K.R. and Becker-Heidmann, P. (1992) Towards establishing the age, location and identity of the inert soil organic matter of a Spodosol. *Zeitschrift für Pflanzenernährung und Bodenkunde* 155, 181–184.

Vinceslas-Akpa, M. and Loquet, M. (1994) ^{13}C CPMAS NMR spectroscopy of organic matter transformation in ligno-cellulosic waste products composted and vermicomposted (*Eisenia fetida andrei*). *European Journal of Soil Biology* 30, 17–28.

Vassallo. A.M., Wilson, M.A., Collin, P.J., Oades, J.M., Waters, A.G. and Malcolm, R.L. (1987) Structural analysis of geochemical samples by solid-state nuclear magnetic resonance spectrometry. Role of paramagnetic material. *Analytical Chemistry* 59, 558–562.

Wilson, M.A. (1987) *NMR Techniques and Applications in Geochemistry and Soil Chemistry*, 1st edn. Pergamon Press, Oxford, 353 pp.

Wilson, M.A. and Hatcher, P.G. (1988) Detection of tannins in modern and fossil barks and in plant residues by high-resolution solid-state ^{13}C nuclear magnetic resonance. *Organic Geochemistry* 12, 539–546.

Wilson, M.A., Barron, P.F. and Goh, K.M. (1981a) Differences in structure of organic matter in two soils as demonstrated by ^{13}C cross polarization nuclear magnetic resonance spectroscopy with magic angle spinning. *Geoderma* 26, 323–327.

Wilson, M.A., Pugmire, R.J., Zilm, K.W., Goh, K.M., Heng, S. and Grant, D.M. (1981b) Cross-polarization ^{13}C-NMR spectroscopy with 'magic angle' spinning characterizes organic matter in whole soils. *Nature* 294, 648–650.

Wilson, M.A., Heng, S., Goh, K.M., Pugmire, R.J. and Grant, D.M. (1983a) Studies of litter and acid insoluble soil organic matter fractions using ^{13}C-cross polarization nuclear magnetic resonance spectroscopy with magic angle spinning. *Journal of Soil Science* 34, 83–97.

Wilson, M.A., Pugmire, R.J. and Grant, D.M. (1983b) Nuclear magnetic resonance spectroscopy of soils and related materials. Relaxation of ^{13}C nuclei in cross polarization nuclear magnetic resonance experiments. *Organic Geochemistry* 5, 121–129.

Zech, W., Haumaier, L. and Kögel-Knabner, I. (1989) Changes in aromaticity and carbon distribution of soil organic matter due to pedogenesis. *The Science of the Total Environment* 81/82, 179–186.

21   Development and Use of a Carbon Management Index to Monitor Changes in Soil C Pool Size and Turnover Rate

G.J. Blair, R.D.B. Lefroy, B.P. Singh and A.R. Till

Department of Agronomy and Soil Science, University of New England, Armidale NSW 2351, Australia

Introduction

Much of agriculture throughout the world has developed, and continues to be developed by opening up new land to production. Initially, productivity is supported through the utilization of nutrients released from the accumulated surface litter and soil organic matter (SOM), which also contribute to the physical fertility of the soil. The release of nutrients from SOM is largely through microbial activity. To accomplish this, the microbial population needs readily usable carbon (C) to provide the energy source.

There are important changes in both the C pool size and turnover rate when natural systems, such as grassland, are converted to crop land and when legume green manures are introduced into the system. In natural grassland systems there is a large pool of C with residues of different ages and quality which are turning over at varying rates. When the land is cultivated and cropped the rate of breakdown of organic debris is increased (Syers and Craswell, 1995) and, although the amount of residue returned may be the same as in the grassland, a major portion of the organic matter addition occurs at the one time, at harvest, and is of a similar quality. Generally such residue has a wide C-to-N ratio and is poor in nutrients. As C is lost from the system the remaining C is more resistant to breakdown. Introduction of a legume green manure into a crop rotation provides a large amount of easily decomposable C and a ready supply of N for the microorganisms. If temperature and moisture conditions are favourable, this results in a rapid breakdown of the added organic matter.

Technological developments in the production of plant cultivars with shorter growing seasons and increased adaptation to adverse soil and climatic conditions has increased the possibilities for multiple cropping systems in which more than one crop is grown per year. This second crop has often been grown at the expense of a fallow or green manure crop. Another consequence of multiple cropping systems has been a decline in the retention of crop residues and an increase in burning of such residues. This results in a low return of C to the system and the potential loss of C and nutrients via volatilization, smoke and losses of ash in surface runoff.

Most attempts to develop models of SOM turnover, and to relate SOM dynamics to soil fertility, have involved the separation of carbon into a number of pools on the basis of their rate of turnover (Parton *et al.*, 1987; McCaskill and Blair, 1989; Swift *et al.*, 1991). Only when a number of pools, with very different turnover rates, are incorporated into these models can the modelled release of nutrients approximate the observed variations in soil nutrient supply. The rapidly cycling labile carbon pool holds the key to understanding carbon dynamics and nutrient supply in agricultural systems, as was found in the model of C, N, S and P in grazed pastures developed by McCaskill and Blair (1989).

Residue Breakdown and Green Manuring

The proportion of total C decomposed in the initial phase of breakdown is similar for a wide range of crop residues, with about two-thirds of their C lost in this initial phase (Jenkinson, 1981). In contrast, the rates of decomposition during this initial stage of decomposition vary widely between different plant materials although with some exceptions, this stage is usually completed within one year (Jenkinson, 1981). The rate of decomposition is controlled by the content of nutrients particularly nitrogen, lignins and polyphenols (Tian *et al.*, 1995 – and other chapters in this volume).

When residues and SOM are broken down, some C is released as CO_2 and some is present in the soil as soluble, or at least mobile, C compounds. The situation for movement of mobile soil C compounds is similar to that for mobile nutrients. The movement of C means a loss of an energy source for microbial activity or at least a change in the position of this microbial activity in the soil profile. In addition to the direct effects of crop residues, green manures and fertilizers on nutrient dynamics and crop growth, considerable interactions can occur. These interactions, which must be taken into account when designing improved management systems include the 'priming' effect of fertilizer to stimulate the release of nutrients from recently added and/or native organic matter.

Fractions of SOM

Swift *et al.* (1991) argue that sustainable management of SOM is based on two assumptions. First, that organic matter can be separated into a number of frac-

tions, each of which is differentially responsive to management and land-use practices, and second, that the decomposition and synthesis of each of the fractions is regulated by definable sets of physico-chemical and biological factors, which in turn may be modified by management. As such, the key to developing improved residue management systems is the ability to make appropriate measurements of SOM. To this end, changes in SOM can be measured as changes in total SOM, chemical fractions of SOM, physical fractions of SOM or combinations of these fractions.

A commonly used method for chemical fractionation of SOM was the separation into fulvic acids, humic acids and humins on the basis of differences in solubilities of organic constituents in acid and alkali (Hayes and Swift, 1978). However, these fractions are not closely related to the functions of SOM. Another approach to fractionation of SOM is to use procedures that involve both physical and chemical aspects to the extraction of SOM. To this end, there is interest in procedures that involve physical separation of SOM, using sieving, density separation or floatation, followed by chemical analysis of the fractions, using any one of a number of analytical techniques, or which involve aspects of the physical and the chemical protection of SOM in the one procedure.

Solutions of potassium permanganate ($KMnO_4$) have been extensively used for the oxidation of organic compounds. The rates and extent of oxidation of different substrates is governed by their chemical composition (Hayes and Swift, 1978), the physical protection and the concentration of permanganate. Oxidation with less than the amount of permanganate required for complete oxidation should reveal the quantity of readily oxidizable components in the SOM. Modification and standardization of the $KMnO_4$ method of Loginow *et al.* (1987) by Blair *et al.* (1995) has increased the precision and simplified the technique to use only the highest concentration of permanganate used by Loginow *et al.* (1987) (333 mM), thereby dividing soil carbon into two fractions: labile (C_L) and non-labile (C_{NL}) carbon. These measurements of labile carbon have been used, in combination with similar data from a soil of an uncropped, reference area, to calculate a carbon management index (CMI), as a measure of the relative sustainability of different agricultural systems (Lefroy and Blair, 1994; Blair *et al.*, 1995). This index compares the changes that occur in the total and labile carbon as a result of the agricultural practice, with increased importance attached to changes in the labile, as opposed to the non-labile, component of the SOM.

Derivation of the Carbon Management Index

Since the continuity of C supply depends on both the total pool size and the lability (an estimate of turnover rate), both must be taken into account in deriving a soil carbon management index.

This can be achieved as follows:

1. Change in total C pool size: The loss of C from a soil with a large carbon pool is of less consequence than the loss of the same amount of C from a soil already depleted of C or which started with a smaller total C pool. Similarly, the more a soil has been depleted of carbon the more difficult it is to rehabilitate. To account for this a C pool index is calculated as follows:

C pool index (CPI)

$$= \frac{\text{sample total C (mg g}^{-1})}{\text{reference total C (mg g}^{-1})}$$

$$= \frac{C_T \text{ sample}}{C_T \text{ reference}}$$

2. The loss of labile C is of greater consequence than the loss of non-labile C. To account for this, since it is the turnover of labile carbon which releases nutrients and the labile carbon component of SOM appears to be of particular importance in affecting soil physical factors (Whitbread, 1995), a carbon lability index is calculated as follows:

Lability of C (L)

$$= \frac{\text{C in fraction oxidized by KMnO}_4}{\text{C remaining unoxidized by KMnO}_4}$$

$$\frac{\left(\text{mg labile C g}^{-1}\text{ soil}\right)}{\left(\text{mg non-labile C g}^{-1}\text{ soil}\right)} = \frac{C_L}{C_{NL}}$$

Lability index (LI)

$$= \frac{\text{Lability of C in sample soil}}{\text{Lability of C in reference soil}}$$

3. The carbon management index (CMI) can then be calculated as follows:

Carbon management index (CMI)
= C pool index × lability index × 100
= CPI × LI × 100

The soil to be used as the reference must be chosen carefully as it has a significant affect on the CMI. Which soil is used depends, to a certain extent, on the use to which the CMI is put. When the objective is to compare a number of experiment treatments or different management systems, then one treatment can be chosen as the relative standard for all other treatments. However, most often we find that, if it is available, the most useful reference soil is from an undisturbed area adjacent to the site or sites of interest. This should represent a stable amount and form of SOM, developed under the same climatic conditions and with the same mineralogy. The introduction of crop and pasture species, cultivation, fertilizers, irrigation etc. can then be compared with this area. The depth, or depths, to which soil samples are taken will vary with the way CMI is to be used, the soil type, the cultivation systems, the rooting depths etc. Clearly the same depth(s) must be used for the sample and reference soils. There is no value of CMI that can be considered good or bad. The index provides a measure of the rate of change of the system relative to a comparatively more stable reference area. It should be the objective of agricultural managers to increase CMI and hence provide the system with greater resilience.

Table 21.1. Labile and total C and carbon management indices for some cropped and uncropped soils in New South Wales, Australia (from Lefroy *et al.*, 1993).

Location	Cropping or grazing history	C_T (mg g^{-1})	C_L (mg g^{-1})	CPI	LI	CMI
Nyngan (Solonized brown earth, Palexeralf)						
	Uncropped (reference)	18.03	4.50	1	1	100
	4 year crop	12.59	2.21	0.70	0.64	45
Gunnedah (Black earth, Pellustert)						
	Grazed (reference)	20.45	3.62	1	1	100
	7 year crop	10.73	1.55	0.52	0.79	41
Warialda (Red earth, Paleustalf)						
	Light grazing (reference)	16.77	3.99	1	1	100
	18 year crop	7.51	1.02	0.45	0.50	23
	16 year crop and 2 year lucerne	9.13	1.62	0.54	0.69	38

C_T, Total carbon.
C_L, Labile carbon.
CPI, Carbon pool index.
LI, Lability index.
CMI, Carbon management index.

Utility of CMI in a range of agricultural systems

Soil samples were analysed from three cropped and non-cropped sites from central New South Wales (Lefroy *et al.*, 1993) using an uncultivated site as the reference. In all three soils, the reduction in C_L due to cropping was proportionally greater than the decline in C_{NL} or C_T (Table 21.1). Averaged over the three soils the declines were 63.3%, 39.3% and 44.9% for C_L, C_{NL} and C_T, respectively. Incorporation of 2 years lucerne into the cropping rotation at Warialda partly restored soil C, with the increases being 58.8%, 15.7% and 21.6% for C_L, C_{NL} and C_T, respectively. These data demonstrate that C_L declines faster and is restored faster than C_{NL} or C_T, hence is a more sensitive indicator of the C dynamics of the system.

Data from the two soils from Mackay, Queensland, cropped to sugarcane (Table 21.2) provide a contrast. The soil from Marian had been cropped for 90 years and has a markedly lower C_L, C_{NL} and C_T than the reference (uncropped) soil. By contrast, the Victoria Plains soil which had

been cropped for 15 years has a higher C_L, C_{NL} and C_T concentration than the adjacent non-cropped reference soil. Green trash management has been practised in this area for some time, replacing the practice of trash burning. As such, although both sites have had similar periods of green trash management, the proportion of green trash to burnt trash management is much lower for the Marian site.

In the data from Brazil (Table 21.3) similar trends in the C fractions are evident, with an increase of 39.7%, 2.4% and 8.5% in C_L, C_{NL} and C_T, respectively, over the 12-month period after mulch return. Ball-Coelho *et al.* (1993) reported no significant increase in C_T in the mulched treatment reported here. As for the other sites, the C_L data indicates a marked change in the amount of labile or active C in the soil.

Measurement of total (C_T) and labile (C_L) carbon from the surface layer of two soils cropped continuously to sugarcane for 60 and 22 years, were compared with an adjacent forest area. Soil samples were collected from the sites described by Cerri *et al.* (1985, 1994). The decline in C_T was

Table 21.2. Labile and total C and carbon management indices from sugarcane cropped and adjacent non-cropped areas of Mackay, Queensland (from Blair *et al.*, 1995).

Location	Cropping or history (years)	C_T (mg g^{-1})	C_L (mg g^{-1})	CPI	LI	CMI
Marian (Yellow podzolic, Haplustalf)						
	0 (reference)	14.99	4.08	1	1	100
	90	8.55	1.54	0.57	0.59	34
Victoria Plains (Black earth, Pelloxerert)						
	0 (reference)	18.78	3.56	1	1	100
	15	23.69	4.00	1.26	0.87	110

C_T, Total carbon.
C_L, Labile carbon.
CPI, Carbon pool index.
LI, Lability index.
CMI, Carbon management index.

Table 21.3. Carbon data and CMI for soil samples from a mulch return treatment of a sugarcane experiment conducted on a red yellow latosolic podzolic (oxic haplustult) soil by Ball-Coelho *et al.* (1993) in Brazil (from Blair *et al.*, 1995).

Time after mulch return (months)	C_T (mg g^{-1})	C_L (mg g^{-1})	CPI	LI	CMI
0 (reference)	7.34	1.21	1	1	100
12	7.97	1.69	1.09	1.36	148

C_T, Total carbon.
C_L, Labile carbon.
CPI, Carbon pool index.
LI, Lability index.
CMI, Carbon management index.

accompanied by an even greater decline in C_L (Table 21.4). The samples from these sugarcane soils in Brazil (Table 21.4), provide an opportunity to use the ratio of ^{13}C-to-^{12}C in assessing the proportion of carbon derived from the forest (predominantly C3) and the proportion derived from sugarcane (C4, δ^{13}C= -12‰). The ^{13}C-to-^{12}C ratio in SOM is comparable to that of the source plant material (Schwartz *et al.*, 1986), and thus every change in vegetation between C3 and C4 plants leads to a corresponding change in the ^{13}C-to-^{12}C value of SOM. This principle has been used by Schwartz *et al.* (1986) to study changes in vegetation in the Congo,

Skjemstad *et al.* (1990) in studying the turnover of SOM under pasture and Lefroy *et al.* (1993) on changes in SOM as a result of cropping. The δ^{13}C values indicate that the reduced amount of carbon in the sugarcane soils is predominantly derived from sugarcane, particularly after 60 years of continuous sugarcane cultivation (Table 21.4). This demonstrates that cultivation not only increases the turnover of recently added residue but of the existing SOM. It is likely that the small amount of 'forest' SOM has a very slow turnover rate, almost to the extent of being inert.

The utility of the CMI has been evaluated in a long term cropping experiment in

Table 21.4. Effect of sugarcane cropping on C status of soil from São Paulo State, Brazil.

	Forest (reference)	22 years sugarcane	60 years sugarcane
C_T (mg g^{-1})	43.34	13.27	16.18
C_L (mg g^{-1})	10.96	2.45	2.73
CPI	1	0.31	0.37
LI	1	0.67	0.60
CMI	100	20	22
$\delta^{13}C(^{o}/_{oo})$	−22.56	−16.51	−12.31
% Forest C	100	43	3

C_T, Total carbon.
C_L, Labile carbon.
CPI, Carbon pool index.
LI, Lability index.
CMI, Carbon management index.

Australia which was established in 1921. It contains seven rotations which are made up of various sequences of maize, spring oats, autumn oats, red clover and fallow. Oaten hay is cut and removed, maize grain removed and stover returned and red clover ploughed in. Data from three of the rotations are included in Tables 21.5 and 21.6. Soil samples were collected from the plots in 1994 after the maize and red clover had been incorporated. The reference soil sample was taken from an adjacent improved pasture which had never been cultivated. The input of recalcitrant material and the lack of cultivation resulted in a reference material with high C content (47 mg g^{-1}), but low lability (0.199). The cropping regimes over the 12 years rotation period and the changes in maize yield from the first cycle (1922–1933) to the fifth cycle (1970–1981) are presented in Table 21.5.

There was a significantly lower C_T concentration in the soil in the maize/spring oats rotation (M/SO) compared with the rotations where red clover was included (M/SO/RC and M/M/SO/RC) (Table 21.6). There was no significant difference in C_T between the M/SO/RC and M/M/SO/RC rotations. Despite there being no difference in C_T

between these two rotations there was a substantially lower C_L in the M/M/SO/RC rotation. These changes mean that the CPI is similar in the rotations containing red clover and these are significantly different from the M/SO rotations. The higher C_L concentrations in M/SO/RC rotation is reflected in the substantially higher LI in this treatment. When CPI and LI are combined to give CMI there is a significant difference between treatments in the order M/SO/RC >M/M/SO/RC >M/SO. These differences in CMI are also reflected in soil total N. It is estimated that in the M/M/SO/RC rotation approximately 27.4 t of maize stover had been incorporated into the plots in the previous 12 years. This compares with 11.53 t in the M/SO/RC rotation and 9.75 t in the M/SO rotations. The higher return of corn stover in the M/M/SO/RC rotation is countered, to some extent, by the shorter period of red clover and longer fallow period in this rotation. Coupled with this are additional cultivations associated with the double maize cropping which have lead to there being no significant difference in C_T between the two red clover rotations but a higher lability of the carbon where a longer period of red clover and fewer cultivations are practiced in the M/SO/RC rotations.

Table 21.5. Cropping sequences used in Glen Innes crop rotation experiment.

	Crops or months in 12 year rotation		
	Maize/ Spring oats (M/SO)	Maize/Maize/ Spring oats/ Red clover (M/M/SO/RC)	Maize/ Spring oats/ Red clover (M/SO/RC)
Maize (crop)	6	6	4
Oats (crop)	6	3	4
Clover (months)	0	45	60
Fallow (months)	72	42	36
Average maize yield (t ha^{-1})			
1922–1933 (cycle 1)	2.05	2.42	2.73
1970–1981 (cycle 5)	0.93	2.00	2.52
% Reduction (1 to 5)	54	17	7

Table 21.6. Effect of crop rotations on C dynamics in the Glen Innes crop rotation experiment.

Rotation			
	M/SO	M/M/SO/RC	M/SO/RC
C_T (mg g^{-1})	13.74 [b]	19.70[a]	20.71[a]
C_L (mg g^{-1})	2.83[c]	3.82[b]	6.28[a]
C_{NL} (mg g^{-1})	10.91[c]	15.88[a]	14.43[b]
CPI	0.291[b]	0.416[a]	0.438[a]
LI	1.30[b]	1.21[b]	2.19[a]
CMI	37.8[c]	50.3[b]	95.6[a]
% N	0.121[c]	0.218[b]	0.247[a]

In each row, numbers followed by the same letter are not significantly different using Duncan's multiple range test ($P < 0.05$).
M, maize; SO, spring oats; RC, red clover.
C_T, Total carbon.
C_{NL}, Non-labile carbon.
C_L, Labile carbon.
CPI, Carbon pool index.
LI, Lability index.
CMI, Carbon management index.

Conclusion

In this chapter we have argued the central role of carbon in sustainability of agricultural systems. What is evident from the literature and from a wide range of experiments is that the use of total carbon as an index of soil health and productivity is of limited value as indeed is the use of total nutrient concentrations in soils. Adaptation of the KMnO$_4$ oxidation method of Loginow et al. (1987) by Blair et al. (1995) enabled a Carbon Management Index (CMI) to be calculated. This has been used in a range of agricultural systems and been demonstrated to be a sensitive indicator of the rate of change of the carbon dynamics in these soils. The C_L

and C_{NL} components of CMI allow a two pool C model to be measured and modelled. Work is progressing on an extension of this to a three pool model using Walkley Black C without the correction factor. This potentially allows a highly labile pool (present C_L), a less labile pool (non-corrected Walkley Black $C-C_L$) and a C resistant pool (C_T by catalytic combustion – non-corrected Walkley Black C).

Aiming to increase soil organic matter results in a quandary. Crops have been selected for increasing yield which is normally associated with increasing harvest index. Crop residues have also generally improved in quality (animal digestibility and nutrient content) along with this yield increase. Such increased offtake of carbon and nutrients in yield and higher quality of the residues that are returned to the agricultural system are mutually exclusive to the accumulation of soil organic matter. Material which will add significantly to this SOM pool is that which is essentially indigestible by animals (i.e. Δ SOM pool = Residue added $\times$ (1 −digestibility)). It is evident that crop types and management strategies in warm moist environments need to be reconsidered. What is needed are crops of high yield but with residues that have slow to moderate breakdown rates. This may mean re-engineering plants to meet these goals. In addition, standing organic matter, and that on the surface of the soil will break down more slowly and contribute a more sustained input of carbon and nutrients to the agricultural system than those residues incorporated into the soil. This emphasizes the importance of minimum cultivation techniques in agricultural systems.

In summary, the CMI developed in this chapter allows a sensitive measure of rates of change in the carbon status of agricultural systems and is recommended for evaluation in a wider range of agro-ecological zones.

References

Ball-Coelho, B., Tiessen, H., Stewart, J.W.B., Salcedo, I.H. and Sampaio, E.V.S.B. (1993) Residue management effects on sugarcane yield and soil properties in Northeastern Brazil. *Agronomy Journal* 85, 1004–1008.

Blair, G.J., Lefroy, R.D.B. and Lisle, L. (1995) Soil carbon fractions based on their degree of oxidation and the development of a carbon management index for agricultural systems. *Australian Journal of Agricultural Research* 46, 1459–1466.

Cerri, C., Feller, C., Balesdent, J., Victória, R. and Plenecassagne, A. (1985) Application du traçage isotopique naturel en [13]C a l'étude de la dynamique de la matiére organique dans les sols. *Comptes Rendus de l'Academie des Sciences de Paris* T.300, Serie II, 9, 423–428.

Cerri, C.C., Bernoux, M. and Blair, G.J. (1994) Carbon pools and fluxes in Brazilian natural and agricultural systems and the implications for the global CO_2 balance. In: *Proceedings of 15th International Congress of Soil Science*. Acapulco, Mexico 10–16 July 1994, Volume 5a. International Soil Science Society, pp. 399–406.

Hayes, M.H.B. and Swift, M.J. (1978) The chemistry of soil organic colloids. In: Greenland, D.J. and Hayes, M.H.B. (eds) *The Chemistry of Soil Constituents*. John Wiley & Sons, Chichester, UK, pp. 179–320.

Jenkinson, D.S. (1981) The fate of plant and animal residues in soils. In: Greenland, D.J, and Hayes, M.H.B. (eds) *The Chemistry of Soil Processes*. Wiley, Chichester, pp. 505–561.

Lefroy, R.D.B. and Blair, G.J. (1994) The dynamics of soil organic matter changes resulting from cropping. In: *Transactions of the 15th World International Congress of Soil Science*. Acapulco, Mexico 10–16 July 1994, Volume 9. International Soil Science Society, pp. 235–245.

Lefroy, R.D.B., Blair, G.J. and Strong, W.M. (1993) Changes in soil organic matter with cropping as measured by organic carbon fractions and ^{13}C natural isotope abundance. *Plant and Soil* 155/156, 399–402.

Loginow, W., Wisniewski, W., Gonet, S.S. and Ciescinska, B. (1987) Fractionation of organic carbon based on susceptibility to oxidation. *Polish Journal of Soil Science* 20, 47–52.

McCaskill, M.R. and Blair, G.J. (1989) A model for the release of sulfur from elemental S and superphosphate. *Fertilizer Research* 19, 77–84.

Parton, W.J., Schimel, D.S., Cole, C.V. and Ojima, D.S. (1987) Analysis of factors controlling soil organic matter levels in Great Plains grasslands. *Soil Science Society of America Journal* 51, 1173–1179.

Schwartz, D., Mariotti, A., Lanfranchi, R. and Guillet, B. (1986) $^{13}C/^{12}C$ ratios of soil organic matter as indicators of vegetation change in the Congo. *Geoderma* 39, 97–103.

Skjemstad, J.O., Le Feuvre, R.P. and Prebble, R.E. (1990) Turnover of soil organic matter under pasture as determined by ^{13}C natural abundance. *Australian Journal of Soil Research* 28, 267–276.

Swift, M.J., Kang, B.T., Mulongoy, K. and Woomer, P. (1991) Organic-matter management for sustainable soil fertility in tropical cropping systems. In: *Evaluation for Sustainable Land Management in the Developing World. Proceedings of International Workshop.* Chiang Mai, Thailand 15–21 Sep. 1991. IBSRAM Proceedings No. 12 (2). IBSRAM, Bangkok, Thailand, pp. 307–326.

Syers, J.K. and Craswell, E.T. (1995) Role of soil organic matter in sustainable agricultural systems. In: Lefroy, R.D.B., Blair, G.J. and Craswell, E.T. (eds) *Soil Organic Matter Management for Sustainable Agriculture.* Ubon, Thailand 24–26 August 1994. ACIAR Proceedings No. 56. ACIAR, Canberra, ACT, pp. 1–14.

Tian, G., Brussaard, L. and Kang, B.T. (1995) An index for assessing the quality of plant residues and evaluating their effects on soil and crop in the (sub-)humid tropics. *Applied Soil Ecology* 2, 25–32.

Whitbread, A.M. (1995) Soil organic matter: its fractionation and role in soil structure. In: Lefroy, R.D.B., Blair, G.J. and Craswell, E.T. (eds) *Soil Organic Matter Management for Sustainable Agriculture.* Ubon, Thailand 24–26 August 1994. ACIAR Proceedings No. 56. ACIAR, Canberra, ACT, pp. 124–130.

22 Long-term Vegetation Management in Relation to Accumulation and Mineralization of Nitrogen in Soils

J.Z. Burket and R.P. Dick*

Department of Crop and Soil Science, Oregon State University, Corvallis, OR 97331-7306, USA

Introduction

Approximately 86% of the nitrogen (N) used for primary plant production in world agriculture comes from the mineralization of N in soil organic matter (Jenny, 1980). How plant litter inputs affect the nature of soil organic N and its N availability for future crop production is an increasingly important area of agricultural research.

Proportional differences among N fractions vary according to climatic zone (Stevenson, 1982a), with amino acid N and amino sugar N having the highest proportions in sub-tropical soils and ammonium N being highest in arctic soils. Although soil organic matter C-to-N ratios vary with geographic moisture regimes (Zinke *et al.*, 1984), widely diverse ecological factors such as climate and vegetation form a fairly similar suite of soil organic matter fractions in terms of humic and fulvic acids (Sowden, 1968; Schnitzer, 1977). Within particular environments, vegetation type can be the ultimate controller of soil C-to-N ratios, as observed in grasslands and pine groves in California (Jenny, 1980) and during succession of natural vegetation in the Netherlands (Berendse, 1990).

In agricultural as well as natural ecosystems, vegetation exerts the primary influence on the character of soil N, so it is expected that crop and soil management history will have a profound effect on organic N dynamics. Cultivation of soils, especially when organic residues are not returned to the soil, results in the loss of soil organic C and N (Rosswall and Paustian, 1984) and decreases in biological and biochemical properties in soils (Dick, 1992).

The focus of this chapter is to consider long-term plant residue management effects on both indigenous soil N and, the retention and mineralization of recently added organic N in agricultural soils.

Accumulation of Total C and N

Within a given climate, the amount of organic C that accumulates in aerobic agricultural soils appears to have less to do with the type of organic residue than with the amount of C that is returned. In semi-arid regions, organic C levels have been closely correlated with plant C inputs, regardless of whether the source of C was wheat straw, legume green manure, or animal manure (see review by Rasmussen and Collins, 1991; Campbell and Zentner, 1993). In Michigan USA, accumulation of soil organic C from a variety of crops in rotations such as corn, oats, sugarbeet, and navy beans after 9 years (Zielke and

* Corresponding author.

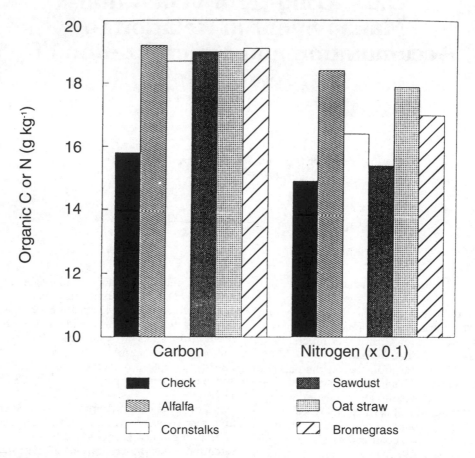

Fig. 22.1. Distribution of soil organic C and N for the check (control) and five types of residues applied at 8 t ha⁻¹ year⁻¹ for 11 years under continous maize production (from Larson *et al.*, 1972).

Christenson, 1986) was closely related to plant biomass additions to soils.

Total N accumulation has been related to biomass inputs but can be moderated by vegetation type and N fertilization. The impact of inorganic fertilizers on soil organic C is not necessarily correlated with total N levels in soil. An example of this is shown in Fig. 22.1 where various organic residues applied to a soil in Iowa USA over 11 years showed no differences in total C accumulation but did show a difference in total N accumulation in soils (Larson *et al.*, 1972).

In general, N rich residues such as legumes will cause an accumulation of more soil organic N than non-N$_2$ fixing plants (Gupta and Reuszer, 1967). However, the growth of non-legumes can be stimulated by N fertilizer (thus enriching plant residues with N and increasing the mass of organic N inputs). Also, addition of N fertilizer can contribute directly to soil organic N pools through N immobilization. This has been shown in Canada (Campbell *et al.*, 1991b) where N fertilizer in combination with continuous wheat, legume green manures, or legume–grass hay crops maintained or increased organic N (Fig. 22.2). Conversely, Gupta and Reuszer (1967), after 7 years of cropping, found that annual alfalfa harvesting without added fertilizer N had equal or slightly higher soil N levels

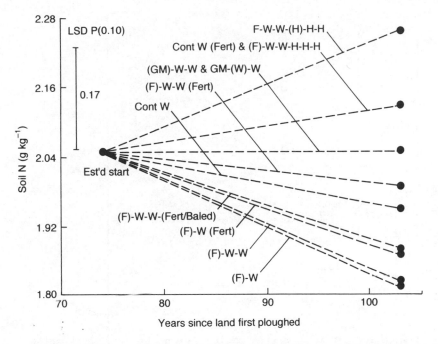

Fig. 22.2. Effects of rotation practices on N concentration in the top 15 cm of soil after 30 years. F = fallow; W = spring wheat; GM = sweet clover turned under for green manure; H = bromegrass–alfalfa cut for hay; Cont. = continuous; ( ) = phase sampled. Treatments not labelled *Fert* were unfertilized. Estimated start is N concentration at start of experiment (1958) derived by interpolation from a plot of literature data (from Campbell *et al.*, 1991d).

than corn or bromegrass that received more than 200 kg N ha^{-1} year^{-1}.

The net stabilization of the fertilizer N in soils is likely to be controlled by climate and soil type. In cooler, moister climates, N reserves tend to increase with N fertilization due to relatively lower rates of organic matter decomposition in these environments. There is relatively little information from long term studies (>15 years) on effects of plant residues on N dynamics in tropical agricultural systems, but some generalizations may be made from several shorter term studies. Goyal *et al.* (1992) found an increase in total soil C after 13 years with only the highest rates of N and P in a pearl millet–wheat rotation in sub-tropical, semi-arid India. But increases in microbial C and N, and total N paralleled yield increases and occurred with N fertilizer alone or with one half the highest rates of N and P. The rapid turnover of

organic N in tropical systems is also indicated in a 2-year study in which N-use efficiency by upland rice in Indonesia was about twice as high with a cowpea residue N source than with fertilizer N (Sisworo *et al.*, 1990).

Distribution of N Fractions

Comparative studies of cultivated and virgin soils have shown about a 30–40% loss of total N in cultivated soils (Smith and Young, 1975; Meints and Peterson, 1977). These losses have been characterized with a fractionation scheme which involves hot acid hydrolysis followed by analysis for various N fractions such as amino acids and amino sugars (Stevenson, 1982b). Keeney and Bremner (1964) examined the N distribution in cultivated and virgin plots in each of ten different soils and

Table 22.1. Effect of cultivation and cropping system on the distribution of the forms of N in soil (adapted from Stevenson, 1994).

Location*	Acid insoluble	Amino NH$_3$	Amino acid	sugar	HUN[†]	Reference
Alberta, Canada (Brenton plots)[‡]						
Rotation of grains and legumes (6)	21.1	15.1	30.9	10.4	22.6	Khan and Sowden (1971)
Wheat–fallow sequence (6)	25.0	17.8	28.8	9.3	19.2	
Germany[§]						
Grass sod	16.4	27.6	27.6	4.2	24.2	Fleige and Beaumer (1974)
Arable (tilled)	16.2	32.1	22.1	4.2	25.4	
Illinois, USA						
(Morrow plots)[¶] Grass border and COCL rotation (2)	20.3	16.6	42.0	10.5	10.7	Stevenson[‖]
Continuous corn and CO rotation (2)	20.2	16.7	35.0	14.4	13.9	
Iowa, USA						
Virgin (10)	25.4	22.2	26.5	4.9	21.0	Keeney and Bremner (1964)
Cultivated (10)	24.0	24.7	23.4	5.4	22.5	
Nebraska, USA[**]						
Virgin (4)	20.8	19.8	44.3	7.3	7.8	Meints and Peterson (1977)
Cultivated (4)	19.3	24.5	35.8	7.0	13.4	

Form of N (%)

* Numbers in parentheses indicate number of soils.
[†] Hydrolyzable unidentified N.
[‡] Treatments for each sequence include a control, manure plot, NPKS plot, NS plot, lime plot, and a P plot.
[§] Average for 0–5, 5–10, and 10–15 cm depths.
[¶] COCL = corn, oats, clover rotation with lime and P additions. CO = corn–oats rotation.
[‖] Unpublished observations.
[**] Soils of the Ustoll suborder.

Table 22.2. Relationship of amino acid content of plant residues to amino acid content of soil under long-term crop rotations at a site in Saskatchewan, Canada, where W is wheat, H is bromegrass–alfalfa cut for hay, F is fallow and letter in parenthesis is rotation phase when soils were sampled (adapted from Campbell *et al.*, 1991d).

| | Rotation soil (kg ha^{-1}) | | | Rotation soil (kg ha^{-1}) | | |
	Cont W	Cont W (fert)	Wheat straw residue (g ha^{-1})	(F)-W-W -H-H-H	F-W-W- (H)-H-H	Alfalfa (g ha^{-1})
Total amino acids	966	1086	4.16	1119	1005	21

found the average losses in the cultivated plots to be 43, 39, 36, 35, 35, 29, and 28% for the N fractions of amino acid, non-hydrolysable, total, total hydrolysable, unidentified hydrolysable, hydrolysable ammonium, and hexosamine-N, respectively. Long-term cultivation may sometimes cause a slight increase in hydrolysable NH_3, but amino acids are the most sensitive N fraction to soil management and generally decreases with cultivation (Table 22.1; Stevenson, 1994). However, neither long-term cropping nor organic amendments greatly affects the relative distribution of N fractions in soils.

Amino acids are prevalent and persist in soils as demonstrated by their presence in paleosols (Goh, 1972). Although persistent, amino acids are also dynamic in soils and laboratory incubations have shown that at least a portion of this fraction is readily mineralized under aerobic conditions (Keeney and Bremner, 1966). The nature of amino acid N in soils may be a useful tool in assessing the influences of soil management on mineralizable N.

Stevenson (1956) showed that the amino acid composition of soils from the long-term Morrow plots in Illinois differed with crop rotation and soil amendments. Soil from non-manured continuous corn and corn–oats rotations contained proportionally higher amounts of basic amino acids such as lysine, ornithine, and arginine than soil from the manured corn–oats–clover rotation, and the grass border soil contained higher amounts of neutral amino acids such as leucine, serine, and alanine (Stevenson, 1956). The polar nature of the basic amino acids may make

them more resistant to complete mineralization, which in turn would explain their predominance in soils with lower organic inputs. Studies of the Morrow plots indicate that soils that have residue removed and no organic amendments are depleted in all but the most resistant organic N compounds such as the basic amino acids and apparently need annual organic inputs for maintaining the more labile amino acids. Amino acid content of the plant residues would seem to be an important factor in determining the amino acid content of soils. This could be indirect by affecting microbial synthesis of soil N fractions including amino acid content or through direct additions of amino acids from plant residues to the soil. Analyses of crop residues have shown that alfalfa has three to four times the amino acid content of grasses and about five times that of wheat straw (Table 22.2) (Campbell *et al.*, 1991d). Long-term application (31 year) of these materials through crop rotations at a site in Saskatchewan, Canada showed no significant effects on the quality of the N fraction distribution in soils. Unlike Gupta and Reuszer, (1967) who found that alfalfa caused significantly higher total amino acids in soil than corn or bromegrass, Campbell *et al.* (1991c) were unable to draw any firm conclusions whether plant quality as indicated by amino N content can have any long-term effect on N fractions.

It seems plausible, as shown by Campbell *et al.* (1991d), that recent additions of plant residues of varying quality can affect amino-N distributions but this may be seasonal or have only short-term effects. This hypothesis is supported by early work of Kuo and Bartholomew (1966) who showed most organic N in soil is of microbial, not plant, origin which would account for the dynamic nature of soil amino acids. Furthermore, current theories indicate that humic substances are formed largely of microbial decomposition products of lignin and cellulose such as polyphenols or quinones that condense with amino groups of amino acids, peptides, and proteins originating from the cell walls of microorganisms (Stevenson, 1994).

Several studies that examined soils from long-term plots of durations between 20 and 40 years report no differences in quality of soil amino acid composition due to cropping system (Young and Mortensen, 1958; Sowden, 1968; Kahn and Sowden, 1971). These differences in findings among studies on the interactions of soil amino acids and cropping systems may be related to differences in environment, soil parent material, or other unknown factors. Further research is needed on methods of fractionation of N that are more sensitive in discriminating N pools that can be related to N mineralization–immobilization processes in relation to soil management.

Microbial Biomass N

Physical properties affecting soils such as pore size distribution and wetting and drying regimes form the primary backdrop for determining the character of soil microbial populations (van Veen et al., 1984), and within these constraints, vegetation management will also influence soil microbial biomass. Higher soil organic matter levels are directly correlated to higher microbial counts, biomass, and activities (Schnürer et al., 1985). Crop rotations that include legumes or manure amendments have higher microbial activities and microbial biomass C levels than do monoculture soils with little or no organic matter inputs (Dick, 1992).

A key component of N mineralization is microbial biomass N (MB_N). It is highly correlated to readily mineralizable organic N in soils (Kai et al., 1973; Myrold, 1987) and to rates of N mineralization (Alef et al., 1988; Bonde et al., 1988; Fisk and Schmidt, 1995), and is a significant source of N nutrition for plants (Lethbridge and Davidson, 1983). Nitrogen availability is likely to be controlled by microbial biomass turnover (Holmes and Zak, 1994). Consequently, long-term effects of plant residues on the activity and size of the MB_N pool is an important consideration in N mineralization.

Crop rotations that include a fallow have significantly lower MB_N than either continuous monocropping or more complex crop rotations (Biederbeck et al., 1984; McGill et al., 1986; Campbell et al., 1991a). It follows that the amount of long-term organic C and N inputs control the MB_N levels. It is less clear whether the quality of plant residues affects MB_N levels. This is partly due to the fact that studies rarely characterize plant residues added to soils for quality beyond C-to-N ratio. However, if one assumes that a legume is a higher quality residue (narrow C-to-N ratio) than a non-legume, some inferences can be made on the long-term effects of 'higher quality' residues such as legumes on MB_N. Studies in Canada (Campbell et al., 1991a) and the southern USA (Franzluebbers et al., 1995) have shown that in the absence of N fertilization, crop rotations that include a legume have substantially higher MB_N. The diminished levels of MB_N with 'lower quality' inputs where non-legumes dominate can, in general, be brought to similar levels as those in rotations that include legumes by the addition of N fertilizer in temperate regions (Campbell et al., 1991a; Harris et al., 1994; Franzluebbers et al., 1995). Similar results have been found in subtropical regions where N fertilization stimulated MB_N (Goyal et al., 1992; Singh, 1995). The added fertilizer N may increase MB_N by increasing N content and amount of the non-legume residue that is returned to the soil and by being immobilized directly into soil MB_N. These results suggest that stimulation of MB_N may have less to do with residue quality than with the total amount of N that is incorporated into the soil as organic or inorganic N.

Nitrogen Mineralization Potential

The ability of a soil to provide plant available N from its store of organic N will be determined by the quantity and quality of organic residue inputs to the soil. The characterization of mineralization was first presented by Stanford and Smith (1972) who developed an aerobic incubation/leaching method to develop cumulative N mineralization curves. Various mathematical models have been fitted to these data (zero order or first order kinetics) in order to calculate N mineralization potential (N_0) and the rate constant (k). The N_0 (mg N kg^{-1}) is defined as the quantity (capacity factor) of the total N in a soil at time zero that is available for mineralization as affected by soil genesis and management whereas k is supposed to be a true rate constant (week^{-1}).

Although cross study comparisons of N mineralization constants are not generally appropriate because of differences in soil pretreatment (e.g. air-dried versus field moist) and use of different models, these constants are sensitive in detecting field treatments. For example, field applications of various manures on a range of soils showed groupings of N_0 as follows: steer manure>poultry manure>sewage sludge (Griffin and Laine, 1983). Recent and continuous additions of organic residues apparently are important in establishing and maintaining N_0 levels. Evidence for this was shown by Janzen (1987) who found that in soils from long-term rotations, the rotation with the highest plant inputs had the highest N_0 values and total N mineralized. This was closely related to the organic light fraction content of the soils but relatively unrelated to total N or C content of the soils. Furthermore, there was a near perfect (negative) correlation between degree of fallowing and N mineralized. Other long-term studies are consistent with this hypothesis where elevated plant contributions from rotations

with legumes, green manures, and hay crops (Campbell et al., 1991c) or soil amendments of steer manure or pea vine residue (Christ, 1993; Dick and Christ, 1995) significantly increased N_0 over systems without these elevated organic inputs. On 19 soil types, cumulative N mineralized and microbial activities were higher in native soil from forest, pasture, wetland, or green manured sites than their pair-wise comparison with the cultivated sites (J.Z. Burket and R.P. Dick, Oregon, unpublished results).

Nitrogen mineralization dynamics are also affected by residue quality. Indirect evidence for this was provided by Beauchamp et al. (1986) who found N_0 to be significantly higher in soils under bromegrass or alfalfa than in soils under corn after 15 years of cropping. Yet from this and the other studies mentioned above, it is difficult to separate the effects of qualitative differences in the organic residue from the effect of the amount of biomass and/or N inputs on N mineralization. This was addressed in a unique study by Bonde et al. (1988) where they compared N mineralization from soils of bare ground, cropped, cropped + N fertilizer (80 kg N ha^{-1} year^{-1}), 1800 kg straw-C ha^{-1} year^{-1} + N fertilizer (80 kg N year^{-1} year^{-1}), and 1800 kg farm yard manure (FYM)-C ha^{-1} year^{-1} + N fertilizer (80 kg N ha^{-1} yr^{-1}). The highest amount of N mineralization and N_0 values occurred in the 1800 kg straw-C or FYM treatments. Using a two-component model they were further able to show that the straw treatment had a lower proportion of N_0 in the recalcitrant fraction than did the FYM treatment. These results are consistent with other studies that have shown FYM to be resistant to mineralization (Castellanos and Pratt, 1981; Fauci and Dick, 1994a).

Long-term addition of inorganic N in the absence of other organic amendments can increase the amount of N mineralized and N_0 values (El-Harris et al., 1983; Bonde et al., 1988). This effect is due to increased N-rich plant biomass additions to soils and N fertilizer being directly immobilized in the soil over the summer when the microbial activity is highest. The extent to which cropping or vegetative history influences soil mineralizable N is a function of both a 'non-living soil mineralizable N pool' and the size and activity of the soil biomass.

Nitrogen Mineralization of Plant Residues

From the above discussion, it is clear that plant residue management can affect N accumulation and N mineralization potential. This section will address mineralization of N from plant residues added to soils where the soils vary in total N content and N mineralization potential because of past long-term residue management. We found only two studies that have specifically addressed this and both were done under greenhouse conditions. Janzen and Radder (1989) conducted a study on soils obtained from long-term plots (initiated in 1951) in Alberta, Canada that had been under different crop rotations that included continuous spring wheat, wheat–fallow, and wheat in rotations with legumes and hay crops. In the greenhouse, soils from the long-term plots were amended with ^{15}N-labelled Tangier flat pea (*Lathyrus tingitanus* cv. Tinga) and plant N uptake was then determined. Total net N mineralized was highest in soils from the rotations of lentil–wheat, continuous wheat (unfertilized) and native grass and lowest in the rotations with fallow–wheat (fertilized with N) and fallow–wheat with 3 years of alfalfa–grass forage (Fig. 22.3). Although there were significant differences in net N mineralization of flatpea as a function of soil history, these differences were small and most of the differences observed in total net N mineralization were accounted for by differences in indigenous soil N mineralization.

In a study by Fauci and Dick (1994a, b), soils were collected from the long-term residue utilization plots (wheat–fallow system; Oregon). The four soils received the following amendments since 1931: steer manure (22 t ha⁻¹ 2 year⁻¹); pea (*Pisum*

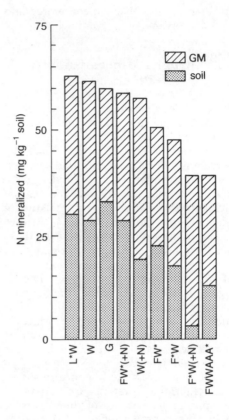

Fig. 22.3. Influence of cropping history on net nitrogen mineralization from indigenous soil and recently added green manure (GM). Effects on soil, green manure, and total mineralization all significant at $P = 0.0001$. Corresponding LSD ($P = 0.05$) values are 9.2, 4.0 and 10.4 respectively. Cropping treatments: F = fallow, W = spring wheat, L = lentil, A = alfalfa/crested wheat grass forage, G = mixed native grasses. The asterisk indicates the phase at which the rotation was sampled. (For example, F*W indicates that the rotation was sampled just before establishment of the spring wheat.) (+N) indicates that fertilizer N was applied in the W phase of the rotation. Soil N was measured at harvest of plants (from Janzen and Radder, 1989).

sativum L.) vine green manure (2.2 t ha⁻¹ 2 year⁻¹); 90 kg N ha⁻¹ 2 year⁻¹; or no amendment (control). Each of these soils was amended with pea vine, composted steer manure, poultry manure, or control

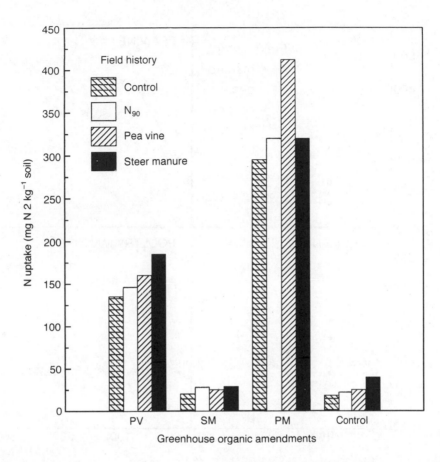

Fig. 22.4. Nitrogen uptake of the first maize crop grown in the greenhouse with the factorial treatments of organic amendments (PV = pea vine, SM = steer manure, PM = poultry manure) and soils from long-term field plots (wheat–fallow system initiated in 1931, Oregon, USA). There was no significant ($P = 0.05$) effect of long-term field history on mineralization of organic residues (adapted from Fauci, 1993).

(organic amendments were added at the rate of 0.5 g N kg^{-1} soil) and had four successive plantings of maize (*Zea mays* L.) over 306 days (Fig. 22.4 shows first maize crop). In the absence of any greenhouse amendment, long-term applications of steer manure had the highest N mineralization which could be related to its higher N content and biological biomass (Dick *et al.*, 1988; Collins *et al.*, 1992). There was no significant effect of field history on N mineralization of any of the organic residues added in the greenhouse (Fig. 22.4). In other words, a soil that had received a particular amendment long-

term did not, in the short-term, have a significant advantage in mineralizing N from the same residue over a soil that had never received that amendment. The type of residue amendment had a much greater effect than did the field history. For example, pea vine and steer manure had similar C-to-N ratios (24 and 21, respectively), but pea vine had much greater short-term N mineralization than steer manure (Fig. 22.4) which was likely due to pea vine having a much lower lignin content (6 vs. 28% in steer manure). This finding was supported by various biological measures such as microbial biomass C

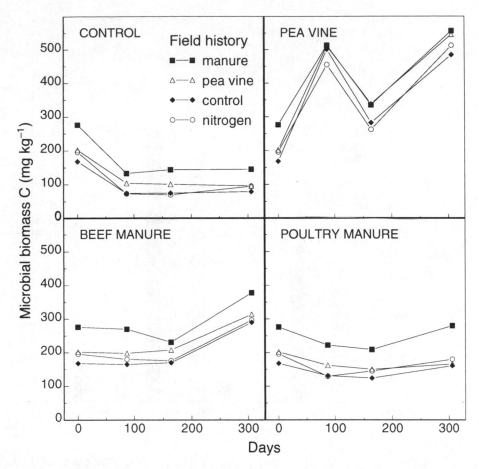

Fig. 22.5. Microbial biomass C (fumigation/incubation method) in soils as a function of previous long-term management history and recent organic amendment, averaged across greenhouse N treatments (n=12). The largest standard error for any treatment combination is 0.15 (from Fauci and Dick, 1994b).

(Fig. 22.5) where pea vine added in the greenhouse had a much larger biological response than did the effect of long-term field residue management.

Both of these studies (Janzen and Radder, 1989; Fauci and Dick, 1994a,b) support a hypothesis that field history through residue or cropping management that results in differences in organic matter or biological levels does not significantly affect rates of N mineralization from recently added organic amendments. Long-term applications of organic residues, though, do appear to affect the mineralization of indigenous soil N.

Clearly, studies from other regions are needed (both of these studies were done on soils from the semi-arid regions of the northern parts of North America) under field conditions to verify these findings.

Perspectives

A review of the literature shows that in general, addition of plant residues can increase soil organic C levels and this increase is much more related to the rates of application than to the type or quality of the residue. Accumulation of total N and various N fractions appears to be more complex. Some types of N-rich plant (e.g. legume and forage species) residues have been shown to increase, maintain, or reduce losses of organic N with intensive cultivation of the soil. But addition of inorganic N in some settings to monoculture cereal or maize plants had the same effect as more complex crop rotations that included hay/legume forage rotations. In the case of inorganic N additions, it is difficult to separate the effects of increased plant biomass production (i.e. greater N uptake in plant residue) from the effect of immobilizing inorganic N directly in the soil organic matter. Furthermore, there were differing effects of plant residues between some locations. Work from Canada by Campbell and co-workers suggested that climate can affect the impact of plant residues, particularly moisture levels, where a drier climate that limits plant growth reduces potential for crop residue management to affect organic N content.

Nitrogen fractionation after acid hydrolysis can differentiate between plant residue management and plant quality (at least with short-term effects on amino N distribution). However, in general this fractionation scheme is relatively insensitive to subtle changes in the distribution of organic N as a function of plant residue management. More sensitive methods are needed to extract the active N fraction that is susceptible to mineralization/immobilization processes in soils.

Clearly, long-term vegetation management has a significant and measurable impact on the activity and size of the soil microbial biomass. It is an important site of N mineralization and source of plant available N. The elevated levels of soil biological properties undoubtedly is related to increases in mineralization of N in soils with long-term organic residue inputs. However, work by Fauci and Dick (1994b) suggests that this elevated level of biological potential resulting from long-term applications has no particular advantage in mineralizing N from new additions of organic amendments.

Measuring N mineralization with the incubation method of Stanford and Smith (1972), combined with calculations of the N mineralization potential (N_0) and the rate constant (k), is relatively sensitive in discriminating between various crop rotations (differing in biomass or N contributions) and other organic amendments such as animal manures. The sensitivity of this method is shown by the fact that there are seasonal differences in N_0 (decreasing through the summer and increasing in the fall) (Bonde and Rosswall, 1987) with potentially large changes over the season (e.g. two-fold or greater decrease in N_0 between fall and spring) (El-Harris et al., 1983). This provides evidence that plant residues (probably in both quantity and quality) do affect N dynamics. Unfortunately, neither N mineralization parameter (N_0 and k) can adequately distinguish between the effects of soil biology from the effect on the readily mineralizable organic N pool. A surprising outcome of preliminary investigations (Janzen and Radder, 1989; Fauci and Dick, 1994a,b) indicates that soils that have higher mineralization potential do not seem to have greater rates of mineralization of N from organic amendments added to soils. If this is a consistent result, it would mean that greater attention can be placed on the quality of the organic residue (e.g. C-to-N ratios, cellulose and lignin content) than on long-term soil management effects in predicting N mineralization from organic amendments of soils (of course it will still be important to predict N mineralization of indigenous N as a function of long-term management). More extensive studies in various climate, cropping systems and soil types are needed under field conditions to confirm these findings.

In conclusion, long-term management and incorporation of plant residues does have a significant effect on N accumulation and mineralization but to understand the mechanisms that are operating and develop more efficient use of plant N, more sensitive methods for identifying labile N pools are needed. Furthermore, it is difficult to separate the long-term vegetation effects on the activity, size, or diversity of the soil biomass in relation to N mineralization of indigenous soil N. A major limitation in answering these questions is a lack of methodologies that can deal with the chemical and biological complexity of soils. Also, studies at long-term sites are needed that characterize the quality (e.g.

amino acid, cellulose, and lignin content, and C-to-N ratios) of the organic residues in order to relate these inputs to measurable N mineralization parameters and soil organic N pools.

References

Alef, K., Beck, T., Zelles, L. and Kleiner, D. (1988) A comparison of methods to estimate microbial biomass and N-mineralization in agricultural and grassland soils. *Soil Biology and Biochemistry* 20, 561–565.

Beauchamp, E.G., Reynolds, W.D., Brasche-Villeneuve, D. and Kirby, K. (1986) Nitrogen mineralization kinetics with different soil pretreatments and cropping histories. *Soil Science Society of America Journal* 50, 1478–1483.

Berendse, F. (1990) Organic matter accumulation and nitrogen mineralization during secondary succession in heathland ecosystems. *Journal of Ecology* 78, 413–427.

Biederbeck, V.O., Campbell, C.A. and Zentner, R.P. (1984) Effect of crop rotation and fertilization on some biological properties of a loam in southwestern Saskatchewan. *Canadian Journal of Soil Science* 64, 355–367.

Bonde, T.A. and Rosswall, T. (1987) Seasonal variation of potentially mineralizable nitrogen in four cropping systems. *Soil Science Society of America Journal* 51, 1508–1514.

Bonde, T.A., Schnürer, J. and Rosswall, T. (1988) Microbial biomass as a fraction of potentially mineralizable nitrogen in soils from long-term field experiments. *Soil Biology and Biochemistry* 20, 447–452.

Campbell, C.A. and Zentner, R.P. (1993) Soil organic matter as influenced by crop rotations and fertilization. *Soil Science Society of America Journal* 57, 1034–1040.

Campbell, C.A., Bowren, K.E., Schnitzer, M., Zentner, R.P. and Townley-Smith, L. (1991b) Effect of crop rotations and fertilization on soil organic matter and some biochemical properties of a thin Black Chernozem. *Canadian Journal of Soil Science* 71, 377–387.

Campbell, C.A., LaFond, G.P., Letshon, A.J., Zentner, R.P. and Janzen, H.H. (1991c) Effect of cropping practices on the initial potential rate of N mineralization in a thin Black Chernozem. *Canadian Journal of Soil Science* 71, 43–53.

Campbell, C.A., Schnitzer, M., Lafond, G.P., Zentner, R.P. and Knipfel, J.E. (1991d) Thirty-year crop rotations and management practices effects on soil and amino nitrogen. *Soil Science Society of America Journal* 55, 739–745.

Campbell, C.A., Biederbeck, V.O., Zentner, R.P. and LaFond, G.P. (1991a) Effect of crop rotations and cultural practices on soil organic matter, microbial biomass and respiration in a thin Black Chernozem. *Canadian Journal of Soil Science* 71, 363–376.

Castellanos, J.Z. and Pratt, P.F. (1981) Mineralization of manure nitrogen-correlation with laboratory indexes. *Soil Science Society of America Journal* 45, 354–357.

Christ, R.A. (1993) Effect of long-term residue management and nitrogen fertilization on availability and profile distribution of nitrogen, phosphorus, and sulfur. MSc. Thesis, Oregon State University, Corvallis OR, USA.

Collins, H.P., Rasmussen, P.E. and Douglas, C.L., Jr (1992) Crop rotation and residue management effects on soil carbon and microbial dynamics. *Soil Science Society of America Journal* 56, 783–788.

Dick, R.P. (1992) A review, long-term effects of agricultural systems on soil biochemical and microbial parameters. *Agriculture, Ecosystems and Environment* 40, 25–36.

Dick, R.P. and Christ, R.A. (1995) Effects of long-term residue management and nitrogen fertilization on availability and profile distribution of nitrogen. *Soil Science* 159, 402–408.

Dick, R.P., Rasmussen, P.E. and Kerle, E.A. (1988) Influence of long-term residue management on soil enzyme activities in relation to soil chemical properties of a wheat-fallow system. *Biology and Fertility of Soils.* 6, 159–164.

El-Harris, M.K., Cochran, V.L., Elliott, L.F. and Bezdicek, D.F. (1983) Effect of tillage, cropping, and fertilizer management on soil mineralization potential. *Soil Science Society of America Journal* 47, 1157–1161.

Fauci, M.F. (1993) Soil biological indices and nitrogen availability during a simulated transition from inorganic to organic sources of nitrogen. MSc Thesis. Oregon State University, Corvallis, OR.

Fauci, M.F. and Dick, R.P. (1994a) Plant response to organic amendments and decreasing inorganic nitrogen rates in soils from a long-term experiment. *Soil Science Society of America Journal* 58, 134–138.

Fauci, M.F. and Dick, R.P. (1994b) Soil microbial dynamics, short- and long-term effects of inorganic and organic nitrogen. *Soil Science Society of America Journal* 58, 801–806.

Fisk, M.C. and Schmidt, S.K. (1995) Nitrogen mineralization and microbial biomass nitrogen dynamics in three alpine tundra communities. *Soil Science Society of America Journal* 59, 1036–1043.

Fleige, H. and Baeumer, K. (1974) Effect of zero-tillage on organic carbon and total nitrogen content, and their distribution in different N-fractions in loessal soils. *Agro-Ecosystems* 1, 19–29.

Franzluebbers, A.J., Hons, F.M. and Zuberer, D.A. (1995) Soil organic carbon, microbial biomass, and mineralizable carbon and nitrogen in sorghum. *Soil Science Society of America Journal* 59, 460–466.

Goh, K.M. (1972) Amino acid levels as indicators of paleosols in New Zealand soil profiles. *Geoderma* 7, 33–47.

Goyal, S., Mishra, M.M., Hooda, I.S. and Singh, R. (1992) Organic matter-microbial biomass relationships in field experiments under tropical conditions: effects of inorganic fertilization and organic amendments. *Soil Biology and Biochemistry* 24, 1081–1084.

Griffin, G.F. and Laine, A.F. (1983) Nitrogen mineralization in soils previously amended with organic wastes. *Agronomy Journal* 75, 124–129.

Gupta, U.C. and Reuszer, H.W. (1967) Effect of plant species on the amino acid content and nitrification of soil organic matter. *Soil Science* 104, 395–400.

Harris, G.H., Hesterman, O.B., Paul, E.A., Peters, S.E. and Janke, R.R. (1994) Fate of legume and fertilizer nitrogen-15 in a long-term cropping systems experiment. *Agronomy Journal* 86, 910–915.

Holmes, W.E. and Zak, D.R. (1994) Soil microbial biomass dynamics and net nitrogen mineralization in northern hardwood forests. *Soil Science Society of America Journal* 58, 238–243.

Janzen, H.H. (1987) Soil organic matter characteristics after long-term cropping to various spring wheat rotations. *Canadian Journal of Soil Science* 67, 845–856.

Janzen, H.H. and Radder, G.D. (1989) Nitrogen mineralization in a green manure-amended soil as influenced by cropping history and subsequent crop. *Plant and Soil* 120, 125–131.

Jenny, H. (1980) *The Soil Resource*. Ecological studies, Vol. 37. Springer-Verlag, New York, 377 pp.

Kahn, S.U. and Sowden, F.J. (1971) Distribution of nitrogen in the black solonetzic and black chernozemic soils of Alberta. *Canadian Journal of Soil Science* 51, 185–193.

Kai, H., Ahmad, Z. and Harada, T. (1973) Factors affecting immobilization and release of nitrogen in soil and chemical characteristics of the nitrogen newly immobilized. *Soil Science and Plant Nutrition* 19, 275–286.

Keeney, D.R. and Bremner, J.M. (1964) Effect of cultivation on the nitrogen distribution in soils. *Soil Science Society of America Proceedings* 28, 653–656.

Keeney, D.R. and Bremner, J.M. (1966) Characterization of mineralizable nitrogen in soil. *Soil Science Society of America Proceedings* 30, 714–719.

Kuo, M.H. and Bartholomew, W.V. (1966) On the genesis of organic nitrogen in decomposed plant residue. In: *The Use of Isotopes in Soil Organic Matter Studies*. Report of the FAO/IAEA Technical Meeting organized by the Food and Agriculture Organization of the United Nations and The International Atomic Energy Commission. Symposium Publications Division, Pergamon Press, London, pp. 329–335.

Larson, W.E., Clapp, C.E., Pierre, W.H. and Morachan, Y.B. (1972) Effects of increasing amounts of organic residues on continuous corn, II. Organic carbon, nitrogen, phosphorus, and sulphur. *Agronomy Journal* 64, 204–208.

Lethbridge, G. and Davidson, M.S. (1983) Microbial biomass as a source of nitrogen for cereals. *Soil Biology and Biochemistry* 15, 375–376.

McGill, W.B., Cannon, K.R., Robertson, J.A. and Cook, F.D. (1986) Dynamics of soil microbial biomass and water-soluble organic C in Breton L after 50 years of cropping to two rotations. *Canadian Journal of Soil Science* 66, 1–19.

Meints, V.W. and Peterson, G.A. (1977) The influence of cultivation on the distribution of nitrogen in soils of the Ustoll suborder. *Soil Science* 124, 334–342.

Myrold, D.D. (1987) Relationship between microbial biomass nitrogen and a nitrogen availability index. *Soil Science Society of America Journal* 51, 1047–1049.

Rasmussen, P.E. and Collins, H.P. (1991) Long-term impacts of tillage, fertilizer, and crop residue on soil organic matter in temperate semiarid regions. *Advances in Agronomy* 45, 93–134.

Rosswall, T. and Paustian, K. (1984) Cycling of nitrogen in modern agricultural systems. *Plant and Soil* 76, 3–21.

Schnürer, J., Clarholm, M. and Rosswall, T. (1985) Microbial biomass and activity in an agricultural soil with different organic matter contents. *Soil Biology and Biochemistry* 17, 611–618.

Schnitzer, M. (1977) Recent findings on the characterization of humic substances extracted from soils from widely differing climatic zones. In: *Organic Matter Studies*, Vol. II. Proceedings of a symposium. International Atomic Energy Agency, Vienna, pp. 117–132.

Singh, H. (1995) Nitrogen mineralization, microbial biomass and crop yield as affected by wheat residue placement and fertilizer in a semi-arid tropical soil with minimum tillage. *Journal of Applied Ecology* 32, 588–595.

Sisworo, W.H., Mitrosuhardjo, M.M., Rasjid, H. and Myers, R.J.K. (1990) The relative roles of N fixation, fertilizer, crop residues and soil in supplying N in multiple cropping systems in a humid, tropical upland cropping system. *Plant and Soil* 121, 73–82.

Smith, S.J. and Young, L.B. (1975) Distribution of nitrogen forms in virgin and cultivated soils. *Soil Science* 120, 354–360.

Sowden, F.J. (1968) Effect of long-term annual additions of various organic amendments on the nitrogenous components of a clay and a sand. *Canadian Journal of Soil Science* 48, 331–339.

Stanford, G. and Smith, S.J. (1972) Nitrogen mineralization potentials of soils. *Soil Science Society of America* 36, 465–472.

Stevenson, F.J. (1956) Effect of some long-time rotations on the amino acid composition of the soil. *Soil Science Society of America Proceedings* 20, 204–208.

Stevenson, F.J. (1982a) Nitrogen-organic forms. In: Page, A.L. (ed.) *Methods of Soil Analysis*, Part 2. American Society of Agronomy, Madison, WI. pp. 625–641.

Stevenson, F.J. (1982b) Organic forms of soil N. In: Stevenson, F.J. (ed.) *Nitrogen in Agricultural Soils*. American Society of Agronomy, No. 22. Madison, WI, pp. 67–122.

Stevenson, F.J. (1994) *Humus Chemistry*. John Wiley & Sons, New York, 496 pp.

Van Veen, J.A., Ladd, J.N. and Frissel, M.J. (1984) Modelling C and N turnover through the microbial biomass in soil. *Plant and Soil* 76, 257–274.

Young, J.L. and Mortensen, J.L. (1958) Soil nitrogen complexes, 1. Chromatography of amino compounds in soil hydrolysate. *Ohio Agricultural Experiment Station Research Circular* 61. Wooster, OH, 18 pp.

Zielke, R.C. and Christenson, D.R. (1986) Organic carbon and nitrogen changes in soil under selected cropping systems. *Soil Science Society of America Journal* 50, 363–367.

Zinke, P.J., Stanenberger, A.G., Post, W.M., Emanuel, W.R. and Olson, J.S. (1984) *Worldwide Organic Soil Carbon and Nitrogen Data*. Publication 2217. Environmental Science Division, Oak Ridge National Laboratory, Oak Ridge, Tennessee, 108 pp.

23 Phosphorus Mineralization and Organic Matter Decomposition: A Critical Review

N. Gressel and J.G. McColl

Division of Ecosystem Sciences, Department of Environmental Science, Policy and Management, 151 Hilgard Hall, University of California, Berkeley, CA 94720, USA

Introduction

Phosphorus is essential for many metabolic processes in living organisms and is a major element in plant nutrition. Current interests in organic amendments and sustainable agriculture and global biogeochemical cycles highlight the necessity for a review of biological cycling of P in litter and soil for both agricultural and non-agricultural research. Its importance is partially a result of its unique chemistry as a phosphate complex (PO_4^{3-}) that is composed of a highly charged cation (P^{5+}) forming a tetrahedral centre for strongly associated oxygen atoms, resulting in three negative sites. This structure enables the formation of molecules with diester links while maintaining structural stability and preventing hydrolysis (Westheimer, 1987). Phosphate is the only stable oxidation state for P under atmospheric and soil conditions, in solution or solid forms (Lindsay *et al.*, 1989). Therefore, in this discussion 'phosphorus' necessarily implies 'phosphate' and 'P' will refer to both, except where quantitative concentrations are reported, where 'P' signifies the phosphorus atom itself.

In this review we critically examine current views regarding P cycling in soils, focusing primarily on forest soils and on the linkages between P transformations and organic matter (OM) decomposition. We discuss the conceptual models and methodology that have developed mainly in conventional agricultural settings, but that typically lack important biotic components, and we propose new biochemically-based approaches.

Current Views of P Cycling

The global P cycle consists of weathering processes, including mineral dissolution, erosion and leaching, which transfer P from the terrestrial environment to oceanic deposits, and of volcanism and uplifting of marine sediments which replenish terrestrial P and complete the cycle. Thus, in terrestrial ecosystems biologically-available P originates from slow weathering of minerals during soil formation. Concentrations of total P are low in soils (usually 0.2–5.0 g kg^{-1}) and soil solutions (<1 mg l^{-1}), there is no measurable gaseous P (e.g. phosphine, PH_3), and atmospheric deposition of P is limited in many environments, especially in forests (Binkley, 1986; Black, 1968; Schlesinger, 1991). As a result, productivity of most forests is limited by P availability, and P is typically conserved by organisms in forest ecosystems (Tate, 1984; Binkley, 1986).

Walker and Syers (1976) provide a basic model of P cycling on a pedogenic time-scale, which describes the relation between P transformations and soil formation across a chronosequence of New Zealand grasslands, and their model has been verified for other ecosystems (McGill

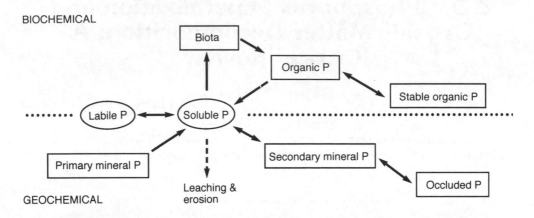

Fig. 23.1. A schematic presentation of phosphorus cycling in soils (modified from Smeck, 1985).

and Cole, 1981; Crews *et al.*, 1995; Cross and Schlesinger, 1995). The main source of P is from weathering of primary minerals, mostly apatites $(Ca_5(PO_4)_3(OH,F))$, which account for about 95% of the total P in igneous rocks (Lindsay *et al.*, 1989). Mineral weathering releases P into the soil solution, where P is taken up by plants and microorganisms, sorbed on to solid surfaces, forms secondary minerals, or is lost through erosion or leaching (Fig. 23.1). The low solubility of P anions $(H_2PO_4^-,$ HPO_4^{2-}, $PO_4^{3-})$ limits the rates of these solution-mediated processes. Litterfall, root decay, exudation and death of organisms return biologically held P creating a complex balance between geochemical and biochemical processes on a pedogenic time-scale, mediated by low P concentrations in the soil solution (Fig. 23.1). During soil formation as soil pH declines, base cations are lost through leaching and P is transformed gradually from primary Ca-phosphates to Al- and Fe-phosphates, while total soil P steadily declines due to leaching and erosion (Walker and Syers, 1976). Thus, the major constraints on P-availability are the release of P from primary minerals in the early stages of soil development, and the extent of leaching losses during the late stages.

On a biological time-scale, mineralization of organic P (P_o) was assumed by McGill and Cole (1981) to be controlled by extracellular phosphatase, produced in

response to demand for available P. Their analysis of trends in C and P contents of soil sequences (chrono-, topo- and climo-sequences) suggest that total C increases steadily with soil development and total P decreases, but P_o increases until primary-mineral P is depleted after which it too

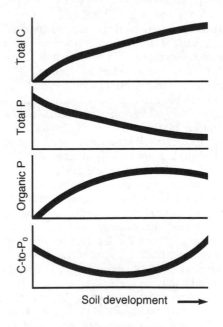

Fig. 23.2. Idealized concentrations of soil C, P and organic P (P_o) contents and the C-to-P_o ratio during soil development as suggested by McGill and Cole (1981).

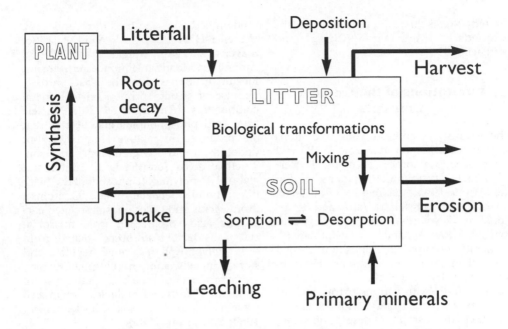

Fig. 23.3. Schematic biogeochemical representation of the major processes and components of the P cycle in plant–soil systems.

declines (Fig. 23.2). Consequently, the C-to-P_o ratio is high during the very early and late stages of soil development, forming a concave-shaped curve (Fig. 23.2). McGill and Cole (1981) attributed this trend to low P-availability at these stages, inducing phosphatase production and subsequent mineralization of P. Their analysis resulted in the current working-paradigm for P cycling in soils: that P availability is independent of OM decomposition on both biological and geological time scales (McGill and Cole, 1981; Smeck, 1985; Tate and Salcedo, 1988; Crews et al., 1995).

Phosphorus cycling through the soil-plant system (Fig. 23.3) will differ from one ecosystem to another depending on the intensity of input, output, transfer and transformation processes determined by interactions between climate, parent material, organisms, topography and time (Jenny, 1941; Schlesinger, 1991). For example, in the Alaskan coastal tundra, where inputs of P from weathering minerals and atmospheric sources are small, decomposition of OM provides most of the

P cycled during the short growing season (Chapin et al., 1978). Phosphorus is available to plants only after crashes in the microbial population, when 40% of the P cycled annually is released to the soil solution in a period of 10 days. In contrast, in desert soils where amounts of Ca-phosphates are high, P-availability is independent of OM accumulation, biological activity, or the degree of soil development (Lajtha and Schlesinger, 1988). In Mediterranean ecosystems, and others receiving aeolian dust from deserts and seas, atmospheric deposition contributes substantial amounts of the requirement for P by the natural vegetation (Singer et al., 1992; Newman, 1995). But in most forests inputs of P from weathering and atmospheric sources are relatively small compared with the annual requirements (Binkley, 1986), and efficient P-conservation has evolved in plants to maintain important metabolic processes. For example, retranslocation of P accounts for 67% of the P requirements in *Pinus nigra* plantations (Miller, 1984), and 70% of total P requirements in a northeastern

United States hardwood forest in the
Adirondack Mountains (Zhang and
Mitchell, 1995).

Limitations of the Current Concepts

Phosphorus research in soils and crops
focuses primarily on inorganic P (P_i) and
the geochemical processes of the P cycle
(Fig. 23.1). We found 60 citations listed in
the 1989–1995 database of *Current
Contents* for 'phosphorus/ate' and 'sorption' as keywords, but the keywords
'phosphorus/ate' and 'transformation',
'mineralization' or 'decomposition' yielded
fewer than 20 related citations. Historic-
ally, there has been much research on
extractable-P (with various extractants) as
an index for P availability, with the understanding that sorption–desorption reactions with mineral surfaces determine
availability in mineral soils, especially in
fertilized systems. Crop and soil P
response to fertilization, calibration of
yields to extraction-based indices, and P
reactions with soil minerals have been the
focus of most studies. Obviously, biochemical assays of P mineralization and the
transformations of P species in litter and
soil OM are necessary to address issues of
long-term soil fertility and land management more rigorously, but these are seri-
ously lacking compared with the large
number and variety of extraction-based
indices (Tate and Salcedo, 1988; Attiwill,
1991; Beck and Sanchez, 1994; Stevenson,
1994).

In agricultural soils, most P_o is in
forms that are considered poor nutrient
sources on a time-scale appropriate for
agricultural production, so that research
regarding P_o was rather limited in countries with fertilizer-based agriculture
(Khasawneh *et al.*, 1980; Magid *et al.*,
1996). However, forests and other non-
agricultural ecosystems rely more heavily
on mineralization of P_o that typically consists of 30–80% of total P in litter and soil
(Stevenson, 1986; Attiwill, 1991). Many
P-containing compounds have been identified, but the majority of P_o remains

unidentified in both litter and soil
(Mueller-Harvey and Wild, 1986;
Stevenson, 1986; Magid *et al.*, 1996).
Among the identifiable monoester P forms,
inositol P persists in the soil, partially
because of strong sorption on to mineral
components (Anderson, 1980). Nucleic
acids and phospholipids are diesters that
are produced in relative abundance, but
undergo rapid degradation in soil.
Teichoic acids (consisting of ribitol and
glycerol P polymers), phosphonates (with
direct bonds between P and C atoms),
uronic acids and other phosphorylated carboxylic acids originating from microbial
cell walls have also been isolated from
soils. Polyphosphates, which regulate and
store P in cells, and pyrophosphates which
are byproducts of metabolic pathways, are
inorganic P forms of biological origin also
present in litter and soil (Hawkes *et al.*,
1984; Stevenson, 1986).

The relationships between decomposition and P mineralization can be
approached from different premises: P can
be considered as a component of OM
'quality', therefore being a factor determining decomposition rates (Stevenson,
1986; Sinsabaugh *et al.*, 1993; Sinsabaugh
and Moorhead, 1994). On the other hand,
mineralization of P_o can be considered as a
function of the decomposition of organic
residues. This second viewpoint has
received little attention during the past 15
years, perhaps because current concepts
consider P mineralization as independent
of decomposition and many models of P
cycling lack essential descriptors for the
decomposition process (Magid *et al.*,
1996). For example, McGill and Cole
(1981) maintained that release of P from
OM depends on demand for P and the
subsequent phosphatase production, and
that their '... conceptual model appears to
provide a rational framework ... over both
geological and biological time scales', and
later models have accepted this basic
premise for both mineral-soil and litter
materials (Smeck, 1985; Sinsabaugh and
Moorhead, 1994).

An example of contradicting evidence
is provided by a study in which radioac-
tively-labelled clover residues were incu-

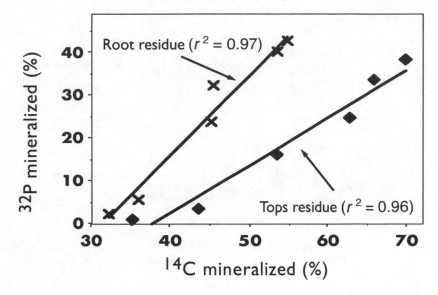

Fig. 23.4. The relationship between release of ^{14}C-labelled CO_2 and uptake of ^{32}P by plants in a pot study with incubated residues of labelled white clover (data from Dalal, 1979).

bated in soil with non-labelled plants growing in it (Dalal, 1979). Uptake of ^{32}P provided a measure for P availability, and release of ^{14}C-labelled CO_2 was the measure for decomposition. Release of labelled CO_2 was strongly correlated with the uptake of ^{32}P (Fig. 23.4), suggesting that P availability was coupled with residue decomposition. Additional contradictions to the hypothesis that phosphatase activity is the sole control of P availability are shown by the lack of correlation, or only weak correlations, between P availability and phosphatase activity found in many field studies (Speir and Ross, 1978; Adams, 1992; Harrison, 1983). A partial explanation why such correlations are not universal is provided by a pot study of Tarafdar and Claassen (1988), in which inorganic orthophosphate and three P_o compounds were added to a P-deficient soil with different plant species. Uptake of P was 300–500% of the non-fertilized control when P_i was added, depending on the plant species (Fig. 23.5). Similar increases in uptake were observed for fertilization with three P_o compounds. Because mineralization of P_o by phosphatase enzymes exceeded the demand, and uptake of P from inorganic and organic

sources were not significantly different, Tarafdar and Claassen (1988) concluded that 'phosphatase activity is not the limiting factor in the use of organic P ...'. Thus, on a biological time-scale, P availability is likely to be limited by its accessibility, rather than by phosphatase activity, as McGill and Cole (1981) suggested. A recent study of a soil chronosequence on volcanic parent material in Hawaii found low decomposition rates, low litter quality and high C-to-P_o ratios in the earliest and latest stages of soil development (Crews et al., 1995). The trend for C-to-P_o ratios is similar to that noted by McGill and Cole (1981) and shown in Fig. 23.2. The fact that low litter quality and low decomposition rates coincide with high C-to-P_o ratios suggests that this trend is controlled by ecological factors such as site conditions and initial litter quality, rather than by demand for P and subsequent activity of phosphatase.

Alternative Approaches

Conservation of P by plants does not begin with active competition for soil P, but rather with retranslocation of P from

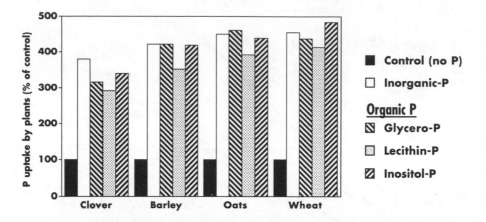

Fig. 23.5. Uptake of P by plants grown in pots with inorganic and organic additions of P (200 mg P kg soil^{-1}), relative to the uptake by plants in control pots with no P addition (data from Tarafdar and Claassen, 1988).

senescent tissue to younger, growing tissue within the plants (Marschner, 1995; Miller, 1984; Salisbury and Ross, 1985). Although the need for P is greater in P deficient plants, retranslocation may decrease because a greater proportion of the nutrient is incorporated in structural tissues (Miller, 1984). As a result, the concentration of P in litterfall will decrease with decreasing availability of P and will also be less accessible for mineralization by extracellular phosphatase enzymes in the litter layer. Under field conditions, P availability in litter and soil OM may therefore depend more on conditions that are conducive for oxidation and hydrolysis of the C structures occluding P, than on optimal conditions for phosphatase activity.

Detailed extraction-based studies, primarily those using the Hedley sequential fractionation (Hedley *et al.*, 1982), have shown important differences in the way P$_o$ is cycled under different conditions, and illustrated its importance especially in weathered soils where other indices of P availability have failed (Tiessen *et al.*, 1984; Beck and Sanchez, 1994; Cross and Schlesinger, 1995). For example, Beck and Sanchez (1994) found that the NaOH-extractable P$_o$ fraction was an important source of plant-available P in a non-fertilized agricultural Ultisol in Peru, but was a

sink for P in the fertilized soil. Despite this success, extraction-based fractions provide only operational information and do not increase our knowledge of the important biochemical mechanisms that drive P dynamics (Magid *et al.*, 1996).

The inability of extraction-based indices and phosphatase activity to explain issues of P availability, and the indication that linkages with C structure may be an important factor controlling P release, prompted us to seek new tools that provide a better understanding of P biochemistry in litter and soil (Gressel *et al.*, 1996). To establish the relation between P$_o$ compounds and the organic moieties with which they are associated during the decomposition process, we chose two spectroscopic analyses that are capable of monitoring P and C functional groups, i.e. ^{31}P and ^{13}C nuclear magnetic resonance (NMR) (for an overview of NMR spectroscopy see Skjemstad *et al.*, Chapter 20, this volume; and Wilson, 1987).

We examined a simple decomposition sequence of a forest soil profile with two well-defined litter layers and two mineral horizons that have high OM content (Table 23.1). This California forest soil profile showed typical decreases of the C-to-N and C-to-P$_o$ ratios with increasing soil depth and little change in total P con-

Table 23.1. Description of litter and mineral horizons of a mixed conifer forest in northern California.

Horizon	Depth (cm)	C-to-N ratio	C-to-P ratio	Description
O_i	7–5	48	834	Conifer needles; desiccated; single particle; loose; acerose; no roots; some insects.
O_e	5–0	30	605	Desiccated; weak, non-compact matted; loose; acerose; few, fine roots; common white mycelia.
A_o	0–3	22	272	Desiccated; very dark brown; granular; loam; friable; plentiful roots and mycelia; abrupt boundary.
A	3–40	22	211	Dry; brown; granular; loam; friable; plentiful roots; gradual boundary.
B_t	40–150	na	na	Dry; reddish brown; subangular blocky; clay loam; friable; common fine roots, few medium and coarse roots; clay films; few weathered rock fragments; diffuse boundary.
R	150–200	na	na	Weathered brownish grey andesite.

na, Not available.

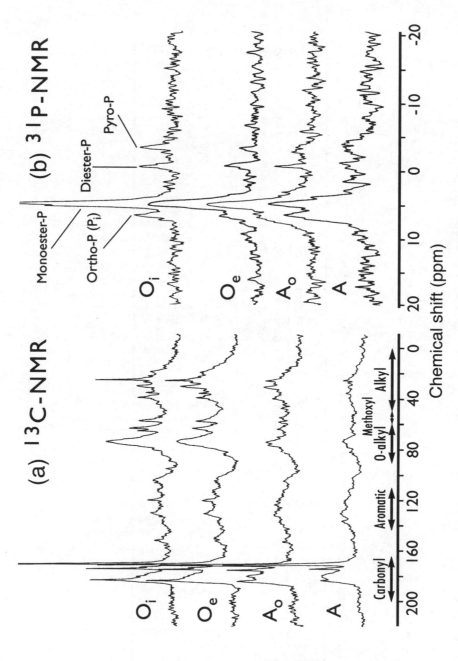

Fig. 23.6. Spectra of solution ^{13}C-NMR (A) and ^{31}P-NMR (B) of the surface horizons from a Californian mixed conifer forest soil (modified from Gressel et al., 1996).

Table 23.2. Significant correlations (r-values) and their probability levels (P-values) for solution (NaOH extracts) ^{13}C and ^{31}P-NMR results and indexes of organic decomposition and P fractions (n=8; only correlations with P<0.10 are shown).

Organic and P fractions	Carbonyl C		Alkyl C		O-alkyl C		Aromatic C		Methoxyl		C/N	
	r	P	r	P	r	P	r	P	r	P	r	P
Decomposition												
Carbonyl C	−0.79	0.016									−0.79	0.016
Alkyl C	−0.80	0.015	0.81	0.012							0.81	0.012
O-alkyl C	0.64	0.089	−0.88	0.003	−0.93	0.0003					−0.63	0.099
P fractions												
P_i	0.82	0.011	−0.87	0.003	−0.76	0.027	0.81	0.012				
Monoester P			0.63	0.095					−0.64	0.088	−0.68	0.064
Diester P					0.63	0.099	−0.74	0.036	0.91	0.001		
Pyrophosphate					0.84	0.006	−0.72	0.043			0.89	0.002

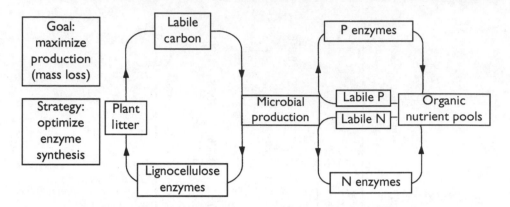

Fig. 23.7. A conceptual model for maximization of plant litter degradation through optimization of extracellular enzyme synthesis and C, N and P availability (modified from Sinsabaugh *et al.*, 1993).

centrations throughout the profile (Gressel *et al.*, 1996). Solution ^{13}C NMR spectroscopy of NaOH extracts (0.5M) revealed a decrease of *O*-alkyl (60–95 ppm) and alkyl (0–50 ppm) fractions with increasing profile depth and degree of decomposition, while carbonyl (165–210 ppm) and aromatic (110–145 ppm) fractions showed relative increases (Fig. 23.6A). Grouping results from soil profiles sampled before and after forest harvesting, we then searched for correlations between ^{13}C-NMR functional groups and other characteristics of the litter and mineral soil horizons. Significant correlations existed between the different C functional groups measured by NMR, and also with the C-to-N ratio (Table 23.2), establishing alkyl, *O*-alkyl, aromatic and carbonyl C as plausible biochemical indices of decomposition. Solution ^{31}P-NMR indicated that inorganic orthophosphate (P_i) increases with increasing depth in accordance with OM decomposition, while the signals from P_o fractions decrease (Fig. 23.6B). A close relationship between P mineralization and litter decomposition is suggested by correlations between P_i and all the C functional groups serving as indices of OM decomposition (Table 23.2). Significant correlations for diester-P and pyrophosphate with *O*-alkyl C and inverse correlations with aromatic C suggest that mineralization of these P fractions coincides with availability of C substrate (e.g. carbohydrates). Correlations between monoester P and

alkyl and methoxyl C suggest that monoester P mineralization is linked to breakdown of plant structural components such as cell membranes, the middle lamella and cuticle tissues (Gressel *et al.*, 1996). In a more recent study, we incubated ponderosa pine litterbags in the forest floor of an 80-year-old California mixed-conifer forest over a period of 18 months, and made periodic measurements of mass loss nutrient content and NMR analyses (Gressel, 1996). The diester P signals, assigned to nucleic acids and initially present in the ^{31}P-NMR spectra (9% of the total signal), were completely absent by the end of the study. In contrast, the amount of monoester P, the dominant P_o form (42–63% of the total ^{31}P-NMR signal), decreased where decomposition rates were high and remained unchanged where they were low (Gressel, 1996). Further research is needed to identify the chemistry of the ^{31}P-NMR functional groups and to elucidate physical compartmentation of P_o in decomposing litter. Manipulation of physical compartmentation may provide means to control nutrient release from litter and agricultural organic residues, and may enable synchronization between P release and plant uptake by crops.

Enzyme-based studies by Sinsabaugh *et al.* (1993), and Sinsabaugh and Moorhead (1994; and Chapter 27 this volume) provide a new approach that may be complementary to ours. They incubated uniform birch wood sticks in different soils for varying periods of time, and then analysed for mass loss, nutrient content and activities of P, N and C-degrading enzymes (Sinsabaugh *et al.*, 1993). Their results indicated that decomposition (i.e. mass loss) was maximized by microbial optimization of P, N and C supplies, through need-based synthesis of extra-cellular enzymes (Fig. 23.7). This approach differs from previous need-based theories in that microbial activity and mass loss are maximized, rather than availability of nutrients which are only components of the general microbial activity. A linear optimization model was then developed to estimate mass loss from the activities of P, N and C-degrading extracellular enzymes (Sinsabaugh and Moorhead, 1994; Chapter 27, this volume). This model sheds light on the relationship between P availability and phosphatase activity, but it also suggests that phosphatase activity alone is the rate-limiting step for P mineralization which is contrary to the findings of Tarafdar and Claassen (1988) presented earlier. To correct this, the conceptual model presented in Fig. 23.7 could be amended; links between the C-degrading enzymes and labile P and N release could be added to illustrate the dependency of nutrient release on breakdown of the C-structure surrounding it. This correction results in a non-linear relationship between mass loss and enzyme activity, but would have the advantage that P availability could be estimated from the activities of both P and C-degrading enzymes. Experiments that investigate the integrated effect of P, N and C-degrading enzymes on transformations of P_o are necessary to link this enzyme-based model with measures of available P.

Conclusions

On a biological time-scale (i.e. the approximate time-frame for both soil management and crop production), P mineralization is associated with decomposition of litter and soil OM, although availability of P on a pedogenic time-scale is ultimately controlled by the release of P from primary minerals and subsequent leaching (McGill and Cole, 1981). Moreover, breakdown of specific P fractions is associated with specific organic moieties, raising an important issue for future research regarding compartmentation of P_o in decomposing OM.

Phosphorus mineralization and OM decomposition must be examined using appropriate temporal and spatial scales and biologically-meaningful fractions. Current conceptual models of P cycling emphasize the importance of geochemical, sorption-based reactions, occurring over pedogenic time-scales. These models usually use arbitrarily defined P fractions, lack appropriate descriptors for biological processes, and are unsuitable for analysis of many processes associated with anthropogenic disturbance. In contrast, biochemically-based models are still in developmental stages, as are the methods associated with measuring biochemically-meaningful fractions of P. These must be identified and better understood to help address the environmental challenges associated with decreasing soil fertility and increasing land utilization.

In a recent review on future research needs for forest management, Attiwill (1991) identifies two points relevant to our topic: (i) 'It is most probable that the availability of phosphorus in forests is largely determined by organic equilibria, but our methodology for studying these equilibria is poorly developed'; and (ii) 'Research in nutrient cycling in forests must increasingly be based on more detailed knowledge of the composition and biochemistry of organic matter in the litter layer and in the soil'. Our review highlights the necessity for integrated conceptual and methodological approaches for examining P and C biochemistry in soils. We believe that this can be achieved through approaches combining analyses of nutrients and OM; e.g. parallel use of ^{31}P- and ^{13}C-NMR techniques (Gressel *et al.*, 1996) or measurement of P, N and C-degrading enzyme activities (Sinsabaugh *et al.*, 1993).

Acknowledgements

We are grateful to our research group members, Angus E. McGrath, Mark Waldrop and Thaïs Winsome for useful discussions and thoughtful comments on earlier versions of this manuscript and to two anonymous reviewers for their help in improving this review.

References

Adams, M.A. (1992) Phosphatase activity and phosphorus fractions in Karri (*Eucalyptus diversicolor* F. Muell.) forest soils. *Biology and Fertility of Soils* 14, 200–204.

Anderson, G. (1980) Assessing organic phosphorus in soils. In: Khasawneh, F.E., Sample, E.C. and Kamprath, E.J. (eds) *The Role of Phosphorus in Agriculture*. ASA, Madison, WI, pp. 411–431.

Attiwill, P.M. (1991) The disturbance of forested watersheds. In: Mooney, H.A., Medina, E., Schindler, D.W., Schulze, E.-D. and Walker, B.H. (eds) *Ecosystem Experiments. SCOPE 45*. Wiley, New York, pp. 193–213.

Beck, M.A. and Sanchez, P.A. (1994) Soil phosphorus fraction dynamics during 18 years of cultivation on a Typic Paleudult. *Soil Science Society of America Journal* 58, 1424–1431.

Binkley, D. (1986) *Forest Nutrition Management*. Wiley, New York, 290 pp.

Black, C.A. (1968) *Soil–Plant Relationships*, 2nd edn. Wiley, New York, 792 pp.

Chapin, F.S., Barsdate, R.J. and Barél, D. (1978) Phosphorus cycling in Alaskan coastal tundra: a hypothesis for the regulation of nutrient cycling. *Oikos* 31, 189–199.

Crews, T.E., Kitayama, K., Fownes, J.H., Riley, R.H., Herbert, D.A., Mueller-Dombois, D. and Vitousek, P.M. (1995) Changes in soil phosphorus fractions and ecosystem dynamics across a long chronosequence in Hawaii. *Ecology* 76, 1407–1424.

Cross, A.F. and Schlesinger, W.H. (1995) A literature review and evaluation of the Hedley fractionation – applications to the biogeochemical cycle of soil phosphorus in natural ecosystems. *Geoderma* 64, 197–214.

Dalal, R.C. (1979) Mineralization of carbon and phosphorus from carbon-14 and phosphorus-32 labelled plant material added to soil. *Soil Science Society of America Journal* 43, 913–916.

Gressel, N. (1996) Linkages between phosphorus and carbon fractions in forest soils. PhD Thesis, University of California, Berkeley, 172 pp.

Gressel, N., McColl, J.G., Preston, C.M., Newman, R.H. and Powers, R.F. (1996) Linkages between phosphorus transformations and carbon decomposition in a forest soil. *Biogeochemistry* 33, 97–123.

Harrison, A.F. (1983) Relationship between intensity of phosphatase activity and physico-chemical properties in woodland soils. *Soil Biology and Biochemistry* 15, 93–99.

Hawkes, G.E., Powlson, D.S., Randall, E.W. and Tate, K.R. (1984) A ^{31}P nuclear magnetic resonance study of the phosphorus species in alkali extracts of soils from long-term field experiments. *Journal of Soil Science* 35, 35–45.

Hedley, M.J., Stewart, J.W.B. and Chauhan, B.S. (1982) Changes in inorganic and organic soil phosphorus fractions induced by cultivation practices and laboratory incubations. *Soil Science Society of America Journal* 46, 970–976.

Jenny, H. (1941) *Factors of Soil Formation: A System of Quantitative Pedology*. McGraw-Hill, New York, 281 pp.

Khasawneh, F.E., Sample, E.C. and Kamprath, E.J. (eds) (1980) *The Role of Phosphorus in Agriculture*. ASA, Madison, WI, 910 pp.

Lajtha, K. and Schlesinger, W.H. (1988) The biogeochemistry of phosphorus cycling and phosphorus availability along a desert soil chronosequence. *Ecology* 69, 24–39.

Lindsay, W.L., Vlek, P.L.G. and Chien, S.H. (1989) Phosphate minerals. In: Dixon, J.B. and Weed, S.B. (eds) *Minerals in Soil Environments*. SSSA, Madison, WI, pp. 1089–1130.

Magid, J., Tiessen, H. and Condron, L.M. (1996) Dynamics of organic phosphorus in soils under natural and agricultural ecosystems. In: Piccolo, A. (ed.) *Humic Substances in*

Terrestrial Ecosystems. Elsevier, Amsterdam, Netherlands, 429–466.

Marschner, H. (1995) *Mineral Nutrition of Higher Plants,* 2nd edn. Academic Press, London, UK, 889 pp.

McGill, W.B. and Cole, C.V. (1981) Comparative aspects of cycling of organic C, N, S and P through soil organic matter. *Geoderma* 26, 267–286.

Miller, H.G. (1984) Dynamics of nutrient cycling in plantation ecosystems. In: Bowen, G.D. and Nambiar, E.K.S. (eds) *Nutrition of Plantation Forests.* Academic Press, London, UK, pp. 379–412.

Mueller-Harvey, I. and Wild, A. (1986) The nature and stability of organic phosphates in leaf litter and soil organic matter in Nigeria. *Soil Biology and Biochemistry* 18, 643–647.

Newman, E.I. (1995) Phosphorus inputs to terrestrial ecosystems. *Journal of Ecology* 83, 713–726.

Salisbury, F.B. and Ross, C.W. (1985) *Plant Physiology,* 3rd edn. Wadsworth Pub. Co., Belmont, CA, 540 pp.

Schlesinger, W.H. (1991) *Biogeochemistry: An Analysis of Global Change.* Academic Press, San Diego, 443 pp.

Singer, A., Danin, A. and Zöttl, H. (1992) *Atmospheric Dust and Aerosol as Sources of Nutrients in a Mediterranean Ecosystem of Israel.* Final report. Joint German–Israeli Research Project 13/610. Bundesministerium für Forschung und Technologie (BMFT), Germany; and Ministry of Science and Development, National Council for Research and Development (NCRD), Israel.

Sinsabaugh, R.L. and Moorhead, D.L. (1994) Resource allocation to extracellular enzyme production – a model for nitrogen and phosphorus control of litter decomposition. *Soil Biology and Biochemistry* 26, 1305–1311.

Sinsabaugh, R.L., Antibus, R.K., Linkins, A.E., McClaugherty, C.A., Rayburn, L., Repert, D. and Weiland, T. (1993) Wood decomposition – nitrogen and phosphorus dynamics in relation to extracellular enzyme activity. *Ecology* 74, 1586–1593.

Smeck, N.E. (1985) Phosphorus dynamics in soils and landscapes. *Geoderma* 36, 185–199.

Speir, T.W. and Ross, D.J. (1978) Soil phosphatase and sulphatase. In: Burns, R.G. (ed.) *Soil Enzymes.* Academic Press, London, pp. 197–250.

Stevenson, F.J. (1986) *Cycles of Soil. C, N, P, S, Micronutrients.* Wiley, New York, 380 pp.

Stevenson, F.J. (1994) *Humus Chemistry: Genesis, Composition, Reactions,* 2nd edn. Wiley, New York, 496 pp.

Tarafdar, J.C. and Claassen, N. (1988) Organic phosphorus compounds as a phosphorus source for higher plants through the activity of phosphatases produced by plant roots and microorganisms. *Biology and Fertility of Soils* 5, 308–312.

Tate, K.R. (1984) The biological transformation of P in soil. *Plant and Soil* 76, 245–256.

Tate, K.R. and Salcedo, I. (1988) Phosphorus control of soil organic matter accumulation and cycling. *Biogeochemistry* 5, 99–107.

Tiessen, H., Stewart, J.W.B. and Cole, C.V. (1984) Pathways of phosphorus transformations in soils of differing pedogenesis. *Soil Science Society of America Journal* 48, 853–858.

Walker, T.W. and Syers, J.K. (1976) The fate of phosphorus during pedogenesis. *Geoderma* 15, 1–19.

Westheimer, F.H. (1987) Why nature chose phosphates. *Science* 235, 1173–1178.

Wilson, M.A. (1987) *NMR Techniques and Applications in Geochemistry and Soil Chemistry.* Pergamon, New York, 353pp.

Zhang, Y.M. and Mitchell, M.J. (1995) Phosphorus cycling in a hardwood forest in the Adirondack Mountains, New York. *Canadian Journal of Forest Research* 25, 81–87.

Part VII

Modelling: Providing the Framework

24 Modelling Litter Quality Effects on Decomposition and Soil Organic Matter Dynamics

K. Paustian[1], G.I. Ågren[2] and E. Bosatta[2]

[1]*Natural Resources Ecology Laboratory, Colorado State University, Fort Collins, CO 80523, USA;* [2]*Department of Ecology and Environmental Research, Swedish University of Agricultural Sciences, Box 7072, S-750 07 Uppsala, Sweden*

Introduction

The concept of litter quality – in its simplest form – deals with the question of why different plant materials, when exposed to identical environmental conditions, decompose at different rates. This question has been the focus of much, if not most, of the experimental work on decomposition. Similarly the issue of litter quality has been a central theme of efforts to develop theoretical and predictive models to describe decomposition processes mathematically.

In this chapter we examine the various components of litter quality and how they can be formalized mathematically. We review and compare how existing models incorporate different aspects of litter quality, how these relate to measurable attributes of litter and how sensitive the models are to changes in litter quality. Finally we extend this analysis to consider models that deal not only with short-term 'primary' decomposition processes but also with how litter quality influences soil organic matter attributes over the long term.

Mathematical representation of litter quality

Components of litter quality

While 'quality' is a useful generic term for characteristics of litter relating to decom-position processes, a more rigorous definition is required for our discussion of modelling approaches. We propose a simple operational definition of quality, i.e. its relative ease of mineralization by decomposer organisms. Unless otherwise specified we refer to quality with respect to C mineralization and specifically its short-term or 'instantaneous' mineralization rate. Thus, glucose would be considered a higher quality substrate compared with cellulose, even though the cumulative amounts mineralized after several weeks might be similar. A more strict definition of quality might be the relative ease of assimilation rather than mineralization; however assimilation by soil organisms is difficult to measure and is not explicitly represented in many decomposition models.

Quality involves intrinsic characteristics of litter that affect its utilization by heterotrophs, chiefly microorganisms. To provide a frame of reference, consider the classical microbial growth model of Monod (1942), i.e.

$$\frac{dS}{dt} = v_m \, B \, \frac{S}{(k_s + S)}$$

$$v_m = \frac{\mu}{e} \tag{24.1}$$

where the decomposition rate of a substrate (S) is related to the specific rate of

Table 24.1. Categorization of litter quality components and their possible interpretation in terms of microbial growth kinetics. See Equation 24.1 for symbol definitions.

Quality component	Quality attribute	Process affected	Parameter interpretation
Chemical	Bond strength/energy	Growth rate, energy yield	μ, e
	Bond complexity	Exoenyzme efficacy	k_s
Physical	Particle size	Soil resources, abiotic	μ
	Physical occlusion	Enzyme accessibility	k_s, e
Inhibitory	Interference	Exoenzyme efficacy	k_s
	Antibiotic	Growth rate	μ

uptake (v_m) per unit microbial biomass (B) and the half-saturation constant (k_s) for microbial uptake. The specific uptake rate (v_m) is a function of the maximum specific growth rate (μ) and the growth yield efficiency (e). From these equations it is apparent that there are two fundamental ways in which the characteristics of an organic substance can affect its decomposition rate; either through effects on the *potential rate* of decomposition (represented in this case by uptake or growth rates) or by influencing its *accessibility* to decomposer organisms (represented by the interaction between k_s and S). We will consider three components of quality – chemical, physical and inhibitory – in relation to how they influence decomposition and how they can be represented in decomposition models (Table 24.1).

Many of the earliest scientific studies of litter decomposition focused on the influence of chemical composition and how rapidly various compounds (e.g. protein, cellulose, hemicellulose, lignin) disappeared from decomposing litters (Waksman and Tenney, 1926). Chemical composition is still generally considered the most significant component of litter quality and is probably the most widely represented quality factor in models. At the biochemical level, the determinants of quality or decomposability are a function of the nature and complexity of the chemical bonds which must be lysed before materials can be assimilated by microorganisms. Characteristics such as the strength of the bond and the catabolic

energy yield of the product would primarily influence the potential rates of dissolution and uptake. Thus this aspect of chemical quality could be expressed in terms of the potential growth rate constant (μ) and/or yield efficiency (e). Structural complexity, in terms of the molecular shape and the diversity of chemical bonds of a material, is another aspect of chemical quality. Lignin is known as one of the more structurally complex plant constituents which is a major factor contributing to its recalcitrance to decomposition (Swift *et al.*, 1979). The effect of structural complexity could be expressed through the substrate availability term (k_s) in Equation 24.1, in that the probability of a lysing enzyme 'finding' the appropriate binding site is smaller in a complex, heterogenous substance like lignin as compared to simpler structures such as cellulose fibres, which are made up of regular repeating units. At the microbial cell level, energetic costs could also be greater due to a poorer yield efficiency for utilizing complex compounds.

Physical components of litter quality have received somewhat less attention in both experimental studies and in decomposition models. Components of physical quality that affect decomposition rate include particle size and surface-area to mass characteristics. Decomposer organisms are most prevalent and active on surfaces. Thus, at the soil–litter interface, the availability of inorganic nutrients and colonizing organisms will be greater for finely divided litter (e.g. twigs) compared with

Table 24.2. Overview of some litter decomposition models and their representation of litter quality affects on decomposability.

Model type/Author	No. of litter compartments	Elements	Decomposer biomass	Litter quality representations/other comments
Single-pool models				
Aber and Melillo (1982)	1	C, N	No	Lignin-to-N
Berg and Ekbohm (1991)	1	C	No	Finite asymptote included
Jansen (1984)	1	C	No	Decomposability defined by 'apparent initial age'
Meentemeyer (1978)	1	C	No	Lignin; Used for regional comparisons
Middleburg (1989)	1	C	No	Decomposability defined by 'apparent initial age', specific rate constant decreases with time
Multiple litter pool models				
Berg and Ågren (1984)	2	C	No	Flux from resistant to labile component to represent solubilization
Jenkinson (1977)	2	C	No	Two pool determination from curve fitting
Minderman (1968)	6	C	No	Proximate analyses to determine initial pool sizes
Moorehead and Reynolds (1991)	4	C, N	Yes	Labile, holocellulose and resistant initial litter fractions; includes lag time due to microbial colonization
Parnas (1975)	2	C, N	Growth rate only	C and N only compounds distinguished
Andrén and Paustian (1987)	4	C	No	Soluble and insoluble fractions used for initial pool sizes; formation of secondary products
Continuous spectrum models				
Carpenter (1981)	na	C	No	Quality parameter derived from curve-fitting
Bosatta and Ågren (1991; 1994)	na	C, N, P, S	Steady-state	Quality parameters have been related to chemical fractionations
Boudreau (1992)	na	C	No	Gamma distribution for reactivities

na, Not applicable.

coarse litter (e.g. large branches).

Particle size is a 'macro-scale' physical component of quality which is manifested indirectly, as a modifier of other factors (e.g. nutrient and water availability, colonization potential). At a finer scale, the arrangement of different compounds within plant tissues can be considered a physical (or physio-chemical) component of quality. One of the best examples of this is the occlusion of cellulose and hemicellulose by the lignification of secondary plant cell walls. In this case the presence of lignin is thought to act as a physical barrier inhibiting the accessibility of enzymes to the more decomposable structural polysaccharides (Swift *et al.*, 1979; Chesson, Chapter 3, this volume). Conceptually, the effects of both particle size and physical occlusion could be represented in models as modifiers of substrate availability.

The presence of inhibitory compounds is yet another component of litter quality. While chemical in nature, we consider this as a separate quality component in that it involves the interference of compounds in enzyme function or decomposer metabolism, resulting in a reduction in decomposition rate of other litter constituents. Some polyphenols such as tannins are believed to inhibit exoenzyme activity (Williams and Gray, 1974) and high concentrations of such components in litter have been correlated with reduced decomposition rates in some studies (Palm and Sanchez, 1990; Tian *et al.*, 1992, 1995). One way to represent the effects of inhibitory compounds mathematically would be to modify the potential rate constants for decomposition of other litter components as a function of the concentration of inhibitory compounds in the litter (Whitmore and Handayanto, Chapter 25, this volume).

Incorporation of quality into models

Most models of litter decomposition are based, at least loosely, on concepts relating to microbial activities as articulated in Equation 24.1. However, these concepts have been abstracted and simplified in a variety of ways in different models of decomposition (Table 24.2).

Among the earliest and still most widely used models is the simple first-order decay model,

$$\frac{dS}{dt} = -kS \qquad (24.2)$$

where the decomposition rate is directly proportional to the amount of substrate (S) remaining and thus mass loss follows an exponential decline. This model treats litter as a uniform, homogeneous substance and makes the implicit assumption that the biological capacity to decompose litter is unlimited and that the specific decay rate ($dS/dt \cdot 1/S$) is constant. Significant departures from a single exponential decay function generally occur well before all of the recognizable plant material has disappeared and thus a simple exponential model is seldom used other than for describing short-term (e.g. first year) mass loss. The only option for including litter quality in this formulation is by allowing k to vary as a function of litter type, in which case quality only has meaning in the sense of *between* litter type comparisons (e.g., different plant species). For comparative studies, the first-order model and litter specific k values are useful primarily as a way of summarizing and comparing data but not for a deeper exploration of the mechanisms governing decomposition processes.

An elaboration of the single first-order model is to explicitly represent different component parts *within* plant litter. The concept of litter pools recognizes that different fractions of plant litter, viewed either as specific compounds or as collections of chemical or morphological constituents, vary in their inherent decomposability and in the factors that influence decomposition rates. Such multiple-pool models assume that each litter constituent has its own potential decomposition rate and typically (but not in all cases) it is assumed that each component decomposes independently according to a first-order relationship. Thus the general form of the decomposition equation is,

$$\frac{dS}{dt} = -k_1 S_1 - k_2 S_2 \dots - k_n S_n \quad (24.3)$$

where

$$S = S_1 + S_2 + \dots + S_n$$

and $S_1, S_2, \dots S_n$ are the different litter constituents, having specific decomposition rate constants, $k_1, k_2, \dots k_n$. In such models quality is explicitly incorporated, in that the composition of the litter material determines the initial amount of material in each pool. If the potential decomposition rate constants are generalized such that they apply to a specific, quantifiable, litter fraction (e.g. decomposition rate for lignin), then the model can be used for evaluating quality influences within litter types (e.g. high compared with low lignin litter of the same species).

One of the earliest applications of a multi-pool decomposition model was by Minderman (1968), who included several chemically defined primary fractions (e.g. sugars, cellulose, lignin). One important conclusion of this work was the realization that in order to predict total mass loss, as well as the decomposition of the primary litter fractions, the formation and turnover of secondary decomposition products need to be included. This led to a class of models which may be best referred to as organic matter turnover models, where organic matter pools representing secondary decomposition products (e.g. microbial biomass, soil organic matter) are included.

In multi-pool decomposition/soil organic matter models (see reviews by Paustian 1994, Ågren et al., 1996), litter quality is expressed mainly through different specific decomposition rate constants for each organic matter pool, and in some instances through effects on substrate availability parameters (e.g. Smith, 1979; McGill et al., 1981). In these models, the most common litter quality attributes are chemically defined, either as collections of specific fractions identified by chemical extraction procedures (e.g. lignin, cellulose and hemicellulose, hot H_2O extractable, non-polar solubles), or as more functional or morphologically-defined components.

An example of the latter type of characterization is the concept of Hunt (1977) and later McGill et al. (1981) and Juma and McGill (1986), defining structural and metabolic components, having slow versus rapid decomposition rates, respectively. For plant litter these fractions represent the 'structural' compounds making up cell walls, i.e. structural polysaccharides, lignin, and wall-bound proteins, versus the 'metabolic' compounds constituting the cell cytoplasm. This functional representation of litter recognizes that individual chemical compounds in litter do not decompose independently and that the physical structure of plant material at the microscale is an important attribute of quality. This type of representation has been adopted in several other models of decomposition and soil organic matter turnover (e.g., Parton et al., 1987, 1994; Verberne et al., 1990; Li et al., 1992).

Some decomposition models, particulary for forest systems (e.g. Pastor and Post, 1986), also incorporate particle size as a physical attribute of litter quality. This is done by including separate litter pools and defining decomposition rate constants on the basis of size (e.g. coarse or fine woody litter).

A novel variation of the multiple first-order pool model, in which the specific decomposition rates are directly related to enzyme activity, was proposed by Sinsabaugh and Moorhead (1994; Chapter 27, this volume). In this model, litter quality is measured by assays of potential enzyme activity.

An alternative way of representing litter heterogeneity and the influence of quality, is to consider litter as a continuous distribution of organic materials rather than a set of discrete pools. This approach can incorporate both the influence of initial litter composition and the transformation of primary litter compounds into secondary materials and subsequent effects on the overall decomposition rate. Relatively fewer models of this type have been developed (Carpenter, 1981; Bosatta and Ågren, 1985, 1991, 1994; Boudreau, 1992). The Q model of Bosatta and Ågren

(1985, 1991, 1994) and Ågren and Bosatta (1987, 1996) will be briefly described here to characterize this approach.

The large number of compounds in plant litter suggests that if a quality (denoted q) is assigned to each of these compounds there could be a virtually continuous range of qualities. Consider first the carbon (C) in one of these compounds. It will be assimilated by decomposer organisms, with C concentration f_c, and the rate of assimilation will depend upon the quality, which can be described by a function, $u(q)$. Part of the assimilated C will be respired away and part will remain as decomposer biomass, until the death of the organism, or be exuded as organic C. The fraction of the assimilated C that remains in organic form until the death of the organism is the efficiency, denoted as $e(q)$. This is the fraction of assimilated C that is returned to the substrate and may be reassimilated by other decomposers. When C is returned to the substrate it will have a different chemical composition from the C which was originally assimilated, i.e. assimilated glucose may be 'returned' as chitin. To account for this process, Bosatta and Ågren (1991) introduced a dispersion function $D(q,q')$ defining the fraction of the C assimilated at quality q' that is returned to the substrate as quality q. They assume further that the rate of assimilation of a substrate is proportional to its availability ($S \ll k_s$ in terms of Equation 24.1) and that the turnover rate of the decomposers is so high that the decomposer biomass can be regarded as in steady-state with respect to the substrate. Let then the amount of C in the quality interval $[q, q+dq]$ at time t be $\rho_c(q,t)dq$. The mass balance equation for $\rho_c(q,t)dq$ is then (Bosatta and Ågren, 1991, 1994),

$$\frac{\partial \rho_c(q,t)}{\partial t} = -f_c \frac{u(q)}{e(q)} \rho_c(q,t)$$

$$+ f_c \int_0^\infty D(q,q')u(q')\rho_c(q',t)dq' \quad (24.4)$$

The total amount of C in the substrate is given by

$$C(t) = \int_0^\infty \rho_c(q,t)dq \quad (24.5)$$

and the average quality of the substrate by

$$\bar{q}(t) = \frac{1}{C(t)} \int_0^\infty q\rho_c(q,t)dq \quad (24.6)$$

They also define the average displacement in quality

$$\eta_1(q) = \int_0^\infty (q-q')D(q,q')dq' \quad (24.7)$$

These equations form a system where decomposition is seen as both a loss of total carbon (Equation 24.5) as well as a continuous change in litter composition along a quality continuum (Equations 24.6 and 24.7). A complete solution of the model requires specification of the decomposer functions $e(q)$, $\eta_1(q)$ and $u(q)$.

In summary, there is a wide array of mathematical descriptions of litter decomposition which include effects of litter quality. The principal differences between model structures lie in whether or not microbial turnover and formation of secondary products are included and whether the array of compounds in plant litter and decomposition products are grouped into discrete pools or are treated as a continuum.

Sensitivity to Litter Quality in Short-term Decomposition Models

Given the differences in representation of the decomposition system and the use of different quantitative indicators of litter quality, it may be useful to compare several decomposition models using a common data set representing litters of different quality. We have compared six different models, applicable to short-term decomposition, ranging from the single exponential model to models with continuous quality distributions (Table 24.3). Time series of mass loss for three plant litters, namely clover roots, barley straw and pine needles, were used to fit the models.

Table 24.3. Comparison of some models of litter decomposition with litters of different quality: clover roots (Berg *et al.*, 1987), barley straw (Wessén and Berg, 1986) and green Scots pine needles* (Berg and Ekbohm, 1991). Duration of incubations were 195, 720 and 1468 days for clover, barley and pine litters, respectively. Clover roots were incubated in the laboratory and barley and pine litter in the field. Model equations are in integral form and the model of Bosatta and Ågren is shown in reduced form since several parameters have been estimated independent of the data. The value k_0 gives the initial specific rate of decomposition, t is time and the asymptote is the mass remaining as $t \to \infty$. EV is the fraction of the variance explained by the model. (For further details about models and definitions see original publications.)

Model	Equation	k_0	Asymptote	Parameter	Clover roots	Barley straw	Pine needles
Berg and Ekbohm (1991)	$1 - m_a(1-e^{-kt})$	km_a	$1-m_a$	m_a	0.81	1.00	0.61
				k	$62.7\,10^{-3}$	$1.21\,10^{-3}$	$1.95\,10^{-3}$
				EV	0.930	0.842	0.989
Bosatta and Ågren (1994)	$(1+pt)^{-1.19}$	$1.19p$	0	p	$42.8\,10^{-3}$	$1.23\,10^{-3}$	$0.90\,10^{-3}$
				EV	0.918	0.787	0.982
Boudreau (1992)	$1-[1-(1+t/a)^{-v}]/v$	$1/a$	$(1-1/v)$	a	11.8	660	810
				v	1.15	1.00	1.13
				EV	0.976	0.776	0.993
Janssen (1984) Middleburg (1989)	$\exp[p\{(a+t)^{-q} - a^{-q}\}/q]$	pa^{-q-1}	$\exp\{-pa^{-q}/q\}$	a	30	3000	600
				p	45	5470	346
				q	0.90	0.90	0.90
				EV	0.975	0.810	0.970
Minderman (1968)	$\sum m_i e^{-k_i t}$	$\sum k_i m_i$	0	k_1	$31.0\,10^{-3}$	$8.04\,10^{-3}$	$2.90\,10^{-3}$
				k_2	$1.99\,10^{-3}$	$2.09\,10^{-3}$	$1.44\,10^{-3}$
				k_3	$1.87\,10^{-3}$	$4.28\,10^{-3}$	$1.58\,10^{-3}$
				EV	0.715	−0.123	−1.27
Exponential	e^{-kt}	k	0	k	$2.54\,10^{-3}$	$1.32\,10^{-3}$	$0.71\,10^{-3}$
				EV	0.725	0.831	0.909

* Initial chemical composition mg g^{-1} ash-free dry matter, using methods described in Wessén and Berg (1986).

	Clover roots	Barley straw	Scots pine forest
Water and ethanol extractable	359	60	270
Acid soluble	574	742	510
Acid insoluble	67	197	220
Total N	20	3.7	1.5

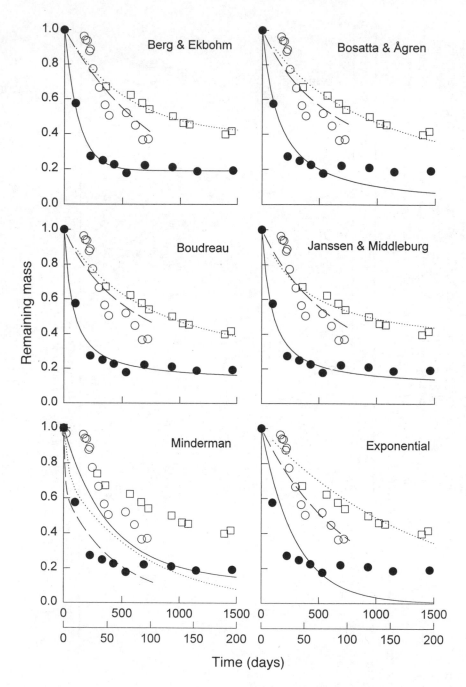

Fig. 24.1. Six decomposition models were fitted to measured mass loss of clover roots (●–), barley straw (○ · · · · ·), and Scots pine needles (□· · · · ·). The lower *x*-axis (0–200 days) applies to the clover root decomposition while the upper *x*-axis (0–1500 days) applies to the straw and pine needle decomposition series. Model equations and values for the fitted parameters are given in Table 24.3.

With one exception, the model equations (shown in the integral form) were fitted to *total* mass loss for each litter type using a non-linear least-squares procedure for the parameters listed (Table 24.3). In some cases, parameters not specific to litter quality have been combined. For the Minderman model, the changes in each of the three measured components, i.e. water plus ethanol solubles, acid solubles, and acid insolubles, were fitted separately and the mass loss was then calculated as the sum of the three components. Since all models contain some characteristic rate constant (see k_0 in Table 24.3), the fitting of this parameter, or the corresponding parameter combinations, accounts for differences in the incubation conditions for the three litter types.

Taking into account that one, two or three parameters were fitted for the different models, there were no large differences in how most of the models fit the observed mass loss curves (Fig. 24.1). The exception was the Minderman model which drastically overestimated the mass loss rate for all three litter types. This indicates the importance of considering transformations between different compounds in the substrate; in other words, total mass loss is not well represented by attempting to predict the loss of individual components, a conclusion already drawn by Mindermann (1968) in his original paper. The decomposition of clover roots was best fit by the model having a single exponential term and an asymptote (Berg and Ekbohm, 1991), which, in effect, assumes that decomposition stops at around 20% remaining mass. This fits well the two-phase decomposition profile of the clover roots – a very rapid initial phase, followed by a low rate of mass loss. Eliminating the asymptote (i.e. the single exponential model) gave a poor fit to clover decomposition and predicted near complete litter disappearance after 200 days. The flattening of the mass loss curves for clover roots and pine needles was better represented by the 'continuum' models; the Boudreau and Janssen/Middleburg models were closest to the observed value, but both these models have more free parameters than the

Bosatta and Ågren model. The poor fit of all models to the initial part of the barley straw curve was a result of that experiment being started in late autumn which was followed by a period of low activity over the first winter. All the models operate with an average climate, hence the lack of fit over the winter period is not unexpected.

Influence of Litter Quality on Soil Organic Matter

Model descriptions

To this point we have considered litter quality primarily in the context of modelling short-term decomposition dynamics, for example, changes in mass loss that might be observed in a litter bag experiment. One of the fundamental questions in soil organic matter research is the extent to which the characteristics of the original plant material influence the longer term soil C balance. In other words, are different plant compounds stabilized differentially in soil, and if so how and by what mechanisms? Early theories of soil organic matter formation, postulated the importance of specific plant compounds, specifically lignin and proteins, as the 'precursors' in the formation of recalcitrant humus compounds (Waksman, 1936). Modern theories focus more on the role of microbial products and metabolites and their interactions with soil physical and textural properties in the formation and maintenance of soil organic matter (Stevenson and Elliott, 1989). However, it is believed that the decomposition products of certain plant compounds, e.g. secondary products of lignin degradation that are not fully metabolized by microorganisms, contribute relatively more to the formation of stabilized soil organic matter (Zech and Kögel-Knabner, 1994).

Several models have been developed to study long-term soil organic matter dynamics, most of which include plant litter quality as an important factor (see reviews by Paustian, 1994; Ågren *et al.*, 1996). We will examine the structure and assumptions of three models, the

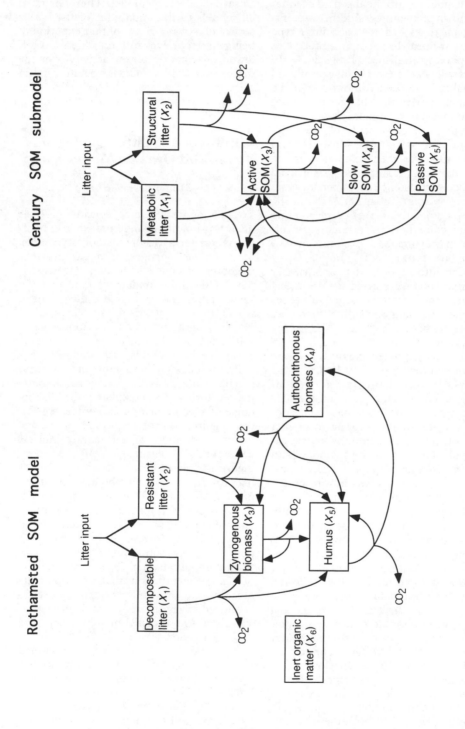

Fig. 24.2. Simplified structures of the Rothamsted (Jenkinson *et al.*, 1987) and Century (Parton *et al.*, 1987) models of SOM dynamics, showing state variables (C pools) and mass transfer pathways. Abbreviated designations for the state variables (i.e. X_i) are used in the steady-state solutions given in Table 24.4.

Rothamsted model (Jenkinson *et al.*, 1987), Century (Parton *et al.*, 1987) and the Q model (Ågren and Bosatta, 1987), which have been used to simulate long-term soil organic matter dynamics. Recent modifications have been made in both Century (Parton *et al.*, 1994) and the Rothamsted model (K. Coleman, personal communication), but their fundamental structures regarding decomposition processes have remained the same. To compare the different behaviours of the models we will examine the steady-state solutions for total C and its relative composition. We will briefly describe the Rothamsted and Century models (Fig. 24.2); the Q model was described earlier in this paper (Equations 24.4–7).

The Rothamsted model (Jenkinson *et al.*, 1987) includes two litter components, 'decomposable' and 'resistant', two microbial biomass pools, representing 'opportunistic' (i.e. zymogenous) and 'basal' (i.e. authochthonous) microorganism populations, and a humus pool, representing the non-living, secondary decomposition products. An inert organic matter pool is included as part of the total soil C, representing very old C which is not involved in turnover at the time scale considered in the model. Litter decomposes from the two primary litter pools (decomposable and resistant), a portion is assimilated by the zymogenous biomass and a portion is respired as CO_2. An important point to note is that a single microbial efficiency is assumed, which applies to both litter pools. The authochthonous biomass assimilates C from the humus pool. Organic matter in both biomass pools is broken down to humus, CO_2 or recycled to the zymogenous biomass. Litter quality is incorporated as differing proportions of the decomposable and resistant components, which are determined from curve-fitting of short-term decomposition data. Typical values used for the proportion of resistant litter in previous model applications, are 40% for most agricultural crops, 60% for grassland and scrub (savanna), 80% for deciduous woodland, and 100% for peat (K. Coleman, personnal communication).

The decomposition submodel of Century (Parton *et al.*, 1987) consists of two litter pools, an easily degradable 'metabolic' component and a more resistant 'structural' component. There are three soil organic matter pools (SOM): active SOM consisting of microbial biomass and metabolites; slow SOM, consisting of partially stabilized organic matter having an intermediate turnover rate; and passive SOM, which is the most recalcitrant pool with the slowest turnover rate. Three mechanisms involving litter quality have an effect on SOM dynamics. First, the partitioning of fresh litter is based on the lignin-to-N ratio of the litter, such that the fraction of newly formed litter going into the structural component increases with increasing lignin-to-N. All lignin is assumed to be part of the structural litter fraction. Second, lignin influences the partitioning of primary decomposition products in that the lignin associated with the decomposing structural fraction flows directly to the slow pool, whereas the remainder of the structural litter and the metabolic component flow through the active pool. The proportional CO_2 loss associated with C flowing to the active pool is also assumed to be greater (i.e. 55–65%) than for lignin-associated C flowing from the structural litter to the slow pool (i.e. 30%). This assumption is based on observations that a relatively small fraction of lignin degradation products are metabolized by microorganisms (Stott *et al.*, 1983). The third effect of litter quality is that the specific decomposition rate of the structural litter fraction is a function of litter lignin content, such that increasing the lignin-to- structural ratio slows the potential rate of decomposition.

Influence of litter quality on amounts and composition of SOM

The state equations and steady-state solutions for total C in the three models are given in Table 24.4. The Rothamsted and Century models consist of sets of equations for each organic matter pool, where rates of change are a function of first-order decomposition constants (k_i) and transfer

Table 24.4. Equations for rates of change of organic carbon components and steady-state solutions for total organic C (X_{tot}). Several rate (k_i) and transfer (f_i) coefficients are functions of other factors (e.g. climate, soil type) which are assumed to be constant for the steady-state solution. Pool designations X_n for the Rothamsted and Century models are given in Fig. 24.3.

State equations	Steady-state solution

Rothamsted

$$\frac{dX_1}{dt} = \gamma I - k_1 X_1$$

$$\frac{dX_2}{dt} = (1-\gamma)I - k_2 X_2$$

$$\frac{dX_3}{dt} = f_1(k_1 X_1 + k_2 X_2 + k_4 X_4) - (1-f_1)k_3 X_3$$

$$\frac{dX_4}{dt} = f_2 k_5 X_5 - (1-f_1)k_4 X_4$$

$$\frac{dX_5}{dt} = f_3(k_1 X_1 + k_2 X_2 + k_3 X_3 + k_4 X_4) - (1-f_4)k_5 X_5$$

$$\frac{dX_6}{dt} = 0$$

$$X_{tot} = I \left\{ \frac{\gamma}{k_1} + \frac{1-\gamma}{k_2} + \frac{f_1}{(1-f_1)k_3} + \frac{1}{\alpha} \left[\frac{f_1}{(1-f_1)k_3} + \frac{1}{k_4} + \frac{(1-f_1)}{f_2 k_5} \right] \right\} + X_6$$

where,

$$\alpha = \left[\frac{(1-f_1)^2(1-f_4)}{f_2 f_3} - 1 \right]$$

γ = fraction decomposable plant material in fresh litter

I = litter input

Century

$$\frac{dX_1}{dt} = \beta I - k_1 X_1$$

$$\frac{dX_2}{dt} = (1-\beta)I - k_2 X_2$$

$$\frac{dX_3}{dt} = f_1 k_1 X_1 + f_2 k_2 (1-\lambda)X_2 + f_7 f_4 X_4 + f_8 k_5 X_5 - k_3 X_3$$

$$\frac{dX_4}{dt} = f_3 \lambda k_2 X_2 + f_4 k_3 X_3 - k_4 X_4$$

$$\frac{dX_5}{dt} = f_5 k_3 X_3 + f_6 k_4 X_4 - k_5 X_5$$

$$X_{tot} = I\left\{\left[\frac{1}{k_3} + \frac{f_4}{k_4} + \frac{f_5 + f_4 f_6}{k_5}\right]\alpha + \left[\frac{1}{k_4} + \frac{f_6}{k_5}\right]f_3\lambda(1-\beta)\right\}$$

$$\alpha = \frac{f_1\beta + [f_2(1-\lambda) + f_3\lambda(f_7 + f_6 f_8)](1-\beta)}{(1 - f_4 f_7 - f_5 f_8 - f_4 f_6 f_8)}$$

$\beta = f(\Lambda/n_l) = $ metabolic fraction of litter input (I)

$\Lambda = $ lignin content of litter

$n_l = $ nitrogen content of litter

$\lambda = \Lambda/(1-\beta) = $ lignin-to-structural ratio of litter input

Q Model

See Equations 24.4 to 24.7 in text

$$X_{tot} = \frac{I_0}{f_c \eta_H \mu_0 q_0^\beta \dfrac{1-e_0}{\eta_H e_0} - \beta}$$

$e_0 = $ microbial efficiency for litter quality q_0

$\beta = $ a constant relating the effect of quality on the microbial growth rate

$f_c = C$ concentration of decomposers

$\eta_H = $ average displacement (Equation 24.7)

$\mu_0 = $ quality modifier for rate of assimilation

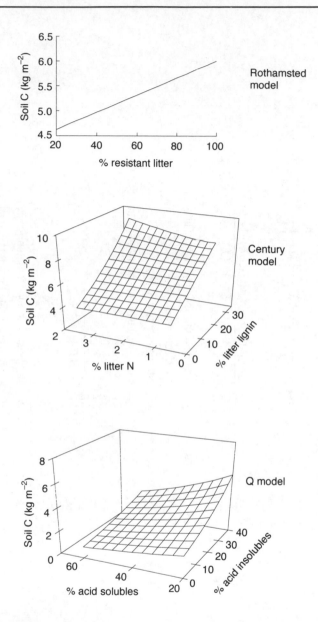

Fig. 24.3. Total soil C levels, at steady-state, across a range of litter qualities for the Rothamsted (Jenkinson *et al.*, 1987), Century (Parton *et al.*, 1987) and Q (Ågren and Bosatta, 1987) models. See text for description of steady-state assumptions.

coefficients (f_i) that govern the flow of C between pools. In both models, decomposition constants are influenced by factors such as temperature and moisture and some of the transfer coefficients are influenced by soil texture and/or clay mineralogy (expressed as inorganic CEC in the Rothamsted model, and by the percentage of sand, silt and clay, in Century). Both models are resolved at a monthly time scale and thus, for the steady-state solutions, temperature and moisture regimes were assumed to be constant and specific decomposition rates were set to an annual basis. The rate parameters in the Q model are already defined for an annual basis.

For the model comparisons, we attempted to equalize as many factors determining soil C levels as possible. The annual rate of C input was set at 300 g m^{-2} year^{-1}, and mean annual temperature at 10°C. Soil moisture was assumed not to be limiting for decomposition rates. A medium textured (loam) soil was assumed since texture is a factor in both the Rothamsted model and Century (but not the Q model).

Total C levels under steady-state conditions for the three models were calculated as a function of litter quality. Because the form of the functional response to litter quality varies between models, the quality values used for model comparisons are not strictly equivalent but rather span the range of values which might be found in most plant litters. In the Rothamsted model, initial litter quality is expressed as the proportion of total litter in the resistant fraction. For Century, litter quality is expressed on the basis of lignin and N contents. For the Q model, the parameter defining the initial litter quality, q_0, has been related to proximate chemical analyses by Ågren and Bosatta (1996) according to a regression analysis,

$$q_0 = c_{ex}q_{ex} + c_{as}q_{as} + c_{ai}q_{ai} \quad (24.8)$$

where c_{ex}, c_{as} and c_{ai} are the fractions of soluble extractives, acid solubles and acid insolubles, respectively (i.e. $c_{ex} + c_{as} + c_{ai} = 1$). Initial quality values, q_{ex}, q_{as} and q_{ai} were estimated to be 1, 1.25 and 0.65, respectively (Ågren and Bosatta, 1996). The higher quality of acid solubles relative to the soluble extractives might be caused by some highly resistant compounds in the non-polar soluble extractable fraction.

Total C varies as a function of litter quality in all three models, but there are differences in the amounts of soil C under the specified conditions (Fig. 24.3). The Century model gave the greatest magnitude of response to the range of litter qualities, ranging from ~ 4 kgC m^{-2} to almost 10 kg C m^{-2}, while the Rothamsted model varied between 4 to 6 kg m^{-2} and the Q model between <1 and 5 kg C m^{-2}.

The Rothamsted model yielded a linear relationship between litter quality and total C while the Century and Q models exibited curvilinear response surfaces (Fig. 24.3). Soil C level predicted by the Rothamsted model increases in direct proportion to the size of the resistant litter fraction. In Q, soil C increases with higher amounts of acid insolubles (e.g. lignin) and decreases as the proportion of acid solubles (e.g. cellulose and hemicellulose) increases. With the Century model, total soil C increases with increasing lignin content but also with increasing N content. The latter response initially appears counterintuitive since larger N contents are generally associated with higher litter quality. However, the response can be understood by recalling the way in which lignin content influences decomposition in the model. At a given lignin content, increasing N contents leads to a lower structural fraction (and a higher metabolic fraction) in the fresh litter. Thus, the lignin-to-structural litter ratio increases, which decreases the decomposition rate of the structural fraction. The net effect is to cause an overall increase in the total soil C due to the increase in the steady-state level of the structural pool.

To interpret the influence of litter quality in these models more fully, we examined the steady-state behaviours of individual pools, in the case of the Rothamsted and Century models, or the integrated quality variable in the case of the Q model.

In the Rothamsted model, differences in litter quality do not affect the size of the secondary organic matter pools, i.e. the humus and microbial biomass (Fig. 24.4). The only change in the composition of the total organic matter, as a function of litter quality, is in the amount of undecomposed litter. This outcome is a result of the assumption that the proportion of decomposed litter which is assimilated by the microbial biomass is the same for both the decomposable and the resistant litter fractions. The resistant fraction has a longer mean residence time than the decomposable fraction and, therefore, the total standing stock of litter at steady-state increases as the litter quality decreases.

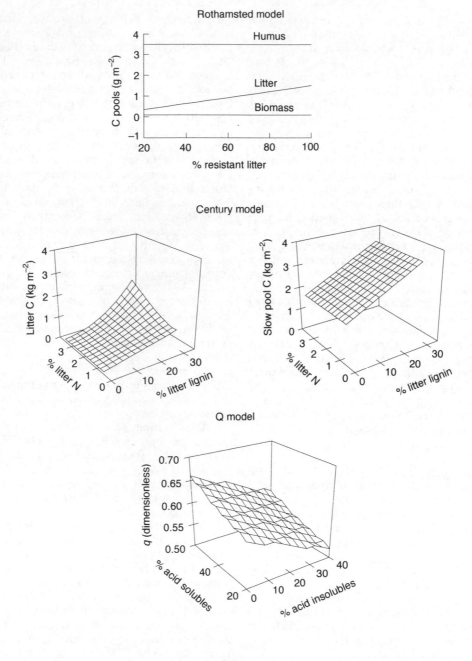

Fig. 24.4. Change in amounts of different C pools, at steady-state, as a function of litter quality parameters in the Rothamsted (Jenkinson *et al.*, 1987) and Century (Parton *et al.*, 1987) models. For the Q model (Ågren and Bosatta, 1987), the effect of initial litter quality on the quality variable (*q*) for the whole soil is shown. Values for the two microbial biomass pools and the two litter pools in the Rothamsted model were combined. Values for the litter pool in Century are for structural plus metabolic fraction (active and passive pools are not shown).

However, the throughput rate to the microbial biomass, and subsequently to the humus pool, is unaffected and thus the size of these pools is constant as a function of litter quality.

The influence of litter quality on SOM composition in the Century model is more complex and the amounts in all the SOM pools change as a function of lignin, N content or both (Fig. 24.4). Steady-state litter levels increase in a non-linear fashion as lignin and N litter contents increase. The increases as a function of N content and the interaction with lignin is mainly associated with the change in the lignin to structural litter ratio as described above. The active (i.e. microbial biomass) fraction (not shown) is least sensitive to litter quality differences. Both the slow and passive fractions show a linear response to lignin content but are unaffected by litter N content. The response of these secondary SOM pools is mainly a consequence of the assumption that the lignin portion of the structural litter is stabilized directly into the slow pool and is not metabolized by the microbial biomass.

In the Q model, changes in the composition or 'quality' of the soil organic matter as a function of initial litter quality is encapsulated in a single quality variable, q, as opposed to changes in discrete pool sizes as in the other two models. At steady-state, the turnover rate of the whole soil organic matter is proportional to q. The value of q is shown to vary linearly with respect to the proximate chemical fractions used to define the initial litter quality (Fig. 24.4). For the range of initial litter qualities included in the analysis, the overall value of q for the SOM varied between about 0.50 to 0.65. As a point of reference this can be compared to the value of q for the acid-insoluble fraction of fresh litter of 0.65.

Both the Century (Paustian *et al.*, 1992) and Q models (Hyvönen *et al.*, 1996) have been applied to a long-term experiment in which SOM was regularly monitored over more than 30 years in cropped field plots. In this experiment the principal treatment factor was addition of different organic materials of varying

decomposability, ranging from green plant matter to sawdust and peat amendments. Inorganic N fertilizers were also added in some treatments. In the organic amendment treatments, amounts of dry matter added were similar. However, there were differences in total C inputs for some treatments due to differences in crop residue production (roots and stem bases were incorporated into the soil) across treatments. Nevertheless the main effects of the differences in litter quality on changes in SOM levels were clearly demonstrated.

The models were applied somewhat differently and not all treatments were simulated by both models. In the Century application, C and N dynamics of the added material as well as organic matter already present in the soil was simulated (Fig. 24.5). In the Q model application, only the turnover of the added material since the start of the experiment (in 1956) was simulated (Fig. 24.6).

Amounts of SOM, both total and residual, were greatest in the plots receiving the most recalcitrant material (e.g. sawdust, peat) and smallest for additions of fresh litter such as green (plant) manure and cereal straw. Both models were able to represent the differences in organic matter changes between treatments successfully, as a function of litter quality (Figs 24.5 and 24.6). In the Century model application, most of the differences in SOM levels in the organic amended treatments could be explained by differences in lignin content and by the indirect effects of N additions (organic and inorganic) on plant growth and subsequent root C inputs. However, neither model was parameterized completely independently of the data. In the Century model application, an analysis of data compared with model output was used to initialize the relative size of the slow pool at the start of the experiment. In the Q model application, the initial quality parameter q_0 was fitted for each treatment to minimize differences between predicted versus observed data.

The question of how SOM composition (versus total amount) is affected by litter quality and whether or not these effects of quality are adequately represented

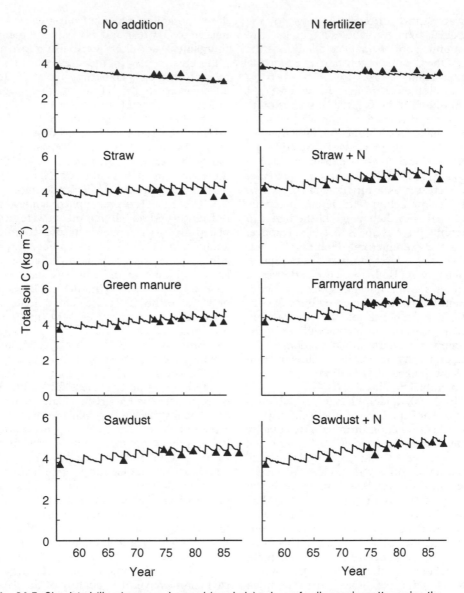

Fig. 24.5. Simulated (lines) versus observed (symbols) values of soil organic matter, using the Century model, for a long-term field experiment (started in 1956) in Uppsala, Sweden. Field treatments included biannual addition of similar amounts (equivalent to c. 185 g C m^{-2}year^{-1}) of different organic materials (except for no-addition and N fertilizer only treatments) (from Paustian *et al.,* 1992).

in existing models is more difficult to address. Experiments such as the one described above would lend themselves to such tests, but in most cases the organic matter data are limited to measurements of total C or short-term activity indices such as microbial biomass or C mineralization (e.g. Schnürer *et al.,* 1985). Analysis of soil organic matter characteristics using techniques such as nuclear magnetic resonance (NMR) (Guggenberger *et al.,* 1994; Skjemstad *et al.,* Chapter 20 this volume), particle and aggregate size fractionations (Oades, 1993; Cambardella and Elliott,

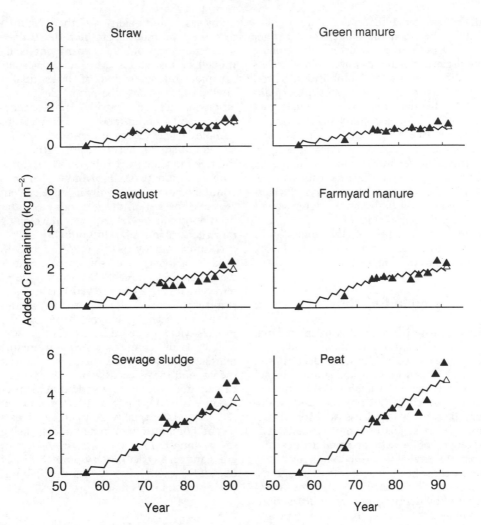

Fig. 24.6. Simulated (lines) versus observed (symbols) values of the amount of organic amendments remaining in the soil, using the Q model, for a long-term field experiment (started in 1956) in Uppsala, Sweden. Field treatments included biannual addition of similar amounts (equivalent to c. 185 g C m^{-2} year^{-1}) of different organic materials (from Hyvönen *et al.*, 1996).

1994) and ^{13}C (Balesdent *et al.*, 1988, Jastrow *et al.*, 1996) and ^{14}C (Jenkinson *et al.*, 1987; Trumbore, 1993) tracers in such experiments could help to elucidate differences in SOM composition and turnover rates. Carefully selected natural sites, where litter quality was the principal difference, and other factors such as climate, soil texture, C input rate were similar, could also provide opportunities to evaluate litter quality effects on soil organic matter.

In summary, there are several similarities as well as differences in the assumptions of how litter quality affects SOM dynamics in these three models. Changes in total amounts of SOM as a function of litter quality were qualitatively similar; total organic matter increases as litter quality decreases. In contrast, the magnitudes and sensitivity of SOM contents to litter quality differed substantially between models. However, exact comparisons are difficult since the definitions of litter quality differ between the models as do other parameters and relationships influencing soil organic matter turnover. The models also differ in how the composition of soil organic matter is influenced by litter quality. The Rothamsted model predicts that the amounts of secondary decomposition products (i.e. microbial biomass and humus) are unaffected by litter quality. In contrast, the secondary soil organic matter pools in Century change as a function of litter quality, due to the inclusion of different stabilization pathways for lignin versus non-lignin components of litter. In the Q model, the quality of the total SOM changes in direct proportion to the quality of the litter entering the soil.

Concluding Remarks

Litter quality is a central issue in research on decomposition and most models of decomposition consider some facet of litter quality. We propose a conceptual framework in which litter quality can be broken down into chemical, physical and inhibitory factors. In the classical Monod formulation these factors can be articulated as modifiers of growth or uptake rates and substrate availability. Operationally, litter quality has been incorporated into most decomposition models in a fairly rudimentary way by defining discrete litter fractions or 'pools'. More recently, the concept of a continuous quality spectrum has been formalized.

Despite substantial differences in structure and assumptions, many different decomposition models can successfully 'fit' the time course of different types of decomposing litter. One difficulty in conducting rigorous testing of existing models is that in many cases litter quality (or other) parameters used in models are not determined independently from the experiment being modelled. Just as there are no universally accepted measurements of litter quality attributes, there are different model formulations of litter quality which make evaluation and inter-model comparisons complex. The recent initiation of multi-site, multi-species decomposition studies in the US and Canada (M. Harmon, personal communication; A. Trofymow, personal communication) will provide an excellent resource for testing the generality of litter decomposition models and for developing more general indices of litter quality.

Litter quality is also considered in models of longer term soil organic matter dynamics. The three SOM models compared in this chapter showed qualitatively similar responses of soil organic matter levels to litter quality. However, the models yield quite different results with respect to the composition of soil organic matter. Well-controlled long-term experiments and cross-ecosystem comparisons can aid in further testing and refinement of theories of soil organic matter dynamics in relation to litter quality. Activities such as a recent comparison of nine models using seven long-term data sets from forest, grassland and agriculture systems (Smith *et al.*, 1996) exemplify promising collaborative research in this field.

Acknowledgements

We thank two anonymous reviewers for constructive comments on earlier reviews of the manuscript. Support for work on this chapter by K. Paustian from a National Science Foundation Grant, DEB-9419854 is acknowledged.

References

Aber, J.D. and Melillo, J.M. (1982) Nitrogen immobilization in decaying hardwood leaf litter as a function of initial nitrogen and lignin content. *Canadian Journal of Botany* 60, 2263-2269.

Ågren, G.I. and Bosatta, E. (1987) Theoretical analysis of the long-term dynamics of carbon and nitrogen in soils. *Ecology* 68, 1181–1189.

Ågren, G.I. and Bosatta, E. (1996) Quality – a bridge between theory and experiment in soil organic matter studies. *Oikos* (in press).

Ågren, G.I., Johnson, D.W., Kirschbaum, M. and Bosatta, E. (1996) Ecosystem physiology – Soil organic matter. In: Breymeyer, A., Hall, D.O., Melillo, J.M. and Ågren, G.I. (eds) *Global Change: Effects on Forests and Grasslands*, J.Wiley (in press).

Andrén, O. and Paustian, K. (1987) Barley straw decomposition in the field: a comparison of models. *Ecology* 68, 1190–1200.

Balesdent, J., Wagner, G.H. and Mariotti, A. (1988) Soil organic matter turnover in long term field experiments as revealed by carbon-13 natural abundance. *Soil Science Society of America Journal* 52, 118–124.

Berg, B. and Ågren, G.I. (1984) Decomposition of needle litter and its organic chemical components – theory and field experiments. Long-term decomposition in a Scots pine forest. III. *Canadian Journal of Botany* 62, 2880–2888.

Berg, B., Müller, M. and Wessén, B. (1987) Decomposition of red clover (*Trifolium pratense*) roots. *Soil Biology and Biochemistry* 18, 589–593.

Berg, B. and Ekbohm, G. (1991) Litter mass-loss rates and decomposition patterns in some needle and leaf litter types. Long-term decomposition in a Scots pine forest. VII. *Canadian Journal of Botany* 69, 1449–1456.

Bosatta, E. and Ågren, G.I. (1985) Theoretical analysis of decomposition of heterogeneous substrates. *Soil Biology and Biochemistry* 17, 601–610.

Bosatta, E. and Ågren, G.I. (1991) Dynamics of carbon and nitrogen in the organic matter of the soil: a generic theory. *The American Naturalist* 138, 227–245.

Bosatta, E. and Ågren, G.I. (1994) Theoretical analysis of microbial biomass dynamics in soils. *Soil Biology and Biochemistry* 26, 143–148.

Boudreau, B.P. (1992) A kinetic model for microbic organic-matter decomposition in marine sediments. *FEMS Microbiology Ecology* 102, 1–14.

Cambardella, C.A. and Elliott, E.T. (1994) Carbon and nitrogen dynamics of soil organic matter fractions from cultivated grassland soils. *Soil Science Society of America Journal* 58, 123–130.

Carpenter, S.R. (1981) Decay of heterogenous detritus: a general model. *Journal of Theoretical Biology* 89, 539–547.

Guggenberger, G., Christensen, B.T. and Zech, W. (1994) Land-use effects on the composition of organic matter in particle-size separates of soil: I. Lignin and carbohydrate signature. *European Journal of Soil Science* 45, 449–458.

Hunt, H.W. (1977) A simulation model for decomposition in grasslands. *Ecology* 58, 469–484.

Hyvönen, R., Ågren, G.I. and Andren, O. (1996) Modeling long-term carbon and nitrogen dynamics in an arable soil receiving organic matter amendments. *Ecological Applications* (in press).

Janssen, B.H. (1984) A simple method for calculating decomposition and accumulation of 'young' soil organic matter. *Plant and Soil* 76, 481–491.

Jastrow, J.D., Boutoon, T.W. and Miller, R.M. (1996) Carbon dynamics of aggregate-associated organic matter estimated by ^{13}C natural abundance. *Soil Science Society of America Journal* 60, 801–807.

Jenkinson, D.S. (1977) Studies on the decomposition of plant material in soil. V. The effects of plant cover and soil type on the loss of carbon from ^{14}C labeled ryegrass decomposing under field conditions. *Journal of Soil Science* 28, 424–434.

Jenkinson, D.S., Hart, P.B.S., Rayner, J.H. and Parry, L.C. (1987) Modelling the turnover of organic matter in long-term experiments at Rothamsted. *INTECOL Bulletin* 15, 1–8.

Juma, N.G. and McGill, W.B. (1986) Decomposition and nutrient cycling in agroecosystems. In: Mitchell, M.J. and Nakas, J.P. (eds) *Microfloral and Faunal Interactions in Natural and Agro-ecosystems*. Martinus Nijhoff/Dr W. Junk Publishers, Dordrecht, pp. 74–136.

Li, C., Frolking, S. and Frolking, T.A. (1992) A model of nitrous oxide evolution from soil driven by rainfall events: 1. Model structure and sensitivity. *Journal of Geophysical Research* 97, 9759–9776.

McGill, W.B., Hunt, H.W., Woodmansee, R.G. and Reuss, J.O. (1981) PHOENIX, a model of the dynamics of carbon and nitrogen in grassland soil. In: Clark, F.E. and Rosswall, T. (eds) *Terrestrial Nitrogen Cycles. Processes, Ecosystem Strategies and Management Impacts. Ecological Bulletins (Stockholm)* 33, 49–115.

Meentemeyer, V. (1978) Macroclimate and lignin control of litter decomposition rates. *Ecology* 59, 465–472.

Middelburg, J.J. (1989) A simple rate model for organic matter decomposition in marine sediments. *Geochimica et Cosmochimica Acta* 53, 1577–1581.

Minderman, G. (1968) Addition, decomposition and accumulation of organic matter in forests. *Journal of Ecology* 56, 355–362.

Monod, J. (1942) Recherches sur la croissance des cultures bactériennes. *Actualités Scientifiques et Industrielles Microbiologie*. Hermann & Cie, Paris, 210 pp.

Moorhead D.L and Reynolds J.F. (1991) A general model of litter decomposition in the northern Chihuahuan Desert. *Ecological Modelling* 56, 197–219.

Oades, J.M. (1993) The role of biology in the formation, stabilization and degradation of soil structure. *Geoderma* 56, 377–400.

Palm, C.A. and Sanchez, P.A. (1990) Decomposition and nutrient release patterns of the leaves of three tropical legumes. *Biotropica* 22, 330–338.

Parnas, H. (1975) Model for decomposition of organic material by microorganisms. *Soil Biology and Biochemistry* 7, 161–169.

Parton, W.J., Schimel, D.S., Cole, C.V. and Ojima, D.S. (1987) Analysis of factors controlling soil organic matter levels in Great Plains grasslands. *Soil Science Society of America Journal* 51, 1173–1179.

Parton, W.J., Ojima, D.S., Cole, C.V. and Schimel, D.S. (1994) A general model for soil organic matter dynamics: Sensitivity to litter chemistry, texture and management. In: *Quantitative Modeling of Soil Forming Processes*. Special Publication 39, Soil Science Society of America, Madison, WI, pp. 147–167.

Pastor, J. and Post, W.M. (1986) Influence of climate, soil moisture and succession on forest carbon and nitrogen cycles. *Biogeochemistry* 2, 3–27.

Paustian, K. (1994) Modelling soil biology and biochemical processes for sustainable agriculture research. In: Pankhurst, C.E., Doube, D.M., Gupta, V.V.S.R. and Grace, P.R. (eds) *Soil Biota: Management in Sustainable Farming Systems*. CSIRO, Melbourne, Australia, pp. 182–193.

Paustian, K., Parton, W.J. and Persson, J. (1992) Modeling soil organic matter in organic-amended and nitrogen-fertilized long-term plots. *Soil Science Society of America Journal* 56, 476–488.

Schnürer, J., Clarholm, M. and Rosswall, T. (1985) Microbial biomass and activity in an agricultural soil with different organic matter conents. *Soil Biology and Biochemistry* 17, 611–618.

Sinsabaugh, R.L. and Moorhead, D.L. (1994) Resource allocation to extracellular enzyme production: A model for nitrogen and phosphorus control of litter decomposition. *Soil Biology and Biochemistry* 26, 1305–1311.

Smith, O.L. (1979) An analytical model of the decomposition of soil organic matter. *Soil Biology and Biochemistry* 11, 585–606.

Smith, P., Smith J.U., Powlson, D.S., Arah, J.R.M., Chertov, O.G., Coleman, K., Franko, U., Gunnewiek, H.K., Jenkinson, D.S., Jensen, L.S., Kelly, R., Li, C., Molina, J.A.E., Mueller, T. and Parton, W.J. (1996) A comparison of the performance of nine soil organic matter models using datasets from seven long term experiments. *Geoderma* (in press).

Stevenson, F.J. and Elliott, E.T. (1989) Methodologies for assessing the quantity and quality of soil organic matter. In: Coleman, D.C., Oades, J.M. and Uehara, G. (eds) *Dynamics of Soil Organic Matter in Tropical Ecosystems*, University of Hawaii Press, pp. 173–241.

Stott, D.E., Kassim, G., Jarrell, W.M., Martin, J.P. and Haider, K. (1983) Stabilization and incorporation into biomass of specific plant carbon during biodegradation in soil. *Plant and Soil* 70, 15–26.

Swift, M.J., Heal, O.W. and Anderson, J.M. (1979) *Decomposition in Terrestrial Ecosystems*. Blackwell, Oxford, UK, 372 pp.

Tian, G., Kang, B.T. and Brussaard, L. (1992) Biological effects of plant residues with contrasting chemical compositions under humid tropical conditions – Decomposition and nutrient release. *Soil Biology and Biochemistry* 24, 1051–1060.

Tian, G., Brussaard, L. and Kang, B.T. (1995) An index for assessing the quality of plant residues and evaluating their effects on soil and crop in the (sub-)humid tropics. *Applied Soil Ecology* 2, 25–32.

Trumbore, S.E. (1993) Comparison of carbon dynamics in tropical and temperate soils using radiocarbon measurements. *Global Biogeochemical Cycles* 7, 275–290.

Verberne, E.L.J., Hassink, J., de Willigen, P., Groot, J.J.R. and van Veen, J.A. (1990) Modelling organic matter dynamics in different soils. *Netherlands Journal of Agricultural Science* 38, 221–238.

Waksman, S.A. (1936) *Humus: Origina, Chemical Composition and Importance in Nature.* Williams and Wilkins, Baltimore, 494 pp.

Waksman, S.A. and Tenney, F.G. (1926) On the origin and nature of the soil organic matter or soil 'humus'. IV. The decomposition of the various ingredients of straw and alfalfa meal by mixed and pure cultures of microorganisms. *Soil Science* 22, 395–406.

Wessén, B. and Berg, B. (1986) Long-term decomposition of barley straw: chemical changes and ingrowth of fungal mycelium. *Soil Biology and Biochemistry* 18, 53–59.

Williams, S.T. and Gray, T.R.G. (1974) Decomposition of litter on the soil surface. In: Dickenson, C.H. and Pugh, G.J.F. (eds) *Biology of Plant Litter Decomposition*, Vol. 2. Academic Press, London , pp. 611–622.

Zech, W. and Kögel-Knabner, I. (1994) Patterns and regulation of organic matter transformations in soils: Litter decomposition and humification. In: Schulze, E.D. (ed.) *Flux Control in Biological Systems*. Academic Press, San Diego, pp. 303–334.

25 Simulating the Mineralization of N from Crop Residues in Relation to Residue Quality

A.P. Whitmore[1,*] and E. Handayanto[2]

[1]DLO Research Institute for Agrobiology and Soil Fertility, PO Box 129, NL-9750 AC Haren, The Netherlands; [2] Faculty of Agriculture, Brawijaya University, Jl. Veteran, Malang, East Java, Republic of Indonesia

Introduction

Different definitions of the quality of crop residues or organic matter in soil can be given depending upon what property is of interest. Looked at purely from the point of view of decomposition, glucose is a high quality substrate because it is used immediately by micro-organisms and is decomposed rapidly in soil. From the point of view of supplying nutrients to plants it is a very poor quality substrate. In decomposition terms, soil humus is of poor quality because it breaks down only slowly. Even so, it is rich in nutrients and can supply much of a plant's needs (e.g. Addiscott *et al.*, 1991). This very property of resistance to decomposition can even be of great value: say in limiting erosion of soils. From an agronomic and environmental point of view, residues whose nutrients remain locked up in soil until the time at which crops need to use them are of much better quality than those that lose their nutrients to the environment too soon or release them too late for the farmers' commercial crops to use (Vallis and Jones, 1973; Whitmore and van Noordwijk, 1995). Thus quality is many-sided, depends on function and in this sense is subjective. Consequently, good, bad or better, high, low or average are not terms that will be used in this chapter with regard to quality.

Models are tools that succinctly sum-marize existing knowledge or, where shown to be reliable, help with decision making and planning. Models need not be complicated: the widely-used lignin-to-N index (Becker and Ladha, Chapter 18, this volume) is a model that predicts quite well the relative availability of N from different crop residues within a given period, for example, a growing season. Such a model is static, however, and the words *relative* and *within* are essential qualifiers of the model. In fact all decomposition processes have a dynamic and many are non-linear. This means that the relationship between lignin-to-N and decomposition may differ with time after incorporation of the residue or with climate (see Meentemeyer 1978 and the section on activation energy below). Worse, Kachaka *et al.* (1993) have shown that young plants contain much less lignin than mature ones; this may not affect crop residues derived from plants of similar age, but it will greatly influence the decomposition of tree prunings taken at different stages of growth. For these reasons dynamic computer simulation models based on sound mechanisms, even if fairly simple, can give a more reliable description of the decomposition of crop residues than empirical, static models. The sections below outline some of the ways in which the quality of crop residues can be handled within models and they explore them with reference to laboratory and field experimental data.

* Corresponding author.

Aspects of Quality and Computer Simulation Models

This chapter describes adaptations only to existing computer simulation models. Leaching is calculated using the capacity approach (e.g. Burns, 1975; Addiscott, 1977) in which incoming rainfall fills consecutive soil layers dissolving or diluting soluble nitrogen already present; when a layer is filled, further rain displaces any water present to the layer below and so on. Layers were 5 cm thick in this study. Organic matter turnover, as described by Bradbury *et al.* (1993) and modified by Whitmore and Groot (1994), expresses the decomposition process by summarizing the state of soil organic matter after each time interval. After any form of organic carbon in soil has decomposed, a proportion of it, α, can be found in microbial biomass, a proportion, β, in the soil humus that is relatively resistant to further decomposition, and the remainder, $(1-\alpha-\beta)$, as CO_2. The rate of decomposition is a (changing) function of quality (see next section). The proportions α and β (cf. *E* in Equation 25.2 below) are fixed as is the C-to-N ratio of both the biomass and humus. The C-to-N ratio of residues as shown below (Equation 25.2) determines whether N is mineralized or immobilized.

Quality and decomposition

Whitmore and Matus (1996) modelled the effect of resistance to decomposition of a substrate in relation to quality as follows. They assumed that the fibre would impede the action of microorganisms or enzymes in decomposing the rest of the crop residues. As the easily decomposable parts of the residues break down, the fibrous parts become more concentrated in the remainder and retard decomposition to a greater and greater extent. A factor f_R, that multiplies the rate of decomposition, is calculated in a manner analogous to that proposed by Parton *et al.* (1987) (see also: Paustian *et al.*, Chapter 24, this volume).

$$f_R = \exp\left(\frac{-k_R F C_0}{C_t}\right) \qquad (25.1)$$

where k_R is a constant and F is the fraction of fibre carbon in crop residues (hemi-cellulose plus cellulose plus lignin in these simulations), C_0 is the initial carbon content of the crop residues and C_t the carbon remaining undecomposed at time t. Table 25.1 shows the difference in quality of the parts of some ^{15}N-labelled clover and wheat plants incorporated in soil (Whitmore and Matus, 1996). Although the two species differed in N content, the main difference was in fibre content and easily decomposable organic matter (calculated by difference). The residues (12.7 g shoots and 3.2 g roots) were added to 6200 g of a sandy soil in open (top and bottom) tubes, buried in soil out of doors and subject to losses by leaching. Lignin and fibre were measured in the residues using the method of van Soest (1967).

Quality and the supply of nitrogen

The carbon-to-nitrogen ratio (C-to-N) has long been asserted to be of great relevance

Table 25.1. The quality of the wheat and clover residues in relation to their soluble and fibrous carbon (data from Whitmore and Matus, 1996).

| Residue | C-to-N ratio | % dry matter | | | |
		Cellulose	Hemi-cellulose	Lignin	Soluble
Wheat shoot	9.8	22.1	14.5	3.7	11.1
Wheat root	18.2	32.4	27.3	4.0	2.7
Clover shoot	7.6	19.0	1.5	3.3	11.3
Clover root	9.1	27.3	4.1	6.9	5.3

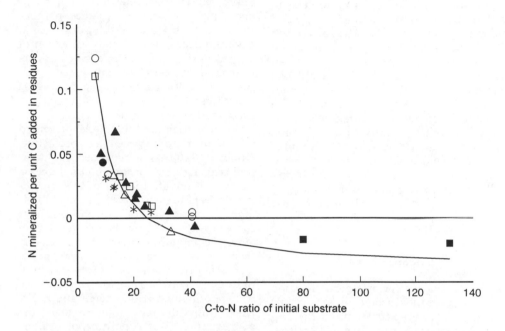

Fig. 25.1. The relationship between nitrogen mineralized (or immobilized) and the carbon-to-nitrogen ratio of added organic matter. The data comes from the following sources: □, Franzluebbers *et al.* (1994); △, Zagal and Persson (1994); ■, Nieder and Richter (1989); ⋆, Jensen (1929); ▲, Chae and Tabatabai (1985); ●, Thorup-Kristensen (1994); ○, Whitmore, unpublished. The solid line is the theoretical relationship derived from Equation (25.2).

to the speed with which N is released in soil, but it is important to realize that this is not the same thing as decomposition (see model case studies below). In general, the smaller or narrower the C-to-N the faster is N released from crop residues (Jensen, 1929). Where residues are uniformly decomposable, the following equation applies:

$$N = C_0 \left\{ \frac{1}{Z} - \frac{E}{Y} \right\} \qquad (25.2)$$

In this equation, N is the amount of nitrogen mineralized once decomposition is complete or almost so, Z is the C-to-N of the substrate (assumed constant), E is a microbiological efficiency factor that can vary for several reasons but can be taken as

about 0.4, and Y is the C-to-N of the end product of the decomposition process, here humus with C-to-N 10. Microbes are assumed to have a demand for nitrogen in order to decompose crop residues. They and their products must maintain a certain C-to-N ratio. This value is almost always narrower than the residues themselves but because microbes are not 100% efficient ($0 < E < 1$) the critical C-to-N ratio, Z_c, where mineralization switches to immobilization, is often wider than in the residues and, with the values assumed for E and Y above, is about 25. Looked at in this way mineralization is less a question of decomposition, but rather of supply and demand. Residues that are highly heterogeneous in terms of quality may deviate from what is predicted by Equation (25.2) which, as

Fog (1988) has argued, is irrelevant or even misleading when applied to recalcitrant substrates. None the less Fig. 25.1 shows that this function is quite general, with data taken from a number of field and laboratory experiments (Jensen 1929; Chae and Tabatabai, 1985; Nieder and Richter, 1989; Franzluebbers *et al.*, 1994; Thorup-Kristensen, 1994; Zagal and Persson, 1994; and Whitmore unpublished). Equation (25.2) also forms the foundation of the mineralization sections of many other models (e.g. Century, Paustian *et al.*, Chapter 24, this volume; DAISY, Magid *et al.*, Chapter 26, this volume); it expresses the release of N after decomposition is complete. Replacing C_0 with $(C_0 - C_t)$ gives the release of N after time t.

Activation energy

Before a chemical (or biological) reaction can take place the reactants must normally acquire a certain activation energy, E_a, (e.g. Moore, 1963); E_a may differ from substance to substance. Interestingly, enzymes function by making this quantity lower than it otherwise would be. Several authors have stated that quality affects the way different crop residues decompose or release N at low temperatures (Meentemeyer, 1978; Nicolardot *et al.*, 1994; Vigil and Kissel, 1995; De Neve *et al.*, 1996). At 28°C, for example, the decomposition of glucose was 4.6 times faster than hollocellulose but at 5°C it was 17.6 times faster (Nicolardot *et al.*, 1994); although these authors eschewed the idea, their data translates into activation energies, E_a, of 45.9 kJ mol^{-1} (5521 °K^{-1}; $E_a \times R$ the gas constant) and 64.0 kJ mol^{-1} (7702 °K^{-1}) for glucose and hollocellulose respectively. The model described in this chapter uses these values of the activation energy to distinguish between lignin-rich and easily-decomposable residues. This is necessarily rather crude but without such a distinction a model based on decomposition in temperate climates would not estimate decomposition in warmer climates correctly.

Binding of organic matter in soil – polyphenolic compounds

Several authors have shown that polyphenolic compounds in crop residues, particularly tree litter and residues found in tropical regions, reduce both the extent of decomposition and release of nutrients from crop residues (Vallis and Jones, 1973; Fox *et al.*, 1990; Palm and Sanchez, 1991; Handayanto *et al.*, 1994). Although the exact mechanism by which polyphenols act is not yet certain and the model presented below must be regarded as speculative, Kuiters (1990) has summarized the literature arguing that polyphenolics increase the rate of humification of organic matter in soil. Measurements of the protein binding capacity (P_{BC}) of polyphenols have shown that some polyphenols react with amino compounds much more readily than others; where soluble polyphenolics were retained in soil this P_{BC} was shown to be better correlated with the rate of decomposition of, and mineralization of N from, tropical tree residues than the concentration of polyphenol alone (Handayanto *et al.*, 1994). The protein binding capacity was measured using the method of Dawra *et al.* (1988). In the model, active polyphenols (or strictly, the protein binding capacity) are assumed to form a separate pool distinct from the others but having the same decomposition characteristics as soil humus. The initial P_{BC} of the residues determines the proportion, u, of residues which supplies the polyphenol pool in soil:

$$u = \frac{f_P P_{BC}}{C_0} \qquad (25.3)$$

where f_P is a constant. The polyphenol pool diverts decomposing substrate to humus; in doing so, however, the polyphenol is itself consumed. Because the polyphenols that are most active are also soluble in water, they readily bind organic matter throughout soil. This very solubility, however, makes them subject to leaching and removal from both field soils and leaching tubes. Nitrogen is bound at the same time as carbon in this enhanced humification process and the C-to-N of

Table 25.2. Quality of the tropical tree prunings (data from Handayanto, 1994).

Pruning mixture % Gliricidia-to-Peltophorum*	C-to-N ratio	Lignin (%)	Protein binding capacity, P_{BC} (µg BSA[†] mg^{-1})
100:0	13.6	20.0	23
50:50	16.5	25.9	55
0:100	20.0	30.0	243

* Dry matter basis.
[†] Bovine serum albumin (Dawra *et al.*, 1988).

the polyphenol pool is assumed to be the same as that of the residues. Nitrogen is lost too, when polyphenols leach from soil. A model containing these ideas was first calibrated against some experiments in pots and in leaching tubes where *Gliricidia* and *Peltophorum* residues and a 50-to-50 mixture of the two were added to soil (Table 25.2; Handayanto, 1994; Handayanto *et al.*, 1994). In the pot experiments the polyphenols were assumed to act as described above but the leaching tubes were periodically flushed as described by Stanford and Smith (1972), removing the active polyphenols. Field simulations were also carried out using the calibrated model and supposing that similar concentrations of the residues to those in the laboratory experiments were applied to soil (mg kg^{-1}) at the beginning of October, just as the rains began; in the field simulations, however, leaching was not as severe as in the laboratory so that the polyphenols remained in soil for longer. Weather data was obtained from the Bah Lias Research Station on Sumatra, latitude 3° 11'S, longitude 99° 20'E. Burns' (1975) leaching equation was used to estimate the effect of rainfall (less evaporation) on both the polyphenols and mineral N in soil.

Model Case Studies

Fibre and decomposition

Simulations made using Equation (25.1) distinguished the rates of decomposition of the clover and wheat residues (Table 25.3). The essential difference in the pattern of decomposition is that the clover released labelled mineral N earlier than the wheat as a consequence of the greater preponderance of easily decomposable organic matter it contained and the greater proportion of fibre (particularly hemicellulose) in the wheat residues. The simulations reflected these differences well although the effect of leaching was slightly underestimated leading to an overestimation of the ^{15}N-labelled mineral N remaining in soil in both cases (Whitmore and Matus, 1996).

C-to-N and immobilization

Figure 25.2 plots the release of C and N from sugarbeet crowns (C-to-N = 40) incorporated in soil (5 mg crowns g^{-1} dry soil; Whitmore unpublished). Although the release of C from these readily decomposable crop residues was complete after 6 or 7 weeks, net mineralization of N only began after this time, exceeding background soil N mineralization after 12 weeks. The mineralization of C is a measure (in this case) of the decomposition of the residues themselves; the mineralization of N is a measure of the decomposition of microbial products formed during the initial decomposition. The total amount of N released can be described well by (25.2) but, as warned above, this differs from the decomposition itself as Fig. 25.2 makes clear. These pot incubations were carried out at 20°C and Whitmore and Groot (unpublished) have found that at ambient temperatures, this period of immobilization could exceed the growing season of a

Table 25.3. Modelled and measured release of ^{15}N-labelled mineral N remaining in soil (mg kg^{-1}) after 40 days incubation in field lysimeters of young crop residues differing in quality (data from Whitmore and Matus, 1996).

Residue	Modelled ^{15}N	Measured ^{15}N	Standard error
Wheat	24.4	12.6	5.67
Clover	41.5	26.4	5.26

spring sown crop. Following a crop such as sugar beet, the strategy for a farmer must be to ensure that the period between incorporation and sowing of the following crop is long enough, if the C-to-N of the residues is wide, to complete the initial decomposition, but not so long, if the C-to-N is narrow, that substantial leaching losses can occur before the crop has a chance to use the nitrogen.

Mineralization or immobilization?

A completely different risk can occur with young green plant residues that are easily decomposable and rich in N; for example, when non-hardy catch crops are killed by the first frosts of the winter. The decomposition of an N-rich catch crop can be contrasted with the decomposition of fibrous, mature maize residues having a much wider C-to-N. Assuming that the catch crop dies on 1st November in a temperate winter in the Netherlands (weather from Wageningen, latitude 51° 58′N, longitude 5° 40′E, during 1989–90 was used), that it leaves 1 t ha^{-1} of dry matter containing 5% N, Fig. 25.3 indicates what could have happened during the following winter and spring. At first, loss of N was tiny but after a mild December following the frost, and heavy rain in January, leaching losses began to be appreciable. Sowing barley in March or sugar beet in April (both deep rooting crops) would probably have prevented further losses and even led to capture of some of the N leached below 60 cm (the depth to which mineral nitrogen measurements are made in advising growers of these crops in the Netherlands; Anonymous, 1986). Sowing maize in May would have missed more and a shallow

rooting crop such as spinach sown in June would have missed the N altogether. Figure 25.3 also traces the loss of N following the incorporation of maize residues in October of the same autumn. The amount of N in residues added to soil has been assumed to be the same, but the C-to-N was much wider (60-to-1) and the presence of fibre retarded decomposition (Equation 25.1). Far less N leached during the winter from under the maize residues; far less of the N it contained was lost. Per hectare some 15 kg less N mineralized from the maize residues during the winter but this continued to become available afterwards, during the growing season. The decomposition of the catch crop was essentially complete before spring. A further advantage of the maize was that it immobilized some native N released from organic matter in soil.

None the less, catch crops have often been used to great advantage by farmers in temperate regions. The benefit is potentially even greater in tropical regions where the warmer temperatures and immense quantities of rain conspire to make the risk of loss very great indeed. However, the results of experiments are sometimes ambiguous because of the many factors that can differ. Not least is the risk that crop residues decompose and their N leach out before a catch crop can grow to take advantage. Catch crops in temperate regions are normally thought of as conserving nitrogen, but in tropical regions the extra carbon they stabilize in soil organic matter is equally valuable. A catch crop is sometimes thought of as a green manure. The extra manure can, however, be supplied in many forms: sometimes as prunings from bushes grown elsewhere especially for the purpose.

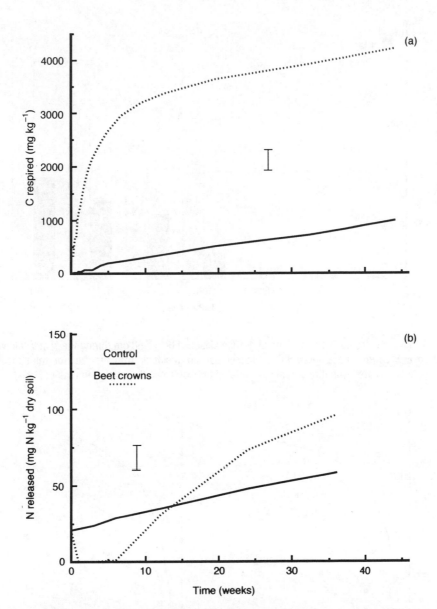

Fig. 25.2. A comparison between the cumulative amounts released and time course of the release of (a) carbon and (b) nitrogen from soils to which sugar beet crowns (C-to-N = 40) were added. In both cases the mineralization from a control soil with no addition is shown for comparison. Data from Whitmore (unpublished). Error bars show the standard error of the measurements.

Polyphenols

The model described above was used to simulate the release of N from tropical tree prunings that are often used in alley-crop-ping systems, and that contain polyphenolic compounds (Fig. 25.4). *Gliricidia* with a narrow C-to-N, less lignin and smaller P_{BC} (Table 25.2) mineralized N immediately; *Peltophorum* with a wider C-to-N,

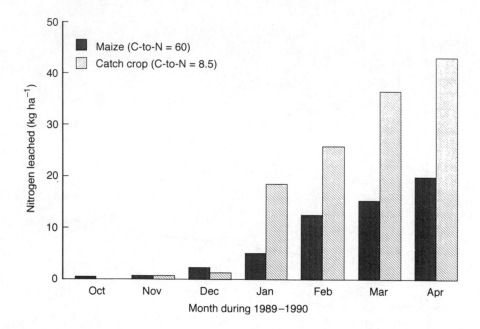

Fig. 25.3. Simulated cumulative amounts of nitrogen leached below 90 cm during the winter following either a cover crop (died-back 1 November, hatched bars) or residues from a maize crop (ploughed-in 1 October, shaded bars). Quantity of residues differed but contained the same amounts of N.

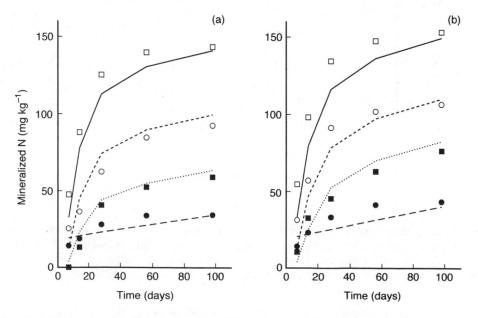

Fig. 25.4. The measured (points) and simulated (lines) amounts of mineralization of N from *Gliricidia* (□——); *Peltophorum* (■······); a mixture of the two, (○-----); background soil (●– – –), from incubations in (a) pots and (b) in leaching tubes. Data from Handayanto (1994).

more lignin and a greater P_{BC} at first immobilized native N from soil but later mineralized N over and above background. Strikingly more N mineralized from the *Peltophorum* residues incubated under leaching conditions than under non-leaching conditions and this is well represented in the simulations. This is because under leaching conditions the polyphenols, which divert extra N to humus, were removed from soil. The agreement between model and measurement is good except that the simulations under leaching conditions released N more slowly at the start than was actually the case (Fig. 25.4b). Release of N from the control soil was similarly retarded in the model compared with measurement which suggests that disturbing the soils to put them in the leaching tubes stimulated microbial activity and mineralization.

Do polyphenols benefit farmers?

In a simulation study the same amount of N (84 kg N ha^{-1}) in *Gliricidia* or *Peltophorum* prunings was added to soil. The total mass added in these simulations differed because the C-to-N of the residues differed. Approximately 35% more dry matter, equivalent to an extra 1.5 t ha^{-1}, was added with the *Peltophorum* prunings. Simulations began on 1 November. The amounts chosen corresponded roughly to the amounts of residues added in the laboratory experiments on which the model was calibrated. The simulations followed the mineral N leached during the following 3 months in a sandy soil containing a maximum of 30% water by volume in the top 15 cm of soil. Mineral N washed below this depth (a typical plough depth) was considered to have leached. During these 3 months the total rainfall was 325 mm, the mean air temperature was 26.6°C and the mean daily evaporation from bare soil was 3.5 mm day^{-1}. Note however, that these are bare soil simulations: no crop was present. The leaching losses from the *Gliricidia* treatments are clearly greater than from the *Peltophorum* (Fig. 25.5a) and this is a con-

sequence of the earlier and greater mineralization of N from residues. The retardation of decomposition of the *Peltophorum* confirms the idea that polyphenol-rich mulches may help to retain N in soil to the benefit of crops, but the amount of N lost in leached-out polyphenols has not been verified. If polyphenols increase humification, the extra secretion of nitrogen will remove mineral N in the short-term but may not be detrimental to soil fertility in the long run. Where *Peltophorum* is added to soil each year for many years, the extra organic matter may provide additional fertility and stability against erosion. Fig. 25.5b shows, however, that greater leaching from under *Gliricidia* is the consequence of greater amounts of mineral N in soil; yields and crop growth depend on the availability of N and although losses were less from *Peltophorum* this does not automatically mean that yields of crops grown in alley cropping systems with *Peltophorum* will be greater than those grown with *Gliricidia*. Once again quality shows an ambiguous face, the full evaluation of which depends on a number of contradictory factors. Vanlauwe *et al.* (1995), studying decomposition in the dry season, have warned that residues of different qualities may respond to rainfall and wetting and drying cycles differently.

Conclusions

Understanding the effect of quality on decomposition and stability of organic matter is vital to all farmers, but particularly to those working in tropical regions where the soils are often depleted of nutrients and vulnerable to erosion. Up to 15% more of the N contained in crop residues may be retained in soil by incorporating *Peltophorum* rather than *Gliricidia* but this N will not necessarily benefit a growing crop. Although sugar beet residues contain a great deal of N, there is a risk, under extreme circumstances, that this may be immobilized and only released after the growing season is complete. Dynamic

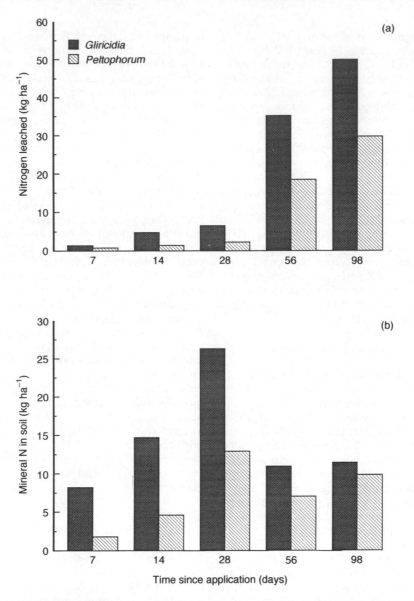

Fig. 25.5. Simulations of (a) cumulative field leaching of N derived from *Gliricidia* or *Peltophorum* prunings added to fallow soil and (b) the amounts of mineral N in soil at the same times after addition. Identical amounts of N were added in both cases but total dry matter additions differ because the C-to-Ns of the additions differ. Precipitation in these three months, 325 mm; mean air temperature, 26.6°C; mean evaporation 3.5 mm day^{-1}.

computer models provide a convenient means to distinguish between immobilization and mineralization of N, and between availability and loss through leaching. This is particularly useful in customizing residues for a particular aim: build up of fertility, prevention of leaching or supply of nutrients. Armed with such information farmers can decide the best strategy for their farms, be they in temperate or tropical regions, aiming for maximum yield or minimum loss to the environment.

Acknowledgements

Both authors acknowledge support from the European Communities Life Sciences and Technologies for Developing Countries Research Programme – STD-3. We are also grateful to J. Hassink, P.E. Smith and H. Terburg who made a number of helpful comments on the manuscript.

References

Addiscott, T.M. (1977) A simple computer model for leaching in structured soils. *Journal of Soil Science* 28, 554–563.

Addiscott, T.M., Whitmore, A.P. and Powlson, D.S. (1991) *Farming, Fertilizers and the Nitrate Problem*. CAB International, Wallingford, UK, 170 pp.

Anonymous (1986) *Adviesbasis voor Bemesting van Bouwland*. Consulent voor Bodem-, Water- en Bemestingszaken in de Akker- en Tuinbouw, Wageningen, The Netherlands, 28 pp.

Bradbury, N.J., Whitmore, A.P., Hart, P.B.S. and Jenkinson, D.S. (1993) Modelling the fate of nitrogen in crop and soil in the years following application of ^{15}N-labelled fertilizer to winter wheat. *Journal of Agricultural Science, Cambridge* 121, 363–379.

Burns, I.G. (1975) An equation to predict the leaching of surface-applied nitrate. *Journal of Agricultural Science, Cambridge* 85, 443–454.

Chae, Y.M. and M.A. Tabatabai (1985) Mineralization of nitrogen in soils amended with organic wastes. *Journal of Environmental Quality* 15, 193–198.

Dawra, R.K., Makkar, H.S.P. and Singh, B. (1988) Protein-binding capacity of micro-quantities of tannins. *Analytical Biochemistry* 170, 50–53.

De Neve, S., Pannier, J. and Hofman, G. (1996) Temperature effects on C- and N-mineralization from vegetable crop residues. *Plant and Soil* (in press).

Fog, K. (1988) The effect of added nitrogen on the rate of decomposition of organic matter. *Biological Reviews* 63, 433–463.

Fox, R.H., Myers, R.J.K. and Vallis, I. (1990) The nitrogen mineralization rate of legume residues in soil as influenced by their polyphenol, lignin and nitrogen contents. *Plant and Soil* 129, 251–259.

Franzluebbers, K., Weaver, R.W., Juo, A.S.R. and Franzluebbers, A.J. (1994) Carbon and nitrogen mineralization from cowpea plants part decomposing in moist and in repeatedly dried and wetted soil. *Soil Biology and Biochemistry* 26, 1379–1387.

Handayanto, E. (1994) *Nitrogen Mineralization from Legume Tree Prunings of Different Quality*. PhD thesis, Wye College, University of London.

Handayanto, E., Cadisch, G. and Giller, K.E. (1994) Nitrogen release from prunings of legume hedgerow trees in relation to quality of the prunings and incubation method. *Plant and Soil* 160, 237–248.

Jensen, H.L. (1929) On the influence of the carbon:nitrogen ratios of organic material on the mineralization of nitrogen. *Journal of Agricultural Science, Cambridge* 29, 71–82.

Kachaka, S., Vanlauwe, B. and Merckx, R. (1993) Decomposition and nitrogen mineralization of prunings of different quality. In: Merckx, R. and Mulungoy, K. (eds) *Soil Organic Matter Dynamics and Sustainability of Tropical Agriculture*. Wiley-Sayce, Chichester, UK, 392 pp.

Kuiters, A.T. (1990) Role of phenolic substances from decomposing forest litter in plant–soil interactions. *Acta Botanica Neerlandica* 39, 329–348.

Meentemeyer, V. (1978) Macroclimate and lignin control of litter decomposition rates. *Ecology* 59, 465–472.

Moore, W.J. (1963) *Physical Chemistry*, 4th edn. Longmans, London, 844 pp.

Nicolardot, B., Fauvet, G. and Cheneby, D. (1994) Carbon and nitrogen cycling through soil microbial biomass at various temperatures. *Soil Biology and Biochemistry* 26, 253–261.

Nieder, R. and Richter, J. (1989) Die Bedeutung der Umsetzung von Weizenstroh im Hinblick auf den C- und N-Haushalt von Löss-Ackerböden. *Zeitschrift für Pflanzenernährung und Bodenkunde* 152, 415–420.

Palm, C.A. and Sanchez, P.A. (1991) Nitrogen release from the leaves of some tropical legumes as affected by their lignin and polyphenolic contents. *Soil Biology and Biochemistry* 23, 83–88.

Parton, W.J., Schimel, D.S., Cole, C.V. and Ojima, D.S. (1987) Analysis of factors controlling soil organic matter levels in great plains grasslands. *Soil Science Society of America Journal* 51, 1173–1179.

Stanford, G. and Smith, S.J. (1972) Nitrogen mineralization potentials of soils. *Soil Science Society of America, Proceedings* 36, 465–472.

Thorup-Kristensen, K. (1994) An easy pot incubation method for measuring nitrogen mineralization from easily decomposable organic material under well defined conditions. *Fertilizer Research* 38, 239–247.

Vallis, I. and Jones, R.J. (1973) Net mineralization of nitrogen in leaves and leaf litter of *Desmodium intortum* and *Phaseolus atropurpureus* mixed with soil. *Soil Biology and Biochemistry* 5, 391–398.

Vanlauwe, B., Vanlangenhove, G., Merckx, R. and Vlassak, K. (1995) Impact of rainfall regime on the decomposition of leaf litter with contrasting quality under subhumid tropical conditions. *Biology and Fertility of Soils* 20, 8–16.

Van Soest, P.J. (1967) Development of a comprehensive system of field analysis and its application to forages. *Journal of Animal Science* 26, 119–132.

Vigil, M.F. and Kissel, D.E. (1995) Rate of nitrogen mineralized from incorporated crop residues as influenced by temperature. *Soil Science Society of America Journal* 59, 1636–1644.

Whitmore, A.P. and Groot, J.J.R. (1994) The mineralization of N from finely or coarsely chopped crop residues: measurements and modelling. *European Journal of Agronomy* 3, 103–109.

Whitmore, A.P. and Matus, F.J. (1996) The decomposition of wheat and clover residues in soil: measurements and modelling. *Proceedings of the 8th Nitrogen Workshop on Soils*, Ghent, (in press).

Whitmore, A.P. and van Noordwijk, M. (1995) Bridging the gap between environmentally acceptable and agronomically desirable nutrient supply. In: *Arable Ecosystems for the 21st Century. 13th Long Ashton International Symposium.* John Wiley and Sons, Chichester, UK, pp. 271–288.

Zagal, E. and Persson, J. (1994) Immobilization and remineralization of nitrate during glucose decomposition at four rates of nitrogen addition. *Soil Biology and Biochemistry* 26, 1313–1321.

26 Modelling the Measurable: Interpretation of Field-scale CO$_2$ and N-mineralization, Soil Microbial Biomass and Light Fractions as Indicators of Oilseed Rape, Maize and Barley Straw Decomposition

J. Magid, T. Mueller, L.S. Jensen and N.E. Nielsen

Department of Agricultural Sciences, Royal Veterinary and Agricultural University, Thorvaldsensvej 40, DK 1871 FC, Frederiksberg, Denmark

Introduction

Several computer simulation models have been developed which allow the prediction of the effects of changes in management and changes in environmental conditions on soil fertility (e.g. Jenkinson *et al.*, 1987; Parton *et al.*, 1987; Hansen *et al.*, 1991). In a comparison of 14 such models de Willigen (1991) concluded that although some models could accurately predict long-term trends, the simulation of short to medium term processes was generally unsatisfactory, especially regarding soil organic matter dynamics. This underscored the need for more detailed experimentation, but further testing of models has been hampered by a lack of methods suitable for measurements of pools and fluxes which can be used for validation of both short-and long-term model predictions of soil organic matter dynamics.

Only after the development of the above cited models was it demonstrated that reliable field measurements of soil microbial biomass can be made in connection with additions of large amounts of substrate (Ocio *et al.*, 1991; Sparling and Zhu, 1993; Voroney *et al.*, 1993; Joergensen *et al.*, 1994). Magid *et al.*

(1996) elucidated the decomposition of ^{14}C labelled plant residues by using different density and size-density fractionation methods. They showed that the residue remaining in soil could be largely recovered from the soil by separations based on size and density. Thus a possibility of quantifying the plant residues remaining in field trials without the use of the litter bag technique was indicated. While field-scale N-mineralization has been used extensively in order to parameterize and test simulation models for short to medium-term dynamics, few if any have been tested against measurements of field-scale C-mineralization. Jensen *et al.* (1996b) compared a static chamber method using alkali-trapping of CO$_2$ and a dynamic chamber method using infrared gas analysis for the estimation of field scale CO$_2$-fluxes from unplanted soils. They concluded that despite considerable limitations and a strong bias at high flux rates, the static method provided an integrated measure, and that it would be necessary to monitor continuously in order to obtain a reliable estimate by use of the less biased dynamic method.

Our aim with this study was to test the concepts of soil organic matter dynamics applied in the 'DAISY' soil–plant–atmosphere model (Hansen *et al.*, 1991). Thus we conducted a series of field trials in which we attempted to measure the temporal variation in field-scale C and N mineralization, soil microbial biomass, and the remaining plant residues as indicated by light particulate organic matter.

A detailed description of the temporal variability of microbial biomass C and N, the soil inorganic nitrogen and the CO_2 flux measurements from the intensive rape straw incorporation trial has been made by Jensen *et al.* (1996a). Mueller (personal communication) made a detailed model interpretation of the 1993–1994 experiment, in which the turnover and physiological state of the microbial biomass was analysed. In this chapter, we will discuss the integration of measurements and model simulations with special reference to the quality of the added plant material.

Methodological Approach

The experiments were conducted on a sandy loam (Højbakkegård). At the end of August 1993 0 and 8 t DM ha^{-1} of oilseed rape straw (0 and 3.5 t C ha^{-1}) was incorporated into the soil. The experimental design was a randomized complete block design with four replicates (see Jensen *et al.* (1996a) for a detailed description). At the end of August 1994 a less intensive field experiment was conducted on the same site, by incorporating 0 and 6 t DM

ha^{-1} barley, and maize residues (0 and 2.7 t C ha^{-1}) in two replicates. Characteristics of the added plant materials are given in Table 26.1. Cellulose and lignin content was analysed by the method of Goering and van Soest (1970). Water-soluble C and N was extracted (1 g finely ground plant material in 100 ml cold water for 45 minutes). Water-soluble and insoluble C and N were determined on an elemental analyser (Carlo Erba).

Soil moisture content was measured once or twice a week, in both experiments. Meteorological data (global radiation, precipitation, air temperature at 2 m height and soil temperature at 10 cm depth beneath short grass) were recorded at the nearby meteorological station (<100 m from the experimental site) of the Experimental Farm. Soil samples (0–15 cm) were taken regularly during autumn, winter and early spring, initially every 2 to 4 weeks and later every c. 4 weeks except in mid winter where soils were frozen. For size-density fractionation of particulate organic matter fractions, c. 30 cores from each plot were sampled, mixed thoroughly and frozen at −18°C.

Soil samples were not sieved, preincubated or further homogenized to avoid any interference before analysis (Ocio and Brookes, 1990). All laboratory analysis were carried out in triplicate (1993–1994) or duplicate (1994–1995) for each soil sample.

Soil microbial biomass was determined by fumigation–extraction (FE)

Table 26.1. Properties of rape and barley straw, and maize (stalks and cobs) materials used in the time-course experiment on a sandy loam.

	Rape	Barley	Maize
Water-soluble (% of DM)	6	12	29
Cellulose (% of DM)	45	43	29
Lignin (% of DM)	13.1	6.8	4.7
Ash (% of DM)	–	4.1	3.5
N (% of DM)	0.55	0.63	1.41
Total C-to-N	80	72	32
C-to-N of water-soluble	19	12	23
C-to-N of water-insoluble	92	110	36
Lignin-to-N	24	11	3.3

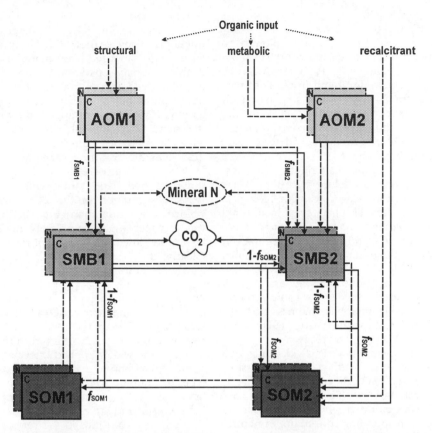

Fig. 26.1. C and N fluxes between the different pools and subpools of organic matter, mineral N and CO_2 in soil. AOM = added organic matter; SMB = soil microbial biomass; SOM = native soil organic matter; f_x=partitioning coefficients; ——— C fluxes; -------- N fluxes.

(Brookes *et al.*, 1985; Vance *et al.*, 1987). Soil samples were rewetted to c. 50% water holding capacity (WHC) before fumigation if soil moisture content was below c. 40% of WHC to ensure fumigation efficiency (Ross, 1989; Sparling and West, 1989). Inorganic N was determined in the unfumigated 0.5 M K_2SO_4 extracts by standard colorimetric methods. Soil surface CO_2-flux as an index for *in situ* microbial respiratory activity was measured at least every time soil samples for chemical and biochemical analysis were taken. A static chamber method with passive trapping of respired CO_2 was used as described in detail by Jensen *et al.* (1996b). Soil samples were analysed in a randomized design, in order to minimize the impact of operator variability, which appeared to be of some magnitude (Magid, personal communication). Soil samples (400 g) were soaked in 35°C NaCl (5% w/w) for c. 20 min before washing over a 100 µm sieve. All materials <100 µm were discarded, whereas the remaining material was decanted several times in water to separate particulate organic matter from heavy sand-sized grains. The particulate organic matter (POM) fraction was further separated by density using an aqueous sodium polytungstate ($Na_6[H_2W_{12}O_{40}]$) solution (40% w/w) at ρ 1.4 g cm⁻³. After 5 min settling time the POM separated into discrete light (ρ<1.4) and heavy (ρ<1.4) POM fractions, and the floating POM was skimmed from the surface with a 100 μm sieve. The heavy POM was stirred once and allowed to resettle, in order to ensure complete separation of light and heavy POM. The 'residue remaining' found in the light particulate fraction, was estimated based on the difference between light POM in the amended and unamended (control) plots.

The DAISY model is a deterministic description of the soil plant atmosphere system, integrating both physico-chemical processes such as solute transport, water movement and heat flow, with biological processes, such as plant growth and soil organic matter dynamics (Hansen *et al.*, 1991). The C- and N pools and fluxes in the soil-organic-matter module are shown in Fig. 26.1. The following fluxes and discrete pools are simulated: mineral N, soil CO_2 respired, added organic matter (AOM), soil microbial biomass (SMB), and native soil organic matter (SOM) not including AOM and SMB.

Each organic pool is divided into two sub-pools: one with a relatively slow turnover (i.e. AOM1) and one with a relatively fast turnover (i.e. AOM2), in order to allow a first order kinetic approximation of the corresponding continuum in nature. The decay rates of the organic pools as well as the maintenance respiration rates of the soil microbial biomass pools are functions of actual soil temperature, soil water potential and the actual clay content except for the turnover rate of the added organic matter and the zymogenous microbial biomass (SMB2) which is not modified by the soil clay content (Hansen *et al.*, 1990). The timestep of the soil-organic-matter submodule is 1 h. The other submodules of DAISY are described elsewhere (Hansen *et al.*, 1990, 1991).

Field Scale Residue Decomposition Measurements

Magid *et al.* (1996) demonstrated that ^{14}C in the large light (>100 μm, $\rho < 1.4$ g cm^{-3}) fraction of ^{14}C amended soil decayed in a way that was consistent with the expected behaviour of fresh plant material, and that the evolution of [14]CO_2 was closely related to the decrease of ^{14}C in the large SOM fractions. They suggested that entrapped air in undecomposed plant material would enable a density separation at normal gravity of undecomposed material from more decomposed material, that would have lost its ability to retain entrapped air, while at the same time it would have interacted with clay particles (Golchin *et al.*, 1994). The dynamics of the unseparated POM fraction (>100 μm) reflected the decay of the amended rape straw (Fig. 26.2c). Density separation of the POM fraction yielded a heavy fraction (Fig. 26.2b) that was very stable, and completely unaffected by ammendment of 8 t ha^{-1} of rape straw. Clearly the heavy fraction represents a slowly decomposable 'native' soil organic matter fraction. The light fraction (Fig. 26.2a) reflects the decomposing rape straw as well as some native SOM, as can be seen from the unamended plots. This approach may provide an interesting alternative to the widely applied litter bag technique, that imposes restrictions on the types of fauna that may enter into the litter, as well as the direct interaction between soil and litter.

The C-to-N ratios in the various fractions (Fig. 26.3) show that the light fraction C-to-N ratio decrease from c. 60 to 20 over 20 months in the 8 t ha^{-1} rape straw treatment, indicating that the decomposing plant residues may be associated with nitrogen rich microbial biomass, or that the residue nitrogen is only very slowly mineralized compared with carbon. The differences in C-to-N ratios in light fractions from the 0 and 8 t ha^{-1} treatments indicate that a nitrogen rich 'native' soil organic matter component is also present in the light fraction. The slight but significant decrease in the heavy fraction C-to-N ratios indicate that they contribute somewhat to the carbon flow but much less to the nitrogen flows in the system. The increase in lignin content and decrease in cellulose content in the light fraction over time (Fig. 26.4) is consistent with the preferential substrate utilization previously reported in the literature (Fog, 1988). Generally it appears that size-density separation may be a useful tool in characteriz-

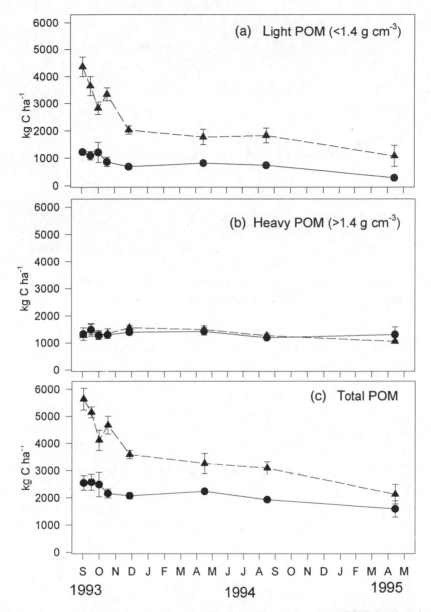

Fig. 26.2. Carbon contents in non-complexed particulate soil organic matter (POM) fractions in rape straw incorporation trial with addition of 0 (●) or 8 (▲) t dry matter ha^{-1}: (a) light POM, (b) heavy POM and (c) total POM fractions.

ing field-scale root and litter decomposition, and as will be discussed below there is a good agreement between the decay of this measurable pool and the predicted decay of plant residues. However, it is obvious that some plant materials will be less suitable for density separation than mature rape straw, due to a lesser retention of entrapped air or an inherent higher density. Recently it was found that a detailed size separation of the POM (>100 µm) fraction was a more reliable means of

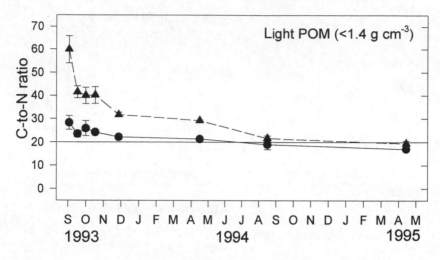

Fig. 26.3. C-to-N ratios in non-complexed light particulate soil organic matter (POM) fractions in the rape straw incorporation trial with addition of 0 (●) or 8 (▲) t dry matter ha^{-1}.

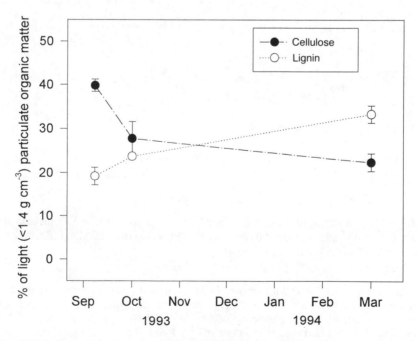

Fig. 26.4. Cellulose and lignin indices in light POM fractions from the 8 t treatment in the rape straw incorporation trial.

elucidating the decomposition of *Vicia faba*, and was comparable or superior to size-density separation for the study of mature barley straw decomposition (Kjærgaard and Magid, unpublished results). This detailed size separation was based on a preliminary examination of the size-distribution of native light and heavy POM. In the soils studied >95% of the native heavy POM was smaller than 400 μm, and using this as a seperation criterion actually yielded better results in the case of *Vicia faba*, because some fresh plant material tended not to float, and because some of the heavy POM would cling to large particles, and thus appear in the light fraction. Apart from being more reliable and less expensive, the detailed size separation has the added advantage over size-density separation that the only significant modification of the decomposing residue during extraction from the soil results from the loss of easily soluble organic and inorganic components. The loss of such easily soluble components during fractionation is most likely to be important during the early stages of decomposition, but of less significance later on.

Parameterization of the DAISY Model Based on 1993–1994 Trials

The data from the first field trial (1993–1994) were used to test and modify the standard setup of the DAISY soil-organic-matter submodule, that was originally parameterized on the basis of available literature data and mainly tested against field measurements of mineral N. The field measurements of soil microbial biomass content (SMB) showed that the DAISY model consistently under-estimated the size of SMB by c. 60% and in order to obtain a reasonable simulation of the unamended plots (Fig. 26.5c) it was necessary to reduce the microbial death rate coefficients and the maintenance respiration rate coefficient (Table 26.2). The model was parameterized for rape straw based on the measured properties. The partitioning of the added organic matter was based on the amounts of C and N in

water soluble (AOM2) and non-soluble (AOM1) components, which were assigned normalized decomposition rates of 0.05 and 0.012 d^{-1} respectively (Table 26.2). It was equally necessary to reduce the substrate utilization efficiencies in order to realistically simulate the response to addition of 8 t ha^{-1} with regard to soil microbial biomass (Fig. 26.5c) and CO_2 surface flux (Fig. 26.5b). The model simulations in the modified setup (Fig. 26.5a–d) were improved markedly compared with those based on the literature derived standard parameterization, resulting in a considerable overall improvement in residual errors and modelling efficiency as defined by Loague and Green (1991). The model prediction of mineral N deviated significantly from the measured values in the initial stages of the experiment. Thus, in the unamended plots (Fig. 26.5a) the mineralization was somewhat under-estimated, until a rather intense rainfall had caused considerable nitrate leaching from the 0–15 cm soil layer investigated. This could indicate that the mechanic perturbation of the field soil made some nitrogen rich material available for decomposition, and such processes are currently not considered in the DAISY-model. In contrast the simulation over-estimated the initial N-mineralization in the amended plots (Fig. 26.5a). This could indicate that the initial decomposition of the rape straw took place using a substrate with a much wider C-to-N ratio than that of the soluble component (Table 26.1) which was assumed to be decomposed most rapidly in the model simulation. In both the unamended and amended plots, the N-mineralization was underestimated in the warm and dry summer period, indicating that the temperature or soil moisture response was somewhat inadequate. The model prediction of remaining plant residues agreed very well with that estimated from light fractions (Fig. 26.5d) during the autumn and winter. However during the subsequent warmer period, the model predicted a rapid decomposition which was in contrast to the measured changes in light fraction. From the temporal changes in the

Table 26.2. Parameterization of the soil organic matter submodule in DAISY, based on literature data (standard) and interpretation of the rape straw incorporation field trial (1993–1994).

Parameterization		Standard	Modified
% of C_t in	SMB1	0.45	1.89
	SMB2	0.15	0.15
C-to-N of	SMB1	6	6.6
	SMB2	10	6.6
Mean C-to-N	SMB	6.7/7.5*	6.6
Death rate coefficient [†]	SMB1 (day^{-1})	0.001	0.00019
	SMB2 (day^{-1})	0.01	0.01
Maintenance respiration rate coefficient [†]	SMB1 (day^{-1})	0.01	0.0018
	SMB2 (day^{-1})	0.01	0.01
% of AOM-C in	AOM1	80	96
	AOM2	20	4
C-to-N of	AOM	80	80
	AOM1	90	92
	AOM2	72	19
Turnover rate coefficient [†]	AOM1 (day^{-1})	0.005	0.012
	AOM2 (day^{-1})	0.05	0.05
Substrate utilization efficiency	SOM1	0.60	0.40
	SOM2	0.60	0.50
	SMB	0.60	0.60
	AOM1	0.60	0.13
	AOM2	0.60	0.69

* Values for 0 t ha^{-1}/ 8 t ha^{-1} rape straw.
[†] = Standard values at 10°C and at field capacity.
SMB, soil microbial biomass.
AOM, added organic matter.
SOM, native soil organic matter.

lignin and cellulose indices in the light fraction (Fig. 26.4) it is obvious that the decay of the light fraction in the long term cannot be described as a single pool with first order kinetics, since the lignin indices increased, while the cellulose indices concurrently decreased. The discrepancy between simulated and measured results could have been corrected by assigning the remaining 30% of the residue directly to the much more slowly decomposing soil organic matter (SOM2) at the outset of the simulation (Fig. 26.1) or changing the conceptual approach, e.g. by using three pools, or second order kinetics. However, as we had no rational way of doing this, based on the measured initial properties of the rape straw (Table 26.1), we declined to use this option to obtain a better fit. From the measurements of field-scale C-mineralization in the unamended and amended plots (Fig. 26.5b) it is clear that there is no considerable difference between the treatments, which would have been expected if the remaining residue had decomposed rapidly over the following summer, as predicted by the model.

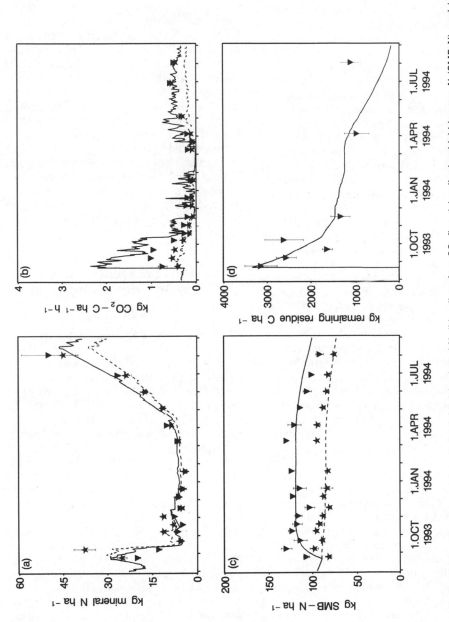

Fig. 26.5. Simulated (lines) and measured data (points) for (a) mineral N, (b) soil surface CO_2 flux, (c) soil microbial biomass N (SMB-N), and (d) remaining residue in the control (0 t) treatment and 8 t dry matter ha^{-1} treatment of the rape straw incorporation trial. Symbols: ★ / dotted line, control treatment; ▼ / solid line, 8 t dry matter ha^{-1} treatment.

Simulation of the 1994–1995 Field Trials

The 1994–1995 field trials were simulated without any further changes in the model parameterization, except for those related to the measured properties (amount of water soluble versus non-soluble C and N) for the barley and maize materials (Table 26.1). As in the case of the previous trial the model underestimated initial N-mineralization in the unamended plot (data for the unamended plot not presented).

The model predictions of mineral N (Fig. 26.6a) were in good agreement with the measurements in the case of barley, but not in the case of maize. The measurements clearly showed an initial immobilization of N in the maize plot, relative to the results from the barley plot. This is remarkable, since the maize material contained large amounts of easily soluble material with a low C-to-N ratio. This indicates that the microbial biomass formed initially by the maize amendment utilized substrate with a wider C-to-N ratio and thus had an increased capacity to immobilize N. This is corroborated by comparing measured and predicted values of microbial N (Fig. 26.6c) where the measured values during the first month after addition of maize were about 40 kg N ha^{-1} higher than those predicted, and by the initially much higher measured CO_2-respiration rates (Fig. 26.6b). Only in one instance did the model clearly underpredict mineral N (mid-October 1994, Fig. 26.6a) which coincided with an unpredictably rapid decline in microbial biomass N (Fig. 26.6c). However, the model generally tended to overpredict mineral N in the maize treatment, and though the differences between predictions and measurements in the later stages may seem small, it should be noted that these differences developed during a period in which leaching occurred frequently.

The model underpredicted the initial formation of microbial N after addition of barley straw, although this difference was less marked compared with the case of maize. The prediction of CO_2 respiration agreed reasonably well with the measured values initially, but since the static chamber method underestimates the CO_2 evolution at high rates, the model probably underpredicted the true initial respiration rates, but overpredicted the true rates at later stages. The light fraction estimates of remaining residues are in a reasonable agreement with the predicted values for barley, but apparently there are some problems connected with the measurements in the case of maize (Fig. 26.6d). Currently we are reanalysing the soil samples, in order to obtain a better replication and more data points.

On the whole, barley amendment was simulated better than maize, which could be expected since the barley straw closely resembled the rape straw (Table 26.1), on which model parameterization had been based.

Residue Quality and Decomposition

C-to-N ratios

We chose to base the parameterization of the added organic matter (AOM) pools on measured properties of the added residues (Table 26.1), rather than arbitrary properties that would fit the measured data more precisely. The measured data on N-mineralization clearly indicates that the subdivision of the residue in a quickly decomposing soluble pool with a narrow C-to-N ratio and a more slowly decomposable pool with a wider C-to-N ratio was inappropriate. A basic assumption in the DAISY model is that once defined, the C-to-N ratio of a pool remains constant, and only the size of the pool will change with time. Considering the change in the C-to-N ratio of the light POM fractions (Fig. 26.3) in the rape straw incorporation trial, this must be a gross simplification, as the only way of reconciling the measured data with the model output would be by claiming that c. 30% of the added residue C was equatable with native soil organic matter already at the time of application, which is allowed by the model (Fig. 26.1). The only

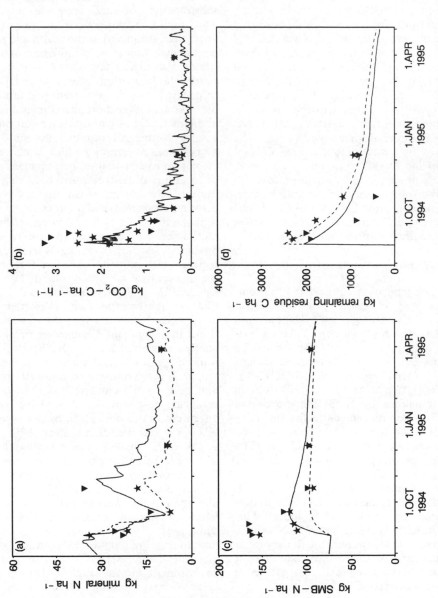

Fig. 26.6. Simulated (lines) and measured data (points) for (a) mineral N, (b) soil surface CO_2 flux, (c) soil microbial biomass N (SMB-N), and (d) remaining residue in the maize and barley treatments (6 t dry matter ha^{-1}). Symbols: ★ / dotted line, barley treatment; ▼ / solid line, maize treatment.

justification we can think of for this claim, though an alluring one, is that it would allow the model a better 'fit'. Even if the model allowed for variable C-to-N ratios in the defined pools, the subdivision of the substrate described above would probably still be inappropriate, as less N was initially mineralized than would be predicted from a rapid decomposition of the water-soluble component. We are not currently able to suggest any better measurable parameterization of the substrates than the one used. For the rape straw simulation the initial lower N-mineralization could easily be obtained by assuming a different C-to-N ratio of the AOM2 component. However this does not apply to the maize residues as both the soluble and non-soluble fraction has a rather narrow C-to-N ratio, and thus a preferential C-rich substrate utilization is clearly indicated for the maize material.

Lignin or lignin-to-N ratio

It has long been recognized that decomposition may be retarded at high lignin contents (Whitmore and Handayanto, Chapter 25, this volume). There is evidence that the lignin-to-N ratio could be indicative of the rate of decay in various types of litter (Melillo *et al.*, 1982; Tian *et al.*, 1992). We used lignin-to-N ratio in our simulations by increasing the overall decomposition rate constants for the barley and maize residue relative to the rape straw (Table 26.1) based on this evidence. The resulting prediction for maize and barley straw (data not shown) clearly overestimated both N-mineralization and initial CO_2 evolution rates, and considerably underestimated the residue remaining. This could indicate that the lignin-to-N ratio is not a critical determinant of the short- to medium-term decomposition rates, but it may still be very important in governing the long-term decay.

Decomposer organisms

Miller and Kjøller (personal communication) studied the changes in enzyme activities on the surfaces of decomposing maize residue particles, and the soil adhering to the residue particles. These changes were indicative of a succession of non-cellulolytic fungi followed by cellulolytic fungi. Little is known about the functioning and substrate use of such organisms in soil. The initial N-mineralization rates are lower than simulated in the maize and rape treatments, indicating assimilation of substrates with higher C-to-N ratios than the water soluble fraction. The large initial discrepancy between measured (by fumigation–extraction) and predicted microbial N in the maize treatment indicates that nitrogen is assimilated rapidly into an early decomposer community with a rather large cytoplasm to cell-wall ratio. The main part of the immobilized N must have derived from organic form, since the difference between measured and predicted mineral N is low (<10 kg N ha^{-1}) compared with the difference between measured and predicted SMB-N (c. 40 kg N ha^{-1}). The apparent lack of reappearance of this nitrogen as excess of measured mineral N compared with simulated mineral N in the later stages, when measured and predicted values of microbial N are in good agreement, could indicate that it had been predominantly transferred into the hyphal walls of a more mature fungal population, with a rather low cytoplasm-to-cell-wall ratio. Similar growth patterns with changes in cytoplasm-to-cell-wall ratios have been interpreted from data given by Schnürer *et al.* (1985) and Wessén and Berg (1986).

A more detailed understanding of the interrelationship between the evolution of the decomposer community and the residue quality may be necessary for predicting both the short as well as the longer term outcome of residue decomposition. Thus Fog (1988) pointed out that the functioning of ligninolytic fungi was markedly depressed in a eutrophic environment and furthermore that the surfaces of decomposing residues could be stabilized by co-polymerization of surface bound phenolic groups with soluble 'browning precursors' (i.e. low molecular amino acid compounds). Our observation that the slowly decomposing, nitrogen rich, heavy particulate organic matter fraction (Fig. 26.3) mainly consisted of small brown-black recognizable plant residue particles is consistent with this hypothesis.

Conclusions

Although the applied methods certainly have limitations we think that the datasets presented are sufficiently detailed and comprehensive to severely restrict the degrees of freedom in the model interpretation of the field scale processes. Certainly, the dynamics of POM appears to be a useful indicator of the decomposition of some plant materials, and may provide an interesting alternative to the litter bag technique for tracing disappearance of certain plant residues. These measurements complemented those of the field-scale CO_2 flux-rates and offset some of the methodological bias and uncertainty in this latter measure of C-dynamics.

With this work we have endeavoured to 'measure the modellable' and vice versa, to test the functioning of the concepts of organic matter dynamics in the 'DAISY' model. Through this interactive modelling and measuring approach it has become increasingly evident that there is no firm relationship between the *standard* set of measurable 'quality' parameters of the added plant materials and an adequate parameterization of the model. This indicates a need to look at other parameters, and it seems likely that a better understanding of the short to medium term decomposition processes may rely on including the decomposer population (the 'demand side' (see also Sinsabaugh and Moorhead, Chapter 27, this volume)), rather than having an exclusive focus on the litter quality (the 'supply side'). The partitioning of water-soluble/insoluble fractions yielded a better description of the short to medium term decomposition than the lignin-to-N ratio. However, none of these measurable quality indices could explain the initial mineral N dynamics during the decomposition, that must have been under the influence of a preferential substrate use by the decomposer population.

Acknowledgements

We thank H. Svendsen and S. Hansen for extensive cooperation on using and modifying the DAISY model. This work was supported by the Danish Environmental Research Program. Torsten Mueller was a collaborator via a fellowship under the OECD project on Biological Resource Management and via a research fellowship of the German Research Society (DFG).

References

Brookes, P.C., Landman, A., Pruden, G. and Jenkinson, D.S. (1985) Chloroform fumigation and the release of soil nitrogen: a rapid direct extraction method to measure microbial biomass in soil. *Soil Biology and Biochemistry* 17, 837–842.

Fog, K. (1988) The effect of added nitrogen on the rate of decomposition of organic matter. *Biological Reviews* 63, 433–462.

Goering, H.K. and van Soest, P.J. (1970) *Forage Fibre Analyses (Apparatus, Reagents, Procedures, and Some Applications)*. Agriculture Handbook, USDA, No. 379, pp. 1–19.

Golchin, A., Oades, J.M., Skjemstad, J.O. and Clarke, P. (1994) Study of free and occluded particulate organic matter in soils by solid state 13C CP/MAS NMR spectroscopy and scanning electron microscopy. *Australian Journal of Soil Research*, 32, 285–309.

Hansen, S., Jensen, H.E., Nielsen, N.E. and Svendsen, H. (1990) DAISY – Soil Plant Atmosphere System Model. *NPo-forskning fra Miljøstyrelsen* A10, 1–269.

Hansen, S., Jensen, H.E., Nielsen, N.E. and Svendsen, H. (1991) Simulation of nitrogen dynamics and biomass production in winter wheat using the Danish simulation model DAISY. *Fertilizer Research* 27, 245–259.

Jenkinson, D.S., Hart, P.B.S., Rayner, J.H. and Parry, L.C. (1987) Modelling the turnover of organic matter in long-term experiments at Rothamsted. *Intecol Bulletin* 15, 1–8.

Jensen, L.S., Mueller, T., Magid, J. and Nielsen, N.E. (1996a) Temporal variation of C and

N mineralization, microbial biomass and extractable organic pools in soil after oilseed rape straw incorporation in the field. *Soil Biology and Biochemistry* (in press).

Jensen, L.S., Mueller, T., Tate, K.R., Ross, D.J., Magid, J. and Nielsen, N.E. (1996b) Measuring soil surface CO_2-flux as an index of soil respiration *in situ*: a comparison of two chamber methods. *Soil Biology and Biochemistry* (in press).

Joergensen, R.G., Meyer, B. and Mueller, T. (1994) Time-course of the soil microbial biomass under wheat: a one year field study. *Soil Biology and Biochemistry* 26, 987–994.

Loague, K. and Green, R.E. (1991) Statistical and graphical methods for evaluating solute transport models: overview and application. *Journal of Contaminant Hydrology* 7, 51–73.

Magid, J., Gorissen, A. and Giller, K.E. (1996) In search of the elusive 'active' fraction of soil organic matter: three size-density fractionation methods for tracing the fate of homogeneously 14C-labelled plant materials. *Soil Biology and Biochemistry* 28, 89–99.

Melillo, J.M., Aber, J.D. and Muratore, J.F. (1982) Nitrogen and lignin control of hardwood leaf litter decomposition dynamics. *Ecology* 63, 621–626.

Ocio, J.A. and Brookes, P.C. (1990) Soil microbial biomass measurements in sieved and unsieved soils. *Soil Biology and Biochemistry* 22, 999–1000.

Ocio, J.A., Brookes, P.C. and Jenkinson, D.S. (1991) Field incorporation of straw and its effects on soil microbial biomass and soil inorganic N. *Soil Biology and Biochemistry* 23, 171–176.

Parton, W.J., Schimel, D.S., Cole, C.V. and Ojima, D.S. (1987) Analysis of factors controlling soil organic matter levels in Great Plains grasslands. *Soil Science Society of America Journal* 51, 173–179.

Ross, D.J. (1989) Estimation af soil microbial C by a fumigation-extraction procedure. Influence of soil moisture content. *Soil Biology and Biochemistry* 21, 767–772.

Schnürer, J., Clarholm, M., and Rosswall, T. (1985) Microbial biomass and activity in an agricultural soil with different organic matter contents. *Soil Biology and Biochemistry* 17, 611–618.

Sparling, G.P. and West, A.W. (1989) Importance of soil water content when estimating soil microbial C, N and P by the fumigation-extraction methods. *Soil Biology and Biochemistry* 21, 245–253.

Sparling, G.P. and Zhu, C.Y. (1993) Evaluation and calibration of biochemical methods to measure microbial biomass-C and biomass-N in soils from Western Australia. *Soil Biology and Biochemistry* 25, 1793–1801.

Tian, G., Kang, B.T. and Brussaard, L. (1992) Effects of chemical composition on N, Ca and Mg release during incubation of leaves from selected agroforestry and fallow plant species. *Biogeochemistry* 16, 103–119.

Vance, E.D., Brookes, P.C. and Jenkinson, D.S. (1987) An extraction method for measuring microbial biomass C. *Soil Biology and Biochemistry* 19, 703–707.

Voroney, R., Winter, J.P. and Beyaert, R.P. (1993) Soil microbial biomass C and N. In: Carter, M.R. (ed.) *Soil Sampling and Methods of Analysis*. Lewis Publishers, London, pp. 277–286.

Wessén, B. and Berg, B. (1986) Long-term decomposition of barley straw: chemical changes and in growth of fungal mycelium. *Soil Biology and Biochemistry* 18, 53–59.

Willigen, P. de (1991) Nitrogen turnover in the soil–crop system; comparison of fourteen simulation models. *Fertilizer Research* 27, 141–149.

27 Synthesis of Litter Quality and Enzymic Approaches to Decomposition Modelling

R.L. Sinsabaugh[1] and D.L. Moorhead[2]

[1]Biology Department, University of Toledo, Toledo, OH 43606, USA; [2]Department of Biological Sciences, Texas Tech University, Lubbock, TX 79409, USA

Introduction

Plant litter decomposition is a rate-limiting process in macronutrient cycles, therefore a mechanistic understanding of litter decomposition has long been a goal of ecosystem science (e.g. Waksman and Tenney, 1928). Many approaches have been pursued, leading to models that relate breakdown rates to measures of climate, litter standing stock, litter composition, and microbial activity (cf. Sinsabaugh and Moorhead, 1994; Moorhead et al., 1995). Despite decades of research, the goal of predictive modelling remains elusive, principally because 'decomposition' is not simply the obverse of primary production. The production of new organic matter is a coherent process, resident within individual organisms; decomposition emerges at the community level, a composite process whose mechanistic components vary among systems and with scale.

Most decomposition models link mass loss rates to measures of litter quality and/or climate (Melillo et al., 1982; Parton et al., 1987; Insam et al., 1989; Taylor et al., 1989; Boulton and Boon, 1991; Moorhead and Reynolds, 1991; Hsieh, 1992; Paustian et al., 1992; Berg et al., 1993). Although widely used, litter quality is difficult to capture in a simple metric. Measures such as C-to-N ratio, lignin-to-N ratio, or lignin-to-lignocellulose ratio often correlate with breakdown rates, but empirical relationships apply only within limited ranges of particle size and composition (Taylor et al., 1991). As an example, the C-to-N ratio and lignin content of fine particulate organic matter in freshwater systems generally decline with particle size while turnover times increase (Sinsabaugh and Linkins, 1990; Sinsabaugh et al., 1994b; Sinsabaugh and Findlay, 1995). Part of the difficulty with indices based on N and fibre composition is that the non-fibre components of plant litter contribute to quality through mechanisms other than N supply. Extractable sugars, extractable and condensed phenolic compounds, terpenoids and lipids, affect decomposition rates by selecting microbial taxa (Savoie and Gourbiere, 1989) and by influencing detritivore feeding (Arsuffi and Suberkropp, 1989; Griffiths, 1994).

Because decomposition rates are also regulated by edaphic and climatic variables, empirical relationships between mass loss and litter composition have to be re-established for each ecosystem or expanded to include these external controls. As an example of the latter, Meentemeyer (1978) predicted litter mass loss as a linear function of lignin content and local evapotranspiration. More mechanistic approaches have variously included general temperature and moisture response functions for decomposition and the impact of nutrient limitation and immobilization (e.g. Parton et al., 1987; Moorhead and Reynolds, 1991).

In principle, some of the limitations of litter quality-based decomposition models could be circumvented by basing decomposition models directly on biotic activity. There have been attempts to do this using general measures of microbial activity, such as respiration (Bunnell et al., 1977; McGill et al., 1981; Howard and Howard, 1993), but the broad physiological characteristics of microorganisms make it difficult to closely link two non-specific processes. This difficulty led Parnas (1975) to propose that plant litter decomposition rates could be specifically related to the macronutrient acquisition activities of saprotrophic communities. Recently, we have extended this approach by developing an explicit model for Microbial Allocation of Resources among Community Indicator Enzymes (MARCIE), which predicts mass loss as a function of extracellular enzyme activity (Sinsabaugh and Moorhead, 1994).

The MARCIE model is based on a series of premises about the regulation of extracellular enzyme activity and the relationships among decomposition rate, microbial productivity, and extracellular enzyme production:

1. Extracellular enzymatic degradation of complex molecules is the rate-limiting step in microbial production.
2. At the transcriptional level, synthesis of extracellular enzymes is regulated by induction and repression–derepression mechanisms that are linked to environmental nutrient availabilities.
3. At the organismal level, the relationships among enzyme expression systems can be described as an optimal resource allocation strategy.
4. At the community level, direct relationships between decomposition rate, microbial productivity, and lignocellulase activity emerge.

In the model, enzymes are grouped into three categories, those involved principally in carbon acquisition (E_C), those involved in nitrogen acquisition (E_N) and those involved in phosphorus acquisition (E_P). Because they share a common regulatory motif (premise 2), we assume that the activities of the many enzymes in each group are correlated such that the activities of one or more enzymes in each group act as an indicators for the whole group. For decomposition, the basic form of the model is:

$$M = k_C E_C \qquad (27.1)$$

where M is mass loss rate, k_C is a first-order rate constant (mass loss per unit E_C activity), and E_C represents relative lignocellulase activity (an average of the relative activities of several indicator enzymes). The model predicts that mass loss from decomposing plant litter is proportional to C flow, expressed in terms of E_C activity. However, the production of C-acquiring enzymes is constrained by the need to enzymatically acquire N and P (premise 3); thus E_C can also be expressed as a fraction of total extracellular enzyme production (E_T, the sum of E_C, E_N, E_P) whose value is dependent on N and P availability (see Sinsabaugh and Moorhead, 1994 for complete derivation):

$$M = k_C E_T /(1 + E_N/E_C + E_P/E_C) \qquad (27.2)$$

Evaluating the model involves testing two hypotheses: mass loss rates are directly proportional to lignocellulase activity; and C, N and P acquisition activities are linked through an optimum resource allocation strategy (specifically that M is directly related to E_C/E_N and E_C/E_P). The null model tests whether M is directly related to E_T, under the hypothesis that mass loss rates reflect total extracellular enzyme activity and that E_T can increase or decrease without reallocating resources allocated among E_C, E_N, and E_P (e.g. through changes in growth efficiencies). We have evaluated these hypotheses by analysing published data (Sinsabaugh et al., 1994a) and by conducting new studies (Jackson et al., 1995; Sinsabaugh et al., 1994b, 1996). In general, both these hypotheses have support, but problems may emerge in nutrient-rich systems, through inappropriate choice of assays, or where non-microbial mass loss mechanisms predominate.

The MARCIE model has potential advantages over litter quality models because it uses enzymic, rather than chemical, terms to express relationships between decomposition and nutrient availability. The bulk abundance of nutrients, even when resolved into multiple pools, is not equivalent to microbial availability (e.g. Sinsabaugh *et al.*, 1993). Enzyme activities are a more direct assessment of microbial perceptions of their environment and therefore provide a more intimate connection between nutrient availability and litter decomposition (Asmar *et al.*, 1994). The enzymic approach also provides improvements in economy and resolution. Assays for a suite (five to ten) of enzymes provide data for estimating both decomposition rates and relative nutrient availability at spatial and temporal resolutions difficult to match with other methods. In some situations, this approach can reduce or eliminate the need for resource-intensive litter bag and nutrient amendment studies. However, the MARCIE model suffers some of the same limitations as litter quality models. Neither approach deals explicitly with external processes such as leaching, invertebrate feeding, or with the effects of temperature and moisture. Also, the rate constant that converts enzyme activities to mass loss may apply to litters only within a limited range of particle size or composition. In general, these problems are somewhat ameliorated because the MARCIE model is biotically based: environmental and litter quality variables regulate extracellular enzyme activities so these controls are to some extent entrained within the model. As a consequence, measurements of enzyme activities can provide instantaneous estimates of decomposition rates, but only after the MARCIE model has been calibrated *a posteriori*. In contrast, it is possible to make *a priori* predictions of relative decomposition rates from litter quality models.

Ideally, the enzymic and litter quality approaches should be merged. A combined model might circumvent some of the intrinsic limitations of each. It would also summarize much of our accumulated knowledge of microbial decomposition and identify questions in need of further investigation. In this chapter, we attempt such a synthesis by proposing enzymic indices of litter quality and by developing a simulation model that makes *a priori* predictions of microbial biomass and decomposition activity from litter quality data.

Enzymic Measures of Litter Quality

The simplest way to link litter quality and enzymic models of decomposition is to develop enzymic indices of litter quality. Because enzyme activities represent microbial responses they are potentially more sensitive decomposition indicators than less direct chemical measures. In previous studies, we have developed two such measures, though their utility has not been extensively investigated.

Sinsabaugh and Linkins (1988) measured the adsorption of *Trichoderma viride* cellulase components by several types of leaf litter. Confirming other studies (Ooshima *et al.*, 1983), they found that exocelluloses sorbed preferentially to holocellulose, while endocellulases had greater affinity for lignin. For deciduous litters, the endocellulose-to-exocellulase activity ratio was positively correlated with the lignin-to-lignocellulase ratio, which has been used as a litter quality metric for decomposition modelling (Melillo *et al.*, 1982). These results confirmed earlier observations that the endocellulose-to-exocellulase activity ratio increased over the course of decomposition of deciduous litter (Sinsabaugh *et al.*, 1981). Studies in aquatic systems suggest that the endo-to-exo activity ratio of fine particles (<1 mm) is inversely related to particle size, and directly related to decomposition rate (Sinsabaugh *et al.*, 1994b; Sinsabaugh and Findlay, 1995). A more broad-based measure, the enzymic index of carbon quality (EICQ) was proposed by Sinsabaugh and Findlay (1995). It can be described as the ratio of cellulase activity to ligninase activity, or more generally, as the ratio of hydrolytic to oxidative enzyme activities. In Hudson River sediments, the differences among sites in EICQ were highly correlated with decomposition rate (Sinsabaugh and Findlay, 1995).

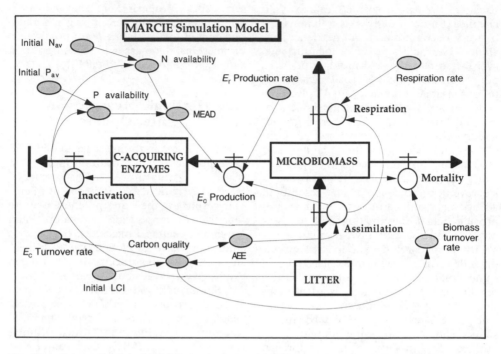

Fig. 27.1. Schematic of the MARCIE simulation model. See text for definitions of stocks (boxes), flows (heavy arrows) and converters (ovals).

The existence of these enzymic trends in relation to litter quality underscore the oversimplification inherent in premise 2 of the MARCIE model. The regulation of cellulolytic and ligninolytic enzymes, either at the cellular level or through community succession, is sufficiently autonomous that the activity of a single enzyme is generally not a reliable indicator of collective ligno-cellulolytic activity. In application, E_C must be estimated from the relative abundance of several classes of enzymes.

Enzymic indices of litter quality are potentially more sensitive than carbon quality measures based on van Soest fibre analyses and are probably more economical. However, to date, there have been no systematic evaluations of applicability or direct comparisons with chemical indices. The use of indices to represent quality illustrates the parallels between litter composition and microbial activity-based approaches to decomposition modelling, but they capture only a small portion of

what we know about the biochemistry of decomposition. A more comprehensive synthesis requires a simulation model.

MARCIE Simulation

As a heuristic exercise, we expanded the MARCIE model into a simulation that links enzymatic activity to microbial biomass and measures of litter quality (Fig. 27.1). The model represents a synthesis of current knowledge of the role of litter quality in the regulation of decomposition, suggests key control mechanisms, and provides foci for future research.

The model consists of three carbon stocks: plant litter, microbial biomass, and a pool of extracellular C-acquisition enzymes. The stocks are linked to one another and to the external environment through five flows: microbial C assimilation, microbial mortality, microbial respiration, extracellular enzyme production, and extracellular enzyme inactivation. The flows are modulated by a series of functions that link each to three measures of litter quality: N availability, P availability and lignin-to-lignocellulose ratio (LCI). The latter is strictly a measure of litter quality while the first two may also reflect the environmental nutrient milieu. In terms of extracellular enzyme production, the simulation is based on the premises of the MARCIE equation (Equation 27.1, Sinsabaugh and Moorhead, 1995). The functions that relate litter quality to enzyme production and enzymatic activity are hypothetical, but are consistent with empirical observations. Other assumptions are described below. The simulation model was developed using STELLA II software (High Performance Systems, Inc.).

Litter dynamics

The standing stock of plant litter is represented in C units. An arbitrary initial value of 10,000 mg m^{-2} was used for simulations. Litter C is calculated daily by:

$$Litter\ (t) = Litter\ (t-dt) - Assimilation \times dt \quad (27.3)$$

where $Litter\ (t)$ is the standing stock of plant detritus at time t and $Assimilation$ is microbial C uptake. In this model all carbon solubilized from the litter is assimilated. Therefore, assimilation is equivalent to gross microbial production. The microbiota are assumed to acquire C from litter as a function of the standing stock of C-acquiring enzymes (E_C):

$$Assimilation = E_C \times AEE \quad (27.4)$$

where AEE is apparent enzymatic efficiency, the increment in litter mass loss per unit of enzyme. Ideally, AEE would be constant among litter types, but field values for lignocellulases, which are calculated as mass loss per unit enzyme activity rather than enzyme C, range over an order of magnitude (Sinsabaugh et al. 1994a). Consistent with these observations, AEE declines with carbon quality:

$$AEE = 10 - (10 \times Carbon\ quality) \quad (27.5)$$

Equation 27.5 predicts that each mole of enzyme C results in the assimilation of 2–8 moles of litter C each day. These values are *ad hoc*: lower values will not sustain decomposition and higher values cause the simulation to crash.

Carbon quality is expressed as the lignocellulose index (LCI). Field studies indicate that LCI increases over the course of decomposition and that decomposition slows as LCI approaches 0.7 (Mellilo et al. 1982, McClaugherty and Berg 1987):

$$Carbon\ quality = 0.75 + (LCI_i - 0.75) \times Litter/Litter_i \quad (27.6)$$

where LCI_i is the initial LCI value for the litter and $Litter_i$ is the initial size of the litter stock (set at 10,000 mg m^{-2} in our simulations). Equation 27.5 predicts that as decomposition progresses carbon quality declines from an initial value to a final value of 0.75. The enzymic index of litter quality (EICQ), might be used here as well; the relevant data would be collected during any field test of the model.

Microbial dynamics

Microbial biomass is represented as C; an arbitrary initial value of 20 mg m^{-2} was used in our simulations. Biomass increases with C assimilation from the litter and decreases with production of C-acquiring enzymes, mortality and respiration:

$$Microbiomass\ (t) = Microbiomass\ (t-dt) + (Assimilation - E_C\ production - Mortality - Respiration) \times dt \quad (27.7)$$

Mortality is calculated as the product of microbial biomass and biomass turnover rate (k_d):

$$Mortality = Microbiomass \times k_d \quad (27.8)$$

As decomposition proceeds, biomass turnover rate declines with carbon quality:

$$k_d = 0.065 - 0.06 \times Carbon\ quality \quad (27.9)$$

For our simulations, k_d ranged from 0.055 to 0.020/day (Table 27.1 shows variation in selected litters). These values were

Table 27.1. Parameter values for litter decay simulations. N_{av} is initial nitrogen availability, P_{av} is initial phosphorus availability, LCI is initial lignin-to-lignocellulose ratio, k_r is respiration rate, k_d is microbial biomass turnover rate, k_c is the turnover rate for carbon-acquiring enzymes. All other model parameters are defined in the text. Dogwood, maple and oak represent fast, medium and slow decomposing leaf litters, respectively.

Litter type	N_{av}	Initial litter quality			Rate variables	
		P_{av}	LCI	k_r	k_d	k_c
Dogwood	6	6	0.25	0.85	0.065–0.04 (C quality)	0.09–0.06 (C quality)
Maple	4	4	0.35	0.88	0.065–0.05 (C quality)	0.09–0.07 (C quality)
Oak	3	3	0.40	0.88	0.065–0.06 (C quality)	0.09–0.08 (C quality)

selected because they resulted in microbial biomass accumulations similar to those reported from field studies.

Respiration is the product of respiration rate (k_r) and microbial assimilation:

$$Respiration = k_r \times Assimilation \quad (27.10)$$

For any particular run, respiration rate was a constant. Among runs, we used values ranging from 0.85 to 0.88/day (Table 27.1) which correspond to a community growth efficiency of 12–15%.

Enzyme dynamics

The standing stock of extracellular C-acquiring enzymes consists largely of lignocellulases. In the model, this stock is represented in units of carbon. In the field, the size of this stock could only be inferred in relative terms by the activity of selected indicator enzymes. For our simulations, the initial value was set at 0.4 mg m^{-2}. Changes in the stock were calculated in daily increments as the difference between E_C *production* and *enzyme inactivation*:

$$E_C(t) = E_C(t-dt) + (E_C \, Production - Inactivation) \times dt \quad (27.11)$$

Thus, only functional enzymes are considered part of the E_C stock.

The production of C-acquiring enzymes (E_C) is the key component of the MARCIE model. Production is calculated as a fraction of the total production of extracellular C, N and P- acquiring enzymes (E_T):

$$E_C \, Production = E_T \, Production \, rate \times Assimilation \times MEAD \quad (27.12)$$

The E_T production rate is fixed at 0.02 day^{-1}. While total secretion of extracellular C by a microbial community can be estimated using radiotracers, estimating what fraction of this total is functional enzymes is difficult. Consequently this number is a guess, equivalent to about one-sixth of the daily community production.

The MEAD function determines the relative allocation of C to synthesis of C-acquiring enzymes on the basis of N and P availability (N_{av} and P_{av}):

$$MEAD = 1 \, / \, (1 + (1/N_{av}) + (1/P_{av})) \quad (27.13)$$

The MEAD (Microbial Enzyme Allocation during Decomposition) function is derived from the assumption that microbial communities maximize their productivity by optimizing their resource allocation among extracellular C, N and P-acquiring enzymes. When either N or P availability is low, the production of N and P-acquiring enzymes increases with a corresponding reduction in the production of C-acquiring enzymes. N availability is expressed as E_C/E_N, the ratio of C-acquisition activity to N-acquisition activity. Similarly, P availability is expressed as E_C/E_P, the ratio of C-acquisition activity to P-acquisition activity. It is hypothesized that these ratios decrease as decomposition proceeds from an initial value to a final value of 1:

$$N_{av} = 1 + (Initial \ N_{av} - 1) \times Litter/Litter_i \quad (27.14)$$

$$P_{av} = 1 + (Initial \ P_{av} - 1) \times Litter/Litter_i \quad (27.15)$$

Among runs, *initial* N_{av} and *initial* P_{av} varied from 3 to 6 (Table 27.1). At present, there are no field data which can be used to estimate these availabilities. Values were chosen solely to generate reasonable decomposition patterns.

Enzymes are lost from the E_C stock through processes such as inhibition, absorption, hydrolysis, denaturation, or leaching. In the model these losses are collectively described as inactivation, which is calculated as the product of E_C stock and E_C turnover rate (k_c):

$$Inactivation = k_c \times E_C \quad (27.16)$$

As carbon quality declines with decomposition, we hypothesize that the rate of turnover of C-acquiring enzymes also declines:

$$k_c = 0.09 - 0.08 \times (Carbon \ quality) \quad (27.17)$$

For our simulations, k_c ranged from 0.075 to 0.030 day^{-1} (Table 27.1 shows variation in selected litters). Like biomass turnover, these values were selected because they resulted in decomposition patterns that appeared reasonable when compared with field studies.

Simulations

We used the extended MARCIE model to simulate the decomposition of three contrasting leaf litter types typical of eastern deciduous forests: flowering dogwood, maple species and oak species. Dogwood leaves have low lignin content and a relatively high N-to-lignin ratio; oak leaves have much higher lignin content and lower N-to-lignin ratio; maple leaves are intermediate. The decay patterns of these litter types have been examined in a number of field and laboratory studies (e.g. Webster and Benfield, 1986; Taylor *et al.*, 1989; Linkins *et al.*, 1990a,b). They represent slow (oak), medium (maple) and fast (dogwood) decomposing litters.

We emphasize that this model is an attempt to synthesize information on enzyme, microbial, and litter quality dynamics into a coherent mechanistic framework to stimulate further research. Based on our general knowledge of decomposition, the model seems plausible. However, values for several model parameters are not known and several key relationships are hypothetical.

Simulation results

We simulated the decomposition of our three litter types by manipulating initial litter quality parameters and rates of biomass and enzyme turnover (Table 27.1). No external drivers, i.e. climate, mineral nutrients or additional litter inputs, were included. In all cases, the simulations indicated an initial acceleration of decomposition until about 50% mass loss, followed by deceleration, leaving a small fraction of litter that resisted further degradation (Fig. 27.2). This mass loss profile deviates somewhat from a simple exponential decay curve because the model describes only fibre degradation. During early decomposition, modelled mass loss was slower than the initial rapid mass loss often observed in field studies. This rapid loss is caused by leaching and rapid mineralization of soluble litter fractions, a process that is not part of the model. Deviations from exponential decay also occur late in the simulations because we explicitly tie decomposition rate to carbon quality which progressively declines.

For all litters, microbial biomass peaked shortly after 50% mass loss at abundances equivalent to 6.3, 2.4 and 1.8% of standing litter mass, respectively, for dogwood, maple and oak litter (Table 27.2). The ratio of biomass peaks was 2.6 : 1.3 : 1.0. Among litter types, peak biomass declines with initial litter quality because biomass turnover varies only slightly through time while growth is directly linked to litter decay rate. Therefore, the higher the decomposition rate the higher the ratio of growth to mortality and the larger the biomass peak. For dogwood leaves decomposing in laboratory

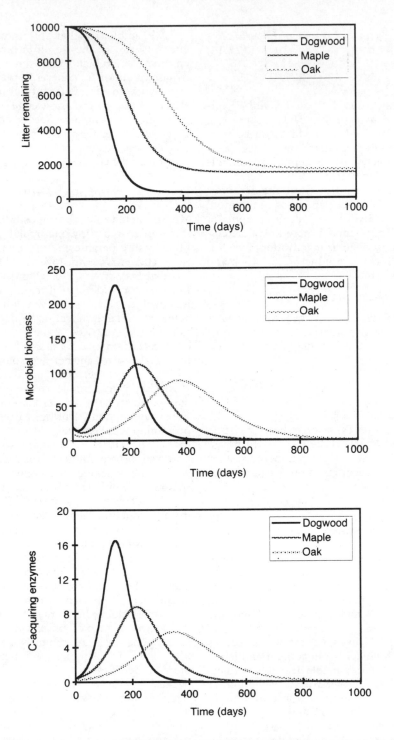

Fig. 27.2. Litter mass loss, microbial biomass, and lignocellulose-degrading enzyme stocks for decomposing dogwood, maple and oak leaf litter, predicted by the MARCIE simulation model.

Table 27.2. Summary of results from MARCIE model simulations. Dogwood, maple, and oak represent fast, medium, and slow decomposing leaf litters, respectively. E_c is the stock of C-acquiring extracellular enzymes. All stocks are represented in carbon units.

Litter type	Peak microbiomass	Peak E_c	50% Mass loss
Dogwood	226 @ day 153	16.5 @ day 142	@ day 134
Maple	111 @ day 233	8.8 @ day 213	@ day 223
Oak	87 @ day 375	5.8 @ day 348	@ day 368

microcosms, Carreiro and Sinsabaugh (unpublished) found that fungal biomass, measured as ergosterol, peaked at about 83 mg g^{-1} (8.3% of standing litter) at a mass loss of 55%; for red oak leaves decomposing in temperate forest plots, peak fungal biomass was 50 mg g^{-1} at a mass loss of 30%. These data suggest that the microbial growth efficiency of 12–15% incorporated into the model may be too conservative. Increasing the growth efficiency will elevate the biomass peaks, but will also increase decomposition rates unless the E_C turnover rate (k_c) also increases and/or enzymatic efficiency (AEE) decreases. AEE and k_c cannot be directly measured, but their values are bound by data on biomass abundance and mass loss rate.

Enzyme stock (E_C) peaked 1–3 weeks before microbial biomass at values that corresponded to 7–8% of peak microbial biomass. The ratio of E_C peaks was 2.8:1.5:1 for dogwood, maple and oak, respectively. These results are comparable to those obtained in field studies. From data presented by Linkins *et al.* (1990a), the ratio of peak cellulase activity for dogwood, maple and oak leaves decomposing in laboratory microcosms was about 3.5:4.6:1, respectively, with activity peaking at mass losses of about 30%, 50%, and 30%, respectively. For the same three litter types decomposing in a deciduous forest plot (Linkins *et al.*, 1990b), the ratio of cellulase activity peaks was 3.1:2.5:1 with peak activity measured at mass loss values of about 30%, 40% and 30%, respectively. For dogwood leaves decomposing in laboratory microcosms, Carreiro and Sinsabaugh (unpublished) found that cel-

lulase activity peaked at about 50% mass loss; for oak and maple leaves decomposing in temperate forest plots, cellulolytic activity peaked at about 30% and 40% mass loss, respectively.

The field data suggest that the initial rate of enzyme turnover or the initial AEE used in the MARCIE simulations may have been somewhat too high; causing the E_C stock to peak at mass loss values near 50%. However, a more fundamental problem is that the MARCIE model contains only one stock for lignocellulose–degrading enzymes. For real leaf litter, cellulolytic activity generally peaks before 50% mass loss while ligninolytic activity peaks later; an observation that led to the development of the EICQ indicator described above. To simulate this pattern with the MARCIE model, it would be necessary to create separate cellulase and ligninase stocks and partition resources between them in relation to carbon quality.

The initial LCI values used for each simulation correspond to published analyses of the fibre composition of dogwood, maple and oak leaf litters (e.g. Taylor *et al.*, 1989; Sinsabaugh *et al.*, 1981). However, the assumption that N and P availabilities decline from dogwood to maple to oak is hypothetical (Table 27.1). What we propose is that the N and P content of the litter becomes physically or chemically less accessible from dogwood to maple to oak, requiring the microbiota to expend a larger fraction of their resources to acquire it. Preliminary results from a study in progress (Carreiro and Sinsabaugh, unpublished) suggest that this hypothesis may be reasonable: mass loss per activity-day (AEE) for phosphatase, peptidase, and

Table 27.3. Apparent enzymatic efficiencies (% mass loss/activity-day) for acid phosphatase, glycine aminopeptidase, and β-1,4-N-acetylglucosaminidase associated with decomposing dogwood (*Cornus florida*), maple (*Acer rubrum*), and oak (*Quercus borealis*) leaves. Because the study is still in progress (Carreiro and Sinsabaugh, unpublished), these values apply only to the early stages of decomposition (cumulative mass losses of 67%, 43% and 35%, respectively, for dogwood, maple and oak). The numbers shown are means for three treatments which received N amendments in the form of ammonium nitrate at doses of 0, 20 and 80 kg N ha^{-1} year^{-1}. N and P acquisition costs, in terms of enzymatic activity, were considerably higher for oak relative to maple and dogwood. The ratio of N-acquisition activity per unit mass loss was approximately 3:4:6 for dogwood, maple, and oak, respectively; for P-acquisition the ratio was about 3:2:27.

	Apparent enzymatic efficiency		
Enzyme	Dogwood	Red maple	Red oak
Acid phosphatase	0.025	0.036	0.0028
Glycine aminopeptidase	0.30	0.056	0.15
β-N-acetylglucosaminidase	0.046	0.070	0.025

β-*N*-acetylglucosaminidase, tends to decrease from flowering dogwood to red maple to red oak litter, even with additions of N fertilizer (Table 27.3).

Discussion

In the course of expanding the MARCIE model into a simulation, several constraints emerged as essential for the generation of reasonable results, given our premises. To get outputs consistent with known plant litter decomposition patterns it was necessary to establish dynamic relationships between litter quality and C flows. Microbial assimilation, microbial mortality and enzyme inactivation were linked to C quality by linear equations. The production of C-acquiring enzymes was linked to N and P availability through the MEAD function. Respiration was the only flow that did not require a direct link to litter quality measures; however for low quality litters, we got more reasonable simulation results by selecting lower respiration rates (Table 27.1).

The requirement for enzyme and biomass turnover to decline with litter quality is one of the interesting hypotheses to emerge from the simulation exercise. It is consistent with models by Parnas (1975) and Moorhead and Reynolds (1991) that link biomass turnover directly to growth. A corollary finding is that, regardless of litter quality, the extracellular enzyme pool must turnover more rapidly than microbial biomass. The relative difference falls within a narrow range. If enzyme turnover is too rapid relative to biomass, decomposition ceases. If enzyme turnover is slower than biomass turnover, the simulation crashes. To get reasonable decomposition patterns, the enzyme turnover is set at 0.01–0.02 greater than the biomass turnover rate. Unfortunately, it is very difficult to estimate the turnover rate for extracellular enzyme pools in the field. However, biomass turnover throughout the course of decomposition can be monitored using established methods for assaying bacterial and fungal productivity (Newell and Fallon, 1991; Michel and Bloem, 1993).

A third hypothesis to emerge from the MARCIE simulation is that the apparent enzymatic efficiency of lignocellulases also varies with litter quality. In the model, AEE declined by two to three fold over the course of decomposition, depending on initial carbon quality. This hypothesis has not been explicitly tested, but it is consistent with data from previous studies which show as much as a ten-fold range of variation among litters spanning a wide range of quality (Sinsabaugh *et al.* 1994a,b).

The MARCIE simulation assumes that N and P availabilities, as well as carbon quality, decline with decomposition leading to a progressive reallocation of resources among extracellular enzyme pools. Many studies have followed N and P dynamics over the course of plant litter decomposition. The general pattern for N is net immobilization during early decomposition, during which the absolute N content of the litter may increase up to several-fold relative to initial content, followed by net mineralization (Harmon *et al.*, 1986; Webster and Benfield, 1986; Prescott *et al.*, 1993). A similar pattern occurs for P except that the absolute P content of the litter typically does not increase beyond that initially present. It is further known that most of the N immobilized by decomposing litter arrives in mineral form, while most of the N released is in organic form (e.g. Hedin *et al.*, 1995); that the production of lignin-degrading enzymes is induced by low N availability (Kirk and Shimada, 1985; Kirk and Farrell, 1987); and that decomposition is stimulated in its early stages and repressed in its later stages by fertilization with mineral N (Fog 1988; O'Connell, 1994). Asmar *et al.* (1994) found that rate of N mineralization from soil was limited by the availability of labile carbon and was directly related to the activity of extracellular proteases. Collectively, these observations are consistent with the MARCIE hypothesis that nutrient availability declines over the course of decomposition, but direct corroboration will require following the activity of N and P-acquiring enzymes in relation to mass loss, litter quality, and microbial biomass.

The field of decomposition research is approaching a state where its various lines of inquiry are converging. The expansion of the MARCIE model into a simulation is a step toward this synthesis. Merging enzymic and litter quality models in effect links the microbial decomposition process with the ecosystem process. What emerges are new relationships that can be applied to facilitate field studies and new hypotheses that can direct further research. Of the latter, the most basic are that: (i) enzyme and biomass turnover rates are functions of litter quality, declining as decomposition progresses; (ii) extracellular enzymes turnover faster than microbial biomass; (iii) the apparent enzymatic efficiency of lignocellulases is linked to litter quality and declines as decomposition progresses; and (iv) for the microbiota, N and P availabilities decline throughout decomposition even though their bulk abundance may increase.

References

Arsuffi, T.L. and Suberkropp, K. (1989) Selective feeding on leaf-colonizing stream fungi: comparison of macroinvertebrate taxa. *Oecologia* 79, 30–37.

Asmar, F., Eiland, F. and Nielsen, N.E. (1994) Effect of extracellular-enzyme activities on solubilization rate of soil organic nitrogen. *Biology and Fertility of Soils* 17, 32–38.

Berg, B., Berg, M.P., Bottner, P., Box, E., Breymeyer, A., Calvo de Anta, R., Couteaux, M., Escudero, A., Gallardo, A., Kratz, W., Madeira, M., Mälkönen, E., McClaughery, C., Meentemeyer, V., Muñoz, F., Piussi, P., Remacle, J. and Virzo de Santo, A. (1993) Litter mass loss rates in pine forests of Europe and Eastern United States: some relationships with climate and litter quality. *Biogeochemistry* 20, 127–159.

Boulton, A.J. and Boon, P.I. (1991) A review of methodology used to measure leaf litter decomposition in lotic environments: time to turn over an old leaf? *Australian Journal of Marine and Freshwater Research* 42, 1–43.

Bunnel, F.L., Tait, D.E.N., Flanagan, P.W. and van Cleve, K. (1977) Microbial respiration and substrate weight loss. I. A general model of the influence of abiotic variables. *Soil Biology and Biochemistry* 9, 33–40.

Fog, K. (1988) The effect of added nitrogen on the rate of decomposition of organic matter. *Biological Reviews* 63, 433–462.

Griffiths, B.S. (1994) Microbial-feeding nematodes and protozoa in soil: their effects on microbial activity and nitrogen mineralization in decomposition hotspots and the rhizosphere. *Plant and Soil* 164, 25–33.

Harmon, M.E., Franklin, J.F., Swanson, F.J., Sollins, P., Lattin, J.D., Anderson, N.H., Gregory, S.V., Cline, S.P., Aumen, S.G., Sedell, J.R., Cromack, K. and Cummins, K.W. (1986) Role of coarse woody debris in temperate ecosystems. *Recent Advances in Ecological Research* 15, 133–302.

Hedin, L.O., Armesto, J.J. and Johnson, A.H. (1995) Patterns of nutrient loss from unpolluted, old-growth temperate forests: evolution of biogeochemical theory. *Ecology* 76, 493–509.

Howard, D.M. and Howard, P.J.A. (1993) Relationships between CO_2 evolution, moisture content and temperature for a range of soil types. *Soil Biology and Biochemistry* 25, 1537–1546.

Hsieh, Y.-P. (1992) Pool size and mean age of stable organic carbon in cropland. *Soil Science Society of America Journal* 56, 460–464.

Insam, H., Parkinson, D. and Domsch, K.H. (1989) Influence of macroclimate on soil microbial biomass. *Soil Biology and Biochemistry* 21, 211–221.

Jackson, C., Foreman, C. and Sinsabaugh, R.L. (1995) Microbial enzyme activities as indicators of organic matter processing rates in a Lake Erie coastal wetland. *Freshwater Biology*, 34, 329–342.

Kirk, T.K. and Farrell, R.L. (1987) Enzymatic 'combustion': the microbial degradation of lignin. *Annual Review of Microbiology* 41, 465–505.

Kirk, T.K. and Shimada, M. (1985) Lignin biodegradation: the microorganisms involved, and the physiology and biochemistry of degradation by white-rot fungi. In: Higuchi, T. (ed.) *Biosynthesis and Biodegradation of Wood Components*. Academic Press, New York, pp. 579–605.

Linkins, A.E., Sinsabaugh, R.L., McClaugherty, C.M. and Melillo, J.M. (1990a) Cellulase activity on decomposing leaf litter in microcosms. *Plant and Soil* 123, 17–25.

Linkins, A.E., Sinsabaugh, R.L., McClaugherty, C.M. and Melillo, J.M. (1990b) Comparison of cellulase activity on decomposing leaves in a hardwood forest and woodland stream. *Soil Biology and Biochemistry* 22, 423–425.

McClaugherty, C.A. and Berg, B. (1987) Cellulose, lignin and nitrogen concentrations as rate regulating factors in late stages of forest litter decomposition. *Pedobiologia* 30, 101–112.

McGill, W.B., Hunt, H.W., Woodmansee, R.G. and Reuss, J.O. (1981) Phoenix, a model of the dynamics of carbon and nitrogen in grassland soils. *Ecological Bulletins (Stockholm)* 33, 49–115.

Melillo, J.M., Aber, J.D. and Muratore, J.F. (1982) Nitrogen and lignin control of hardwood leaf litter decomposition dynamics. *Ecology* 63, 621–626.

Meentemeyer, V. (1978) Macroclimate and lignin control of hardwood leaf litter decomposition rates. *Ecology* 59, 465–472 .

Michel, P.H. and Bloem, J. (1993) Conversion factors for estimation of cell production rates of soil bacteria from [³H]thymindine and[³H]leucine incorporation. *Soil Biology and Biochemistry* 25, 943–950.

Moorhead, D.L. and Reynolds, J.F. (1991) A general model of litter decomposition in the northern Chihuahuan Desert. *Ecological Modelling* 59, 197–219.

Moorhead, D.L., Sinsabaugh, R.L., Linkins, A.E. and Reynolds, J.F. (1996) Decomposition processes: Modelling approaches and applications. *Science of the Total Environment* 183, 137–144.

Newell, S.Y. and Fallon, R.D. (1991). Toward a method for measuring instantaneous fungal growth rates in field samples. *Ecology* 72, 1547–1559.

O'Connell, A.M. (1994) Decomposition and nutrient content of litter in a fertilized eucalypt forest. *Biology and Fertility of Soils* 17, 159–166.

Ooshima, H., Sakata, M. and Harand, Y. (1983) Adsorption of cellulase from *Trichoderma viride* on cellulose. *Biotechnology and Bioengineering* 25, 3103–3114.

Parnas, H. (1975) Model for decomposition of organic material by microorganisms. *Soil Biology and Biochemistry* 7, 161–169.

Parton, W.J., Schimel, D.S., Cole, C.V. and Ojima, D.S. (1987) Analysis of factors controlling soil organic matter levels in Great Plains grasslands. *Soil Science Society of America Journal* 51, 1173–1179.

Paustian, K., Parton, W.J. and Persson, J. (1992) Modeling soil organic matter in organic-amended and nitrogen-fertilized long-term plots. *Soil Science Society of America Journal* 56, 476–488.

Prescott, C.E., Taylor, B.R., Parsons, W.F.J., Durall, D.M. and Parkinson, D. (1993) Nutrient release from decomposing litter in Rocky Mountain coniferous forests: influence of nutrient availability. *Canadian Journal of Forest Research* 23, 1576–1586.

Savoie, J.-M. and Gourbiere, F. (1989) Decomposition of cellulose by the species of the fungal succession degrading *Abies alba* needles. *FEMS Microbiology Ecology* 62, 307–314.

Sinsabaugh, R.L., Antibus, R.K., Linkins, A.E., McClaugherty, C.A., Rayburn, L., Repert, D. and Weiland, T. (1993) Wood decomposition: nitrogen and phosphorus dynamics in relation to extracellular enzyme activity. *Ecology* 74, 1586–1593.

Sinsabaugh, R.L., Benfield, E.F. and Linkins, A.E. (1981) Cellulase activity associated with the decomposition of leaf litter in a woodland stream. *Oikos* 36, 184–190.

Sinsabaugh, R.L. and Findlay, S. (1995) Microbial production, enzyme activity and carbon turnover in surface sediments of the Hudson River Estuary. *Microbial Ecology* 30, 127–141.

Sinsabaugh, R.L., Findlay, S., Franchini, P. and Fisher, D. (1996) Enzymatic analysis of riverine bacterioplankton production. *Limnology and Oceanography* (in press).

Sinsabaugh, R.L. and Linkins, A.E. (1988) Adsorption of cellulase components by leaf litter. *Soil Biology and Biochemistry* 20, 927–932.

Sinsabaugh, R.L. and Linkins, A.E. (1990) Enzymic and chemical analysis of particulate organic matter from a boreal river. *Freshwater Biology* 23, 301–309.

Sinsabaugh, R.L. and Moorhead, D.L. (1994) Resource allocation to extracellular enzyme production: a model for nitrogen and phosphorus control of litter decomposition. *Soil Biology and Biochemistry* 26, 1305–1311.

Sinsabaugh, R.L., Moorhead, D.L., and Linkins, A.E. (1994a) The enzymic basis of plant litter decomposition: emergence of an ecological process. *Applied Soil Ecology* 1, 97–111.

Sinsabaugh, R.L., Osgood, M. and Findlay, S. (1994b) Enzymatic models for estimating decomposition rates of particulate detritus. *Journal of the North American Benthological Society* 13, 160–169.

Taylor, B.R., Parkinson, D. and Parsons, W.F.J. (1989) Nitrogen and lignin content as predictors of litter decay rates: a microcosm test. *Ecology* 70, 97–104 .

Taylor, B.R., Prescott, C.E., Parsons, W.F.J. and Parkinson, D. (1991) Substrate control of litter decomposition in four Rocky Mountain coniferous forests. *Canadian Jornal of Botany* 69, 2242–2250.

Waksman, S.A. and Tenney, F.G. (1928) Composition of natural organic materials and their decomposition in the soil: III. The influence of nature of plant upon the rapidity of its decomposition. *Soil Science* 26, 155–171.

Webster, J.R. and Benfield, E.F. (1986) Vascular plant breakdown in freshwater ecosystems. *Annual Review of Ecology and Systematics* 17, 567–594.

Part VIII

Outlook

28   A Minimum Dataset for Characterization of Plant Quality for Decomposition

C.A. Palm[1] and A.P. Rowland[2]

[1] *Tropical Soil Biology and Fertility Programme, PO Box 30592, Nairobi, Kenya;*
[2] *Institute of Terrestrial Ecology, Merlewood Research Station, Grange-over-Sands, Cumbria LA11 6JU, UK*

Introduction

Decomposition and nutrient release patterns of organic materials are determined by the organic constituents and nutrient content of the material, the decomposer organisms present, and the environmental conditions (Tenney and Waksman, 1929). The relative importance of each of these biological or environmental controls was reviewed by Swift, Heal and Anderson (1979) where they introduced the concept of resource quality to convey the chemical and physical composition of organic materials that influence the rate of decomposition. Aber and Melillo (1991) further differentiated the effects of organic constituents and carbon quality from that of nutrient content and nutrient quality on decomposition.

Several plant quality parameters and indices have been proposed for predicting decomposition and nitrogen release patterns, many of which are discussed in other papers in this volume. Why are there many parameters, rather than one? Several factors must be considered when choosing parameters to describe plant quality. First of all, the processes of decomposition, nutrient release or soil organic matter formation may be controlled by different parameters or at least the order of importance of the parameters might change depending on the process. Likewise, the critical parameters will depend on the time

frame under consideration as indicated by Melillo *et al.*(1989). It is also apparent that the importance of certain parameters changes with the type of plant material (Palm, 1995). Therefore, a range of indices is likely to be meaningful given the objectives, time frame and plant materials of a particular study.

This chapter provides guidelines for choosing which parameters to measure based on the objectives of the study. Methods for measuring those parameters are reviewed and compared and standardized methods are then recommended. The ultimate aim is to identify robust plant quality indices that provide improved prediction of decomposition, nutrient release, and soil organic matter formation. These indices coupled with decomposition models (Parton *et al.*, 1987; Whitmore, Chapter 25, this volume, Paustian *et al.*, Chapter 24, this volume), could be used to assess organic materials for decomposition, nutrient release, and soil organic matter formation. Measurement of a few quality parameters could then replace the need for detailed decomposition studies for each plant material in each location.

Plant Quality Minimum Dataset

Decomposition and plant quality research include studies that test the usefulness of current parameters and models and studies

Table 28.1. Parameters and methods to characterize plant input quality for decomposition and soil organic matter studies, those parameters included in the Plant Quality Minimum Dataset (PQMD) are specified with an X. * indicates if the parameter is considered important for the process.

Parameters	Plant quality minimum dataset	Recommended method	Short term decomposition/ nutrient release	Long term decomposition/SOM formation
Carbon quality				
Lignin	X	ADF-H₂SO₄ (Van Soest, 1963; Rowland and Roberts, 1994)	*	*
Soluble carbon	X	Hot water extraction (TAPPI, 1988) followed by weight loss or simple sugars (Dubois et al., 1956)	*	
Soluble phenolics	if %N >1.8	Aqueous methanol (50%) extraction followed by Folin-Ciolcalteu assay, (Constantinides and Fownes, 1994)	*	?
α-Cellulose		ADF-residue (Van Soest, 1963; Rowland and Roberts, 1994)	*	*
Nutrient quality				
Total nitrogen	X	Kjeldahl (Anderson and Ingram, 1993)	*	*
Total phosphorus	X	Same digest as N, (Anderson and Ingram, 1993).		
Total carbon		Nelson Sommers (Anderson and Ingram, 1993) or CHN		
Ash-free dry weight	X	Ash 3 h 500°C	*	
Physical quality				
Specific leaf area		Weight per area, or penetrometer (Choong et al., 1992)	*	

that investigate new parameters and methods for analysing plant quality as it relates to decomposition. A minimum set of parameters should be measured to characterize plant quality for all decomposition studies to allow cross-site comparisons and synthesis of results from a broad range of studies. Even though some of the parameters may not be essential for the objectives of a particular study, researchers are encouraged to provide this basic list of plant quality characteristics, in addition to information on the climate, ecosystem and soil type where the study is conducted.

The quality parameters in Table 28.1 include this minimum dataset, in addition to some parameters that are recommended for specific objectives and plant types. This list was chosen based on current indices and decomposition models. The minimum dataset includes: lignin, soluble carbon, total nitrogen, total phosphorus and ash-free dry weight; if the nitrogen content is greater than 1.8% then soluble phenolics are also recommended. No parameters have been included for describing physical quality because of lack of agreement as to what is important. Nevertheless, the fact that physical characteristics have basically been ignored but may be important for controlling early rates of decomposition is addressed.

Standardized Methods

For minimum datasets to be collected, methods for measuring them must be appropriate for laboratories with varying levels of sophistication and the equipment and supply requirements must not be excessive. If the methods do not fit these criteria, the results obtained may not be comparable and few laboratories will adopt them because of lack of access to materials or prohibitive cost.

The most important consideration to keep in mind, however, is the aim to establish comparability between studies. Several methods exist for measuring many of the key plant quality parameters but the results obtained from the different methods do

not always compare, hindering the identification of quality indicators. For a plant quality index to be useful there needs to be some standardization of methods or a means of comparing the results from different methods. In cases where different methods are used, conversion factors can be established for converting the results to that of the recommended procedure. Laboratories should use standardized plant materials with known values for the parameters for verifying results. There is a need for plant material with certified values (Certified Reference Material) for the quality parameters identified here. Organizers of International Sample Exchange Schemes should add these parameters to their list of determinations. Researchers that have not yet established methods for characterizing plant quality are encouraged to adopt procedures listed in Table 28.1.

There are many steps in processing and analysing plant materials that can lead to results that are not comparable if they are not standardized: (i) sample collection and preparation; (ii) type of extractants and tissue-to-extract ratios; (iii) sequential extractions versus individual analyses; and (iv) the methods themselves. In all cases it is essential to report the details of the procedures used so that other researchers can follow the procedures and determine if the methods were comparable to theirs. The same sample collection and preparation methods must be used for all parameters while extractions and analytical procedures are specific and discussed separately for each parameter.

Plant Sampling and Preparation

If plant samples are not prepared in a similar manner, then subsequent standardization of analytical methods may be meaningless. First, it is important to collect a representative sample of the plant, plant community, or litter. The composition of materials varies with the part of plant, age of plant and season so it is important to note all of these factors when sampling. The material can include parts with differing quality, such as leaves and their associated twigs and stems or manures mixed with crop residues. In such cases it is best to separate the plant parts and analyse them individually but also to obtain the percentage composition of the each of the parts. Composite samples from several plants growing in the same location should be made to ensure that the natural field variability is included.

The manner in which plant samples are dried is important. If soluble compounds are to be analysed it is often recommended to analyse materials fresh from the field or freeze-dried (Allen 1989; Waterman and Mole, 1994). In many situations, neither of these options is possible because of the remoteness of sites from laboratories or lack of necessary equipment. Others have air dried material (Mafongoya et al., Chapter 13, this volume) but the temperature and sunlight conditions will be different at most sites, making standardization difficult. A compromise standardized methodology of drying plant materials in a forced air oven at 35–40°C is recommended (Allen, 1989). The sample must be spread out or small amounts placed in ventilated bags to assure circulation of air and quick drying to avoid enzymatic and microbial degradation of the material. If soluble carbohydrates or phenolics are not of interest a drying temperature of 60–80°C can be used. It is necessary to express all results on a dry weight basis (dried at 100°C) when using material dried at lower temperatures.

The particle size, or fineness of grinding, will also have an effect on chemical analysis of plant material. Vanlauwe et al. (Chapter 12, this volume) found that ball milling produces the best correlations with measured parameters but such machines are not widely available. For standardization purposes, plant materials should pass a mesh size of 1 mm. A representative sample of the ground material must be taken by thoroughly mixing prior to taking a sample for analysis. The subsampling becomes more critical as the sample weight needed for analysis becomes smaller; in such cases increasing the number of laboratory replicates is recommended.

Parameters to Characterize Carbon Quality

Proximate analysis of groups of carbon compounds, rather than detailed analysis of individual carbon compounds, is normally used for characterizing plant quality. Many procedures involve sequential extractions, first by removing nonpolar waxes and fats, then proceeding with water and then acid extractions (Waksman and Stevens, 1928; King and Heath, 1967; Schlesinger and Hasey, 1981; Ryan et al., 1990). Although these sequential methods are more specific in isolating components they are tedious and because of the many steps are likely to be less precise and less comparable between laboratories. If separate extractions are used for each group of compounds rather than sequential extractions, a portion of some of the constituents may be extracted by each extraction. In other words, there may be overlap among the constituents and the sum of the components is greater than 100%. In this section the most frequently used proximate analyses for lignin, cellulose, soluble carbohydrates and soluble phenolics are presented. The methods in turn are discussed in terms of the advantages and disadvantages for use as standardized methods for characterizing plant quality.

Lignin

Lignin, as such, does not have a precise chemical formula, but is a complex molecule composed of repeating phenylpropane units composed of an aromatic ring with three carbon side-chains. The phenolic groups that make the aromatic structure varies between plants, confounding even more the definition of lignin (Stevenson, 1994). Lignin is interwoven and covalently bonded to other cell wall constituents, imparting the rigid, woody nature to plants. Only a few organisms are known to produce the enzymes necessary for lignin degradation (Hammel, Chapter 2, this volume); some degrade the lignin completely, while others attack only the side chains or cleave certain bonds to expose the more easily decomposed cellulose (Martin and Haider, 1980; Paul and Clark, 1989; Chesson, Chapter 3, this volume). The carbon associated with lignin, at least from the aromatic ring, does not appear in the microbial biomass but shows up in the soil organic matter (Martin and Haider, 1980; Paul and Clark, 1989). Some consider lignin to be a major precursor to soil organic matter, as direct input from slightly modified, residual lignin or through the oxidation of by-products of lignin decomposition (Stevenson, 1994) but evidence is inconclusive to date (Melillo *et al.*, 1989).

Lignin is considered by many to be the most important component determining the rate of decomposition (Meentemeyer, 1978); it is therefore included in the plant quality minimum dataset. Melillo *et al.* (1989) suggest that the lignin-to-N ratio predicts the early stage of decomposition, but the lignin content alone or the lignocellulose index [lignin/(lignin+holocellulose)] predicts the longer term decomposition rate.

Lignin content varies widely, increasing with senescence of plant materials and as litter decomposition proceeds. Values in fresh, nonsenescent leaves range from 5 to about 20% lignin, while that of senesced litter range from 10 to 40% (Constantinides and Fownes, 1994a). Information on animal forage suggest that once lignin content surpasses 15% decomposition is impaired because lignin is covering and thus protecting the cellulose from attack (Chesson, Chapter 3, this volume). Melillo *et al.* (1989) suggest that quality no longer affects decomposition when the lignocellulose index reaches 0.7.

Lignin has been measured by various methods but is generally considered to be the acid-insoluble fraction (Waksman and Stevens, 1928). The major methods used for determining lignin are those developed for ecological materials (Allen, 1989), forest products (Effland, 1977) and those for assessing animal forage quality (Van Soest, 1963). The forest products technique, also known as the Klason lignin, involves treatment with 72% H_2SO_4 followed by boiling in 3% H_2SO_4. Several variations of this method are followed, differing in the diges-

tion times and the pre-extraction steps. In addition to lignin, the isolated fraction may contain small amounts of ash, proteins, and cutins (Theander and Westerlund, 1993). The forage fibre method first destroys the protein with an acid detergent solution (cetyltrimethyl ammonium bromide (CTAB)) followed by treatment with 72% H_2SO_4 (Van Soest, 1963). This method was later modified by treating the residue with potassium permanganate (Van Soest and Wine, 1968). The fraction measured using the acid-detergent fibre is often referred to as ADF-lignin. A recent comparison of the 1963 and 1968 ADF methods shows that the permanganate modification is not good for measuring lignin for litters and leaves with ADF greater that 35% (Rowland and Roberts, 1994). The method using ADF followed by acid digestion also proved to be more reproducible and involves fewer, although more corrosive, reagents.

Studies by both Rowland and Roberts (1994) and Ryan *et al.* (1990) indicate that the lignin (and cellulose) determined by the forage fibre technique is simpler and more precise than the Klason method and adequately measures lignin. A drawback of both methods is that other compounds are included in this fraction.

Given that both the Klason and ADF-lignin methods are widely used, conversion factors are needed to allow cross laboratory correlations. Ryan *et al.* (1990) developed regression equations for this from ten litters; these equations need validation using a broader range of materials.

Cellulose

α-Cellulose and hemicellulose are structural polysaccharides that combined are referred to as holocellulose. α-Cellulose, a relatively pure compound of glucose units as the building block, is the main component of the primary cell wall. External enzymes, produced by a few organisms, are required to break down the complex polymer. Hemicellulose, also important in cell wall structure, refers to a group of compounds. The structural polysaccharides are of intermediate quality as an energy source to decomposers, cellulose being degraded a bit more readily than hemicellulose (Tenney and Waksman, 1929). The close association of lignin with the cell wall constituents can decrease access by enzymes for degrading cellulose (Theander and Westerlund, 1993; Chesson, Chapter 3, this volume).

α-Cellulose and hemicellulose combined constitute over 25% and are often over half of the plant material (Tenney and Waksman, 1929; Theander and Westerlund, 1993; Rowland and Roberts, 1994). Despite the fact that they are the major component of the plant, cellulose is surprisingly not used as an indicator of decomposition so is not included in the plant quality minimum dataset. Melillo et al. (1989) include holocellulose in their lignocellulose index for estimating plant quality and long term decomposition trends. This index has not been widely adopted but perhaps merits more consideration particularly for studies concerned with long term trends and soil organic matter formation.

Methods for measuring cellulose and hemicellulose can be found in Allen (1989), though they are quite intensive and therefore not recommended given the lack of correlation of these parameters to decomposition. The acid-soluble fractionation method of Effland (1977) can be used for estimating cellulose plus hemicellulose. α-Cellulose can be estimated accurately using the ADF-lignin method (Van Soest, 1963; Rowland and Roberts, 1994) but not hemicellulose since the acid detergent extracts other compounds along with the hemicellulose (Rowland and Roberts, 1994).

Soluble carbon compounds

The soluble polar compounds, in general, are metabolic carbohydrates, amino acids and phenolics. These substances can be readily leached from plant material. They may also serve as an energy rich source for microbes and are the first compounds to decompose (Tenney and Waksman, 1929; Schlesinger and Hasey, 1981; McClaugherty et al., 1985). The amount of soluble car-

bon compounds are important in determining early rates of mass loss and carbon mineralization (Reinertsen et al., 1984; Collins et al., 1990). The amount of plant material in this soluble fraction varies considerably depending on the plant type, plant part, and stage of maturity (Amato et al., 1984) but is usually less than 15% (Tenney and Waksman, 1929; Collins et al., 1990).

As the readily degradable forms of carbon are consumed the growing microbial population requires nutrients to grow. If these nutrients are not available from the decomposing plant materials then the nutrients must be taken from the soil and net immobilization will occur. In this sense the ratio of soluble carbon compounds to nutrient content would be important in determining immediate rates and patterns of nutrient release/immobilization.

Given the still uncertain role and importance of soluble carbon compounds to decomposition and nutrient dynamics it should perhaps be included in the plant quality minimum dataset. As yet there have been too few measurements of this parameter; its inclusion in the list may provide the information necessary to clarify its role.

Soluble polar carbon compounds can be extracted by cold water, hot water, ethanol, and various other extractants. The carbon compounds extracted by each procedure are a bit different (TAPPI, 1988) so methods are not directly comparable. The forest products method of extraction by hot water (TAPPI, 1988) is perhaps the most frequently used method. The amount of soluble compounds in this extract can be variously determined by the difference in weight of the initial plant tissue and residue (TAPPI, 1988), total C (Collins et al., 1990), or simple sugars (Dubois et al., 1956). These three methods offer a range in degree of analytical requirements for laboratories with differing capabilities; how the results compare is not known. An alternative method involves extracting with 50% methanol followed by the anthrone procedure for the analysis of reducing sugars (Schlesinger and Hasey, 1981). Again, there are no comparisons between the carbon extracted and analysed by the different methods.

Soluble phenolics

Phenolics are compounds that have a hydroxyl group bonded to an aromatic ring. They include a range of compounds differing in size, complexity and reactivity. Caffeic acid, coumarins, flavonoids and tannins are all phenolic compounds (Waterman and Mole, 1994; Harborne, Chapter 4, this volume). The condensed and hydrolysable tannins are perhaps the most important in terms of decomposition and nutrient dynamics.

Soluble phenolics can serve as a carbon substrate (Martin and Haider, 1980) but many of them, particularly tannins, can also inhibit the growth or function of the decomposer organisms by binding to enzymes, or render N unavailable to organisms by chemically binding to proteins (Martin and Haider, 1980; Harborne, 1989; Waterman and Mole, 1994). Although it is recognized that binding of soil or plant N by soluble phenolics could be an important pathway in the formation of recalcitrant soil organic nitrogen (Handley, 1961; Martin and Haider, 1980; Stevenson, 1994), their role in decomposition remains unclear (Swift et al., 1979; Anderson and Swift, 1983). Recent studies (Palm and Sanchez, 1991; Constantinides and Fownes, 1994a; Handayanto et al., 1994) showing that phenolics slow the release of N but not necessarily the C mineralization (P. Gibbs, 1995, unpublished results), indicating that phenolics merit further attention. Tian et al. (1995) have proposed a Plant Residue Quality Index for predicting decomposition rates that includes the C-to-N ratio, lignin and polyphenol content. Values for soluble phenolics in plant materials are generally less than 10% tannic acid equivalents (TAE) and may therefore seem unimportant but levels above 3% seem to be enough to affect nitrogen dynamics (Palm, 1995). The longer term availability of the nitrogen bound to phenolics is unknown.

The possible role of phenolics in soil organic matter formation and both long and short term nitrogen availability provides reason for inclusion in the plant quality minimum dataset. It may not be justified for all materials but is important for fresh materials with N contents higher than 1.8% but not necessarily senesced litter (Constantinides and Fownes, 1994a).

Several methods exist for the extraction and analysis of phenolics (Hagerman and Butler, 1989; Waterman and Mole, 1994). The amount of soluble phenolics extracted from plant material depends on the type of extractant, the plant-to-extract ratio, drying temperature and exposure to UV (Mueller-Harvey, 1989; Cork and Krockenberger, 1991; Waterman and Mole, 1994; Constantinides and Fownes, 1994; Mafongoya et al., 1996, Chapter 13, this volume). The amount present in an individual plant also varies considerably with the time of day, position in the plant canopy and often the maturity of the leaf (Owen-Smith, 1993). Constantinides and Fownes (1994b) found the phenol content was considerably less in materials dried at 50°C compared to 23°C. So, plant sampling and preparation to obtain representative values for phenolics are perhaps more critical than for other constituents.

Aqueous acetone, aqueous methanol and hot water are the most frequently used extractants. While some researchers strongly recommend aqueous acetone because it is a more efficient extractant, some compounds, particulary condensed tannins, are less stable in acetone (Cork and Krockenberger, 1991) and methanol may be preferred. The plant tissue to extract ratio is extremely important for extraction efficiency. A procedure outlined by Anderson and Ingram (1993) recommends 37.5 mg of plant material per ml of extractant but a ratio of 2 mg ml^{-1} has since been shown to be more effective (Constantinides and Fownes, 1994b).

Total soluble phenolics in the extract are most commonly measured by the Folin-Denis, or more recently the Folin-Ciocalteu assays (Waterman and Mole, 1994). The Folin-Ciocalteu reagent is more robust than the Folin-Denis reagent and it can now be purchased pre-prepared, eliminating the previous cumbersome

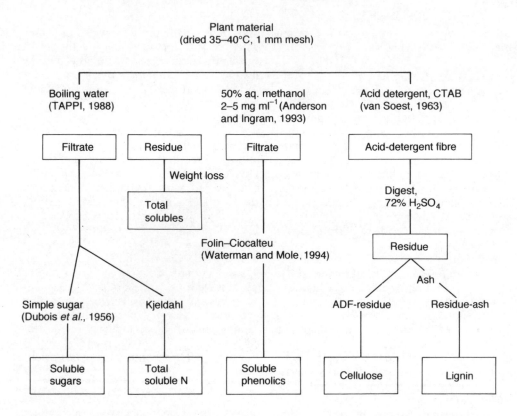

Fig. 28.1. Recommended procedures for analysis of carbon proximate fractions using three separate extractions.

methods of preparing the Folin reagents. It has been argued that this assay is crude, extracting a variety of phenolics, including non-tannin types (Hagerman and Butler, 1989). Despite this, consistent and significant negative correlations have been obtained for total soluble phenolics measured by the Folin-Denis assay and net nitrogen mineralization (Palm, 1995).

Correlations between phenolics and nitrogen dynamics might be improved if the extract is analysed for condensed tannins, the major group of phenolics that binds proteins. Several tannin assays exist including the acid butanol and vanillin assays for condensed tannins, also known as proanthocyanidins (Waterman and Mole, 1994), and assays to assess the protein-binding capacity of the phenolics, including the bovine serum albumin (BSA) precipitation assays (Dawra *et al.*, 1988;

Waterman and Mole, 1994; Handayanto *et al.*, 1995). Unfortunately there is no apparent correlation among those different assays (Martin and Martin, 1982) and the presence of tannins is not always associated with a functional response or inhibition of organisms (Owen-Smith, 1993). Handayanto *et al.* (1994) did find the relationship with phenolics and nitrogen dynamics increased when the extract was analysed for its protein-binding capacity, particularly under non-leaching conditions.

Recommended procedures for carbon quality

Based on the above discussions and keeping in mind accuracy, precision, economy of methods and adoptability, a recommendation is made here involving three sepa-

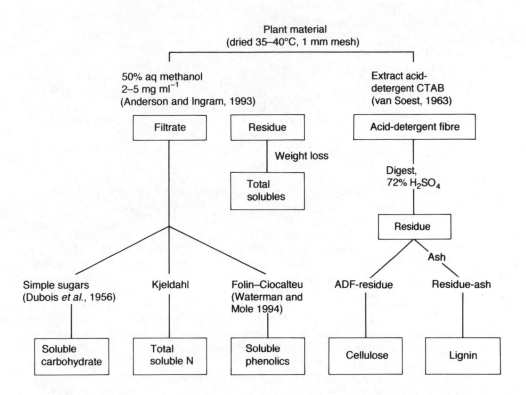

Fig. 28.2. Alternative procedure for analysis of carbon fractions using two separate extractions.

rate extractions that will allow analysis of soluble carbon, phenolics, cellulose and lignin (Fig. 28.1). An alternative recommendation requires only two extractions (Fig. 28.2). The first is more consistent with currently used procedures while the second has fewer steps and provides a means of analysing the soluble carbon and phenolics from the same extract. There is an urgent need to compare the results from the two procedures.

These parameters are usually measured on the initial plant material. Repeated analyses through the process of decomposition should also be made to show the composition of the material that remains to help in clarifying the role of the various parameters in soil organic matter formation.

Parameters to Characterize Nutrient Quality

Total nitrogen, the C-to-N ratio, and lignin-to-N ratio have all been related to decomposition rates (Iritani and Arnold, 1960; Melillo *et al.*, 1982) therefore, nitrogen is included in the plant quality minimum dataset. It is considered to be important in decomposition because it is essential but often limiting to the growth of decomposer communities. The concentration of nitrogen in the plant material also affects nitrogen dynamics. Values less than 1.7% to 2.0% are generally associated with net immobilization (Jensen, 1929; Constantinides and Fownes, 1994).

Total carbon in itself is not related to decomposition but the distribution of carbon into the various carbon compounds is important, as discussed above. Carbon is not included in the plant quality minimum dataset. Carbon values of plant materials do not vary widely and if needed for a particular index can be estimated as 45% of the mass.

The role of phosphorus in controlling decomposition is not as clearly defined as that of nitrogen (Schlesinger and Hasey, 1981; Vogt et al., 1986). This is perhaps due to the fact that the majority of decomposition studies have been conducted on soils that are limiting in nitrogen and not phosphorus. As more studies are conducted in the tropics on phosphorus limited soils the role may become more clear (Vitousek et al., 1994). For these reasons phosphorus is recommended for the plant quality minimum dataset. Phosphorus concentrations in plant materials do affect phosphorus dynamics, although the threshold value below which net phosphorus immobilization occurs is not well established.

Methods for the analysis of plant nutrients are more accurate and precise than those for the carbon compounds, so there is less of a need to standardize as long as laboratories are cross calibrated or reference materials are used. Nitrogen is usually analysed by micro-Kjeldahl procedures (Anderson and Ingram, 1993) or by carbon–hydrogen–nitrogen (CHN) analysers. Analysis of the nitrogen in the readily decomposable (water soluble) fraction and more decomposition resistant (acid-insoluble) fraction may help explain short and long term nitrogen dynamics. Nitrogen can be readily measured in the soluble fraction by Kjeldahl procedures (Fig. 28.1) and ninhydrin assays (Amato and Ladd, 1988). It may not be possible to obtain a reliable estimate of N in the ADF-lignin fraction because the acid detergent reagent, CTAB, contains nitrogen. The acid-insoluble residue of the Klason procedure can be analysed for nitrogen. Phosphorus can be analysed from the same digest as nitrogen. Carbon is analysed by Walkley Black or Nelson Sommers methods (Anderson and Ingram, 1993) or by CHN analysers. As mentioned above,

carbon can also be estimated as 45% of the mass.

Ash-free dry weight is also determined and data reported for the carbon constituents on an ash-free basis. This also allows for correction of contamination by the mineral soil in decomposition studies. Samples are usually ashed at 450–500°C for 3 h.

Parameters to Characterize Physical Quality

Decomposition studies to a large extent have ignored the effects of physical properties. While no parameters are recommended here for the plant quality minimum dataset, attention is called to certain parameters that may help explain decomposition patterns. In cases where they seem important their measurement is encouraged.

The physical nature of plant materials that could affect decomposition and nutrient release include surface properties, toughness and particle size (Swift et al., 1979), although some consider these physical parameters merely a reflection of the chemical parameters. Surface features that can deter fungal growth and feeding include waxiness and pubescence. The wax content, which again relates to chemical properties, can be assessed by using nonpolar solvents such as dichloromethane (Ryan et al., 1990) or ether (Schlesinger and Hasey, 1981).

The texture of leaves has been variously described as toughness, hardness, or stiffness and has been related to the degree of schlerophylly (Grubb, 1986). The schlerophylly index can be estimated by the crude fibre-to-crude protein ratio (Loveless, 1961) and as such could be estimated by the ADF fibre and lignin content. Toughness can also be measured as the penetrometer resistance (Choong et al., 1992).

The particle size of the material may be important in determining the initial rates of decomposition through comminution, particularly for materials with a high schlerophylly index. The combined particle area and its weight, known as the specific leaf area in g cm^{-2}, may be a simple means of estimating toughness. This measurement is determined by taking the dry weight and area of the leaf particles. Though not included in minimum dataset, measurement of the specific leaf area is encouraged for future decomposition studies.

New Approaches

Proximate analyses have provided considerable insight into the controls on decomposition. New approaches with near-infrared reflectance (NIR) and ^{13}C nuclear magnetic resonance (NMR) spectroscopy have been shown to identify groups of compounds more accurately and at the same time have shown discrepancies with the interpretation of results from proximate analysis.

NIR spectra may provide a rapid and economical means of determining the concentration of certain chemical constituents of plant materials. McLellan *et al.* (1991) combined NIR spectra with proximate analysis data from various plant litters to calculate multiple regression equations for estimating nitrogen, lignin, and cellulose. They conclude that this method can be widely adapted for measuring these parameters on fresh and decomposed plant materials and that it is more easily standardized. This rapid, less expensive method could be used for repeated measurements on materials during decomposition.

Data from NMR spectra provide information on the relative abundance of functional groups in plant materials that can be related to the proximate components. Through such analyses compounds extracted as acid-insoluble lignin were found to contain as much as 50% alkyl carbon rather than the expected predominace of aromatic carbon compounds (J.A. Trofymow, 1995, personal communication). The significance of such findings to the interpretation of plant quality as it relates to decomposition perhaps will change as more such information is gathered.

NMR can also be used for comparing the quality of plant materials entering the soil with the resulting composition of the soil organic matter (Skjemstad *et al.*, Chapter 20, this volume). Such approaches were not possible using proximate analyses because the components are difficult to isolate in soil. Analyses by NMR are expensive so will only be used by a few research groups. It will prove to be a valuable tool to further our understanding of the plant decomposition to soil organic matter continuum.

Other analytical techniques are available to enable researchers to fractionate individual compounds from the broad groups of phenolics, lignin and carbohydrates (Theander and Westerlund, 1993). Kuiters and Sarink (1986) used gas chromatography (GC) to identify 18 phenolic compounds from aqueous leachates from deciduous leaves and conifer needles and decomposed litter. Kogel (1986) developed a method to characterize the intact lignin structures by oxidative degradation of the lignin and analysis of the products by reversed phase high performance liquid chromatography (HPLC). This detailed study provided information on the modifications of the material during microbial decomposition.

None of these powerful techniques are recommended in the minimal dataset as they are time consuming, costly and offer the risk of embarking on yet new protracted studies to define in detail the chemical composition of plant material. The results generated by a few research groups, however, may take us the next step in understanding plant quality as it relates to decomposition and soil organic matter formation.

Conclusions

In this chapter we have presented recommendations on a basic list of parameters for describing the chemical quality of plant materials that should be included in all decompositions studies. This list is not exclusive or intended to curtail other measurements that will further our understanding of the decomposition process but is aimed at progressing beyond merely descriptive studies of decomposition. There are numerous decomposition studies that report mass loss in the absence of plant quality, climate, and ecosystem data that are needed for explaining the process and used for synthesis and modelling purposes. Students of decomposition are hereby encouraged to move from the descriptive to the predictive phase.

Acknowledgements

The authors would like to thank Ken Giller and Georg Cadisch and the working group on Methods for Quality Characterization held at the conference Driven By Nature for providing the basis for this paper. Special thanks go to J. Anderson, T. Trofymow and L. Greenfield for their assistance on specific topics.

References

Aber, J.D. and Melillo, J.M. (1991) *Terrestrial Ecosystems*. Rinehart and Winston Inc., PA, USA. 430 pp.

Allen, S.E. (ed.) (1989) *Chemical Analysis of Ecological Material*. 2nd edn. Blackwell, Oxford, UK, 368 pp.

Amato, M. and Ladd, J.N. (1988) Assay for microbial biomass based on ninhydrin-reactive nitrogen in extract of fumigated soils. *Soil Biology and Biochemistry* 20, 107–114.

Amato, M., Jackson, R.B., Butler, J.H.A. and Ladd, J.N. (1984) Decomposition of plant material in Australian soils. II. Residual organic ^{14}C and ^{15}N from legume plant parts decomposing under field and laboratory conditions. *Australian Journal of Soils Research* 22, 331–341.

Anderson, J.M. and Swift, M.J. (1983) Decomposition in tropical forests. In: Sutton, S.L., Whitmore, T.C. and Chadwick, A.C. (eds) *Tropical Rain Forest: Ecology and Management*, Blackwell Scientific Publications, London, UK, pp. 287–309.

Anderson, J.M. and Ingram, J.S.I. (1993) *Tropical Soil Biology and Fertility: A Handbook of Methods*. 2nd edn. CAB International, Wallingford, UK, pp. 221.

Choong, M.F., Lucas, P.W., Ong, J.S.I., Pereira, B., Tan, H.T.W. and Turner, I.M. (1992) Leaf texture toughness and sclerophylly their correlation and ecological implications. *Phytology* 121, 597–610.

Collins, H.P., Elliott, L.F., Rickman, R.W., Bezdicek, D.F. and Papendick, R.I. (1990) Decomposition and interactions among wheat residue components. *Soil Science Society of America Journal* 54, 780–785.

Constantinides, M. and Fownes, J.H. (1994a) Nitrogen mineralization from leaves and litter of tropical plants: relationship to nitrogen, lignin and soluble polyphenol concentrations. *Soil Biology and Biochemistry* 26, 49–55.

Constantinides, M. and Fownes, J.H. (1994b) Tissue-to-solvent ratio and other factors affecting determination of soluble phenolics in tropical leaves. *Communications in soil science and Plant Analysis* 25, 3221–3227.

Cork, S.J. and Krockenberger, A.K. (1991) Methods and pitfalls of extracting condensed tannins and other phenolics from plants: insights from investigations on *Eucalyptus* leaves. *Journal of Chemical Ecology* 17, 123–134.

Dawra, R.K., Makkar, H.S.P. and Singh, B. (1988) Protein binding capacity of microquantities of tannins. *Analytical Biochemistry* 170, 50–53.

Dubois, M. Gilles, K.A., Hamilton, J.K., Rebers, P.A. and Smith, F. (1956) Colorimetric method for determination of sugars and related substances. *Analytical Chemistry* 28, 350–356.

Effland, M.J. (1977) Modified procedure to determine insoluble lignin in wood and pulp. *TAPPI* 60, 143–144.

Grubb, P.J. (1986) Sclerophylls, pachphylls and pycnophylls : the nature and significance of hard leaf surfaces. In: Juniper, B.E. and Southwood T.R.E. (eds) *Insects and the Plant*

Surface. Edward Arnold, London, UK, pp. 137–150.

Hagerman, A.E. and Butler, L.G. (1989) Choosing appropriate methods for assaying tannin. *Journal of Chemical Ecology* 15, 1795–1810.

Handayanto, E., Cadisch, G. and Giller, K.E. (1994) Nitrogen release from prunings of legume hedgerow trees in relation to quality of the prunings and incubation method. *Plant and Soil* 160, 237–248.

Handayanto, E., Cadisch, G. and Giller, K.E. (1995) Manipulation of quality and mineralization of tropical legume tree prunings by varying nitrogen supply. *Plant and Soil* 176, 149–160.

Handley, W.R.C. (1961) Further evidence for the importance of residual protein complexes on litter decomposition and the supply of nitrogen for plant growth. *Plant and Soil* 15, 37–73.

Iritani, W.M. and Arnold, C.Y. (1960) Nitrogen release of vegetable crop residues during incubation as related to their chemical composition. *Soil Science* 89, 74–82.

Jensen, H.L. (1929) On the influence of the carbon:nitrogen ratios of organic material on the mineralization of nitrogen. *Journal of Agricultural Science* 19, 71–82.

King, H.G.C. and Heath, G.W. (1967) The chemical analysis of small samples of leaf material and the relationship between the disappearance and composition of leaves. *Pedobiolgia* 7, 192–197.

Kogel, I. (1986) Estimation and decomposition pattern of the lignin component in forest humus layers. *Soil Biology and Biochemistry* 18, 589–594.

Kuiters, A.T. and Sarink, H.M. (1986) Leaching of phenolic compounds from leaf and needle litter of several deciduous and coniferous trees. *Soil Biology and Biochemistry* 18, 475–480.

Loveless, A.R. (1961) A nutritional interpretation of sclerophylly based on differences in the chemical composition of sclerophyllous and mesophytic leaves. *Annals of Botany* 25, 168–184.

McClaugherty, C.A., Pastor, J. and Aber, J.D. (1985) Forest litter decomposition in relation to soil nitrogen dynamics and litter quality. *Ecology* 66, 266–275.

McLellan, T.M., Aber, J.D. and Martin, M.E. (1991) Determination of nitrogen, lignin and cellulose content of decomposing leaf material by near infrared reflectance spectroscopy. *Canadian Journal of Forest Research* 21, 1684–1688.

Martin, J.P. and Haider, K. (1980) Microbial degradation and stabilization of ^{14}C-labelled lignins, phenolics, and phenolic polymers in relation to soil humus formation. In: Kirk, T.K., Higuchi, T. and Chang, H.M. (eds) *Lignin Biodegradation: Microbiology, Chemistry, and Potential Applications.* Vol. 2. CRC Press. West Palm Beach, FL, USA, pp. 78–100.

Martin, J.S. and Martin, M.M. (1982) Tannin assays in ecological studies: Lack of correlation between phenolics, proanthocyanidins and protein-precipitating constituents in mature foliage of six oak species. *Oecologia* 54, 205–211.

Meentemeyer, V. (1978) Macroclimate and lignin control of litter decomposition rates. *Ecology* 59, 465–472.

Melillo, J.M., Aber, J.D. and Muratore, J.F. (1982) Nitrogen and lignin control of hardwood litter decomposition dynamics. *Ecology* 63, 621–626.

Melillo, J.M., Aber, J.D., Linkins, A.E., Ricca, A., Fry, B. and Nadelhoffer, K.J. (1989) Carbon and nitrogen dynamics along a decay continuum: plant litter to soil organic matter. In: Clarkholm, M and Bergström, L. (eds) *Ecology of Arable Land.* Kluwer Academic, Dordrecht, The Netherlands, pp. 53–62.

Mueller-Harvey, I. (1989) Identification and importance of polyphenolic compounds in crop residues. In: Chesson, A. and Orskov, E.R. (eds) *Physio-chemical Characterization of Plant Residues for Industrial and Feed Use.* Elsevier Science Publishing Co. Inc., New York, pp. 88–109.

Owen-Smith, N. (1993) Woody plants, browsers and tannins in southern African savannahs. *Suid-Afrikaanse Tydskrifvir Wetenskap* 89, 505–509.

Palm, C.A. (1995) Contribution of agroforestry trees to nutrient requirements of intercropped plants. *Agroforestry Systems* 30, 105–124.

Palm, C.A. and Sanchez, P.A. (1991) Nitrogen release from the leaves of some tropical legumes as affected by their lignin and polyphenolic contents. *Soil Biology and Biochemistry* 23, 83–88.

Parton, W.J., Schimel, D.S., Cole, C.V. and Ojima, D.S. (1987) Analysis of factors controlling soil organic matter levels in Great Plains grasslands. *Soil Science Society of America Journal* 51, 1173–1179.

Paul, E.A. and Clark, F.E. (1989) *Soil Microbiology and Biochemistry*. Academic Press Inc., San Diego, CA, USA, 273 pp.

Reinertsen, S.A., Elliott, L.F., Cochran, V.L. and Campbell, G.S. (1984) Role of available carbon and nitrogen in determining the rate of wheat straw decomposition. *Soil Biology and Biochemistry* 16, 459–464

Rowland, A.P. and Roberts, J.D. (1994) Lignin and cellulose fraction in decomposition studies using acid-detergent fibre methods. *Communications in Soil Science and Plant Analysis* 25, 269–277.

Ryan, M.G., Melillo, J.M. and Ricca, A. (1990) A comparison of methods for determining proximate carbon fractions of forest litter. *Canadian Journal of Forest Research* 20, 166–171.

Schlesinger, W.H. and Hasey, M.M. (1981) Decomposition of chaparral shrub foliage: Losses of organic and inorganic constituents from deciduous and evergreen leaves. *Ecology* 62, 762–774.

Stevenson, F.J. (1994) *Humus Chemistry: Genesis, Composition, Reactions*. 2nd edn. John Wiley and Sons, NY, USA, 496 pp.

Swift, M.J., Heal, O.W. and Anderson, J.M. (1979) *Decomposition in Terrestrial Ecosystems*. Studies in Ecology. Vol. 5. University of California Press, Berkeley, CA, USA, 372 pp.

TAPPI (1988) *Water Solubility of Wood and Pulp, T 207 OM-88*. Technical Association of the Pulp and Paper Industry, Atlanta, GA, USA, 2 pp.

Tenney, F.G. and Waksman, S.A. (1929) Composition of natural organic materials and their decomposition: IV. The nature and rapidity of decomposition of the various organic complexes in different materials, under aerobic conditions. *Soil Science* 28, 55–84.

Theander, O. and Westerlund, E. (1993) Quantitative analysis of cell wall components. In: *Forage Cell Wall Structure and Digestibility*. ASA-CSSA-SSSA, Madison, WI, USA, pp. 83–104.

Tian, G., Brussaard, L. and Kang, B.T. (1995) An index for assessing the quality of plant residues and evaluating their effects on soil and crop in the (sub-) humid tropics. *Applied Soil Ecology* 2, 25–32.

Van Soest, P.J. (1963) Use of detergents in analysis of fibrous feeds. II. A rapid method for the determination of fibre and lignin. *Association of Official Agricultural Chemists Journal* 46, 829–835.

Van Soest, P.J. and Wine, R.H. (1968) Determination of lignin and cellulose in acid-detergent fibre with permanganate. *Association of Official Agricultural Chemists Journal* 51, 780–785.

Vitousek, P.M., Turner, D.R., Parton, W.J. and Sanford, R.L. (1994) Litter decomposition on the Mauna Loa environment matrix Hawaii: Patterns, mechanisms and models. *Ecology* 75, 418–429.

Vogt, K.A., Grier, C.C. and Vogt, D.J. (1986) Production, turnover and nutrient dynamics of above- and below-ground detritus of world forests. *Advances in Ecological Research* 15, 303–377.

Waksman, S.A. and Stevens, K.R. (1928) Contribution to the chemical composition of peat: I. Chemical nature of organic complexes in peat and methods of analysis. *Soil Science* 28, 113–135.

Waterman, P.G. and Mole, S. (1994) *Analysis of Phenolic Plant Metabolites*. Methods in Ecology. Blackwell Scientific Publications, London, UK, 238 pp.

29   Driven by Nature: A Sense of Arrival or Departure?

K.E. Giller and G. Cadisch
*Wye College, University of London, Wye, Ashford,
Kent TN25 5AH, UK*

Introduction

The title *Driven by Nature* is perhaps ambiguous, yet it captures the essence of the debate on which this book is based. Our aim was to explore to what extent the nature (i.e. 'the essential qualities of a thing'; Little *et al.*, 1988) of litter determines its decomposition and this is explored from a wide variety of standpoints in the various chapters.

Litter quality undoubtedly regulates the potential rate and outcome of decomposition, but the actual rate and degree of decomposition are moderated by the activity of the decomposer organisms and the environmental conditions. This is equally true of litter decomposition in soil as described by Heal *et al.* (Chapter 1) as of digestion in the rumen (Chesson, Chapter 3). In fact, Heal's review of all aspects of decomposition is so comprehensive that a detailed summary is not warranted here. A particular interest of ours relates to the application of the outputs of research and we focus on that in this summary chapter.

Although we can only ascribe the role of governing potential decomposition to litter quality, it is perhaps the factor most amenable to manipulation and intervention in agricultural systems. By contrast, the options for modification of the decomposer organisms or the environment under which they can function are rather limited, although placement of crop residues is one simple tool for manipulation of the environment for decomposition (see Jones *et al.*, Chapter 19).

Recent Research Highlights and Future Priorities

A questionnaire was circulated at the *Driven by Nature* conference requesting participants to indicate both the advances made over the last 20 years or so and priorities for future research. A number of themes emerged from this survey, many of which are of course highlighted elsewhere in detail in other chapters in this book. Some topics, however, deserve particular mention and these are discussed further below.

Environmental controls

The importance of environmental controls on decomposition is covered thoroughly by Heal *et al.* (Chapter 1), and was highlighted by Meine van Noordwijk in summarizing a discussion session at the conference as the 'nurture' aspect of decomposition. The importance of the environments in regulating actual decomposition is addressed by simulation models (Paustian *et al.*, Chapter 24) which through a variety of approaches are able to combine effects of litter quality and the environmental controls to predict the outcomes of decomposition.

Litter quality and biodiversity in soil

The other side of the Driven by Nature debate, that is the extent to which processes are regulated by the soil organisms is elegantly illustrated in the theory of Wardle and Lavelle (Chapter 8) which suggests that fungi are more strongly influenced by substrate quality, whereas bacterial populations are largely regulated by predation.

The role of soil fauna in turnover of high quality plant litter is relatively minor, but becomes increasingly important with low quality litters (Tian *et al.*, Chapter 9). A more diverse soil faunal community leads to accelerated early decomposition, but has little effect on the final extent of decomposition (Couteaux *et al.*, 1991), and this may be of relevance in regulating decomposition rates in agricultural systems. The

observation that an increased diversity of soil functional groups tends to dampen responses to environmental change (Heal *et al.*, Chapter 1), suggests that with ongoing problems with global climatic shifts and pollution we will become increasingly aware of such positive interactions of biodiverstity on ecosystem stability.

Plant quality and synchrony: in search of the ideal plant residue

Despite the increasingly common use of the term 'synchrony' (Myers *et al.*, Chapter 17) we still lack successful demonstrations in which this concept has been translated into practice with plant litters. An interesting exception is the study in wetland rice systems of Becker and Ladha (Chapter 18) where improved

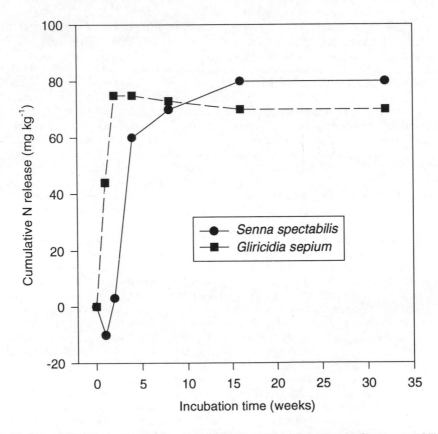

Fig. 29.1. Cumulative net mineralization of leaf litter from *Gliricidia sepium* or *Senna spectabilis* in a soil from Malawi (O.A. Itimu, unpublished results).

synchrony is demonstrated by mixing *Sesbania* and rice residues. On the other hand, mixing of residues with different quality can result in rather unpredictable strong interactions in litter quality and the resulting decomposition patterns (Handayanto *et al.*, Chapter 14). Use of polyphenol-rich residues in mixtures reduced the initial rate of N release, at least over the short-term, but at the expense of a reduced amount of N mineralization which could potentially lead to crop N shortfall. Other results with materials of low polyphenol content are more promising where the rates of N release can differ markedly without having strong differences in the overall amount of N released (e.g. Fig. 29.1).

The future potential for manipulation of plant litter quality by selection or genetic manipulation has been clearly demonstrated (Bavage *et al.*, Chapter 16), although we must bear in mind the other functions of plant quality characteristics such as physical strength and defence mechanisms. Alternatively, the range of litter types studied to date is limited compared with the diversity which exists in the plant kingdom.

Methodologies

As ever when scientists meet, considerable emphasis was focused on developments in methodologies. Much discussion considered whether the frequent use of litter bags had actually been misleading in that emphasis was placed on what remained in the bags and ignored the components which had been lost. The use of soil organic matter (SOM) fractionation to isolate large, light organic matter can be viewed as an *in situ* litter bag method in which the soil and litter are not spatially separated (see Magid *et al.*, Chapter 26). However, similar criticisms can be levelled at SOM fractionation methods as with litter bags unless the fractions which become enriched during decomposition (the finer, heavier SOM fractions) are also considered (e.g. Magid *et al.*, 1996). The use of ^{13}C-NMR methods provides another alternative for the assessment of changes in chemical composition of subtsrates during decomposition (Baldock *et al.*, Chapter 5; Hopkins and Chudek, Chapter 6; Wachendorf *et al.*, Chapter 10) and of SOM characteristics under various management treatments (Skjemstad *et al.*, Chapter 20) but, as emphasized by Baldock *et al.* (Chapter 5), the use of several methods together gives much more insight into processes and serves as a check for reliability of new approaches.

There is little progress in the search of measurable pools in both plant residues and in soil organic matter which can be used in simulation models. Magid *et al.* (Chapter 26) and Marstorp (Chapter 7) tried to use water soluble fractions as the metabolic component of litter with little success, but they showed promising results when light fractions separated from the soil were used to follow litter transformations. The use of 'decomposition days', analogous to degree days used in plant physiology and pathology is another useful suggestion to assist in data comparison and modelling (Vanlauwe *et al.*, Chapter 12).

The 'right' quality index?

Various indices (ranging from the simple C-to-N ratio, to combinations including lignin-to-N, polyphenol-to-N, (lignin+ polyphenol)-to-N and the lignocellulose index have been highlighted for prediction of decomposition and nutrient release (see chapters by Vanlauwe *et al.*, Mafangoya *et al.*, Handayanto *et al.*, Myers *et al.* and Palm and Rowland). It is now widely recognized that the different conclusions reached in each case often result from the differences in the range of materials tested. Heal *et al.* (Chapter 1) described the sequential pattern of resource utilization which results in a shift in the relative importance of different quality parameters in regulating further attack as decomposition progresses (Fig. 1.2). However, it is generally not feasible to perform detailed quality analysis at different stages of decomposition. An approximation can be achieved by observing the change in correlation between a quality parameter and decomposition or N mineralization over time (Fig. 29.2). Using this approach, at

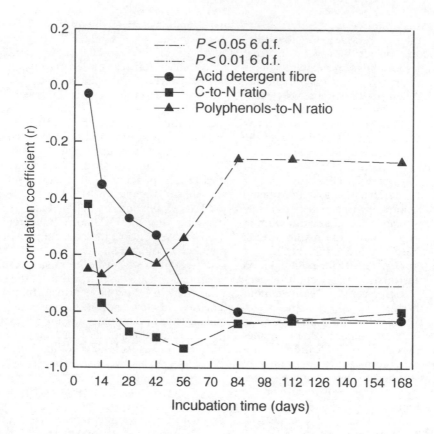

Fig. 29.2. Temporal changes in the correlation coefficients between chemical composition and net N mineralization of indigenous tree forages from Mexico (I. Almendariz-Yanez, unpublished results).

early stages the polyphenol-to-N ratio showed the best correlation with N mineralization of indigenous trees from Mexico, but as decomposition progressed acid-detergent fibre (ADF) became the dominant parameter. Thus shifts in the importance of different quality factors with time must be recognized when results are compared between studies.

Recommendations are made by Palm and Rowland (Chapter 28) for standardizing of methods for description of chemical quality of plant litters. This should in no way restrict researchers from pursuing new ideas and approaches but hopefully will assist us in the future to make proper comparisons of results between studies conducted in different laboratories and systems. It seems likely that such quality parameters will also be useful for predict-

ing phosphorus release from plant litters (Gressel and McColl, Chapter 23).

Chesson (Chapter 3) highlighted the importance of plant cell wall architecture in animal digestion which has received relatively little attention in decomposition studies. If lignin contents reach 15%, cellulose and other degradable cell wall components tend to be physically protected from enzymic attack due to their inaccessibility. Little research has been conducted to evaluate the physical quality of plant litters in relation to decomposition and few promising methods are currently available (Palm and Rowland, Chapter 28). From an experimental perspective it is desirable to measure the surface area of litter available for enzymic attack, but an enzyme-sized molecule which could be used in absorption/desorption studies remains elusive.

Does plant litter quality control the quality of soil organic matter?

It is well established that the amount and quality of organic residues added to a given soil regulate the total soil organic matter content. Our current understanding is sufficient to develop models which successfully predict the outcome of various long-term organic residue amendments to soil (Paustian *et al.*, Chapter 24). Incorporation of more complex chemical quality factors such as polyphenols into simulation models are described for the first time in this volume by Whitmore and Handayanto (Chapter 25). Recent and continuous additions of organic residues are also important in establishing and maintaining the N mineralization potential of soils (Janzen, 1987), and long-term high quality residues and green manures including legumes give consistently higher N mineralization rates compared to organic amendments of poor quality or fallow periods (Burkert and Dick, Chapter 22). Thus there is good evidence that the quality of plant litter residue additions affects N dynamics over the short-term.

The question whether, and to what extent, the chemical nature of the inputs influences the chemistry of soil organic matter has long been debated with little hard evidence (see Paustian *et al.*, Chapter 24). Recent developments in analytical techniques such as ^{13}C-NMR allow the evaluation of SOM constituents in chemically undisturbed soil samples (Skjemstad *et al.*, Chapter 20) . Using NMR methods to examine the nature of SOM fractions separated by physical methods seems a promising way to extend our knowledge of SOM formation. For example (Golchin *et al.*, 1994) showed how the high alkyl carbon content of a brigalow vegetation was preserved in soil organic matter fractions. There is little evidence, however, that aromatic groups survive in the soil environment in the long term unless they originate from fire and are preserved in charcoal (J.A. Skjemstad, personal communication).

The long-term field studies reported by Burket and Dick (Chapter 22) demonstrate the difficulty of separating residue quality effects from effects due to differences in the quantity of organic matter (or N) added, but provide no evidence of large effects of litter quality on the quality or functional properties of soil organic matter. These experiments examined a range of agricultural inputs with differing C-to-N ratio or lignin-to-N ratio ranges. The question thus remains if continuous inputs of residues with more diverse qualities such as polyphenol rich material (see Wardle and Lavelle (Chapter 8) who discuss effects of the ericaceous shrub *Empetrum hermaphroditum*) would give substantial residual effects on the quality and function of SOM. Are polyphenols a 'fast-route' to soil organic matter (see Handayanto *et al.*, Chapter 14) due to their ability to complex proteins and carbohydrates? The stability of protein-polyphenol complexes formed during early decomposition is a focus of ongoing research in our own laboratory at Wye, as our results have indicated that the protein-N is not released over periods of several months once it has been complexed by polyphenols. The degradative capacity of basidiomycetes in attacking polyphenol- and lignin-rich litter in forest systems is highlighted below, but such fungi are not abundant in many agricultural systems.

One problem raised in discussions on effects on soil organic matter quality was the lack of long-term experiments with a range of residues for study, particularly in the tropics. For such investigations increasing use of soils with contrasting long-term agricultural management is one option for comparisons. Whereas most arable systems are complicated due to concurrent differences in soil properties, in many plantation systems spatial distribution of organic matter inputs over quite small distances can lead to pronounced differences in soil organic matter replicated from tree to tree across fields (e.g. Fairhurst, 1996).

Some Omissions

Whilst we have attempted to address all aspects of the relationships between litter quality and decomposition, inevitably there are some areas which have not received sufficient emphasis.

In systems where virtually all the nitrogen is locked up in organic matter, such as boreal forests, a concerted effort is required to unlock the nutrients for plant uptake. The utility of basidiomycete fungi in degrading the most recalcitrant of substrates, in terms of both lignin and polyphenolic contents, comes to the fore in such systems. Many of the fungi are ecto-mycorrhizal so that degradation of the poor quality litters liberates and translocates N directly for tree growth in an environment which is otherwise nutrient limited (Bending and Read, 1995). The evolutionary significance of the trees being provided with a direct route to their 'own supply' of nutrients is a source of current speculation (Northrup et al., 1995).

Given estimates that litter from roots may often comprise a greater input than above-ground litter surprisingly little emphasis was placed on root decomposition at this meeting, with the notable exception of the paper by Arp et al., (Chapter 15) and some poster presentations at the conference. The effects of living roots in priming decomposition is a further area which is complex experimentally, but demands further investigation.

During the hard task of selecting topics for inclusion as spoken papers in the conference on which this book is based, it became clear that there was a strong bias in activities towards temperate or boreal forests on one hand or agricultural systems on the other. Although the sample of abstracts was limited (~100 submitted) this perhaps indicates the need for greater research attention to natural ecosystems in the tropics.

Future Outputs

Many agricultural systems depend increasingly on biological inputs, both in temperate agriculture due to environmental concerns and in the tropics from necessity and there is a need 'to be able to use a handful of organic matter with as much precision as … a handful of compound fertilizer' (Pedro Sanchez quoted by Heal et al., Chapter 1). But is it possible to distil simple tests which can be used by agronomists or farmers to evaluate litter quality and thereby predict the utility of a given litter in terms of rates of breakdown and nutrient supply? Leaf colour can give a clue to the nitrogen content (Collinson quoted by Heal et al., Chapter 1), but can a simple test such as crushing or tearing leaves be developed to indicate the degree of lignification? Farmers introduced to a range of multipurpose legume trees on a demonstration field in Zimbabwe assessed their suitability as cattle fodder simply by chewing leaves (P. Mafongoya, personal communication). The tree species with strong protein-binding capacities due to reactive polyphenols were readily excluded in this way presumably due to the strong binding of polyphenols to salivary proteins (Harborne, Chapter 4). Thus the development of simple tests which could be used in the field by agronomists or farmers is perhaps a possibility for the future.

An Endnote

Finally, to repeat the plea of Heal et al. (Chapter 1), decomposition is a process of equivalent status to photosynthesis and we need to understand it just as fully! We cannot claim to have arrived in our search for a full understanding of the role of plant litter quality in regulating decomposition, but hopefully this book will provide a departure point for a new phase of research on this fascinating topic.

Acknowledgements

We thank Bill Heal for summarizing the questionnaire responses and John Darbyshire, Fred Palmer and Robin Sen for their letters and suggestions which were included in this summary.

References

Bending, G.D. and Read, D.J. (1995) The structure and function of the vegetative mycelium of ectomycorrhizal plants V. Foraging behaviour and translocation of nutrients from exploited litter. *New Phytologist* 130, 401–409.

Couteaux, M.M., Mousseau, M., Clerier, M.L. and Bottner, P. (1991) Increased atmospheric CO_2 and litter quality: decomposition of sweet chestnut leaf litter with animal food webs of different complexities. *Oikos* 61, 54–64.

Fairhurst, T.H. (1996) Management of nutrients for efficient use in smallholder oil palm plantations. PhD Thesis, Wye College, University of London.

Golchin, A., Oades, J.M., Skjemstad, J.O. and Clarke, P. (1994) Soil structure and carbon cycling. *Australian Journal of Soil Research* 32, 1043–1068.

Janzen, H.H. (1987) Soil organic matter characteristics after long term cropping to various spring wheat rotations. *Candian Journal of Soil Science* 67, 845–856.

Little, W., Fowler, H.W. and Coulson, J. (1988) *The Shorter Oxford Dictionary.* Guild Publishing, USA.

Magid, J., Gorissen, A. and Giller, K.E. (1996) In search of the elusive 'active' fraction of soil organic matter: three size-density fractionation methods for tracing the fate of homogenously ^{14}C-labelled plant materials in soil. *Soil Biology and Biochemistry* 28, 89–99.

Northrup, R.R., Yu Zengshou, Dahlgren, R.A. and Vogt, K.A. (1995) Polyphenol control of nitrogen release from pine litter. *Nature* 377, 227–229.

Index